Astronauts perform servicing functions in the cargo bay of the Space Shuttle. Satisfactory servicing is a combination of both manned and robotic activities. *Courtesy of NASA.*

ON-ORBIT SERVICING OF SPACE SYSTEMS

Donald M. Waltz

with Foreword by
Frank Cepollina

AN ORBIT SERIES BOOK

KRIEGER PUBLISHING COMPANY

MALABAR, FLORIDA
1993

Original Edition 1993

Printed and Published by
KRIEGER PUBLISHING COMPANY
Krieger Drive
Malabar, Florida 32950

Copyright © 1993 by Krieger Publishing Company

All rights reserved. No part of this book may be reproduced in any form or by any means, electronic or mechanical, including information storage and retrieval systems without permission in writing from the publisher.

No liability is assumed with respect to the use of the information contained herein.

Printed in the United States of America.

Library of Congress Cataloging in Publication Data

Waltz, Donald M.
   On-orbit servicing of space systems / by Donald M. Waltz.—
Original ed. 1989.
    p. cm.
   ISBN 0-89464-002-X (alk. paper)
   1. Space vehicles—Maintenance and repair. I. Title.
TL915.W35   1990
629.47—dc20
                                                  89-31729
                                                     CIP

10 9 8 7 6 5 4 3 2

Series editor
Edwin F. Strother, Ph.D.

Contents

Foreword vii
Preface ix
1. **On-Orbit Satellite Servicing Background** 1
2. **Status of Satellite Servicing** 21
3. **Mission Operations** 51
4. **Orbital Maneuvers** 87
5. **Steps to Spacecraft Design** 113
6. **Serviceable Spacecraft** 139
7. **Manned Servicing** 177
8. **Automated Servicing** 193
9. **Benefits and Economics** 229
10. **Road Map to On-Orbit Servicing** 249
    Appendix A. Acronyms and Glossary 263
    Appendix B. SAMS Study Summary 269
    Appendix C. On-Orbit Servicing Checklist 275
    Appendix D. Add-a-Pod Concept 277
    Index 281

Foreword

Consider the development of transportation technology in the context of human exploration of this planet. Significantly greater levels of technological maturity have been reached whenever our ability to travel from point to point has been enhanced by our ability to maintain and repair our mode of transportation—whether animal or vehicle—as we travel. The Greeks, the Romans, the Vikings, Columbus, the Pilgrims, and the pioneers of the American West all developed unique approaches for "on the way" repairs and as a result accelerated the great exploration discoveries of the past.

In similar fashion, the journey of manned and unmanned space exploration reached a new plateau with the repair of the unmanned Solar Maximum spacecraft. NASA combined the manned Space Shuttle's two-way capability with the scientific and technological needs of the space solar physics community to bring about a multitude of scientific benefits. This much too brief, but all-important, union led the way to significant economic benefits for the U.S. commercial communications industry. The subsequent commercial missions of Westar, Palapa, Syncom IV, and Intelsat have resulted in significant revenue gains. New data transmission capability has allowed the expanded television coverage of such events as the Olympic Games—a capability enhanced by the repair of the crippled Intelsat communication satellite.

Necessity is indeed the mother of invention! Space repair and servicing are necessitated by our desire to succeed as pioneers confronting the great mysteries of space. As we attempt to build challenging and complicated spacecraft such as the Hubble Space Telescope, the necessity to correct, upgrade, renew, and improve our systems results in ever more challenging and scientifically rewarding endeavors.

This book is the first all-encompassing look at on-orbit servicing. It addresses technical, programmatic, and economic servicing issues. Donald Waltz suggests that three things are central to building a successful national servicing program: the ability to access the space system with a servicing capability; the ability of space systems to be serviced; and the ability to create a national plan encompassing the interests and needs of NASA, DOD, and the commercial world.

The author traces the development of Shuttle servicing concepts from both manned and unmanned perspectives. Space Station Freedom will provide a permanent repair base in orbit and is highlighted as yet another step in the evolution and maturation of space exploration. Robotics usage and orbital maneuvering are explored in depth. Concepts of common interfaces, standard services and tools, and common hardware for cost-effective logistics are all discussed at length.

On-orbit servicing offers a cost-effective alternative to satellite replacement. In the final analysis, servicing of our expensive assets in space, whether for life extension or commercial and scientific renewal, will *survive* because of *economic necessity*.

FRANK J. CEPOLLINA
Project Manager
HST Flight Systems and Servicing
NASA/Goddard Space Flight Center

Preface

This is the era of the on-orbit serviceable satellite.

With access to space via the Shuttle and the rapid development of technologies in space servicing, maintenance, and assembly operations, program managers now have a choice of orbit repair versus satellite replacement with ground launch of another vehicle. This book traces the history of orbital servicing; shows the status of servicing today; then presents information on mission operations, spacecraft design, servicing equipment, and benefits/economics. The final chapter depicts a possible road map to future orbital servicing as a national capability. Four appendixes supplement the chapter contents. The book, overall, makes the points that:

1. On-orbit satellite servicing is technically feasible.
2. Space servicing will support and benefit a wide range of NASA, DoD, commercial, and international missions.
3. The Space Shuttle and the Space Station Freedom can and must play vital roles in developing a national satellite servicing capability.
4. Satellite servicing, to be successful, must be a two-way street. The space vehicle and the space worker crew providing the servicing must impart extensive assembly, maintenance, resupply, and repair functions at reasonable costs; and the satellite itself must be designed to efficiently accept a variety of servicing actions.

Although many excellent technical symposium papers and contract reports have been written on satellite servicing, this publication, to my knowledge, is the first book to address the topic. Writing a book of this kind in the face of a vast amount of satellite servicing literature presented a major problem of information selection. I have used a large number of references, listed at the end of the chapter where they were used, to try to integrate as much up to date information as possible into the text.

In writing the text and assembling the figures, I assumed the reader has some knowledge of space programs, spacecraft engineering, and orbital operations. The book is directed at working spacecraft program managers, aerospace system engineers, satellite designers, and space operations planners, and to students seeking a career in these disciplines. Servicing technology is relatively new and very dynamic. But it is gaining maturity and, therefore, sophistication. Today satellite servicing features manned operations in low Earth orbit; but at some point in the late 1990s we should see the addition of multiple automated servicing activities in high altitude orbits. Implementation of servicing techniques for Earth orbiting satellites will expand to use on manned and automated lunar and planetary exploration and explotation programs. The servicing concepts and hardware developed for today's satellites will grow and be usefully employed for tomorrow's space systems. Space servicing is not a dead-end idea.

In writing this book many people provided valuable information and advice. I am grateful to Dr. Edwin F. Strother, executive editor, for his astute recommendations, to Mr. and Mrs. Robert Krieger for publishing a book the subject of which is considered controversial in some space program circles, and to Mary Roberts, of Krieger Publishing Company, for her patience and guidance and for recognizing, in the first place, the potential need in the aerospace community for a book on this subject. My enduring thanks goes to my wife, Mary Lou, for her tireless persistence in word processing the entire manuscript.

Finally, this book is for our grandchildren, Christine, Brian, Michael, Bob and Kristen, all currently students, who, if they want to, can make great space operations a part of their future.

Chapter 1

On-Orbit Satellite Servicing Background

On-orbit servicing is work in space. The work, performed by men, machines, or a blend of both, relates to space assembly, maintenance, and servicing (SAMS) tasks to enhance the operational life and capability of satellites, platforms, space station attached modules, and space vehicles. In the broadest context of its definition, satellite servicing also includes the in-space launching, reboosting, and retrieval of space systems. Growth versions of some servicing functions involve space debris capture or containment and emergency operations for crew rescue and return to an in-space safe haven or to the Earth.

On-orbit servicing is a relatively new technology. As such, it has spawned a new set of terms, definitions, and acronyms. The reader is urged to consult the list of acronyms and glossary provided in Appendix A.

Satellite Servicing Scope

Satellite servicing includes a very broad set of functions within three operations: assembly, maintenance, and servicing. Assembly is the fitting together of manufactured parts into a structure, a subsystem, or elements of a subsystem. It is the on-orbit joining, construction or fabrication of spacecraft, space systems, or space structures and includes the deployment of solar arrays, antennas, and other appendages into their operational configurations. On-orbit assembly occurs *before* a space system becomes operational. The basic limited forms of space assembly are initial spacecraft checkout, alignment, calibration, and deployment. The next level of space assembly is the mating of large self-contained modules in orbit such as the attachment of a spacecraft to an upper stage placed in orbit on a separate launch vehicle. A higher order of space assembly involves either the construction of large space structures from small component parts which require welding or similar joining processes, or it involves manufacturing of components from unique materials in the space environment.

Maintenance is the upkeep of facilities or equipment either as necessitated or as directed by a scheduled program. It includes any on-orbit activity performed for the purpose of extending the operational life of a space system, but it does not include replenishment of consumables. On-orbit maintenance is performed *after* a space system has become operational. Preventive space maintenance can include observation, inspection, surface restoration, realignment, recalibration, repair, replacement of modules/payloads/subassemblies, contamination removal, test, and checkout. Corrective space maintenance includes all actions performed as a result of a system failure to restore a system to a specified condition.

The broad view of servicing encompasses the replacement of expended consumables and the logistics required to strategically locate supplies. It includes the on-orbit replenishment of consumables and expendables. On-orbit servicing is normally performed *after* a space system is operational; however, initial supplies of consumables and expendables loaded during the space assembly process can be considered servicing. Space servicing may involve direct transfer of fluids including both liquids and pressurized gases, exchange of fluid tankage, and exchange of packaged consumables such as batteries or film.

It should be noted that the word *servicing* is often used to depict any or all of the functions named above under the three operations.

Evolution of Orbital Servicing

Past spacecraft programs have usually resulted in unique, custom tailored system designs which were then developed into specific hardware components and vertically integrated. These systems were designed for high reliability, long life, redundancy, no on-orbit maintenance or servicing, and, if required, spacecraft replacement via another launch. This approach using a nonserviceable design was driven by several factors: access to space, technology obsolescence, and total system cost.

Many present and future systems are designed as multiple platforms with horizontal integration and standardization between the platforms. As space systems have evolved and become more complex with greater capability, the United States has become increasingly reliant on them to support its military operational forces and its national science programs. This increased reliance has been accompanied by an evolving need to ensure that space systems are operationally available. This evolution, in conjunction with improved operational capabilities and technology advances, provides an opportunity for the development of a space assembly, maintenance, and service capability as an alternative to spacecraft replacement.

Significant servicing events since 1984 have moved satellite servicing from concept to reality:

- The capture and repair in space of the Solar Maximum Mission (SMM) spacecraft.
- Space retrieval and Earth return and repair of the Palapa B2 and Westar VI communications spacecraft.
- On-orbit retrieval, repair, and redeployment of the SYNCOM IV-3 communications spacecraft.
- The Experimental Assembly of Structures in Extravehicular activity/Assembly Concept for Construction of Erectable Space Structures (EASE/ ACCESS) in-space demonstrations.
- The in-space launch and deployment of the SYNCOM 4-F5 military communication satellite.
- The capture, retrieval to the Space Shuttle and return to Earth of the Long Duration Exposure Facility (LDEF).
- The astronaut extravehicular activity to deploy the high gain antenna on the NASA Gamma Ray Observatory.
- The in-space evaluations of hardware and methods of moving astronauts from one location to another on large structures via carts on monorails. Assessment of these Shuttle bay tests suggest that in space, as on Earth, simple is often better.
- The on-orbit retrieval, attachment of a new booster stage, and relaunching of the Intelsat 6/F3 communications satellite during the first flight of the Shuttle *Endeavour*.

The inherent feasibility of space assembly, maintenance, and servicing functions were proven on these NASA conducted missions.

Prior to the spring of 1984 and the successful orbital capture, repair, and redeployment of the NASA Solar Maximum Mission spacecraft, when a spacecraft encountered a massive failure, the spacecraft owners and operators on Earth were almost helpless. Typically, the opportunity to extend human knowledge was lost, to say nothing of the considerable investment of talent and funding resources.

The risk of equipment failure has always shadowed any space system project. The best insurance has always been redundancy, the expensive incorporation of backup components into spacecraft subsystems and payloads. Now the era of the National Space Transportation System (NSTS) Space Shuttle introduces new options for ensuring the success of scientific, applications, military, and commercial missions. As access to space becomes more convenient with the Space Shuttle, it is now possible to send repair crews and spare parts on service calls to ailing satellites. Therefore, it makes sense to design refuelable, modular satellites for on-orbit servicing repair and service.

Initially, space servicing was not a factor in spacecraft design or operations, but the complexity and size of custom space systems in the 1960s drove costs up. Commonality was introduced into the design of space systems to focus resources on function. Then, modularity was incorporated in the 1970s as a means of implementing commonality. Arrival of the United States Space Shuttle in the early 1980s baselined the idea of capitalizing on modularity for on-orbit maintenance and repair. New technology and favorable economics fostered space servicing. Larger payloads and evolving Space Shuttle capabilities have now promoted the concept of orbital assembly. Future technology and the possibility of unmanned, man-tended, or permanently manned space bases, NASA and military, in the 1990s suggest the routine use of space servicing. The space based crew of repair/maintenance specialists could be aided by an orbital maneuvering vehicle (OMV), diagnostic devices, standard and mission unique servicing tools, fault tolerant systems, automation, robotics, and artificial intelligence. These could make satellite servicing an attractive business as a commercial space venture.

The goal of NASA and the military is that this new satellite SAMS capability will be operationally available in the late 1990s. This capability will utilize a first-generation space based servicing facility, namely the International Space Station Freedom, as the collection point or base for payloads on unmanned launch vehicles and Space Shuttle delivered payloads. An orbital maneuvering vehicle could be available to move payloads or spacecraft assembled at the Space Shuttle or Space Station Freedom to other Earth orbit locations. With this capability, it will be possible to boost fully checked-out optimum spacecraft configurations to planetary exploration trajectories from the Space Shuttle or Freedom, that would have previously exceeded the size for a single Space Shuttle direct launch and orbital deployment.

This space assembly approach could result in lower space exploration costs and higher reliability. The servicing capability from orbital space stations can also permit refueling, repair, retrieval, changeout replacement, resupply, maintenance, or modification to free-flying satellites via an orbital maneuvering vehicle, either at the space station or the satellite's orbital location. This capability will also permit the modifications in orbit to the early space-based servicing facilities so that they can be improved, based on future mission requirements and early operational experience.

The NASA manned Space Station Freedom, which is now in active design and development, and is projected to be in initial operation in the late 1990s, will greatly extend on-orbit servicing capabilities by virtue of (1) constituting a permanent operations base in low Earth orbit (LEO), (2) its greater and more highly developed resources, and (3) the presence of crewmembers operating without the time constraints inherent in all Space Shuttle missions. Of particular relevance are unique human cognitive, sensing, and manipulative skills, and especially, the ability to react to new and unforeseen situations. Given appropriate tools, resources, and facilities, the crew can perform on-orbit operations, such as satellite servicing, of greater scope and complexity than would be feasible on board the Space Shuttle. However, certain crew assigned satellite servicing functions can be automated such that the best of human abilities and automation capabilities are combined to achieve the highest degree of productivity in satisfying user needs. Numerous spacecraft systems engineering and design studies and related mission analyses have been performed to establish principal requirements, constraints, and technology needs of on-orbit servicing. The driving considerations are (1) cost economy attainable through extension of spacecraft life by correcting unexpected malfunctions, exchange of defective units, and resupply of depleted consumables, notably propellants, and (2) mission flexibility by on-orbit payload changeout.

From the present to the mid-1990s, on-orbit servicing operations will be very much a Space Shuttle based, manned, labor intensive, hands-on activity. Gradually as the technology develops, automation and robotics will be added in the late 1990s to where a productive mix of man and machine will be employed to accomplish servicing events. Satellite servicing must be approached in a two dimensional iteration of:

1. The ability of the space system to be serviced.
2. The ability of the host vehicle to provide service. Here the host vehicle is the Space Shuttle, a space based facility like the Space Station Freedom or an orbital maneuvering vehicle.

Space Assembly, Maintenance, and Servicing (SAMS) Timeframes

Before 1980 spacecraft were nonserviceable. Inflight demonstrations of servicing had been limited to manned spacecraft such as Skylab and Apollo. Support structures consisted of expendable launch vehicles and upper stages. Trends were toward satellites that were not accessible, and the possibilities of SAMS were considered by only a few NASA program office pioneers in the Solar Max Mission, Hubble Space Telescope, and Gamma Ray Observatory programs.

Because of the Space Shuttle, many spacecraft by 1985 were partially serviceable, either because they were designed for servicing or because innovative techniques were created that circumvented operational impediments.

The ground and space servicing support structure expanded dramatically after 1985 and was repeatedly demonstrated on-orbit. It included reusable as well as expendable launch vehicles, specialized servicing tools, and a push for the development of teleoperation and robotics. Trends were toward limited spacecraft accessibility for repairs and possible retrieval.

Decisions now need to be made between satellite replacement or orbital servicing. There is a requirement for an established national policy based on satellite servicer system feasibility studies and space demonstrations of promising servicer concepts. For the Department of Defense, (DoD), the future will be heavily influenced by SAMS decisions on each individual military program. NASA has already made its SAMS decisions, as exemplified by Space Station Freedom, the Hubble Space Telescope, the Gamma Ray Observatory, Advanced X-Ray Astrophysics Facility, and other programs.

Potential Trends of SAMS

Given a decision to implement SAMS techniques, the space program from now until the late 1990s, Figure 1.1, could include the development and deployment of many orbiting spacecraft and platforms designed for on-orbit assembly, maintenance, and servicing. A strong support structure could be in place, including the initial vehicles in the advanced launch system (ALS), incorporating significant reduction in the cost of transportation to LEO. In addition, the space program may have an operational OMV, an expendable orbital transfer vehicle (OTV), and an international space station in place.

Possible trends include consideration of assembly, maintenance, and servicing needs in spacecraft design. Concurrent to this would be demonstrations of developed SAMS equipment designs and a continuing assessment of SAMS feasibility.

The post-1995 era will then build on the implementation of SAMS technology to include

- Fully modularized, larger spacecraft
- An advanced satellite servicer system combined with an OMV for remote servicing
- Large constellations of spacecraft
- More complex and expensive space systems

Extended spacecraft life will be allowed by the on-orbit maintenance and preplanned product improvement options. The support structure could include advanced launch vehicles, reusable upper stages, a growth space station, on-orbit servicing platforms and depots (manned and man-tended), tank farms to support on-orbit refueling, a high degree of commonality for all interfaces and tools, and orbital warehouses.

Possible trends include fully serviceable spacecraft, the ability to access all Earth orbits in cislunar space, space architecture designed to exploit SAMS on an operational basis, and the existence of a national pricing policy and pos-

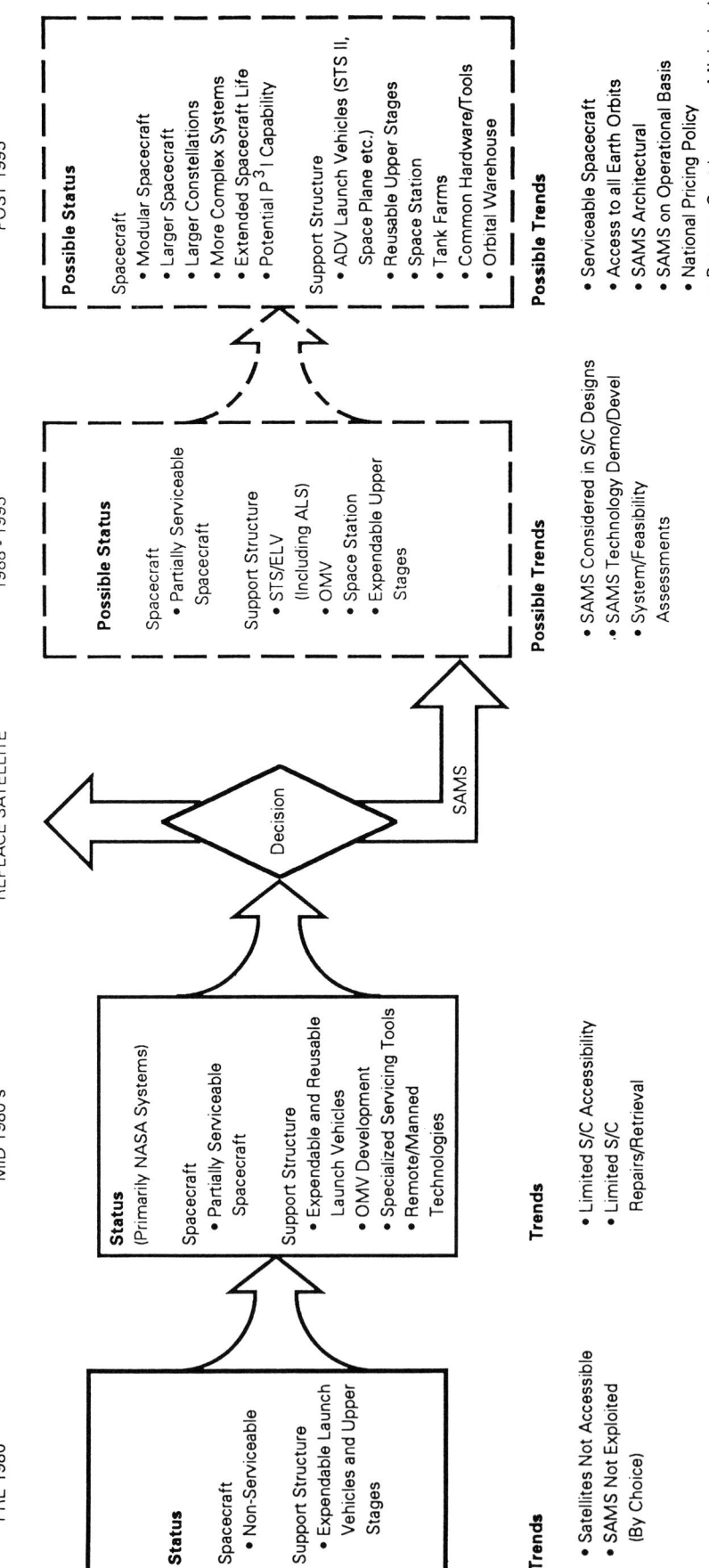

Figure 1.1 Evolution of space assembly, maintenance, and servicing (SAMS). *Courtesy of TRW Space and Technology Group.*

sible commercial competition for controlling costs of space operations and servicing hardware.

Satellite servicing—past, present, and future—can be summarized in the following ten points:

1. Technologically, satellite servicing is an evolving space operations activity in early development.
2. The NASA experience base provides for future capability.
3. Human participation is an essential ingredient today.
4. Telerobotic and robotic development will be essential tomorrow.
5. A servicing host vehicle base is required; NSTS, Space Station Freedom, and an OMV are vital ingredients to the servicing architecture.
6. Servicing is an enabling activity for many major space projects.
7. SAMS is not universally applicable or accepted by all program managers.
8. Early planning for servicing is essential.
9. NASA is working with DoD to achieve a set of workable standards and policies on which to build a national on-orbit servicing capability.
10. Planned on-orbit servicing, maintenance, and upgrading of certain satellites could enhance the national investments in space systems.

SAMS as a Program

Space assembly, maintenance, and servicing (SAMS) viewed as a program is a joint effort between the Department of Defense (DoD), the Strategic Defense Initiative Office (SDIO), and the National Aeronautics and Space Administration (NASA). Its primary objective is to define and establish, where cost-effective, SAMS capabilities to meet requirements for improving space systems capability, flexibility, and affordability. The secondary objectives are to define the SAMS architecture to support evolving space systems and identify common SAMS requirements, cost methodology, life cycle cost benefits, and a technology roadmap.

Included in this program is the near-term demonstration of a spacecraft servicing capability based predominantly on NASA efforts. The cost-effectiveness of an unmanned on-orbit spacecraft and space platform maintenance and servicing capability has been clearly proven in many NASA flight demonstrations and studies during the past decade. The ability to change out failed or wornout spacecraft orbital replacement units (ORUs) and to replenish fuels and other expendable commodities leads to greatly reduced operating costs for spacecraft programs. The design, development, and cost impacts involved in producing compatible spacecraft can be minimal compared to the benefits of on-orbit maintenance and servicing.

In September 1984 the Office of the Secretary of the Air Force documented the need to address potential benefits to be gained from space assembly, maintenance, and servicing. In addition, SDIO has required a space system servicing evaluation to support its operational objectives in the SDIO Supportability Research policy, Work Package Directive B233, October 1985.

SAMS Background

The program consists of three phases, developed under the auspices of the Joint DoD-NASA On-orbit Maintenance Working Group.

To institutionalize commitment and preclude duplication of effort, the Joint DoD-NASA On-orbit Maintenance Memorandum of Agreement (MMA) was signed in June 1986. The program's three phases (study, concept development, and implementation) span a 7 year period.

The Phase I SAMS 16 month study was performed from February 1986 through June 1987 by two large contractor teams—one headed by the TRW, Inc., Space and Defense Sector, Redondo Beach, California; the other led by Lockheed Missiles and Space Company, Sunnyvale, California [1].

The SAMS Phase I Study teams consisted of:

- *TRW*, supported by Grumman Space Systems, McDonnell Douglas Space Systems Company, Booz Allen & Hamilton, and Advanced Technology, Inc.
- *Lockheed*, supported by Boeing, Honeywell, Illinois Institute of Technology, Carnegie-Mellon University, and Life Support Systems

The SAMS Phase I results are found in Appendix B. This effort assessed the impact of SAMS operations on space systems between 1985 and 2010. A top level system analysis was performed on selected design reference missions (DRMs) and identification of benefits was presented.

Phase II will generate national SAMS architecture requirements. It will also structure the testing phase to evaluate present technology shortfalls.

Phase III will initiate the development of the SAMS infrastructure and will culminate in the demonstration of SAMS hardware. The SAMS program elements are shown in Figure 1.2

Many parts of this book are extracts from the documentation from the TRW and Lockheed Phase I SAMS Study performed under contract to the Department of the Air Force, Headquarters Space Division (Air Force Systems Command), Los Angeles, California [1].

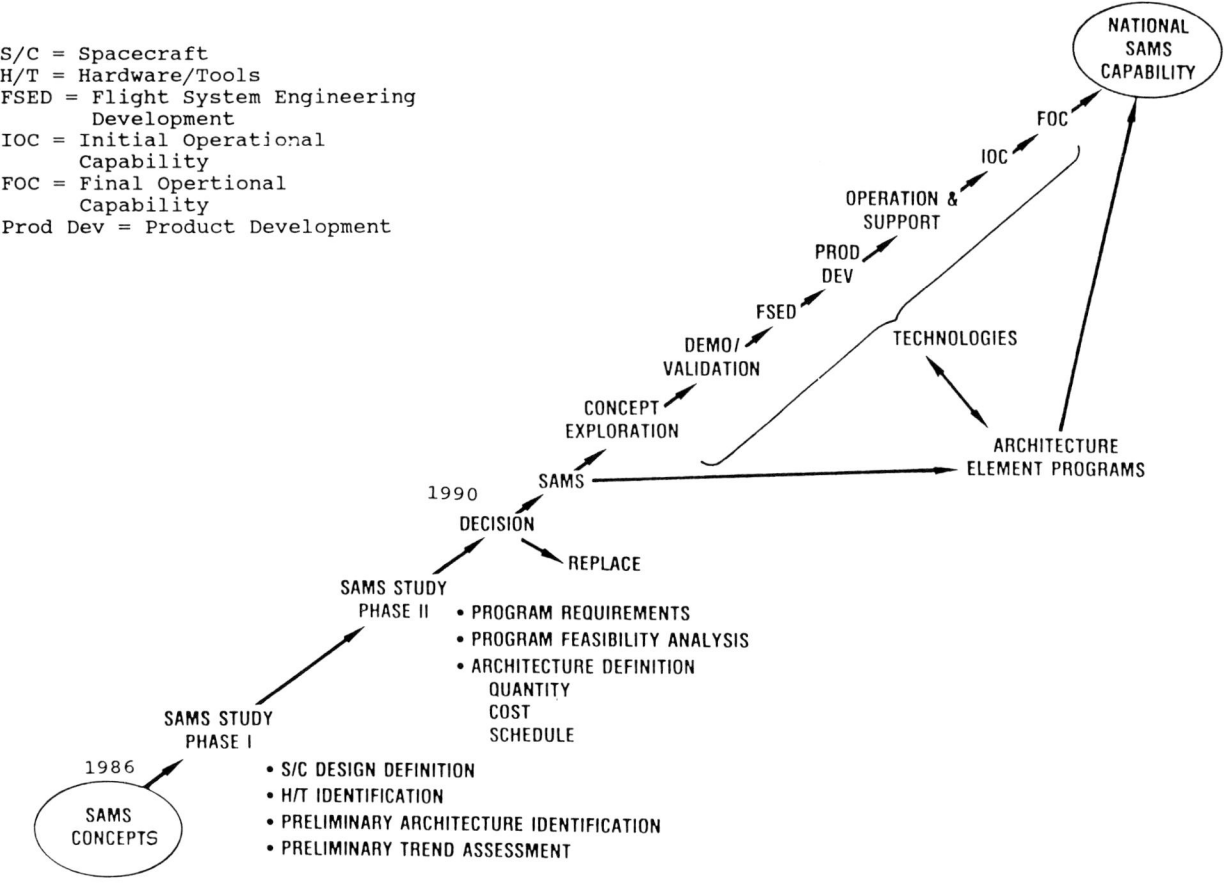

Figure 1.2 SAMS program elements. *Courtesy of TRW Space and Technology Group.*

Servicing at Space Station—A New Era

The Space Station Freedom will open up a new era in on-orbit servicing [2]. It will provide the opportunity to use space in more rational, economical, and imaginative ways than have been possible previously. With the orbiting Space Station Freedom, Figure 1.3, and an associated OMV, Figure 1.4, it will be possible to reach, retrieve, and service the satellites, attached payloads, and platforms in an unprecedented manner. By performing periodic servicing, replenishing consumables, changing or enhancing instruments, and updating components and subsystems on orbit, the life and scientific and applications utility of the satellite missions will be lengthened from a few years to up to 20 years. Another important capability that will be made possible with the Space Station Freedom is the assembly of large structures in space. The ability to assemble an extremely large detector array in orbit, for example, could have a profound impact on astrophysics investigations for generations to come.

Certain significant attributes of the Space Station Freedom and its servicing facility will make it particularly attractive as a servicing base:

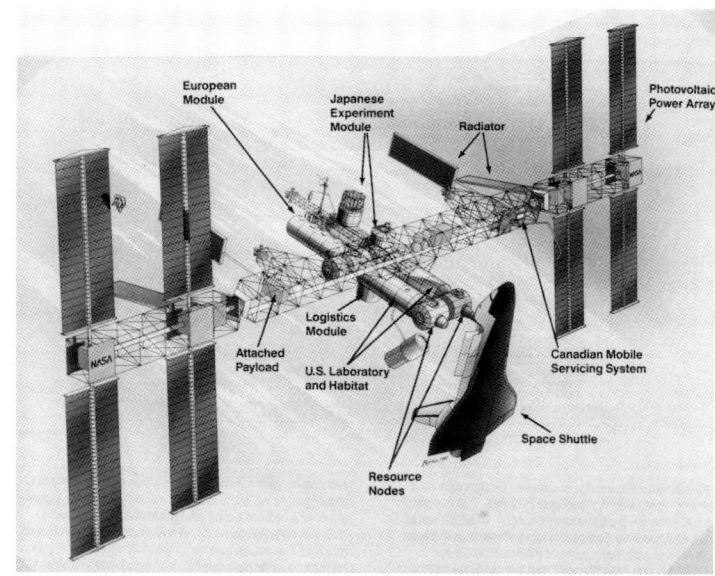

Figure 1.3 Space Station Freedom. *Courtesy of NASA.*

Figure 1.4 Concept of an orbital maneuvering vehicle to aid satellite servicing missions. *Courtesy of TRW Space and Technology Group.*

- The Space Station Freedom will be a permanent facility, available at virtually all times, and subject only to restrictions of traffic and other operational considerations such as crew time, customer spare parts, and/or ORU availability. In contrast, the Shuttle, shown in Figure 1.5, is limited by its nominal 7 to 14 day on-orbit stay time.
- The Station's servicing facility will be able to provide a controlled environment for the servicing of payloads in terms of thermal, contamination, solar impingement, and electromagnetic interference and contamination (EMI/EMC). The Space Shuttle capabilities in these areas are very limited.
- Because the Station is permanently manned, there is an inherent capability to respond to situations in near real time. This may prove to be the single most important attribute as Freedom reaches full maturity.

On-orbit servicing of some satellites will require either the use of the Space Shuttle or servicing in situ. Included in this category are polar orbiting satellites, platforms not in the same plane with the Space Station, and other free-flying satellites in orbits outside the reach of the Space Station or Station-based OMVs and orbital transfer vehicles (OTVs). Although not directly involved with the Space Station servicing facility, the operational techniques and hardware items used in the Shuttle based servicing and in situ servicing will, in many instances, be identical to those used in Space Station-based missions. Equipment commonality is an objective for cost effectiveness.

To allow the full realization of Freedom's potential as a servicing base, NASA has incorporated specific servicing capabilities into its future design options. The synthesis of these capabilities is necessarily driven by two factors: first, the requirements of all foreseeable users must be accounted for, and second, the end product must have a cost-to-benefit ratio consistent with overall budgetary guidelines.

The servicing capability consists of two elements: a multipurpose unpressurized servicing facility and a "work-

Figure 1.5 Space Shuttle with fixed and rotating service structures on the left. *Courtesy of NASA.*

bench," located in one of the Space Station's pressurized modules. These elements are supported by a manipulation and transportation system (the mobile servicing system or MSS), an OMV, standard tools, EVA support equipment, and a telerobotic servicing system. Taken together, they make up a Space Station capability that can fully accomplish the goals ascribed to it. Figure 1.6 is a pictorial of some representative servicing missions.

Historical Perspective

The concept of space maintenance dates from the first satellite failure. In the early days of spaceflight, the first concern was merely to get the satellite into orbit. Once that barrier had been passed and orbiting satellites began returning data, there was a firm desire to reestablish the satellite performance, if lost.

Satellites, while not inexpensive initially, became increasingly more expensive to build and launch as they became more complex and capable. The Space Shuttle concept was envisioned as a means of cutting launch and support costs. On-orbit maintenance of satellites was investigated as one

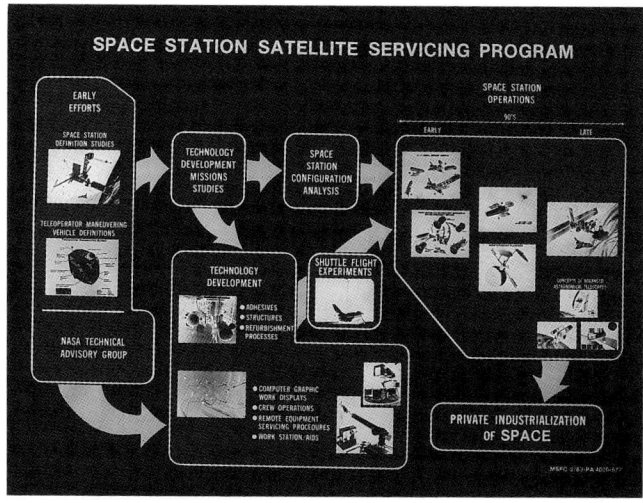

Figure 1.6 Representative Space Station servicing missions. *Courtesy of NASA.*

Figure 1.7 Skylab space station in Earth orbit. Astronaut Garriott performs Skylab 3 EVA.

capability to help justify the acquisition of the Space Shuttle. Payloads were redesigned to take advantage of weight and volume growth permitted by use of the Shuttle. Also, the new concept of on-orbit module replacement and return of failed or used modules to Earth for refurbishment and reuse was investigated. It was concluded that up to 50 percent of program cost could be saved by using refurbishable and modularized spacecraft.

Further cost reduction approaches for payload programs were examined. It became apparent that if the Space Shuttle were used to maintain a large quantity of different low cost spacecraft, the next step would be in the standardization of spacecraft subsystem or module hardware. Cost impact of standardization was analyzed for a complete space program by The Aerospace Corporation and NASA. The costs of future expendable payload programs were used as a baseline. Then payloads representing low cost refurbishment and reuse as well as standardization of spacecraft subsystem modules were extrapolated from point designs and their costs substituted for the baseline.

Space maintenance has been practiced since the beginning of manned spaceflight in 1961. Few missions were completed without crew intervention to correct malfunctions. In some cases extensive effort was required to ensure mission completion. Since the astronaut was originally cast in the role of passive experiment, no specific training and few specialized tools were available to perform space maintenance. On-orbit maintenance was successful due to crew skills and knowledge combined with good training and ground support. The number of examples of on-orbit servicing and repair is too extensive to recount but a few of the most well-known examples will show the range and variety of functions accomplished. These include examples from the experience of the United States and the former Soviet Union, now the Commonwealth of Independent States (CIS).

Skylab Program

The Skylab missions, Figure 1.7, included scheduled maintenance activity, both to maintain vital systems and to get experience with on-orbit servicing. In addition, major unscheduled maintenance was performed on the six subsystems below during the four 1973–74 Skylab missions to correct failures which occurred during launch/deployment:

- Release of orbital workshop solar array
- Deployment of parasol sun shield
- Deployment of twin-pole sun shield
- Installation of rate gyro package
- Coolant system servicing
- Major microwave antenna repair

NASA's first experience with on-orbit servicing and repair was during the 1973 Skylab mission. The unmanned Saturn Orbital Workshop (Skylab) was launched May 14, 1973, atop a Saturn V launch vehicle from Pad A of Launch Complex 39 at Kennedy Space Center (KSC), Florida. The workshop's initial orbit was 500 km (270 nmi) circular with an inclination to the equator of 50 degrees.

An hour after launch, ground controllers still were waiting for confirmation that the workshop's solar arrays had deployed, a signal they never received.

Analysis of launch data showed a failure of the meteoroid shield some 63 seconds into the flight. Slight deployment of one of the two solar array wings, which provided about half of the electrical power used in Skylab, also was indicated.

NASA quickly convened an investigative team to assess the extent of the failure and to recommend corrective actions. The team decided that of several possible failure modes of the meteoroid shield, the most probable was excessive internal pressurization of its auxiliary tunnel which acted to force the forward end of the meteoroid shield away from the shell of the workshop and into the supersonic air stream. The breakup of the meteoroid shield, in turn, broke the tiedowns that secured one of the solar array systems. Complete loss of this solar array system (SAS) occurred at 593 seconds when the exhaust plume of the S-11 stage retrorocket impacted the partially deployed solar array system.

In the hours that followed, NASA and contractor personnel worked to salvage the mission in the face of mounting trouble. Skylab was maneuvered so its telescope mount solar arrays faced the Sun to provide as much electricity as possible. But in this attitude Skylab, without the meteoroid shield that was to protect against solar heating as well, got too warm—up to 52.2°C (126°F) inside.

During the 11 days between the Skylab launch and the subsequent manned launch, repair procedures and equipment to effect the repair of the SAS and the deployment of a backup thermal shield were designed, developed, and tested on the ground.

Several NASA centers designed various thermal shields of reflective cloth to protect the workshop's exposed areas from direct sunlight. Three shields were decided upon—a parasol type to be deployed through an experiments airlock in the Skylab was the primary device; a "sail" to be drawn up over a twin-pole frame and a similar sail to be deployed from the command module were alternatives.

Astronauts Charles "Pete" Conrad, Joseph Kerwin, and Paul Weitz lifted off the KSC Complex 39 Pad B on a Saturn 18 on May 25, 1973, after twice being rescheduled for launch. Rendezvous with Skylab occurred in the fifth revolution and, after an hour and a half of station keeping, the crew docked and finished preparations for a fly-around inspection and standup extravehicular activity (SEVA). Weitz stood in the open hatch while Kerwin held him by the legs and Conrad maneuvered the command-service module. The scientific airlock was reported free of debris, one solar array system completely gone, the other deployed 5 to 10 degrees and jammed there by an aluminum strap.

In the 75-minute extravehicular activity (EVA), Weitz attempted but was unable to cut or pry loose the strap. After five attempts, the crew redocked with Skylab. They spent the night in the command module.

The next day, following procedures worked out only 2 days before, the crew deployed the parasol sunshade. By June 4, the temperature inside the orbital workshop was down to 24°C (75°F).

Another power problem occurred May 30 when 4 of 18 battery packs in the telescope mount power supply system showed they were taking less than one-half charge from the solar arrays, a result of overheating during the unmanned period. While the crew continued a power-limited schedule of experiments and observations, mission support personnel worked out and tested procedures to free the jammed solar wing.

Radioed to Skylab one day, practiced inside the workshop the next, the procedures were used on Day 14 of the mission, June 7.

Conrad and Kerwin spent about 4 hours and 10 minutes in extravehicular activity. They freed the array, and within hours the electric power supply was such that a mission close to the original plan was authorized. In fact, after a series of successful manned missions, Skylab remained in orbit until 1979 when it reentered the Earth's atmosphere.

On June 19, 1973, the 26th mission day, Conrad and Weitz undertook a 96 minute EVA to retrieve film from the telescope mount. Conrad also reactivated a battery regulator relay—he simply tapped the case with a hammer!

The second Skylab manned mission lifted off at On July 28, 1973, with a crew of Alan Bean, Owen Garriott, and Jack Lousma. The trio splashed down 59 days 11 hours and 9 minutes later, September 25, near the U.S.S. *New Orleans*, 417 km (225 nmi) southwest of San Diego, California.

Between those two events was a list of conducted experiments that exceeded the planned workload by 50 percent. The crew also experienced hardware difficulties beyond those that had plagued their predecessors.

Early in the mission, one cluster of four reaction control system (RCS) rockets, then a second, developed leaks and had to be deactivated. The RCS quads are used to "steer" the command-service module in flight. The problem potentially was serious enough that around-the-clock activity was ordered to prepare a modified command module for a rescue mission. But when the two remaining quads stayed healthy, and simulations on the ground demonstrated that adequate control was available with just the two units, rescue preparations were curtailed.

Three EVA sessions were undertaken—the first a marathon 6 hour 31 minute excursion to deploy an experiment, install solar telescope film, and augment the parasol with the twin-pole sunshade. The second was 4 hours 31 minutes to change film, deploy experiments, and replace the faltering rate gyro "6-pack" with a new unit. Garriott and Lousma conducted those EVAs, the first on August 6 and the second on the 24th.

Garriott and Bean spent 2 hours and 45 minutes outside the workshop on September 22, retrieving film and experiment samples. The film canisters plucked from the telescope mount contained some 77,600 pictures of the Sun. Fresh film loaded into the solar cameras during that last EVA enabled three experiments—SO52 White Light Coronagraph, SO54 X-Ray Spectrographic Telescope, and SO55 UV Scanning Polychromator Spectroheliometer—to operate through the 6 weeks of unattended operation, just as they had between the first two missions. The crew did experience one situation not encountered by the first crew—motion sensitivity. It bothered them for the first few days of the mission but as they adapted to weightless flight, the astronauts recovered with no aftereffects.

By the 10th mission day, the crew was putting in about 19 man-hours a day on scientific experiments, but a week to 10 days later they had increased this to 27 to 30 man-hours of experiments each day.

The third and final manned mission in the program got underway on November 16, 1973, after a 6 day delay to replace cracked stabilizing fins on the launch vehicle. With the experience of the successful first two Skylab missions to guide them, crewmen Gerald Carr, Edward Gibson, and William Pogue were prepared to stay as long as 84 days in space, nearly as long as the previous missions combined.

The first extravehicular activity by the crew occurred on Thanksgiving Day, November 22, when Pogue and Gibson put in 6 hours and 33 minutes to deploy experiments, load fresh film into the cameras, and repair a jammed antenna. On February 3, Carr and Gibson teamed again to retrieve the last of the film and experiments in a 5 hour and 19 minute EVA.

Almost all of the 53 Skylab experiments experienced various degrees of maintenance activity during the mission.

Solar Maximum Repair Mission

On April 11, 1984, with the successful completion of the Solar Maximum Mission spacecraft repair, via Space Shuttle EVA, and subsequent spacecraft checkout, the era of the throwaway spacecraft ended. This first on-orbit spacecraft repair was a milestone for the concept of on-orbit servicing and is the cornerstone on which we build that experience and lay our plans for the 1990s [3].

The NASA Goddard Space Flight Center Solar Maximum Mission (SMM) spacecraft was launched by an expendable Delta vehicle on February 14, 1980. For 10 months the observatory collected spectacular new data on solar flares, sunspots, magnetic fields, and solar energy output. Then a problem appeared in the form of a generic deficiency in the design of the momentum wheel drive electronic circuits. Within a month, three wheels had failed, precluding the observatory from accurately pointing at specific parts of the Sun. Two months prior to this, the coronograph/polarimeter instrument started to show a pronounced deterioration in its performance. The problem was traced to its main electronics box, called the MEB. At this time, the Solar Max spacecraft was placed in a backup slow-spin mode 1 degree per second with a slight wobble of about 5 to 7 degrees. This mode allowed the spacecraft to collect sufficient energy on its solar arrays to sustain life indefinitely, but precluded the use of the three remaining fine-pointing instruments. This was a perfect setup to demonstrate the Space Shuttle's capabilities to rendezvous, repair, check out, and redeploy a free-flying spacecraft (see Figure 1.8). The lessons learned from this mission proved without a doubt that on-orbit servicing was not only feasible, but practical. Additional lessons were learned on the mission that set the stage for on-orbit servicing in the 1990s:

- Spacecraft lifetimes of a decade or more are achievable with servicing.
- Properly designed spacecraft can be upgraded in orbit.
- Proper preparation and training will allow more intricate repairs than replacing modules.
- Critical spacecraft replacement hardware must be factored into future programs.
- Powered tools and a flexible, multiposition work station are essential for efficient on-orbit repairs.

Preparation for the repair mission presented unique challenges. Since the "wounded" spacecraft was in orbit, it was not available for checkout of the critical modular attitude control system. It was therefore necessary to create a high fidelity simulation of SMM on the ground to verify the module both mechanically and electrically. This simulator was built up using SMM's structural test model and its spare flight harness. Configuration records were carefully reviewed to assure the accuracy of the simulation. Further fidelity was achieved by use of the spare Landsat communications and data handling and power modules which are virtually identical to those modules on the SMM spacecraft. Verification of the compatibility of the Flight Support System (FSS) with the Shuttle Orbiter's systems was required as well as the integrity of communications and data flow between the Orbiter, the launch site, the Mission Control Center, and the Payload Operational Control Center (POCC). A test program of over a year's duration at NASA/GSFC was completed in December 1983 and the flight hardware, simulators, and ground support equipment (GSE) shipped to the Kennedy Space Center (KSC) for cargo integration test equipment (CITE) testing and installation into the Shuttle Orbiter. Final verification of communications and data links was accomplished in the Orbiter bay utilizing a "suitcase simulator" specifically developed for this purpose.

On April 8, 1984, the Shuttle Orbiter *Challenger* made rendezvous with the Solar Maximum Mission spacecraft in orbit at 491 km (265 nmi). Astronaut George Nelson rode the Man Maneuvering Unit (NMU) 61 meters (200 feet) through space to the Solar Max but was unable to latch onto it when his docking device refused to snap shut on a trunnion pin on the spacecraft. His actions put the Solar Max into a near-catastrophic tumble that almost doomed the repair mission. Only a desperate series of commands sent to the Solar Max from the Goddard Space Flight Center

Figure 1.8 Solar Maximum repair mission. *Courtesy of NASA.*

brought it under control and allowed it to be captured on April 10 by the Orbiter's Remote Manipulator System (RMS) operated by astronaut Terry Hart. Shuttle Commander Robert L. Crippen's precision rendezvous with minimal fuel, supported by extensive analysis by personnel at the Johnson Space Center, succeeded in positioning the *Challenger* next to the spacecraft for the pickup by the RMS. Crippen feared he would run out of forward reaction control system propellant. Predictions showed he would have no more than 3–6 percent fuel remaining at the critical grapple point. His piloting technique left twice that amount. Hart then skillfully deposited the Solar Max spacecraft which was on the end of the RMS onto its berthing cradle in the Orbiter cargo bay.

NASA believes that a small grommet that holds the Solar Max's insulation in place may have been too close to the trunnion pin, preventing Nelson's attaching device from going far enough to trigger its jaws. His locking device, which was strapped to the front of his space suit, was to have snapped shut when placed around a three-inch trunnion pin protruding from the satellite. Despite three attempts, the jaws did not close, though they had worked in the cargo bay before he went to the satellite and again after he returned. The space agency said the grommet, which protrudes between one-quarter and one-half of an inch from the satellite, may have inhibited the locking device and kept the trunnion pin from reaching the device's trigger.

Nelson and Astronaut James van Hoften exited the Shuttle's airlock and in their 7 hour 18 minute single EVA replaced the faulty attitude control module and repaired the main electronics box for the chronograph polarimeter. They achieved all of the work that had originally been planned for two separate EVAs extending over 12 hours. The only longer U.S. EVA occurred on one Apollo 17 lunar surface activity that lasted an additional 19 minutes.

Replacing the attitude control module that had limited solar pointing observations for 3 years consumed only 60 minutes. Within 15 minutes of the repair's completion, Goddard began receiving data from the new module, indicating the changeout had worked. The EVA crew also placed a baffle over a Solar Max vent port that was allowing space plasma to enter the satellite and disrupt detectors in the x-ray polychromator instrument.

After the astronauts finished their EVA repair, the RMS was moved away from the Solar Max. The Solar Max was then checked out and released back into space to continue its scientific mission. *Challenger* went home. The mission was a success even though events did not follow one-to-one the mission plan. Retrieval and repair of the Solar Maximum spacecraft met the Space Shuttle program goal to develop the capability of servicing satellites in orbit, and in the process demonstrated the ability of the U.S. space program to recover from a potentially serious setback.

The original cost of the Solar Maximum Mission spacecraft was $230 million. The cost of the repair mission was estimated by Goddard Space Flight Center at $60 million. Certainly it was cost effective for NASA to opt for parts repair over total spacecraft replacement to complete this important project.

Space Shuttle Missions

There have been numerous examples of actual or attempted unscheduled maintenance on Space Shuttle missions to date. Examples include: The development flight instrumentation (DFI) recorder failure attempted repair, fuel cell workarounds, video cassette repairs, keyboard repairs, and text and graphics system (TAGS) maintenance.

The operational history of the Shuttle for on-orbit service and repair is, of course, still developing, but it is apparent from the flights to date that it will be similar to Apollo and Skylab. A sufficient number of examples exist to show this. Figures 1.9 and 1.10 indicate the pre-Shuttle and Shuttle-based major historical milestones in servicing capability development.

Hubble Space Telescope

A major NASA on-orbit servicing activity centers around the Hubble Space Telescope (HST), Figure 1.11, launched April 24, 1990, from the Space Shuttle [4,5]. This satellite, designed and built by Lockheed Missiles and Space Company and Hughes Danbury Optical Systems, Inc. for NASA/MSFC, will last for two decades. A consortium of 26 companies contributed a broad array of state of the art components. It is designed with replaceable spacecraft subsystems. The five instruments are also on-orbit replaceable for observatory upgrading over the life of the mission. The original plan called for on-orbit repair and update at approximately 5 years, and to return to Earth at 10 years for major refurbishment. Present plans are to routinely service it at the Space Station Freedom. Any servicing required prior to the Station activation will be conducted by the Space Shuttle Orbiter [4].

Future NASA Science Spacecraft Programs

The Advanced X-Ray Astrophysics Facility (AXAF), shown in Figure 1.12 and now under design/development by TRW Space and Technology Group for NASA/MSFC, is a major observatory comparable in size and scope to the Hubble Space Telescope. Planned for launch in the late 1990s, the AXAF is the first scientific satellite to take advantage of lessons learned during the Hubble Space Telescope experience [4]. General similarities in the design of the two observatories suggest that some elements of the Hubble Space Telescope servicing concept are applicable to AXAF. For example, access to the AXAF science instruments probably will be through large doors similar to those on the Hubble Space Telescope, and use of a limited selection of fasteners and connectors chosen for the Hubble Space Telescope will recycle tools already in inventory, decrease their variety, and streamline training.

As a later generation observatory, however, AXAF will benefit from new capabilities not available when the Hubble Space Telescope was being designed. For example, computer-aided design (CAD) and manufacturing will become a valuable tool in designing future spacecraft for servicing. Since the HST design started before CAD was widely available, it evolved through a series of mockups on which engineers and astronauts tested successively refined concepts.

Advanced NASA programs are also influencing the design of future serviceable spacecraft. For instance, it is assumed that the Space Station Freedom will be the principal servicing site for AXAF, with the Space Shuttle available for contingency repair missions. This new option will influence the selection and size of AXAF orbital replacement units (ORUs), which may be standardized to conform to other ORUs used on the Space Station common platforms.

The impact of servicing the Hubble Space Telescope and AXAF with robot spacecraft is being evaluated. Replacement of ORU modules by a telerobotic satellite servicer system on an orbital maneuvering vehicle would eliminate some trips to the Space Station and reduce the observatories' out of operation time. Such a technique is within near-term technology. This option could affect the design of other ORUs.

Repair work inside the Space Station is being considered for some elements of AXAF and the Hubble Space Telescope to reduce the amount of hardware brought up for a servicing call. However, planners must decide whether troubleshooting components within a black box, often a time-

MISSIONS	MILESTONES & ACCOMPLISHMENTS	MISSION DATES	SERVICING ELEMENTS & CAPABILITIES DEMONSTRATED						
			EVA	IVA	RETRIEVING	REPAIRING	RESUPPLYING	UPGRADING	OTHER ELEMENTS & SUBELEMENTS*
GEMINI 4	• FIRST U.S. EVA	JUN 3-7, 1965	•						
GEMINI 6/7	• RENDEZVOUS & STATION KEEPING	DEC 15-16/4-18, 1965							•
GEMINI 8	• RENDEZVOUS & DOCKING	MAR 16-17, 1966							•
GEMINI 10	• EXPERIMENT HARDWARE RETRIEVED FROM DOCKED AGENA WITH EVA	JUL 18-21, 1966	•		•				•
GEMINI 12	• DOCKINGS & 3 EVA'S	NOV 11-15, 1966	•						•
APOLLO 7	• LIVE TV BROADCAST FROM SPACE	OCT 11-22, 1968							•
APOLLO 9	• CREW TRANSFER BETWEEN DOCKED SPACECRAFT	MAR 3-13, 1969	•	•					•
APOLLO 11	• EXTENSIVE EVA ACTIVITIES & SAMPLE RETURN	JUL 16-24, 1969	•	•					•
APOLLO 12	• RETURN OF ELEMENTS OF DEPLOYED SPACECRAFT (SURVEYOR 3)	NOV 14-24,1969	•	•	•				•
APOLLO 13	• EXTENSIVE ON-BOARD PROBLEM SOLVING & RESOURCE MANAGEMENT	APR 11-17, 1970		•					•
APOLLO 15	• USE OF LUNAR ROVER IMPROVED SPACE SUIT	JUL 26-AUG 7, 1971	•	•					•
APOLLO 17	• SATELLITE DEPLOYED IN LUNAR ORBIT	DEC 7-19, 1972	•	•					•
SKYLAB 2	• SUNSHIELD DEPLOYED & SOLAR ARRAY RELEASED DURING EVA • EXTENSIVE IVA	MAY 25-JUN 22, 1973	•	•		•			•
SKYLAB 3	• SUNSHIELD & RATE GYROS REPLACED WITH EVA • EXTENSIVE IVA	JUL 28-SEP 25, 1973	•	•		•			•
SKYLAB 4	• COOLANT SUPPLIES REPLENISHED • ANTENNA REPAIRED WITH EVA • SINGLE EVA DURATION RECORD	NOV 16, 1973- FEB 8, 1974	•	•		•	•		•

* INCLUDES RENDEZVOUS, DOCKING, ELEMENTS OF TELEROBOTICS, ASSEMBLING, DEPLOYING, AND RETURNING

16354 010

Figure 1.9 Major historical milestones in servicing capability development (Pre-Shuttle Experience). *Courtesy of NASA.*

MISSIONS	MILESTONES & ACCOMPLISHMENTS	MISSION DATES	SERVICING ELEMENTS & CAPABILITIES DEMONSTRATED						
			EVA	IVA	RETRIEVING	REPAIRING	RESUPPLYING	UPGRADING	OTHER ELEMENTS & SUBELEMENTS*
STS-7	• RMS DEPLOYMENT & RETRIEVAL OF SPAS-01 FREE FLYER	JUN 18-24, 1983			•				
STS-9	• SPACELAB 1 WITH EXTENSIVE IVA	NOV 28-DEC 8, 1983		•					•
41-B	• MMU & FLUID PUMPING DEMONSTRATED • SMM REHEARSAL	FEB 3-11, 1984	•						•
41-C	• SMM RETRIEVED, REPAIRED & REDEPLOYED	APR 6-13, 1984	•		•	•			•
41-D/F	• 31-M EXTENDED SOLAR ARRAY DEPLOYED & RETRIEVED • ICICLE REMOVED FROM SHUTTLE SURFACE WITH RMS	AUG 30-SEP 5, 1984	•				•		•
41-G	• ON-ORBIT FUEL TRANSFER & FUEL VALVE RETROFIT DEMONSTRATED	OCT 5-13, 1984	•				•		•
51-A	• PALAPA B2 & WESTAR VI RETRIEVED & RETURNED	NOV 8-16, 1984			•			•	•
51-B	• SPACELAB 3 WITH EXTENSIVE IVA	APR 29-MAY 6, 1985		•	•	•			
51-F	• SPACELAB 2 RMS DEPLOYMENT & RETRIEVAL OF PDP	JUL 29-AUG 5, 1985			•	•			•
51-I	• SYNCOM IV-3 RETRIEVED, REPAIRED AND REDEPLOYED	AUG 24-SEP 1, 1985	•		•	•			•
61-B	• EASE/ACCESS ASSEMBLY & DISASSEMBLY WAS FIRST SPACE STATION ASSEMBLY REHEARSAL	NOV 26-DEC 3, 1985	•						•

* INCLUDES RENDEZVOUS, DOCKING, ELEMENTS OF TELEROBOTICS, ASSEMBLING, DEPLOYING, AND RETURNING

Figure 1.10 Major historical milestones in servicing capability development (Shuttle-based experience). *Courtesy of NASA.*

Figure 1.11 Hubble Space Telescope. *Courtesy of NASA and Lockheed.*

Figure 1.12 The Advanced X-Ray Astrophysics Facility. *Courtesy of TRW Space and Technology Group.*

consuming and difficult task, is the most effective use of a very precious resource—an astronaut's time.

Not every orbital servicing lesson to be learned from the Hubble Space Telescope is yet available; more will be learned as the observatory acquires operational experience and the maintenance and repair (M&R) plan moves from paper to reality. Furthermore, future scientific instruments for the HST will require two operations that will be important for many future spacecraft: gas venting and fluid replenishment. The pathfinders are the TRW/NASA/GSFC Gamma Ray Observatory for propellants, and both AXAF and the second-generation HST instruments for cryogens. Other aspects of servicing will be recognized as the Space Station Freedom design matures. The designed-for-servicing aspects of the HST, AXAF, GRO, and an OMV spacecraft are discussed in Chapter 4.

Soviet (now C.I.S.) Program

Soviet maintenance and servicing experience on Salyut has necessarily been extensive because of the long Salyut mission duration. Replenishment of expendables has been accomplished via the unmanned Progress and manned Soyuz spacecraft. There have been numerous subsystem repairs, including, for example, repair and rewiring of electronic systems; disassembly, repair, reassembly, and alignment of the global navigator; and contingency EVA to jettison the 10 meter radio antenna. Several technologies have been advanced by the Soviets in their scheduled servicing program on Salyut: automatic docking and fuel transfer refurbishment and replacement of mission instruments.

Although the Soviets have had several manned programs, the most important of these is the Salyut, the first of which was launched in April 1971. The Salyut program developed an effective support operation that routinely supplies the spacecraft with crews, fuel, life support, parts, instruments, and subsystems for refurbishment. This has meant the development of new vehicles and techniques for servicing, giving the Soviets the ability to dock automatically and transfer fuel from the supply spacecraft to the Salyut. The crews conduct experiments that require frequent equipment changes. On-board technology work results are routinely returned to Earth. The crews continually perform repair and refurbishment tasks.

Transfer of expendables from ground to orbit and return of products and trash have progressed significantly since Salyut 1. Three vehicles have been used for this purpose: the Soyuz 1 until about 1977, the Soyuz 2 during 1978, and the Progress since 1980. Initially the crews transferred all supplies and fuel to the Salyut during EVA. Docking was accomplished by point maneuvers from mission control and Salyut. Subsequently, automatic docking was developed and with Progress vehicles, fuel was transferred automatically. The Soviets have become very proficient at resupply and there have been no reported accidents during dock and transfer although there have been several examples of launch failures and incomplete rendezvous or docking.

During the extended flight of the Salyut 6 spacecraft, maintenance and repair were important. The crews have made numerous repairs to the vehicles, space suits, navigation instruments, television, scientific and technical equipment, and communication equipment. An example was the repair of the attitude sensing system used for navigation. The crew disassembled, cleaned, repaired, reassembled, and aligned this instrument with assistance from ground control. The instrument functioned correctly without subsequent failure.

Continual changes have been made to the technical, scientific, and military equipment that have flown on Salyut. The required modifications have been made on-orbit by the crews. The Soviets report all modifications have worked as expected and good results obtained.

U.S. Military Satellite Servicing Considerations

During the last 25 years military dependence on space systems has increased dramatically. This increased dependence has generally been justified on the basis of cost and mission effectiveness. Space systems were either the least expensive way or the only way to provide an essential military capability. Military space systems are not, however, inexpensive in absolute terms. Some spacecraft hardware now costs more than $200,000 per pound and the cost of replacing that pound in orbit can vary from $5,000 per pound for a low Earth orbit to over $50,000 per pound for a geostationary orbit. Those relatively high unit costs, coupled with the sometimes short or limited spacecraft lifetimes, have constrained military space systems to a few critical force enhancement roles that do not require a large number of vehicles. However, if the costs of space systems can be reduced and/or their performed capabilities increased, there are a number of other military missions that could be effectively accomplished by space systems. In the past it has been possible to lower the total program cost by investing in sophisticated technology that increased the longevity, reliability, and capability of the space vehicles. Those investment opportunities have been highly exploited and are becoming marginal. New approaches that have large potential payback margins need to be developed.

On-orbit servicing of a satellite is one approach that might be made to operate effectively on both the spacecraft and the space transportation costs while relaxing certain performance and technical constraints. This was successfully demonstrated in principle with the on-orbit repair of NASA's Solar Maximum Mission spacecraft.

Satellite servicing does, however, require that military space program managers commit to a fundamental change in the way they do business. Today, most military long term programs maintain their systems by launching replacement spacecraft as required. In effect, the spacecraft is the minimal orbital replacement unit (ORU). This approach allows the program manager to limit dependence on third parties for support and limits exposure to "unproven" technology. It also allows reasonably stable long term budgetary plans. If program managers are to be persuaded to adopt a new

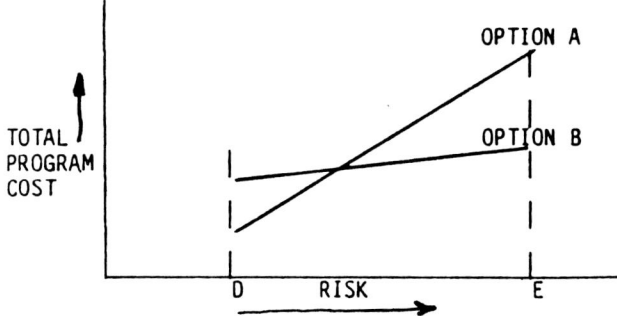

Figure 1.13 Typical total program cost versus risk.

approach, they must be convinced they can afford to make the change. In particular, the manager must be shown that a change in program strategy will not expose the program to unacceptable cost growth risks. Such risks, if they were to materialize, could significantly alter the choice of strategy. Unfortunately, in most cases, once a strategy is selected, it is not easy to change because of technical, cost, and political ramifications. It is important, therefore, that program managers be able to select the most cost-effective and stable option or strategy.

That is a reasonably straightforward problem if program managers have stable, well-defined cost models and a mature technology base for every potential strategy. At the present, those conditions are not met. Therefore, program managers need an analytical tool to select a strategy, given less than perfect knowledge.

Servicing models are based on the premise that program managers want the most cost-effective strategy within a credible envelope of risks. Figure 1.13 illustrates the effect of risk on relative total program cost (TPC). At very low risk, option A is more cost-effective than option B; however, since option A is more sensitive to cost growth, because of technical uncertainties, it quickly becomes less optimal as the risk level rises. If the credible risk envelope extends from point D to point E, then the more stable solution or strategy from the program manager's point of view is option B. Two risk elements have been used to date. They are (1) space transportation cost risk and (2) ground refurbishment versus on-orbit servicing cost risk.

The models usually feature four scenarios:

1. The spacecraft is expendable. If it fails, another is launched.
2. The spacecraft is retrieved, serviced, refurbished, and relaunched, all in space, after each cycle of its operational life.
3. The spacecraft is serviced on-orbit after one cycle, then retrieved and delivered to the ground after its second cycle for *factory* refurbishment and relaunch.

4. The spacecraft is serviced on-orbit after one cycle of mission operations and then abandoned.

As noted earlier, two types of risks are accounted for by the model; namely those associated with space transportation and those associated with the unproven factory refurbishment and on-orbit servicing design changes.

Specific cost estimates of projected dollars are not always realistic. They are dependent on input assumptions of the model being used. Therefore, the various program costs associated with on-orbit servicing of space systems should be normalized for each major cost subcategory, then used as parameters in any multidimensional analysis of satellite life cycle costs. One cost subcategory frequently used is the recurring cost of a baseline space vehicle.

Total program costs by scenario and risk level are then obtained by multiplying the subcategory unit cost by the scenario use factor and the subcategory risk factor and then summing within the scenario. High and low bounds for space transportation are obtained by reentering the summation loop at the space transportation cost point and doubling all costs.

Perception and Reality—Applied to Both NASA and Military Space Systems

Space-based systems either directly or indirectly support all of our major military weapon systems. The capabilities and force enhancements provided by these space-based systems are so important that extreme measures are taken to ensure their performance and survivability. Being able to access spacecraft on-orbit and perform servicing functions has the potential to increase performance, improve survivability, and lower total program cost.

A NASA committee on spacecraft maintenance and repair was chartered in the early 1980s to determine if a NASA policy on spacecraft maintenance and repair was warranted. When affirmative recommendations resulted, the Air Force in 1984 directed that a similar study be conducted within the DoD. Consequently, an ad hoc group was convened from various Air Force/Navy commands and the Air Force Secretariat to determine if the USAF should pursue planned maintenance of spacecraft and, if so, provide the recommended policy for implementation.

To arrive at its recommendation, the Air Force study group evaluated the past efforts of the DoD, NASA, and aerospace contractors and reviewed current and planned spacecraft programs.

This Air Force study concluded that the Air Force must act to posture future spacecraft programs to take advantage of the potential provided by maintainable designs. The report recommended space program managers be required to consider spacecraft on-orbit maintenance options on their new program and to justify their selection or nonselection of a maintainable design prior to the first preliminary requirements review (PRR)—usually 6 months after contract go-ahead.

What We Know

Two things are required to accomplish on-orbit maintenance and servicing of spacecraft: first, the ability to access the spacecraft with a maintenance capability; second, the ability of the spacecraft to be maintained. Since we are theoretically able to rendezvous with any spacecraft which we put into orbit, the real issue is—can we perform servicing and maintenance functions on the orbiting spacecraft once docked with it? In the past, there has been no validated military requirement for on-orbit maintenance. Thus existing and near-term DoD spacecraft programs have not included on-orbit maintainability as a design requirement. The fact that some manufacturers have designed systems to facilitate pre-launch integration, test, and assembly has provided possibilities for on-orbit application as well.

In the future, if NASA and the DoD direct their space systems designs to include component accessibility, modularity, and standardization, there will be considerable potential for mission enhancement through maintenance and servicing. This philosophy is only good, however, if it provides benefits within acceptable levels of cost and risk. The initial NASA/DoD/SDIO SAMS study [1] established the methodology, and generated some generic examples, of this cost benefits analysis. But, the cost benefits and risk analysis, sensitivities, trends, break points and parametric decision trees must be made on a *real program-by-program basis*.

On-orbit maintenance has not been consciously considered to date during the design of existing and programmed DoD spacecraft. It has been indirectly accomplished as a function of satellite health monitoring and station-keeping operations. Consciously designing for servicing and maintainability, applied programmatically where warranted, will provide NASA, USAF, and commercial spacecraft a capability and flexibility previously unavailable.

Increased spacecraft maintainability will continue to evolve as a result of NASA efforts and industry Independent Research and Development (IR&D). The Air Force must take advantage of this evolution and insert its concerns and requirements.

Survey of Future Missions

A survey of scientific and commercial missions planned for the 1990–2010 time period reveals that almost all missions could potentially benefit from some form of on-orbit servicing. The possible services range from simple cryogen resupply to on-orbit assembly and deployment. When the International Space Station Freedom becomes operational, a significant number of satellites that require servicing will already be on orbit at 28.5 degrees—a very probably inclination for the initial Space Station and others that will be launched over the next two decades. The list of candidate NASA satellites starts with the Hubble Space Telescope and continues with the Gamma Ray Observatory, Advanced X-Ray Astrophysical Facility, Space Infrared Telescope Facility, Orbiting Solar Laboratory, Large Deployable Reflector, X-Ray Timing Explorer, Extreme Ultraviolet Explorer, Far Ultraviolet Spectroscopy Explorer, a Solar Corona Diagnostic Mission, and the Equatorial Science Platform of the Space Station program. At geosynchronous orbit the communications satellites, such as the Land Mobile Communications Satellite, will be candidates for remote servicing via the Space Station-based NASA orbital maneuvering vehicles. The larger of these communication satellites may also require assembly at the Station with subsequent deployment via an OMV.

The probability that other satellites requiring servicing will be clustered at other inclinations may be a driver in establishing requirements for servicing facilities at these locations also. For example, near 57 degrees inclination free-flyers on-orbit may include an Upper Atmosphere Research Satellite, the Topography Experiment, and a Solar Terrestrial Observatory. The satellites in near polar orbit will not be closely grouped, primarily because of the particular orbits required for various Sun-synchronous missions, but the list of free-flyers within probable range of a 90 degree space station includes the Polar Orbiting Platforms which support the Earth Observing System, Windsat, the Polar Meteorological Satellite, the Ocean Microwave Package, the Data Collection and Location System, the Advanced Land Observing System, the Magnetic Monitoring Mission, the Navy Remote Ocean Satellite System, the Geopotential Research Mission, the Lower Atmosphere Research Satellite, and the Free Flying Imaging Radar Experiment. In addition, the small Earth-viewing satellites at low altitude and near 90 degrees inclination are especially strong candidates for servicing because they may have film packs that require replacement and because their low altitude orbits degrade quickly thus making a reboost service desirable.

At any of the Space Station Freedom inclinations, the station system may include one or more multipayload, unmanned platforms orbiting near the manned station. The platforms would provide support such as power, heat rejection, communications, and data capability to a variety of payloads. Changeout of these payloads at intervals of 6 months to 5 years, as well as general tending of the platforms and payloads, would be a station servicing function.

The Air Force and Strategic Defense Initiative Organization (SDIO) satellite classes that are candidates for on-orbit servicing are discussed in Chapter 3.

Operating Philosophy

Spacecraft maintenance is viewed by some as a major change of philosophy or redirection that will, for the first time, require the application of new and unique logistics concepts to space systems. This leads to the misrepresentation of these logistics concepts as requirements unto themselves rather than as means of achieving cost-effective support of operational systems. Further, the perception is that maintenance planning and concepts have not been a part of satellite development and acquisition programs. In reality, past equipment design had been driven to emphasize reliability/redundancy for performance over maintainability considerations to achieve operational availability. Now, due to the evolution of technology which provides on-orbit access to spacecraft, maintenance has become a significant factor for consideration. Perceptions are that this is revolutionary. Reality is that it has evolved. Fortunately, on-orbit servicing allows the application of existing logistics support methods. This comes at the very time when cost and technical constraints may require these methods to satisfy increasingly demanding mission requirements.

The perception that maintenance considerations have not always been an integral part of spacecraft design and system engineering is a matter of semantics not reality. There are tasks in the development and operation of equipment that are fundamentally similar regardless of their application in support of space, air, sea, or ground missions. For example:

1. Launch processing, which is identified as part of operations in the Space Shuttle program, consists of all tasks from touchdown to re-launch. Parallel tasks in aircraft programs are identified as maintenance.

2. In-flight servicing performed by a technician or a crewmember, with or without advice from the ground over a communications link, is correction of a malfunction. This is a maintenance action whether it is accomplished on a USAF B-1 aircraft or on a Space Shuttle mission.

3. Going out to acquire deployed/failed equipment, taking it to another location to repair it, and returning it to service is logistics support, whether maintenance is accomplished in a ship at sea or the Space Shuttle payload bay in space.

4. The Air Force Satellite Control Facility applies complex software programs through their telemetry tracking and control (TT&C) capabilities to monitor health and status of spacecraft and utilize built-in redundancies and engineering expertise to return equipment to operational status. The mission avionics of high technology aircraft, such as the Advanced Warning Aircraft Communication System (AWACS), includes built-in redundancies allowing graceful degradation and equipment recondition upon failure while continuing to meet mission requirements. Complex AWACS software programs include automated diagnostics to assure health and status and take corrective actions not dissimilar to space applications.

In a number of the above examples, the distinction between which tasks in the scenarios are operations or maintenance becomes a matter of individual perception. The reality is that all tasks contribute to continued mission support. The key is recognizing the utility of the tasks early in the planning and design stages to take advantage of the advances of space technology and application of appropriate levels of logistics support techniques when they make opational and business sense.

References

1. *Space Assembly, Maintenance and Servicing Study (SAMS) Final Report*, Volume I, Executive Summary, TRW No. SAMSS-196, Volume II, System Analysis, TRW No. SAMSS-195, Volume III, Design Concepts, TRW No. SAMSS-197, Volume IV, Concept Development Plan, TRW No. SAMSS-198, Volume V, Neutral Buoyancy Simulation, TRW No. SAMSS 199. TRW, Redondo Beach, CA, July 6, 1988.
and
Space Assembly, Maintenance, and Servicing Study (SAMS) Final Report, Volume I, Executive Summary, Volume II, System Analysis, Volume III, Design Concepts, Volume IV, Concept Development Plan, Volume V, Simulation Report. Lockheed Missiles and Space Company, Inc., Sunnyvale, CA, July 6, 1988.
2. LaVigna, T. A., and H. P. Cline. *Satellite Servicing In the Space Station Era*, NASA/Goddard Space Flight Center, Greenbelt, MD, 20771.
3. Cepollina, F., and L. Spratt. *On-Orbit Maintenance and Repair. New Era In Space Research and Industrialization*. IAF-84-372, October 7–13, 1984.
4. *Designing An Observatory For Maintenance In Orbit*. Published by the Space Telescope Project Office, NASA/Marshall Space Flight Office, Huntsville, AL., (MSFC 1186), 1986.
5. J. S. Childs. *The Making of Hubble*, Market Supplement, Aviation Week and Space Technology, June 11, 1990.

Chapter 2

Status of Satellite Servicing

Although it receives increasing emphasis, satellite servicing is still a research and development activity for NASA and DoD. Before discussing servicing missions operations in Chapter 3, it seems appropriate to survey the current status of satellite servicing. This chapter offers a broad view of systems thinking on:

- Servicing functions and requirements
- Servicing prerequisites
- Space Station assembly
- Today's satellite servicing technology and hardware
- Servicing decisions
- Major satellite servicing issues

Servicing Functions and Requirements

Satellite servicing in its fullest context includes a very broad set of functions. Figure 2.1 shows 20 servicing functions placed in five task categories together with the designated location in space where they will be performed. Satellite servicing is multitasked with functions performed at multi-locations in space.

The assembly task pertains to the initial construction as well as the upgrade and growth modifications to a space station. It also includes the assembly, test, and checkout of a spacecraft on-orbit at a space station, the deployment of the spacecraft's appendages, and its launch from the station. The orbital transfer task involves the in-space checkout of a spacecraft and the possible mating of a propulsion stage or a transfer vehicle to it for delivery to its final orbit. It also includes retrieval from orbit via a transfer vehicle and preparation of a space system for return to Earth in the Space Shuttle cargo bay.

Resupply involves the on-orbit refueling of spacecraft, replacement of other fluids and gases, exchange of raw materials stock in materials processing in space (MPS) payloads for finished or processed products, and the replacement of fresh film and tape in spacecraft or payloads. The maintenance and repair task is the most varied in its content. It accounts for (1) planned and unplanned changeout to spacecraft modules, subsystems, and payloads; (2) refurbishment/retrofit and modifications to space systems; (3) decontamination and cleaning/resurfacing (of solar arrays, optics, instruments); and (4) inspection/test/checkout of whole space systems as well as component parts. Maintenance and repair, together with refueling, will dominate satellite servicing work in the early years of the Space Station Freedom.

The special task category includes the functions of controlling space debris by capture for Earth return deorbit and Earth entry burnup or reboost to safe orbits. Other tasks in this category are the emergency activities associated with user needs or space rescue as well as the retrieval and return to Earth of commercial products made in space [1].

Concepts of several servicing functions performed at or near the Space Shuttle are displayed in Figures 2.2 through 2.8. Figure 2.2 shows satellite servicing capabilities provided by the Space Shuttle. Crew involvement in servicing functions is emphasized in Figures 2.3 through 2.6. Figure 2.7 shows the launching from the Shuttle of an Air Force Inertial Upper Stage (IUS) with an attached satellite to be delivered to a higher orbit, while Figure 2.8 depicts the idea of on-orbit satellite buildup by joining a satellite to a propulsion stage.

Satellite servicing functions remote from the Shuttle or Space Station will be carried out by servicing vehicles similar to the one shown in Figure 2.9.

Figure 2.10 is an artist's concept of two servicing activities at the Space Station: OMV operations and AXAF refueling from a fuel tank facility.

System Requirements

An analysis of satellite population and location from the present to about the year 2010 indicates five orbital regimes where satellite servicing operations will be conducted [2]. These primary regimes vary from low altitude, less than 1,853 km (1,000 nmi), to geosynchronous orbit; and from inclinations that vary from 27 to 100 degrees. Generic design reference missions (DRMs) can be constructed to create servicing scenarios for the various satellite systems in each regime. Then DRMs specific to a given space system

SERVICING TASKS	TASK FUNCTIONS	SERVICING LOCATIONS			
		SERVICING TO SPACE STATION	SPACE SYSTEM BERTHED AT SPACE STATION	SPACE SYSTEM IN LEO 1	SPACE SYSTEM IN GEO 2
ASSEMBLY	1. SPACE STATION ASSY.	●			
	2. SPACE STATION UPGRADE/MODIFICATION	●			
	3. LARGE SPACECRAFT ASSEMBLY		●	●	
	4. DEPLOYMENT OF APPENDAGES	●	●	●	
ORBIT TRANSFER	1. DELIVERY TO FINAL ORBIT		●	●	●
	2. RETRIEVAL FROM ORBIT		●	●	●
	3. EARTH RETURN		●	●	●
RESUPPLY	1. FLUIDS	●	●	●	●
	2. MATERIALS	●	●	●	●
	3. FILM/TAPE	●	●	●	●
MAINTENANCE AND REPAIR	1. MODULE CHANGEOUT/REPLACEMENT	●	●	●	●
	2. REFURBISHMENT/RETROFIT	●	●	●	●
	3. MODIFICATION	●	●	●	●
	4. DECONTAMINATION	●	●	●	●
	5. CLEANING/RESURFACING	●	●	●	●
	6. TEST AND CHECKOUT	●	●	●	●
	7. UNPLANNED REPAIR	●	●	●	●
SPECIAL	1. SPACE DEBRIS CONTROL	●	●	●	●
	2. EMERGENCY OPS.	●	●	●	●

1 LEO - Low Earth Orbit
2 GEO - Geosynchronous Earth Orbit

Figure 2.1 Examples of servicing functions and orbital locations.

can be generated from which mission and spacecraft requirements are derived for servicing operations. Figure 2.11 is a layout of the five orbital locations. Examples of space systems in each location are indicated. The DRMs for each satellite under each location could be combined into one generic DRM for the orbit location.

The estimated spacecraft distribution by orbit location is shown in Figure 2.12. Programs are defined as groups of satellites performing the same mission, usually developed and controlled by the same NASA or DoD program office. The program numbers on Figure 2.12 were derived from a 1985–87 Air Force sponsored Space Transportation Architecture Study and the NASA Civil Needs Data Base [3,4].

There are no overall documented operational requirements which dictate on-orbit spacecraft servicing functions and procedures across the spectrum of all candidate space systems. There are, however, NASA, DoD, and commercial programs in the concept validation or development phase which have operational requirements for which planned on-orbit servicing and maintenance could provide increased mission capability, improved availability, and lower life cycle costs. The decision to incorporate on-orbit maintenance capabilities into spacecraft programs should be made when it is the most economical or the best method to satisfy program requirements.

The requirements structure [2] shown on Figure 2.13 subdivides servicing missions into an early timeframe from 1986 to 1995—before the International Space Station Freedom and a later timeframe from 1996 to 2010—after deployment of the Station. Five user organizations are identified: DoD, NASA, NOAA, commercial, and foreign users. Regardless of the timeframe or organization, however, servicing requirements fall into the following three basic categories: mission, system, and spacecraft. Figure 2.13 lists the various types of requirements that are relevant to each category.

Status of Satellite Servicing

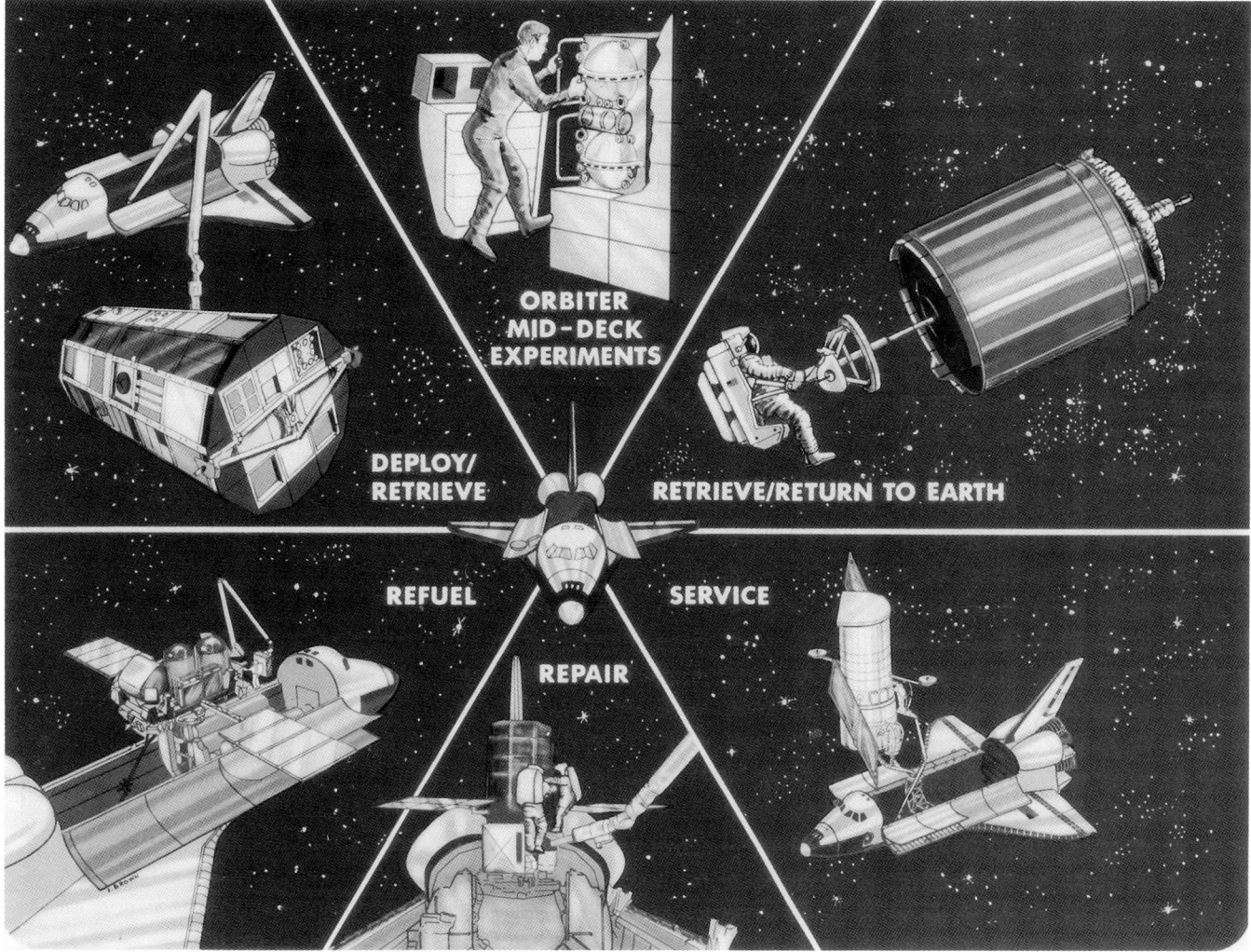

Figure 2.2 Shuttle satellite servicing capabilities. *Courtesy of NASA.*

Figure 2.14 shows the satellite servicing mission data base requirements for NASA and Air Force programs that are candidates for on-orbit servicing.

How To Develop System Requirements

Satellite on-orbit servicing is an excellent example of space logistic systems operations in action. Logistics and operations integration requirements are central to developing the activities needed to conduct on-orbit satellite servicing, Figure 2.15. If the servicing event requires manned operations, the capabilities of the crew to perform IVA or EVA tasks impact the requirements. Note the extensive interaction and feedback shown on Figure 2.15 between the five major activity categories and the central role of logistics and operations integration. Servicing logistics is discussed in detail in the next chapter.

There are seven major parts to the total systems engineering process for synthesizing satellite servicing concepts:

1. Derivation of servicing requirements
2. System design and analysis
3. System integration
4. Mission operations planning
5. System effectiveness
6. System test planning and audit
7. Software system design

Systems engineering, in turn, is part of the program management whole. The other parts are manufacturing, assembly, test operations, systems safety, product assurance, program controls, business operations, administration, and support services.

The achievement of excellence in satellite servicing results is strongly dependent on the accuracy and realism of the mission, system, and spacecraft functional and performance requirements. Mission and system requirements data base activities should be directed toward the structured col-

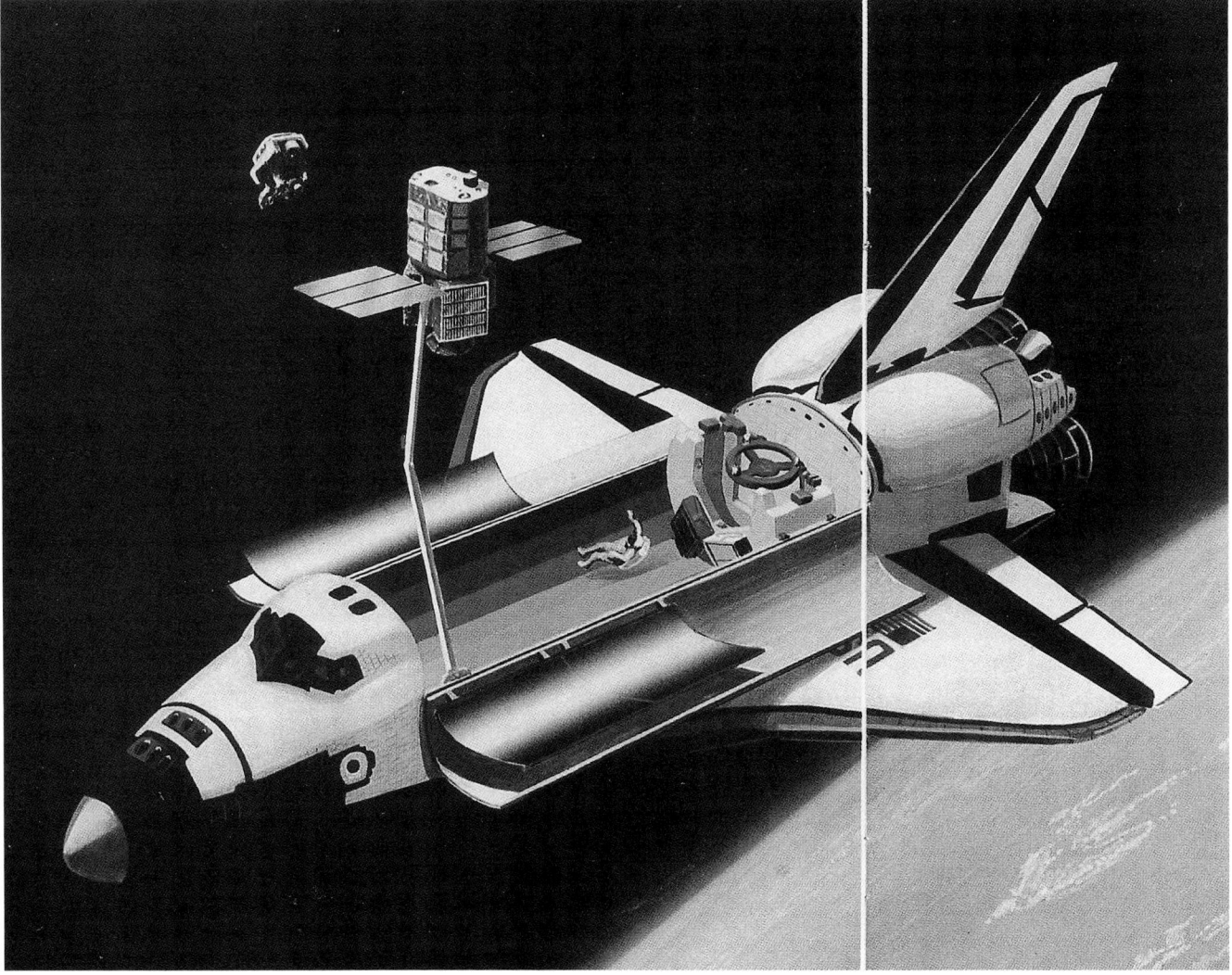

Figure 2.3 Shuttle remote maneuvering system holding satellite above cargo bay prior to bringing it into the bay for servicing. *Courtesy of NASA.*

lection of basic space operations data and policy information. Next, the data must be compiled into usable formats and categories, with rationale directly traceable to mission needs and characteristics. Spacecraft servicing requirements can then be operated and integrated with the mission and system requirements. The recommended approach to development of integrated requirements and decision criteria includes the conduct of significant mission analysis work on all phases of the servicing mission.

Figure 2.16 indicates the major steps in the development of integrated requirements and decision criteria. Mission analysis flow encompasses the steps of processing: mission definition, mission planning, critical parameters, mission functions, system requirements, and requirements allocation.

Complex mission timelines and critical events place significant focus on where to allocate requirements for implementation functions. The orbit and size of a satellite, together with its cost and reliability estimates, dictate the allocated function requirements of servicer systems. Technology readiness is a major driver of mission strategy and the resultant flow-down of requirements.

Overall mission assurance is another top issue for satellite servicing program implementation. Trip times, on-orbit stay times, delta V (ΔV) for transit phases, and variability with opportunity are parameters that enter into the mission assurance analysis.

Servicing Prerequisites

A number of prerequisites are now being defined by NASA and DoD servicing studies to take full advantage of on-orbit spacecraft maintenance and logistics support methods. Some prerequisites exist; others will evolve. They are not

Figure 2.4 Satellite inspection in Shuttle cargo bay prior to servicing tasks. *Courtesy of NASA.*

categorized as servicing functions, although many will be required to permit successful servicing actions. Their application is mission scenario dependent.

It is important to view spacecraft maintenance as merely another service to be provided to the total satellite program, very much like launch or security. It should be a fleetwide service, within economic/technical capabilities, for use by specific projects which need it.

It is also important to realize that many of the prerequisites which enable spacecraft maintenance to be routinely performed are themselves useful for things other than spacecraft maintenance. This includes, for example, satellite placement or retrieval, on-orbit assembly, ground processing flow, and checkout. Servicing prerequisites include:

- *Orbital Access:* Operations involved in transfer of a spacecraft from the surface of the Earth to a stable target orbit. In this context, the mechanism for accomplishing this operation is a space launch vehicle.

- *Orbital Transfer:* Operations involved in transfer of a spacecraft from one orbit to another orbit.

- *Proximity Operations:* Operations involving two or more spacecraft conducting sustained joint activities within 93 km (50 nmi) of each other. It also includes those operations that require monitoring of joint activities, such as:
  — Navigation Control—in the context of proximity operations, those operations involved in bringing two or more space vehicles within a grappling, capture, or docking envelope.
  — Safe/Safing—actions resulting in equipment being placed in a configuration considered safe to perform maintenance or actions which preclude an unsafe/or failure event from occurring.
  — Docking—the act of physically attaching or connecting two space vehicles on-orbit.
  — Thermal Control—in the context of proximity operations, the ability of two or more spacecraft to oper-

Figure 2.5 Astronaut operating satellite servicing equipment in the Shuttle cargo bay. *Courtesy of NASA.*

ate in concert without exceeding the thermal constraints of either vehicle.
  — Observation—the ability to visually acquire and interpret data about a spacecraft.
  — Deployment—physical release or movement of equipment to and from the Space Shuttle (or other launch vehicle or space platform) by device, vehicle, or EVA, as part of a mission or maintenance scenario.
  — Retrieval—grapple, capture, and return of equipment into the Shuttle (or other space platform) bay to perform maintenance or Earth return.
- *Emergency Operations:* An operation or set of operations that is not part of a nominal set or sequence, and which must be accomplished immediately.
- *Jettison:* The act of separating subsystems or structure from a space vehicle with disposal of the separated element on orbit.
- *Orbital Assembly/Modification/Upgrade:* Physical (mechanical and electrical) combination/configuration of modularized assemblies resulting in initial, changed, or enhanced equipment performance.
- *Earth Return:* Design option describing retrieval of equipment for ground processing to include teardown, failure analysis, repair/refurbishment, and modification/upgrade prior to redeployment as a cost-effective alternative to replacement with new equipment.
- *Space Debris Control:* Operations which minimize contamination of an orbit with physical debris, and/or the removal of debris previously left on orbit.
- *Safety Monitor:* A system of continuous assessment and reporting of critical equipment data to identify and alarm unsafe conditions.
- *Support Equipment Control:* The management of devices used to perform equipment maintenance functions such as servicing. Includes a range of technology from simple in-bay fixtures to robotics.

Space Station Assembly

Space assembly is perhaps the most technically challenging of the three operations associated with SAMS (space assembly, maintenance, and servicing). Space assembly oper-

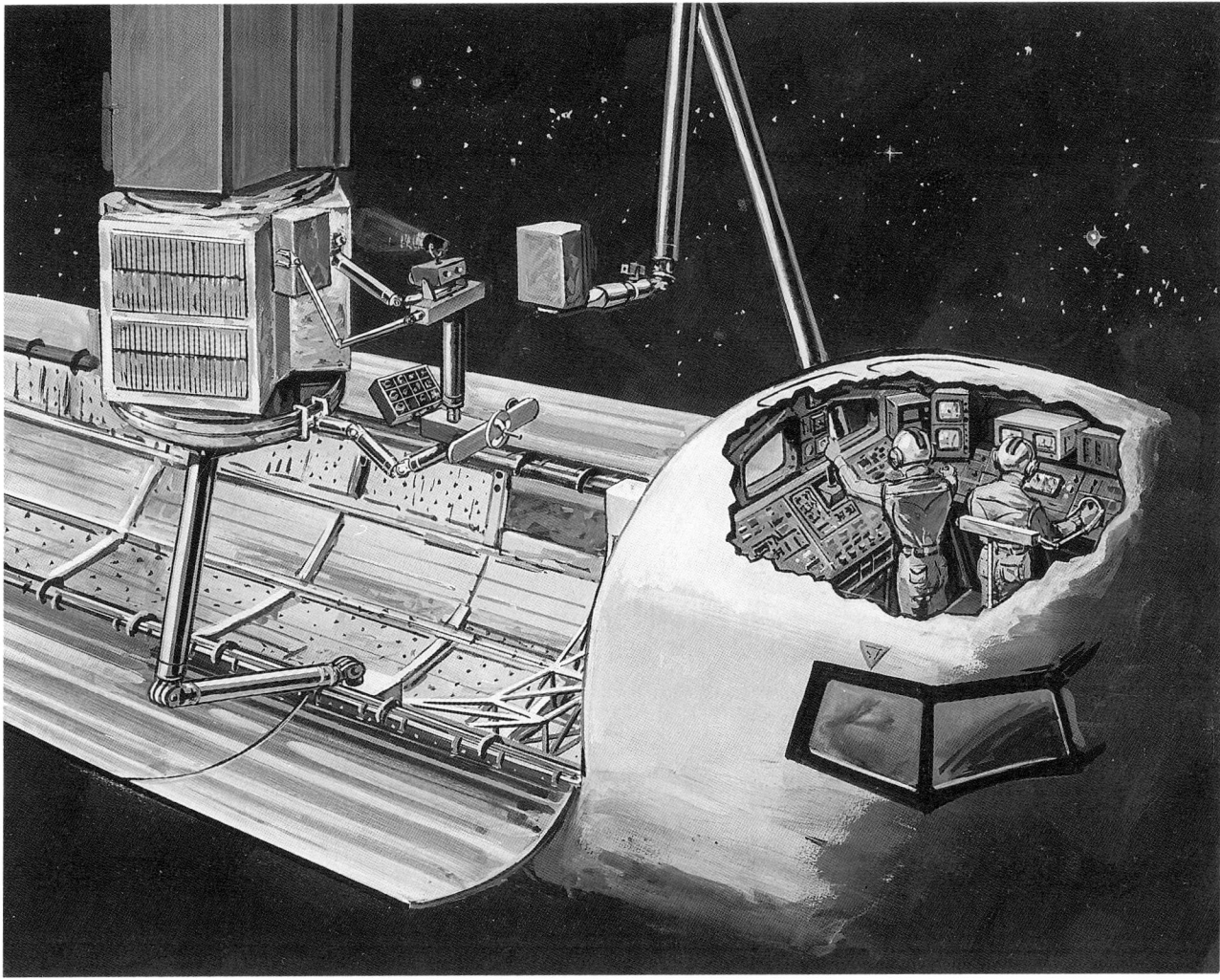

Figure 2.6 Crew in the Shuttle aft flight deck performing satellite module replacement with the Remote Manipulator Systems. *Courtesy of NASA.*

ations bring together the same functions of management, labor, materials, machines, processes, logistics, cost/schedule risks, environment, finished products, checkout, and customer acceptance found in Earth-based assembly activities.

The on-orbit assembly of the NASA Space Station Freedom (SSF) will be a huge endeavor. It will require 16 to 20 Shuttle launches starting in the late 1990s, manned and robotic operations, and a forward leap in the magnitude of space logistics.

The Space Shuttle will transport the Station's elements into space, where astronauts will assemble the Station in orbit. The Space Station will serve well into the 21st century, evolving technologically and expanding to house additional people.

The Space Station construction effort requires novel feats of design and engineering. The orbital environment is unlike any place on Earth and, thus, is a promising location for scientific research and manufacturing processes. However, the same traits that make the space environment beneficial pose unusual problems for builders. Structures, techniques, and work schedules must be carefully designed to fit this unique construction site.

The Space Station's components must be lightweight for transport into orbit, yet durable and strong. Since the Station will be maintained permanently in orbit, it must be made with removable parts for easy servicing and repair by astronauts. The astronaut builders ferried to work in the Shuttle will be wearing space suits and drifting in weightlessness; thus, the Station's components and tools for assembling and servicing the first Space Station must be designed for easy manipulation by a suited astronaut. Construction methods must be efficient because crews can work "outside" in space for only a limited time before they must return to the shelter of a spacecraft with a controlled environment. These aspects of working effectively in microgravity must be thoroughly engineered and practiced before any large structure can be erected in orbit. The SSF assembly sequence is discussed in Chapter 3.

Figure 2.7 Shuttle launching an Air Force inertial upper stage with an attached satellite which will be delivered to a higher orbital altitude. *Courtesy of NASA.*

Initial Capability

When Space Station Freedom is completely assembled, a broad spectrum of research in all the disciplines of life sciences and planetary sciences will be conducted. This will be accomplished with both manned and unmanned elements. The manned facility in a low Earth orbit will consist of four pressurized modules. Three of these modules—one each from the United States, Europe, and Japan—will serve as laboratories. The U.S. laboratory is designed to handle projects that need a stable microgravity environment for materials research as well as R&D in basic physics, chemistry and biology. The European and Japanese modules are designed primarily for research in fluid physics, life sciences, and materials processing. The fourth module provides a habitation area for rest, recreation, and health for the four person crew [5].

The unmanned elements of the program include free-flying platforms in polar or high-inclination orbits as well as attached payloads on the Space Station truss. The platforms will initially be used for Earth observations in a variety of climatology and oceanographic studies. In summary, there will be a variety of manned, man-tended and unmanned user opportunities for science in, on, and around the Space Station [5].

Future Configuration—Although no decision has been made this far in advance, Space Station Freedom is being considered for enhancement sometime after the 16th assembly flight. The long transverse boom will be enhanced by two vertical keels about 105 meters (344 feet) long, and two 45 meter (148 foot) horizontal trusses at top and bottom. This dual keel configuration will add greater stability to the manned base, provide for many additional attached payloads, and offer a wide field of view for scientific instruments. Also included is a solar dynamic electric power system with an additional 50 kW. The mobile servicing center (MSC) will be enhanced to handle heavier payloads [5]. It is noted that this scenario represented NASA's thinking very early in the SSF planning. This thinking will change many times as the program evolves. Funding will drive station configuration growth.

Lunar and Mars Mission Support

NASA scientists and technicians are developing scenarios on how Space Station Freedom can support other explora-

Status of Satellite Servicing

Figure 2.8 Satellite build-up on-orbit using Shuttle RMS and other servicing capabilities. *Courtesy of NASA.*

tions. The dual keel configuration lends itself naturally to the function of a transportation node where spacecraft can be assembled, fueled, and checked out for manned missions to the Moon or Mars. Subsequently, such a spacecraft could be berthed, refueled, and repaired at the Station upon its return. Space Station Freedom could conduct much-needed research in bioregenerative life support systems and artificial intelligence. The Station could define the limits of human endurance for long duration manned spaceflights in a weightless and hostile environment. The dual keel further lends itself to experimentation and as a quarantine facility before lunar and Martian samples are returned to Earth. Consequently, as space policy shifts, Congressional intent emerges, user demands change, and humans find new projects for outer space exploration, Space Station Freedom will evolve to meet these and yet unheard-of uses for a 30-year, multipurpose facility in low Earth orbit [5].

The EASE/ACCESS Experiment

NASA used the Orbiter *Atlantis* on Space Shuttle mission 61-B in 1985 to try out a space construction technique in orbit. The demonstration was called EASE/ACCESS, an acronym standing for Experimental Assembly of Structures in Extravehicular activity/Assembly Concept for Construction of Erectable Space Structures. The EASE/ACCESS experiment was a success.

During the 7 day mission, astronauts Jerry Ross and Woody Spring, working in the cargo bay, assembled ACCESS, a 13.7 meter (45 foot) tower, in 25 minutes, and EASE, a clumsy-looking multiple-sided structure, in 9 to 13 minutes. EASE took so little time to put together that the astronauts assembled and disassembled it several times. Ross and Spring worked from several positions: clamped in foot restraints in the cargo bay, tethered to the bay's sides, and strapped onto the Space Shuttle's robot arm. "The assembly work was very easy," Ross said, though he admitted that the muscles in his hands and forearms tired from trying to hold his body in place. Suited up, the astronauts each weigh about 181 kg (400 lb) on Earth; in orbit they're weightless, but they do have mass and therefore expend a lot of energy controlling their movement. Still, they accomplished their EASE/ACCESS chores in space 18 percent faster than they did during training in a neutral-buoyancy water tank. See Chapter 3 for details on the EASE/ACCESS mission operation.

The NASA astronaut office has established some guidelines for space construction work. It is clear that space suits

Figure 2.9 Example of satellite servicing away from the Shuttle or Space Station. Here a remote servicing vehicle and servicer system approaches a satellite. *Courtesy of NASA.*

need further improvements. New gloves were provided for EASE/ACCESS, but the astronauts still got numb thumbs an hour into the job, and the numbness persisted for more than a month. Another problem is low temperatures, which render the suits inflexible. For health reasons, sessions outside the Shuttle are slated for 6 hours every other day for one week, but not during the first 2 days of a mission, when space sickness might occur. Finally, manned maneuvering units, the jet-powered backpacks, will not be used because they are designed for transport and are too cumbersome to wear while stationary.

Today's Satellite Servicing Technology and Hardware

Government and the aerospace industries space systems engineers and planners agree on the reasons for orbital servicing:

- In-space assembly due to transportation size/weight-to-orbit limitations of current launch vehicle systems
- Scheduled replenishment of expendables
- Planned replacement of degraded hardware
- Planned changes in spacecraft's mission objectives
- Planned technology update for performance improvement
- Unplanned equipment damage or failure; repair or replacement
- Orbit adjustment
- Spacecraft retrieval for return to Earth for repair/refurbishment or commercial product retrieval

With the above eight reasons in mind, a discussion follows on what is feasible today and what constitutes the future satellite servicing technology base.

What Is Feasible Today?

On-orbit servicing is a capability in an early stage of evolution. To date, servicing has been entirely dependent on the presence and support of humans in space. This people dependence has allowed great flexibility in the servicing tasks

Status of Satellite Servicing   31

Figure 2.10 Servicing activities at the Space Station. *Courtesy of TRW Space and Technology Group.*

attempted since astronaut crews are inherently dexterous, adaptable, and innovative. However, crew servicing limits the number of programs that have been used or can use such a capability, since such programs must be associated with or have the capability to achieve close proximity to the manned Space Shuttle vehicle.

This brings into focus the consideration of on-orbit accessibility of the satellite to be serviced. Accessibility is connected with capability and both are a function of time.

Now it is fundamental that the capabilities which put a spacecraft into an orbital location could also access that spacecraft if needed later in its life cycle. Consequently, accessibility could be considered 100 percent regardless of altitude. Thus a graphic representation of accessibility versus support capability shows a straight line along the horizontal axis, depicting a very low level of support capability. See Figure 2.17A.

It is essential in the framework of spacecraft maintenance

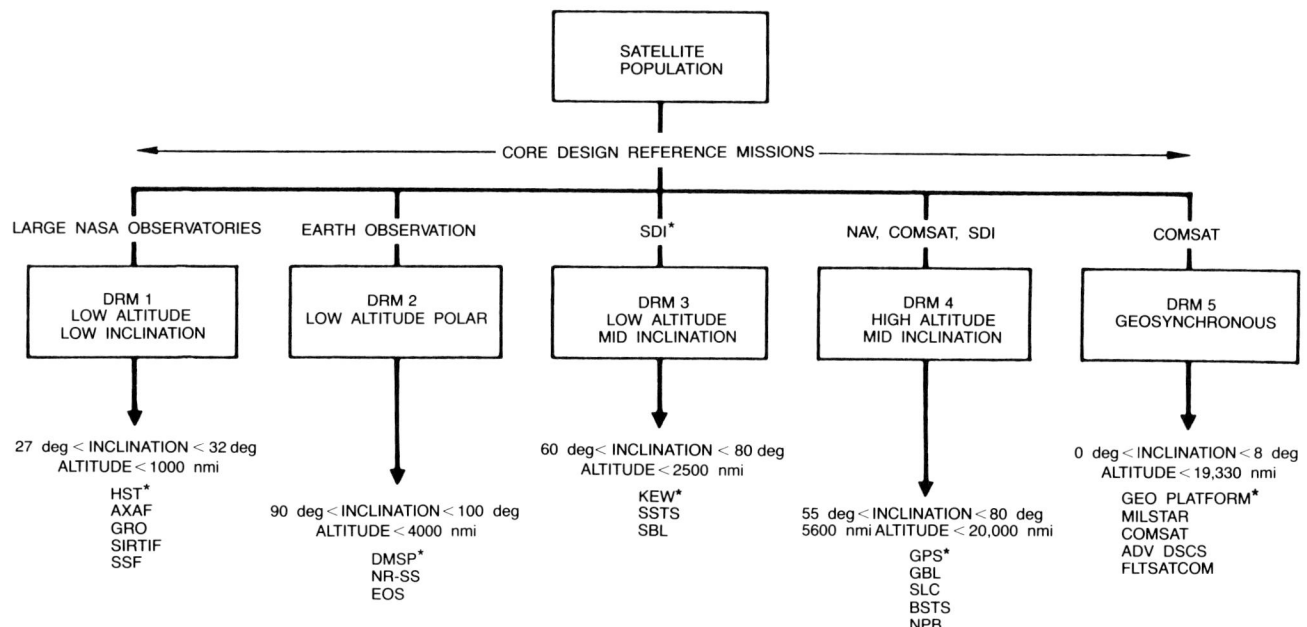

Figure 2.11 Orbit locations to consider for satellite servicing operations.

to have access with capability. Therefore, accessibility is related to support capabilities to establish the potential for spacecraft maintenance at any given time. Figure 2.17A depicts schematically the time phased increases envisioned in support. This depiction includes servicing equipment technology developments and improved spacecraft design features, many of which can be specifically forecast based on program implementation plans.

As space systems evolve, capabilities will continue to improve, always with more capability at the lower altitudes and at the commonly used inclinations. These capabilities generally improve smoothly and gradually over time, as equipment is routinely exercised and enhanced. Periodically there will be significant events or developments which constitute quantum improvements in capability. These improvements in support capability are shown schematically, by altitude, and plotted against time, in Figure 2.17B.

The profiles of Figure 2.17 A and B can be combined to form a more complex graphic, Figure 2.17C, which depicts schematic expectations of spacecraft maintenance capability throughout the various regimes of concern.

Some of the support capability increases (vertical axes of Figure 2.17 B and C) will occur gradually—but consistently—as routine Space Shuttle missions take place and equipment is exercised, extended, and enhanced. These routine increases in support capabilities will include:

- Improved EVA/power tools
- Improved space suit and glove performance
- Communications and enhancements, including television, data management, and computational upgrades
- Remote manipulator system (RMS) enhancements
- Equipment and servicing tools stowage improvements
- Astronaut and equipment restraint capability improvements
- Improved interface standards and specifications
- Improved assembly, maintenance, and repair techniques

On the other hand, some capability increases may occur as quantum jumps—shown on Figure 2.17C—as essentially straight vertical lines. The successful initial operational capability (IOC) or demonstration of an especially significant servicing tool, technique, or operation can be seen as a major increase. Examples are:

- Manned maneuvering unit demonstration/operations with the "open cherry-picker" (OCP)
- Demonstration of fuels/fluids transfer and orbital refueling systems, both by astronaut EVA and by autonomous remote operations
- Integration of handling and positioning aids
- IOC for an enhanced (8 psi) space suit
- Development of orbital control work stations on Shuttle and Space Station Freedom
- On-orbit assembly capability enhancements in techniques, tools, and software
- Demonstration of a satellite servicer systems (SSS) with advanced autonomy and robotics

Status of Satellite Servicing

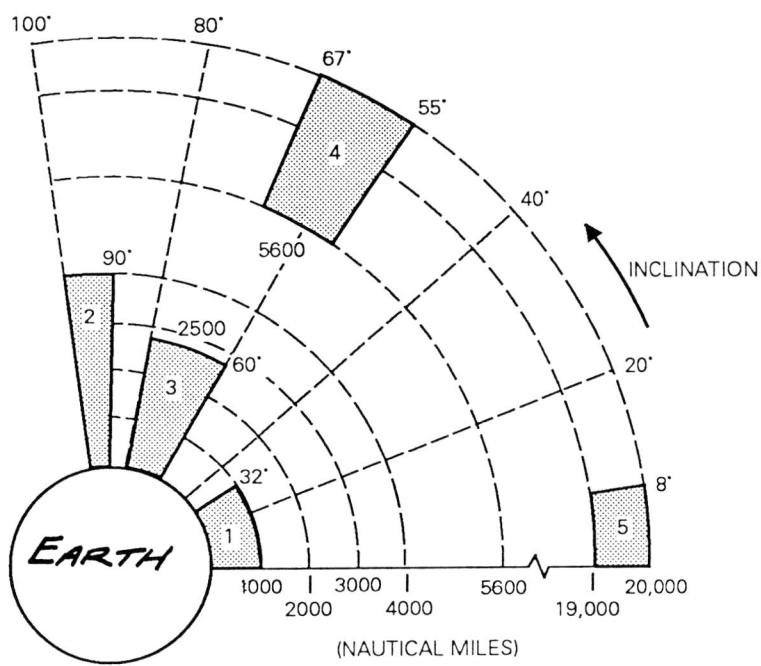

	DRM 1 LOW ALT LOW INC	DRM 2 LOW ALT POLAR	DRM 3 LOW ALT MID INC	DRM 4 HIGH ALT MID INC	DRM 5 GEO	OTHERS	TOTALS		
							NO. PROGRAMS	EARLY BEFORE 1995	LATE AFTER 1995
NUMBER OF PROGRAMS	209	46	2	5	108	12	382	151	231
	210	48	4	7	108	17	394	153	241
	227	48	7	7	108	17	414	153	261
	247	49	10	7	110	24	447	154	293
TOTAL	893	191	23	26	434	70	1637	611	1026

Figure 2.12 Spacecraft population distribution. *Courtesy of Lockheed Missiles and Space Co.*

- Expansion of the Space Station Freedom program to include space-based man-tended servicing facilities, maintenance depots, tank farms, and refueling stations in the servicing architecture
- Artificial intelligence and expert systems for advanced remote servicing
- Development of an in-space assembly/servicing (ISAS) facility that could be manned, man-tended, or totally automated as future requirements dictate

On the Figure 2.17C accessibility axis, the real concern again is accessibility with capability. Maintenance capability is a function of access with the required set of capabilities on board. Improved access will be provided, for example, by:

- Advanced expendable launch systems
- Vandenberg AFB launch site IOC (for access to space system in polar orbits)
- Improved Space Shuttle payload bay "toolbox" and servicing kit availability
- An orbital maneuvering vehicle IOC, followed by a manned OMV system
- A space-based orbital transfer vehicle
- An integrated combined NASA/Air Force satellite servicing logistics system for both space and ground operations
- Satellite integral propulsion system enhancements
- Computerized mission planning and transportation management systems

It is axiomatic that any specific program demands a review of the specific factors from which it will derive support and sustain its operational availability. It is therefore necessary to know, for example, if an OMV would be available by the time the spacecraft will need it to be accessible for maintenance. Figure 2.17 does not show such specific

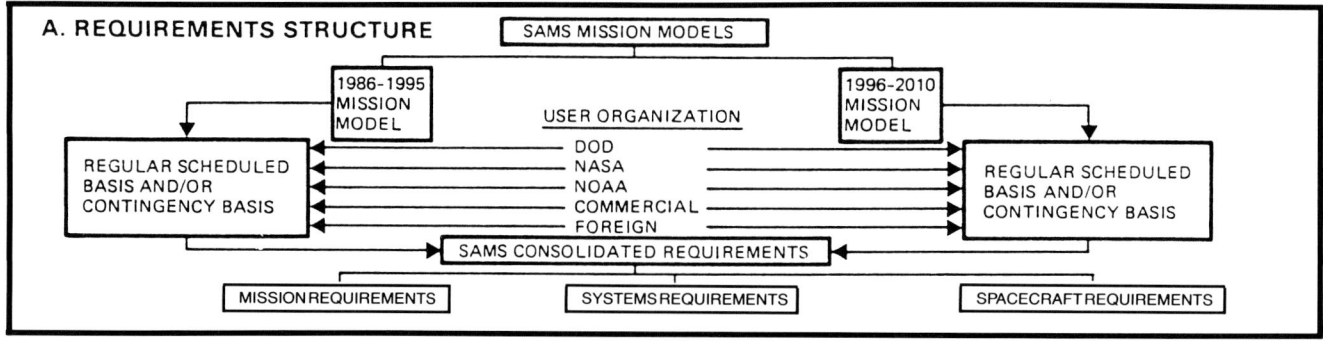

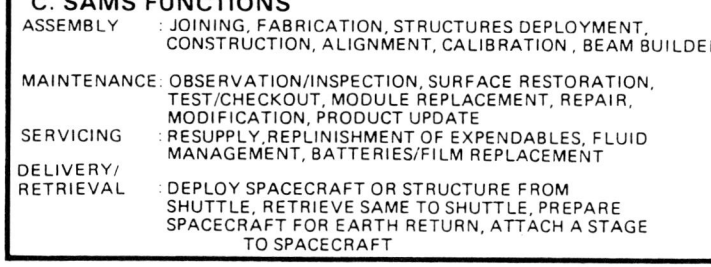

Figure 2.13 Compilation of consolidated requirements considerations. *Courtesy of TRW Space and Technology Group.*

dates, IOCs, and capabilities, but instead shows the general trend. It would be possible, even desirable, for a graphical representation much like this to be made up periodically for use by program managers and planners—with the application of specific dates and capabilities assigned to the quantum jumps and even to points on the routine increase line.

Spectrum of Current Status

In 1988 NASA indicated that the scope of orbital servicing candidate programs included those shown on Figure 2.18. Military and SDI programs will add to this spectrum. Data on transportation vehicle applicability to orbital regions (Figure 2.19) and existing Shuttle-based hardware applicability to servicing capabilities (Figure 2.20) add to the spectrum definition.

The use of the Space Shuttle and Space Station Freedom in current, near, and long term servicing application is projected on Figure 2.21.

It is interesting to note from Figure 2.22 that 70 percent of the servicing events between 1990 and 2010 are slated to occur at low altitude and low to mid orbital inclinations. This percentage will probably change after 2010 when the Strategic Defense Initiative (SDI) servicing architecture is in place, assuming an operational SDI program. By then, advances in robotics and artificial intelligence will have proceeded on a path to economic servicing operations at satellite locations in the higher altitude and inclination regions.

In summary, at present, satellite servicing is essentially a low altitude, Space Shuttle supported, manned/EVA, extensive ground command/control activity. The Space Station Freedom, OMV type vehicles with attached satellite servicer systems, next generation space suits, advanced servicing

Status of Satellite Servicing

Mission	User	Approx Total Weight	Orbit Altitude	Orbit Incl.	Launch Date	First Service	Service Interval	Mission Duration	Servicing Requirements
Hubble Space Telescope	NASA	24,970 lb 11,350 kg	600 km	28.5°	1990	1993	3 Years	15 Years	ORU Replacement, Instrument Changeout, Cryo Replenishment
Gamma Ray Observatory	NASA	35,000 lb 15,909 kg	350 – 450 km	28.5°	1990	Contingency Servicing		10 Years	ORU Replacement, Hydrazine Refueling
Advanced X-Ray Astrophysics Facility	NASA	29,700 lb 13,500 kg	600 km	28.5°	1997	2002	5 Years	15 Years	ORU Replacement, Instrument Changeout, Cryo Replenishment
Large Deployable Reflector	NASA	39,600 lb 18,000 kg	700 km	28.5°	2000	2000	2 Years	15 Years	ORU Replacement, Liquid Helium Replenishment
Upper Atmosphere Research System	NASA	15,000 lb 6,818 kg	600 km	57°	1993	Contingency Servicing		10 Years	ORU Replacement, Refueling
Earth Observatory System	NASA	24,200 lb 11,000 kg	800 km	98.7°	1996	Contingency Servicing		15 Years	ORU Replacement
Orbiting Solar Laboratory	NASA	7,400 lb 3,360 kg	510 km	97.4°	1996	1997	3 Years	4 – 6 Years	ORU Replacement, Instrument Changeout, Hydrazine Refueling
Space Surveillance & Tracking System	SDIO	10,000 lb 4,550 kg	Mid	High	TBD	Contingency Servicing		TBD	ORU Replacement, Fluids Replenishment
Space Based Interceptor	SDIO	TBD	Low	High	TBD	Contingency Servicing		TBD	ORU Replacement, Fluids Replenishment
Zenith-Star	SDIO	(2 Parts) 49,000 lb 22,300 kg each	327 km	28.5°	1995	Contingency Servicing		TBD	ORU Replacement, Cryo and Fuel Replenishment

Figure 2.14 Satellite servicing mission data base. *Courtesy of TRW Space and Technology Group.*

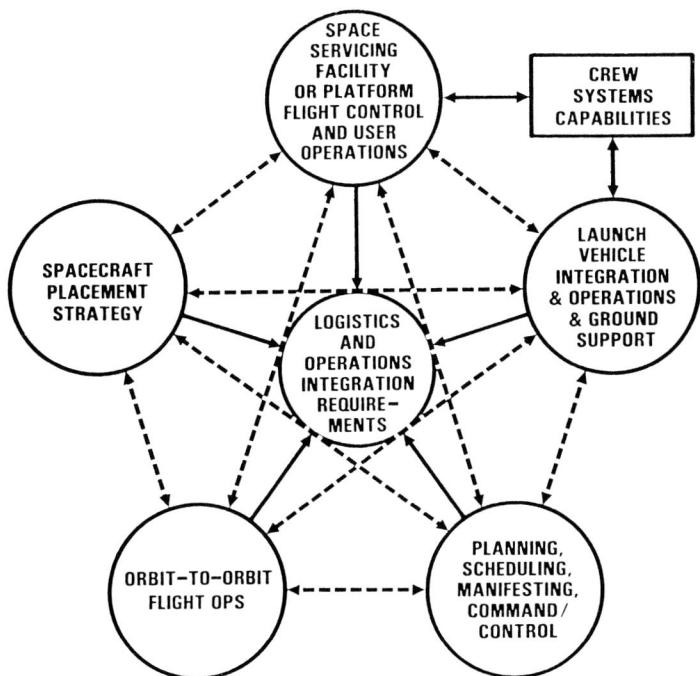

Figure 2.15 Servicing requirements needed to conduct on-orbit satellite servicing. *Courtesy of TRW Space and Technology Group.*

hardware/tools, and an in-space assembly/servicing facility will push orbital servicing to the next plateau of capability. This assumes, of course, that space systems will be designed to accept servicing.

Building the Satellite Servicing Technology Base

Technology is one of the three requirements that impact satellite servicing decisions. The other two are economics and user needs.

Servicing technologies and resultant capabilities will come in steps. Blocks of time are good indicators of when development translates into real applications. Figure 2.23 suggests the present and projected SAMS technology steps. The functions of assembly, maintenance, and servicing are laid out in 5 year time blocks from 1985 through 2010 with an indication of the SAMS capability predicted for each block.

Figure 2.24 is another look at technology readiness—this time from the viewpoint of the flow of development from the present day EVA to the future robotic techniques. Typ-

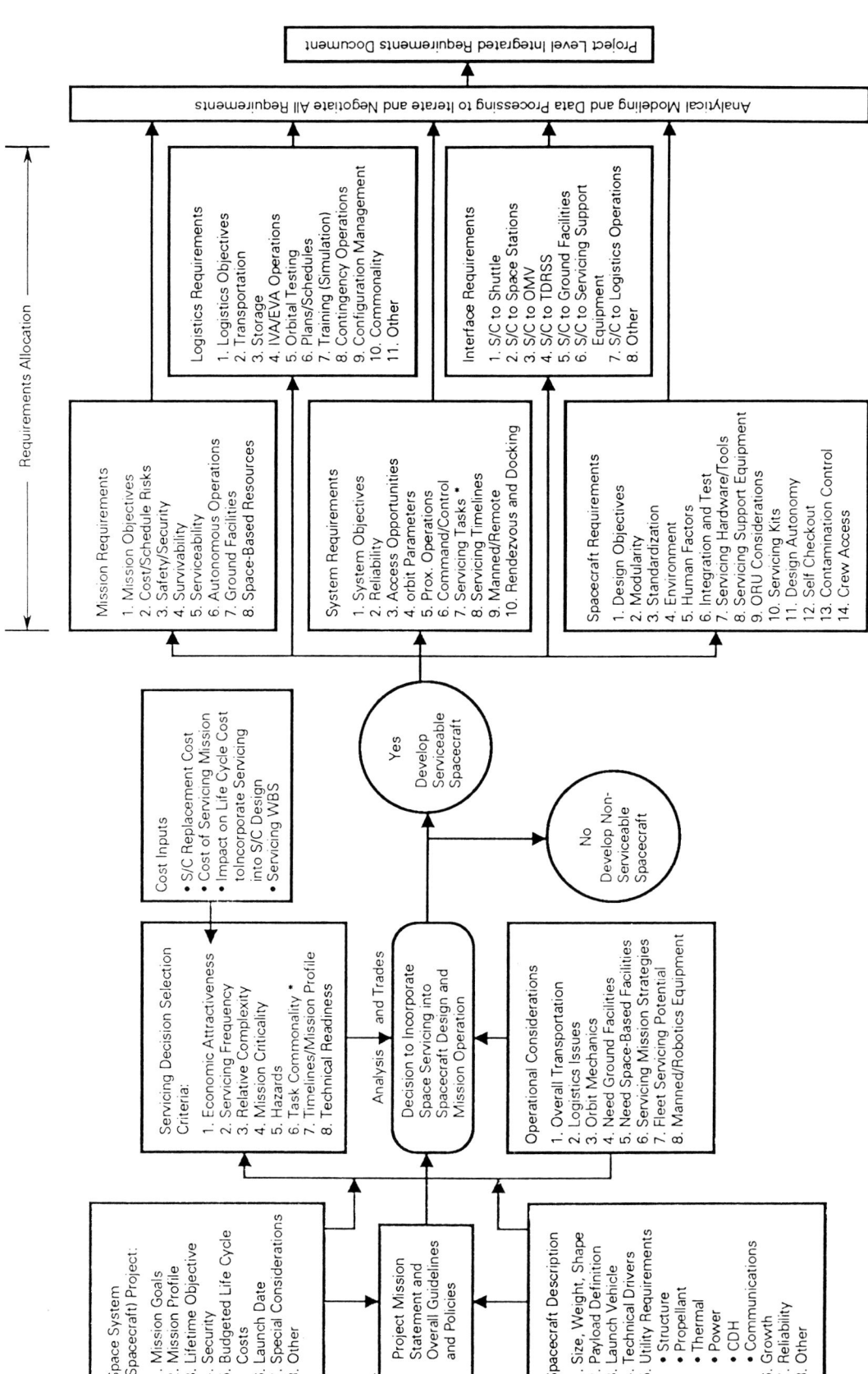

Figure 2.16 Steps in developing integrated servicing requirements. *Courtesy of TRW Space and Technology Group.*

* Tasks Relate to Assembly, Maintenance, or Servicing Work or Combinations of All Three That Apply to a Specific Project

Status of Satellite Servicing

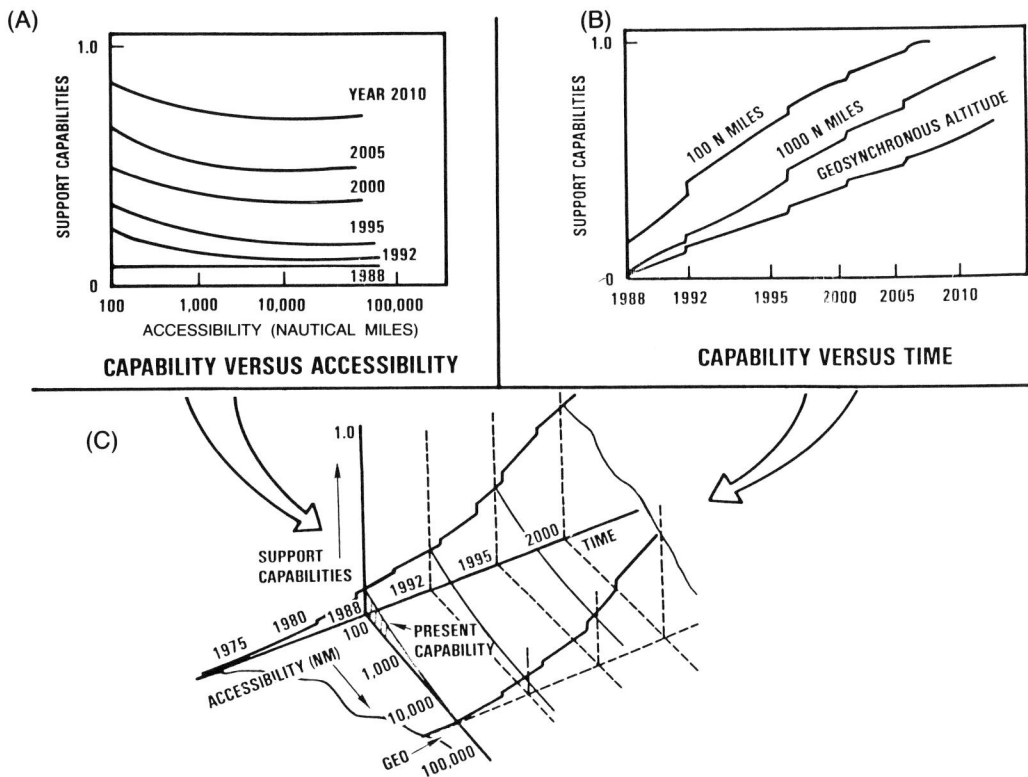

Figure 2.17 Projection of generalized U.S. satellite servicing supportability and excessibility as a function of time.

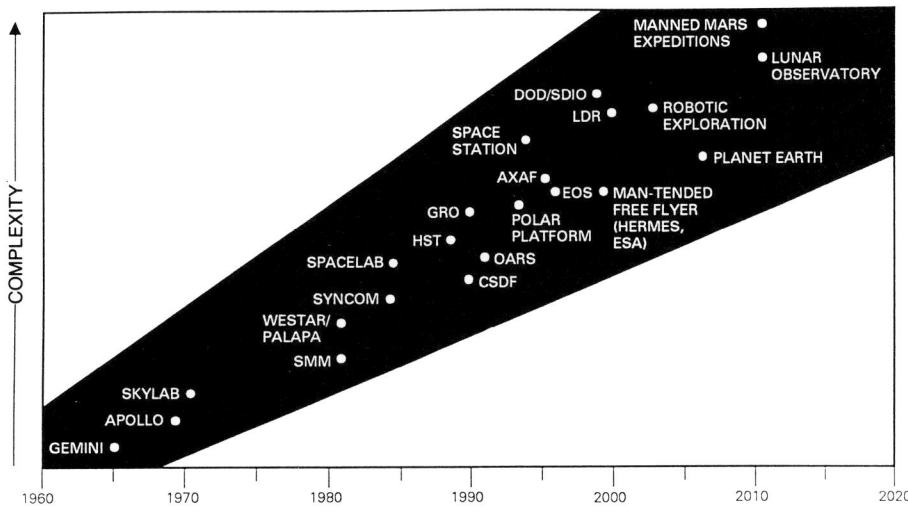

Figure 2.18 On-orbit servicing spectrum. *Courtesy of TRW Space and Technology Group.*

ical NASA programs are shown under the four blocks as EVA transitions to autonomous servicing missions.

Technology Plan

During the SAMS study [2], a range of mission functions and operations was established. An extensive industry/government survey then determined the required technologies and their levels of readiness. The survey results are summarized in Figures 2.25 through 2.28. The technologies are broken into four major categories:

1. Man-in-space technologies
2. Spacecraft design technologies
3. Servicing equipment technologies

VEHICLE APPLICABILITY TO ORBITAL REGIONS

ORBITING REGION / VEHICLE	LOW ALTITUDE/ LOW INCLINATION	LOW ALTITUDE/ MID INCLINATION	LOW ALTITUDE/ POLAR ORBIT	HIGH ALTITUDE/ MID INCLINATION	GEO-SYNCHRONOUS
STS	●	●	ASSUMES VAFB LAUNCH		
SS	●				
OMV	FROM THE STS OR SS	FROM THE STS OR SS	FROM THE STS ONLY	FROM ELV	
OTV				FROM THE STS OR SS	FROM THE STS OR SS
ELV	●	●	●	●	●
COMMENTS	(1) SERVICING OF FREE FLYERS / PLATFORMS AT THE SS INVOLVES RETRIEVAL BY OMV / OTV (2) POLAR ORBITING SPACECRAFT CAN BE SERVICED FROM STS (VAFB LAUNCH) OR ELV's ONLY				

Figure 2.19 Vehicle applicability to orbital regions. *Courtesy of NASA.*

	RETRIEVING RESUPPLYING	REPAIRING	UPGRADING	DEPLOYING	ASSEMBLING CONSTRUCTING
FLIGHT QUALIFIED HARDWARE					
●EVA					
MODULE SERVICE TOOL	●	●	●	●	●
SMART POWER RATCHET TOOL	●	●	●	●	●
EVA FLUID COUPLINGS (MONOPROP.)	●				
ELECTRICAL CONNECTORS	●	●	●	●	●
FLIGHT SUPPORT SYSTEM (FSS)	●	●	●	●	●
HAND TOOLS	●	●	●	●	●
MMU	●		●		●
●RMS	●	●	●	●	●

Figure 2.20 Existing Shuttle-based satellite servicing hardware and capabilities [6]. *Courtesy of NASA.*

4. Mission operations technologies

The survey assessment of technology readiness used the NASA Space Systems Technology Report ratings. There are eight levels of readiness as shown below:

1. Basic principles have been observed and reported.
2. Conceptual designs have been formulated.
3. Conceptual designs have been tested (analysis/experiment).

Status of Satellite Servicing

	STS	SS	OTHERS
NEAR TERM CURRENT - 1994	• FREE-FLYER SERVICING	• N/A	• N/A
MID TERM 1994 - 2000	• FREE-FLYER SERVICING • SS ASSEMBLY • RESUPPLY OF SS	• SERVICING OF SS ATTACHED PAYLOADS • SERVICING OF FREE FLYERS & PLATFORMS	• USE OF OMV TO DEPLOY AND RETREIVE FREE FLYERS
LONG TERM > 2000	• FREE-FLYER SERVICING • RESUPPLY OF SS	• SERVICING OF SS ATTACHED PAYLOADS • SERVICING OF FREE FLYERS & PLATFORMS • ASSEMBLY OF LARGE PAYLOADS	• USE OF OMV/OTV TELEROBOTIC SERVICING • USE OF OTV TO TRANSFER TO AND FROM GEOSYNCHRONOUS • USE OF ELV'S

Note: Many of the mid and long term payloads are still in the planning and development stage and have not yet been approved for development

Figure 2.21 Use of Space Shuttle, Space Station Freedom, and other vehicles for servicing* [6]. *Courtesy of NASA.*

4. Critical functional characteristics have been demonstrated.
5. Component brassboards have been tested in operational environment.
6. Prototype engineering model has been tested in operational environment.
7. Engineering model has been space-tested.
8. Full operational capability is achieved.

A mathematical rating survey of technologies required for the early years of SAMS implementation was performed during the SAMS study. The 6 technology areas, out of the 30 in Figures 2.25 to 2.28, that rated highest from the survey are listed in Figure 2.29. The criteria for rating the technologies included early year payoff, relevance to a real or planned DoD/NASA/SDIO space programs, and belief that "good" proof-of-concept efforts could be developed to advance the technology.

The hardware items that have been developed under NASA sponsorship are listed on Figure 2.30 and scheduled on Figure 2.31.

Servicing Decisions

A national planner of a major national space program or a project manager of one of its elements will want to consider a number of parameters in arriving at a decision to implement or not implement space assembly, maintenance, or servicing. The important parameters are:

1. *Necessity.* Is it absolutely essential to basic mission success?
2. *Satellite mission durability.* If the basic mission of the satellite becomes obsolete before any benefit or extended service life is realized, then servicing to extend life has no merit.
3. *Required near-term investment.* It appears that, in the long term, SAMS is highly cost-effective, but there is a near-term question of affordability. There will not be offsetting cost savings during the buildup to an operational capability, and a national decision to invest in satellite servicing in the future has to be made.
4. *Guaranteed SAMS availability.* The servicing system must be sufficiently robust so there is no significant interruption in the availability of servicing or maintenance due to limited scope casualties.
5. *Total life cycle cost/cost-effectiveness.* Total life cycle cost of a particular system may well be lowered if SAMS equipment user charges are set artificially low, but the national viewpoint must consider full cost recovery for the development of the user equipment to properly assess cost-effectiveness.

ORBITAL LOCATION			TYPICAL MISSIONS	WITHOUT SDI ARCHITECTURE		OPERATIONAL SERVICING CAPABILITY
DESCRIPTION	ALTITUDE N. MI	INCL. DEG.		PERCENT OF TOTAL SPACECRAFT (%)	PERCENT OF SERVICING EVENTS (%)	
LOW ALTITUDE, LOW INCLINATION	<1000	ZERO TO 32	• NASA SPACE STATION • NASA's LARGE OBSERVATORIES • ASTROPHYSICS • SPACE TEST/DEMO • COMMERCIAL	POTENTIAL 54	40	1990 TO 1992

LOW ALTITUDE, POLAR INCLINATION	<4000	90 TO 100	• EARTH OBSERVATION
			  – NASA
			  – DoD
			• METEOROLOGY
			• COMMERCIAL
				13	14	1992 TO 1995

LOW ALTITUDE, MID INCLINATION	<2500	60 TO 80	• SPACE TEST/DEMOS • SDI PROJECTS • NAVIGATION • ASTROPHYSICS	2	30	1995 TO 2000
HIGH ALTITUDE, MID INCLINATION	5600 TO 20,000	55 TO 80	• SDI PROJECTS • NAVIGATION • METEOROLOGY	3	11	1995 TO 2000
GEOSYNCHRONOUS	19,330	0 TO 8	• SPACE TEST/DEMOS • SDI PROJECTS • COMMUNICATIONS • DoD SUPPORT • METEOROLOGY • COMMERCIAL	28	5	2000 TO 2010

Figure 2.22 SAMS data base missions, 1988–2010 serving summary. *Courtesy of TRW Space and Technology Group.*

6. *Technical feasibility.* We must satisfy ourselves that the planned equipment and mission scenarios are technically achievable. Assessment to date indicates that no fundamental enabling technology remains to be "invented."

The above are not necessarily listed in order of importance. Each specific project office must judge the priorities of its mission, then rank this list accordingly. The decision parameters, trends, and cost drivers, Figure 2.32, must then be integrated into the conclusion as to whether to incorporate satellite orbital assembly/maintenance/servicing into a project or to simply replace the satellite at the end of its lifetime or at a major failure point.

Trends

Several trends provide impetus to servicing implementation because they make basic mission operational cost-effectiveness easier to achieve. These are:

1. *Satellite system cost/complexity.* The trend from the earliest days has been for more complex, demanding missions, with increased spacecraft weight and complexity as a consequence. If the trend continues, basic affordability may come into question.

2. *Transportation costs/launch costs.* The trend has been toward a steady increase. The development of new systems such as Heavy Lift Launch Vehicle (HLLV), Advanced Launch Systems (ALS), and STS II *may* reduce launch costs, but that is by no means guaranteed. The higher the cost of transportation, the more attractive the decision to service.

3. *Guaranteed access to space.* If SAMS users must worry about whether access to their systems for servicing may be interrupted for lengthy periods, they might opt to pursue a different strategy and pay the premium price for "silent spares" on-orbit.

4. *Non-DoD users/commercialization.* NASA has, thus far, been the only user agency for satellite servicing

Status of Satellite Servicing

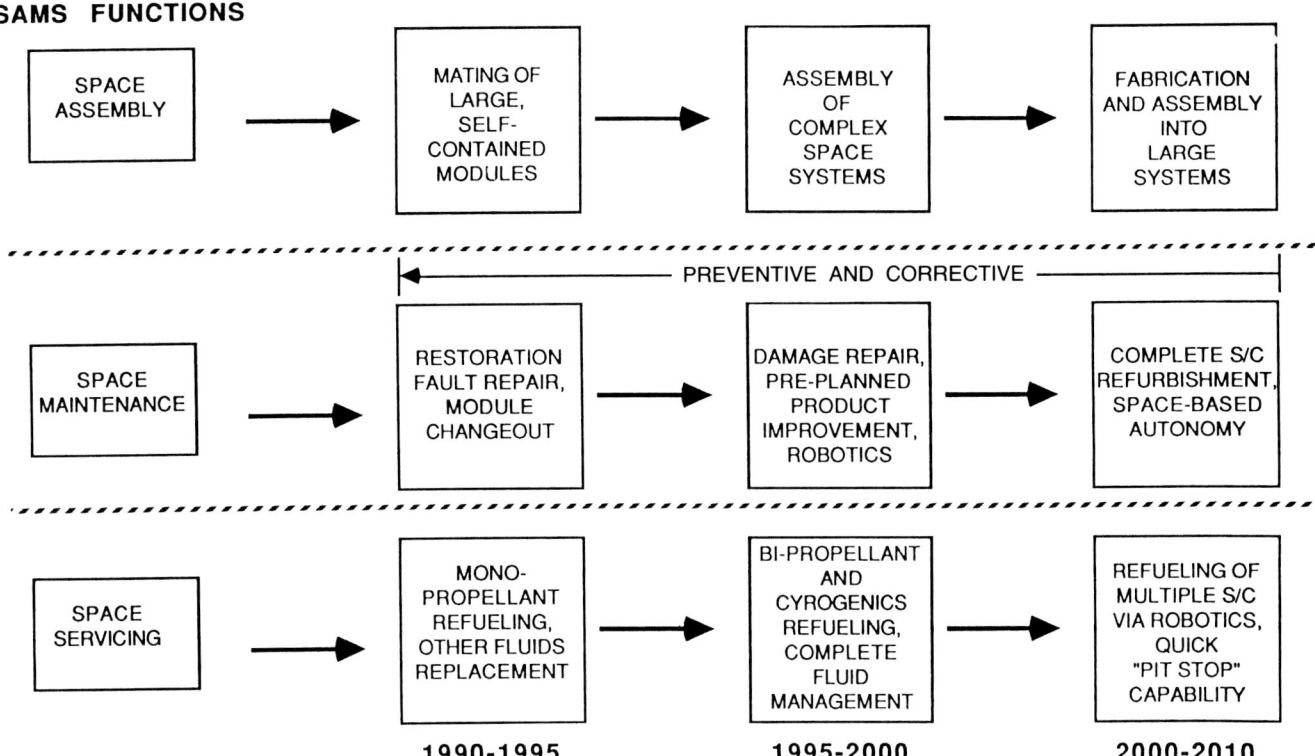

Figure 2.23 Technology evolution. *Courtesy of TRW Space and Technology Group.*.

and has demonstrated the operational feasibility and, to some extent, the economic benefits. The beginnings of commercialization can be seen, and as the lure of profit spurred the industrial revolution of the 18th and 19th centuries, so will the lure of profit spur the industrialization of space.

Cost Drivers

A number of items drive the total cost of any space system. Some operate uniformly on all applications to raise cost and are not discriminators in the decision relative to servicing. Other factors are both total cost drivers and cost discriminators, in that they act differently depending on the strategy chosen. The two major cost drivers are:

1. *Satellite system complexity, size, and weight.* Discussed above.
2. *Transportation costs.* Transportation is a significant fraction of total system servicing cost; the effects of the variation are shown in Figure 2.33. The data on this figure depicts cost of maintaining a three satellite constellation over a 20 year period, after initial deployment. The "replace" line is an estimate of the variation in cost of replacing the constellation on 7 year centers, at $250 million per satellite, assuming transportation costs to low Earth orbit ranging from a low of $100/lb ($45/kg) to a high of $2,000/lb ($907/kg). The three "service" lines show the variation in cost of servicing the same constellation over 20 years, assuming the replacement ORUs cost the specified percentage of the satellite replacement cost. The error bars on Figure 2.33 illustrate the variation in all costs due to changing the size of the user amortization base, from a low of 50 uses, to a high of 300 uses over the 20 year life of the equipment. The curves for 33.3 percent ORU costs and 50 percent ORU costs have the same basic shape and "knees" but exhibit differing break-even points and levels of savings.

Breakpoints for Servicing.

Several identifiable breakpoints, defined as places in the curve where a significant slope change occurs, emerge from consideration of the cost-effectiveness trades, and the study of the curves given in Figures 2.33 and 2.34. Major breakpoints are observed where the:

- ORU costs are not greater than 50 percent of total satellite replacement cost.
- Servicing equipment user charges are less than 50 percent of total satellite replacement cost.
- Servicing time intervals are at least 4 to 5 years.
- Servicing time intervals are at least one-third of the time required for satellite replacement.

Consideration of the number of servicings per "mission" (the numbers in parentheses in Figure 2.35) is driven by

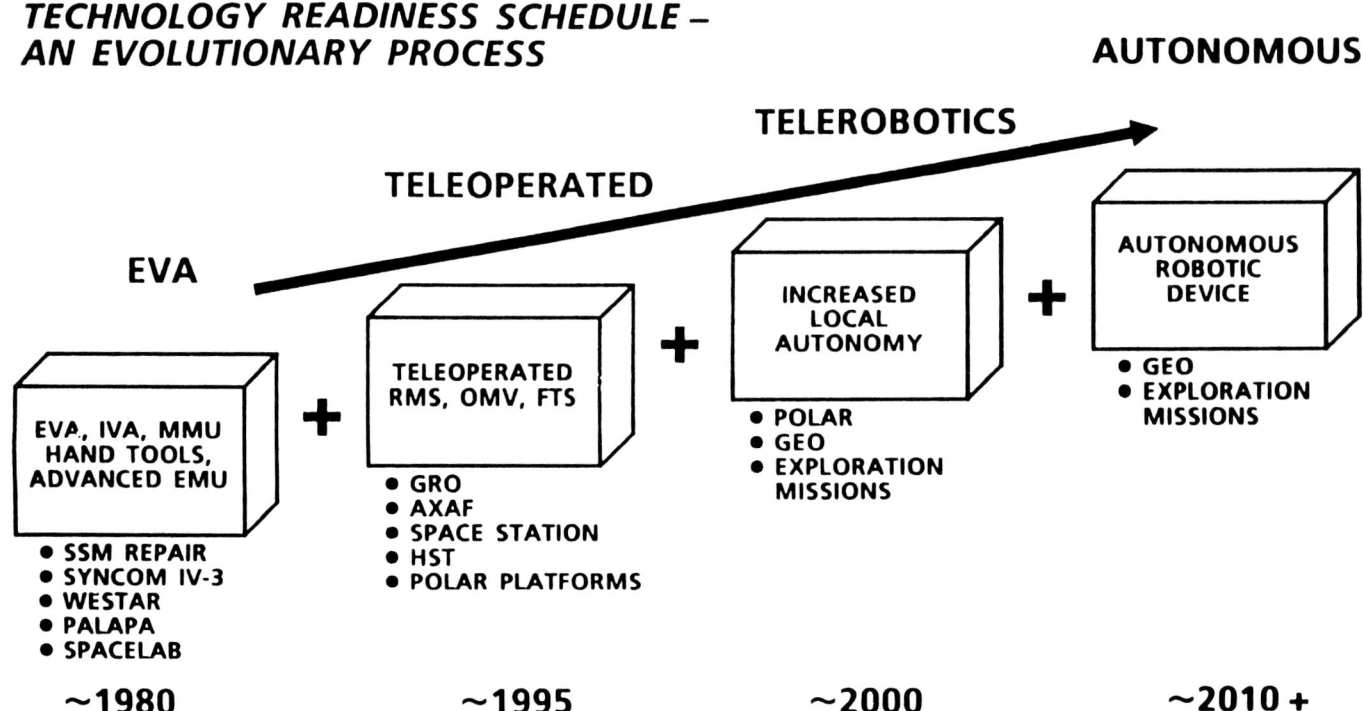

Figure 2.24 On-orbit servicing EVA/robotics prediction. *Courtesy of NASA.*

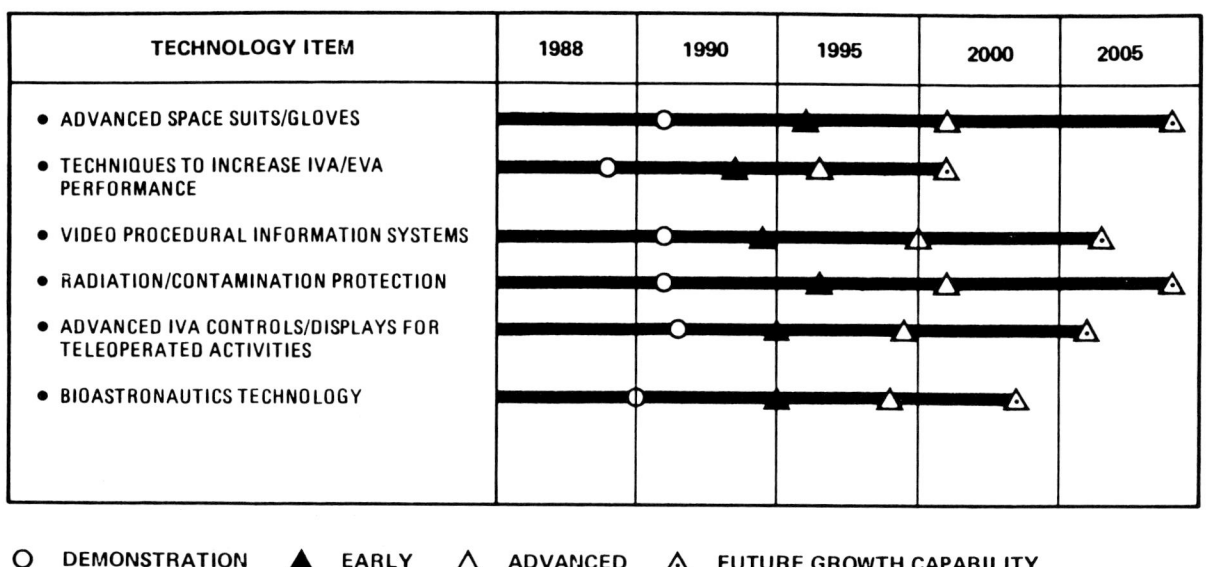

Figure 2.25 Human-in-space technologies readiness requirements. *Courtesy of TRW Space and Technology Group.*

ORU costs as a percentage of satellite replacement costs as well as by the servicing interval. As illustrated, if ORU costs approach 50 percent, anything less than two satellites serviced per mission or a service interval of less than 4 years makes servicing economically unattractive.

Figure 2.35 presents savings as a percentage of replacement costs for varying servicing assumptions. The circled points are calculated data to which the curves have been fitted. The solid lines depict 20 percent ORU costs, the dotted broken lines depict 33.3 percent, and the dashed lines show 50 percent. The numbers in parentheses at the end of the curves denote the assumed number of satellites in the constellation serviced per mission. Three servicings per mission were assumed for the "baseline," but significant savings can still be achieved for two, even one, servicings per mission, if the interval is sufficiently long and ORU costs are controlled.

Status of Satellite Servicing

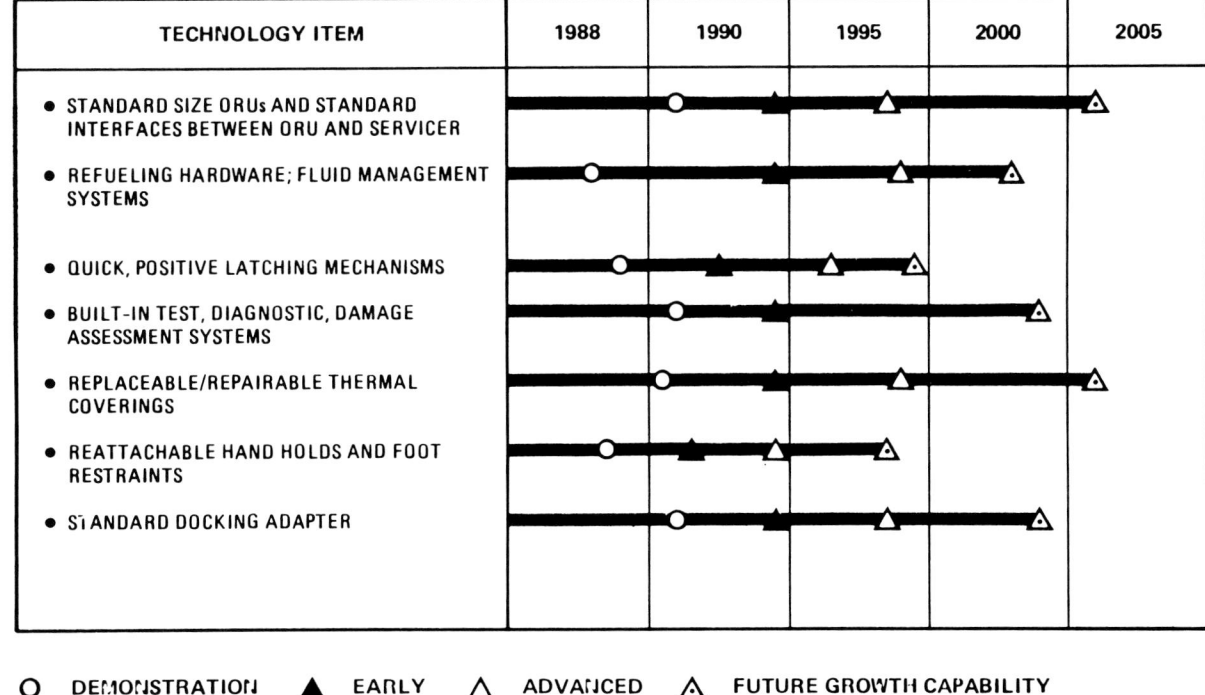

Figure 2.26 Spacecraft design technologies readiness requirements. *Courtesy of TRW Space and Technology Group.*

The basic shape of these curves describes the behavior of costs in relative terms. The actual "position" of these curves will vary depending on the cost estimating relationships used, but the interrelated behaviors described should remain invariant.

Once again, the "knee" in the cost curves appears at about the 4 year service interval point. The ideal service interval would equal the satellite replacement interval.

Satellite Servicing Issues

On-orbit satellite servicing is an emerging concept in the field of space operations. Each servicing study performed by a government or industry organization during the past 5 years included a list of attention grabbing issues.

An issue is defined as a problem or concern needing resolution before servicing can be fully implemented into space systems design/development and/or incorporated into routine space operations.

Servicing issues deemed important enough to attract DoD or NASA review for possible near-term study are listed below under three categories—Engineering, Operations, and Programmatic [6].

Engineering Issues

These issues are grouped under the categories of EVA, automation and robotics, fluid storage and transfer, payload handling, checkout and diagnostics, and satellite design.

The extravehicular activity (EVA) issues encompass a variety of problems associated with:

1. Servicing tasks in high energy orbits
2. Crew protection and decontamination during orbital fluid transfers and system maintenance of toxic/corrosive fluid systems
3. Tool enhancements for more efficient EVA work
4. Robotic/EVA interface compatibility
5. EVA tool catalog update/revision
6. EVA man-machine interface criteria
7. Crew induced loads for more productive use of EVA hours
8. Advanced space suit development
9. Crew training, crew selection criteria, and number of EVAs permissible per Shuttle mission to assure crew health and safety

Automation and robotics issues are being attacked to reduce the crew EVA hours needed to complete major projects such as Space Station Freedom assembly and payload changeout of a large space science observatory. These issues are:

1. Compatibility of robotics activities with EVA tasks to enable an optimum blend of both
2. Assessment of the current status and future needs of critical automation and robotic technologies

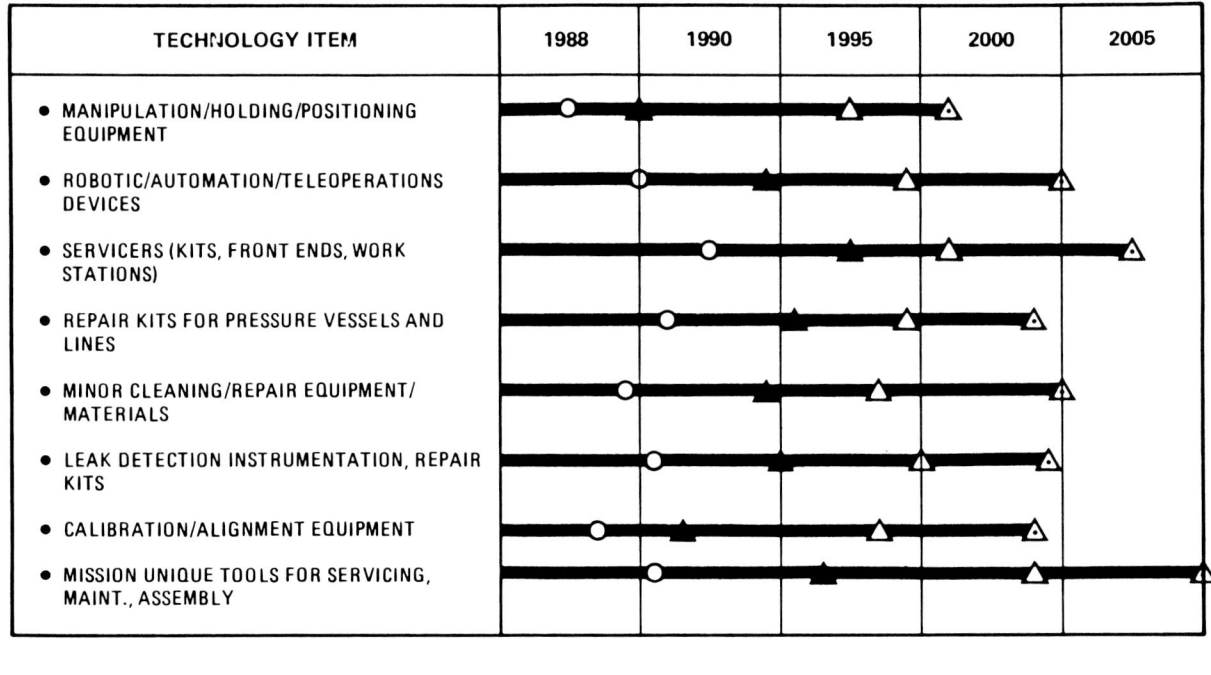

Figure 2.27 Servicing equipment technologies readiness requirements. *Courtesy of TRW Space and Technology Group.*

3. Compatibility of robotic system interfaces with an OMV and with spacecraft being serviced
4. Automation and robotics servicing requirements integrated into the system requirements of new spacecraft design
5. Standard tool interfaces for automated/robotic systems
6. Availability of suitable robotic hardware and software for specific satellite servicing
7. Viewing/cameras/lighting/visualcoding/sensors and display
8. Stereo video for depth perception
9. Enhanced telepresence
10. Role of artificial intelligence

These fluid storage and transfer issues are under study as they are central to extending satellite on-orbit lifetime:

1. Fluid resupply low gravity heat transfer correlations
2. Orbital resupply of high pressure gases
3. Zero "g" liquid tank venting for orbital fluid resupply
4. Bipropellant neutralization for orbital fluid resupply
5. Automatic fluid and gas resupply couplings
6. Pumps and variable pressure regulator/relief valves for orbital fluid resupply
7. Superfluid helium coupling
8. Oxidizer diaphragms
9. Orbital fluid connection leak detection
10. Zero "G" fluid quantity measurement techniques
11. Combination fluid and ORU changeout servicer design

Payload handling issues are key to reducing the time on orbit it takes to service a space systems. The major issues are:

1. Telerobotic interfaces with the Space Shuttle's RMS
2. Payload capture, manipulating, and positioning hardware
3. Hardware to hold, rotate, cradle, and support the payload

Checkout and diagnostic issues are as important in space servicing as they are in Earth servicing of complicated machinery. The issues are associated with:

1. Space Shuttle-based satellite test and diagnostic equipment
2. Satellite/Space Shuttle communications
3. Ground-based decision process, including command and control
4. Extent of spacecraft autonomy to accommodate self-checkout
5. Spacecraft built-in test equipment
6. Use of simulators and mockups to evaluate checkout and diagnostic techniques
7. User friendly software to operate test, checkout, and

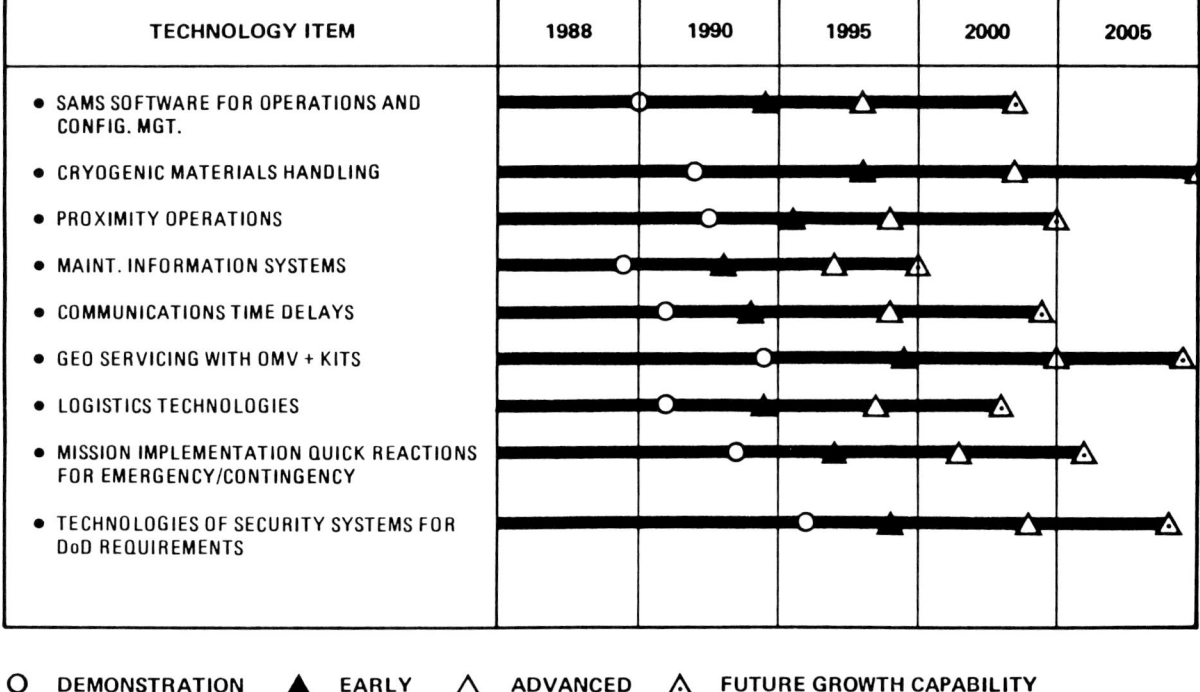

Figure 2.28 Mission operations technologies readiness requirements. *Courtesy of TRW Space and Technology Group.*

Early (1988–1995) Technology Emphasis	SAMS Functions		
	Assembly	Maint.	Servicing
• Man-in-Space IVA/EVA technologies	X	X	X
• Robotic Devices/ Automation Hardware Telepresence Systems	X	X	X
• Built in Test/Diagnostic/ Checkout Equipment		X	
• Fluid Management/ Refueling Systems			X
• On Orbit SAMS Operations related to:			
– P^3I Implementation (ORUs, Standardization, Config. Mgt.)		X	X
– Contamination Control	X	X	X
– Manipulation, Handling, Positioning Equipment	X	X	X
– Logistics Systems	X	X	X
• Mission Unique Hardware/Tools/Support Equipment	X	X	X

Figure 2.29 SAMS technology drivers. *Courtesy of TRW Space and Technology Group.*

diagnostic hardware on the spacecraft and on the servicer

Finally, servicing must be incorporated into the spacecraft's early design layouts. The issues here are:

1. Extent of mission autonomy built into the spacecraft
2. Module partitioning and interfaces and EVA or robotic access to the spacecraft's ORUs
3. Standards and standardization to be employed in the design and operation of a serviceable spacecraft
4. Configuration control documentation and methods to be enforced including closeout photos of all equipment and interfaces as flown
5. Maintainability/supportability requirements
6. Spacecraft designed in concert with servicing hardware/tools support equipment design
7. EVA friendly considerations (crew handholds, markings, color codes, safety)

Operations Issues

The operational considerations relate to proximity maneuvers, Space Station activities, safety, tactical operations, and logistics.

These are the proximity operations issues that are important to a satellite as it approaches or departs the Space Shuttle, Space Station Freedom, OMV, servicing platforms/depots, or fuel tankers:

- **Orbital Maneuvering Vehicle (OMV)** - Reusable propulsion stage capable of transporting spacecraft and other payloads between low earth orbit trajectories.
- **Cryogenic Transfer Couplings** - EVA model for incorporation in near-term spacecraft.
- **Dexterous End Effector** - Mechanical hand for RMS to allow more refined RMS operations than currently achievable.
- **Flight Telerobotic Servicer (FTS)** - Robotic system designed to assist in Space Station assembly and maintenance; to be structured for evolutionary capabilities.
- **Helium On-orbit Transfer Experiment (Shoot)** - Shuttle-based, in-bay demonstration of superfluid helium transfer under operational conditions.
- **Hydrazine Tanker** - Designed for spacecraft refueling.
- **Laser Docking Sensor** - Hardware designed to track passive orbital target spacecraft with sufficient accuracy to enable soft docking near target vehicle.
- **Module Service Tool (MST) Derivatives** - End effectors modified to be mounted on the end of the RMS lightweight MST-modified to reduce weight and facilitate ease of operation to operate as a front-end tool on teleoperated or remote servicing facilities.
- **Optical Communications** - Point-to-point communication capability for operations from the STS aft flight deck to local servicing sites.
- **Standard I/F Connectors and Umbilical** - Remote mateable/demateable flexible umbilical connector to provide remotely operated umbilical connector capability for electrical, gas, and fluid services; umbilical carrier mechanism controlling the connection of umbilicals to minimize disturbances to the satellite and Shuttle.
- **Superfluid Helium Tanker** - Resupply cryogenic coolants on board the SIRTF and possibly second-generation Hubble Space Telescope and AXAF instruments.

Figure 2.30 Hardware items that NASA now has or did have under development. *Courtesy of NASA.*

1. Plume impingement
2. Safety standards
3. Command and control
4. Ground-to-orbit-to-ground RF signals timelag
5. Remote operator training via simulators

The satellite servicing operations at the Space Station issues are:

1. Contamination control
2. Space Station facilities and crew availability
3. Manned/remote optimum mix
4. Commonality and standards
5. Optimum orbital mechanics

Safety is the most important item in any on-orbit activity. The two issues requiring paramount attention are:

1. Crew safety with procedures and equipment in place for assured crew egress from nonsafe conditions
2. Mission and hardware protection

Tactical operations issues are a function of mission planning, hardware capability, on-orbit hardware performance, and crew safety. The principle issues are:

1. How to cope with environment (natural, induced, enemy threat)
2. Prediction and staffing and conduct of contingency servicing missions
3. Servicing strategies including abort potential
4. Placement in space of manned-tended/unmanned work platforms, warehouses, fuel/ORU storage depots
5. Mission support and backup planning

Space servicing is a logistics operation. Logistics planning must be integrated with mission analysis. The principle issues are:

1. Architecture, integrated facilities, transportation scheduling and cost-effective flow of hardware and crews
2. Ground facilities—how many, where, who owns?
3. Space-basing strategies
4. Design inputs to spacecraft and servicing equipment to optimize logistics

Programmatic Issues

The principle issues associated with the planning and execution of on-orbit servicing missions are related to costs,

Status of Satellite Servicing

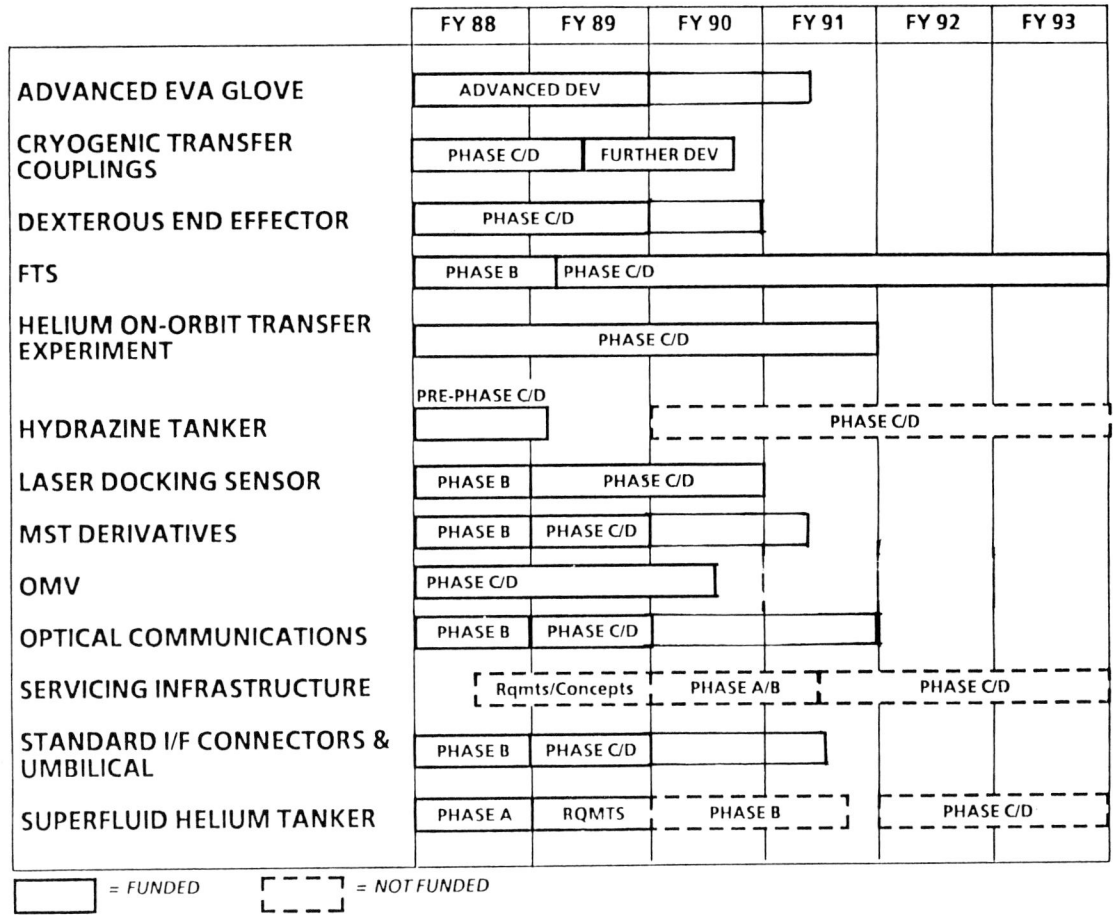

Figure 2.31 Satellite servicing hardware development status [6]. *Courtesy of NASA.*

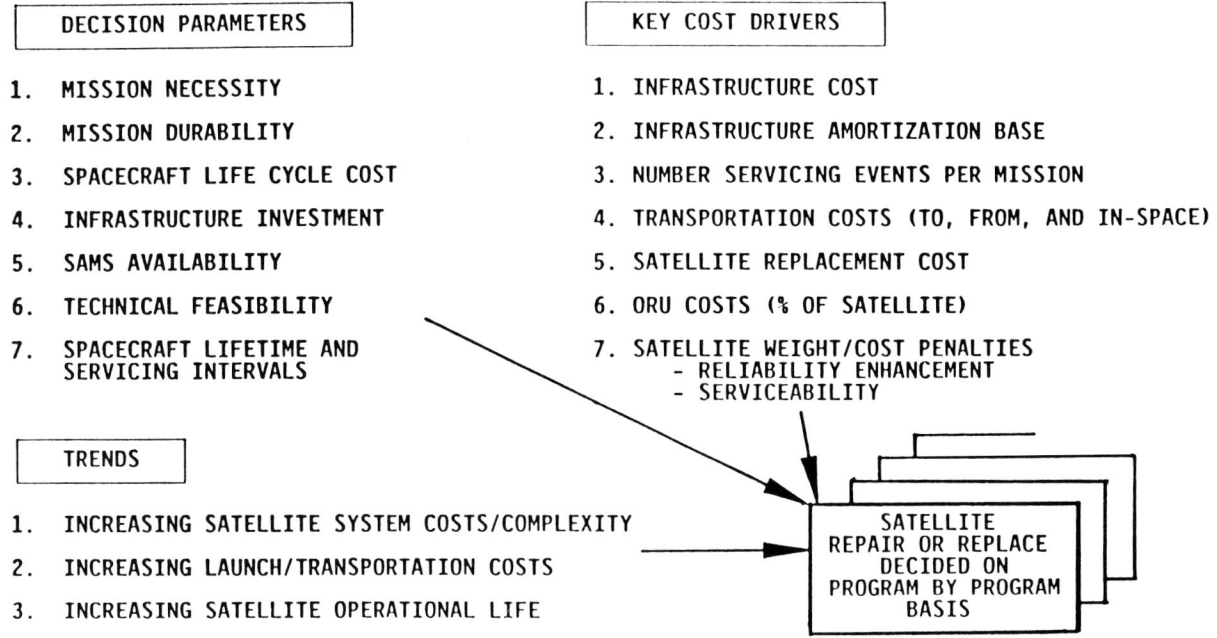

Figure 2.32 Satellite on-orbit servicing considerations.

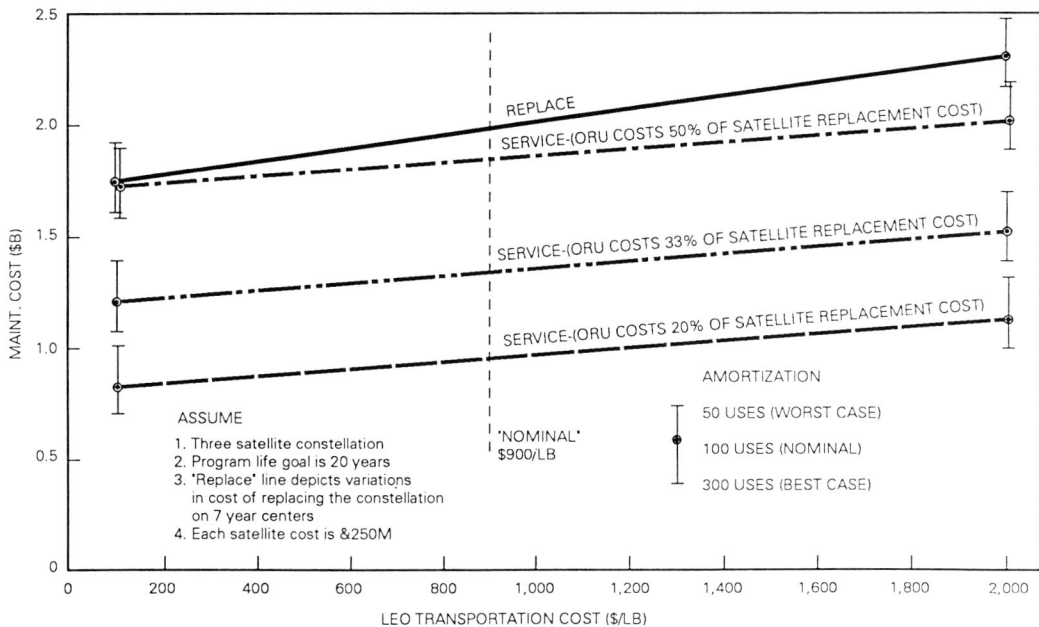

Figure 2.33 Maintenance cost versus LEO transportation cost as amortized over three potential use cases. *Courtesy of TRW.*

EVA, automation and robotics, and decisions related to in-flight operations.

Costs, including space system life cycle costs, spacecraft and satellite servicer system development and first article delivery costs, mission operational costs, and all support systems costs are important issues. In addition, these issues must be considered:

1. Cost to make spacecraft serviceable
2. Access policy
3. Cost traceability within NASA
4. Servicing prices
5. Role of STS during Space Station ERA
6. Orbiter servicing availability and policy
7. Cost and schedule risks
8. Technology readiness versus government budget constraints

The extravehicular activity issues that pertain to programmatics are:

1. EVA R&D laboratory equipment for ground-based simulations
2. NASA satellite servicing commitment
3. Interface standardization policy that pertains to the EVA hours per mission and crew safety

The programmatic automation and robotics issues are:

1. Teleoperations, control station locations for remotely operated systems
2. Extent of commonality employed on the various robotics systems onboard the serviceable spacecraft and the servicer vehicle
3. Costs of using automation and robotics as traded against use of EVA

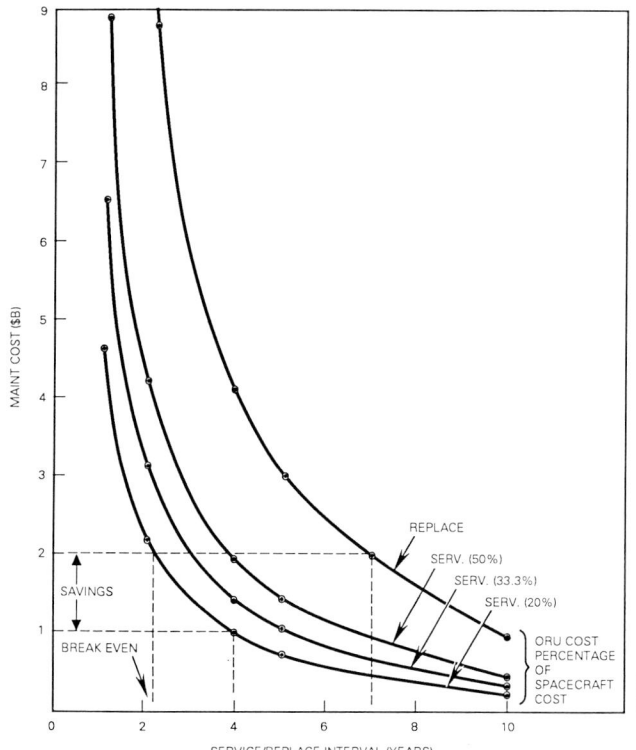

Figure 2.34 Cost versus replace/service interval and ORU percent cost. *Courtesy of TRW.*

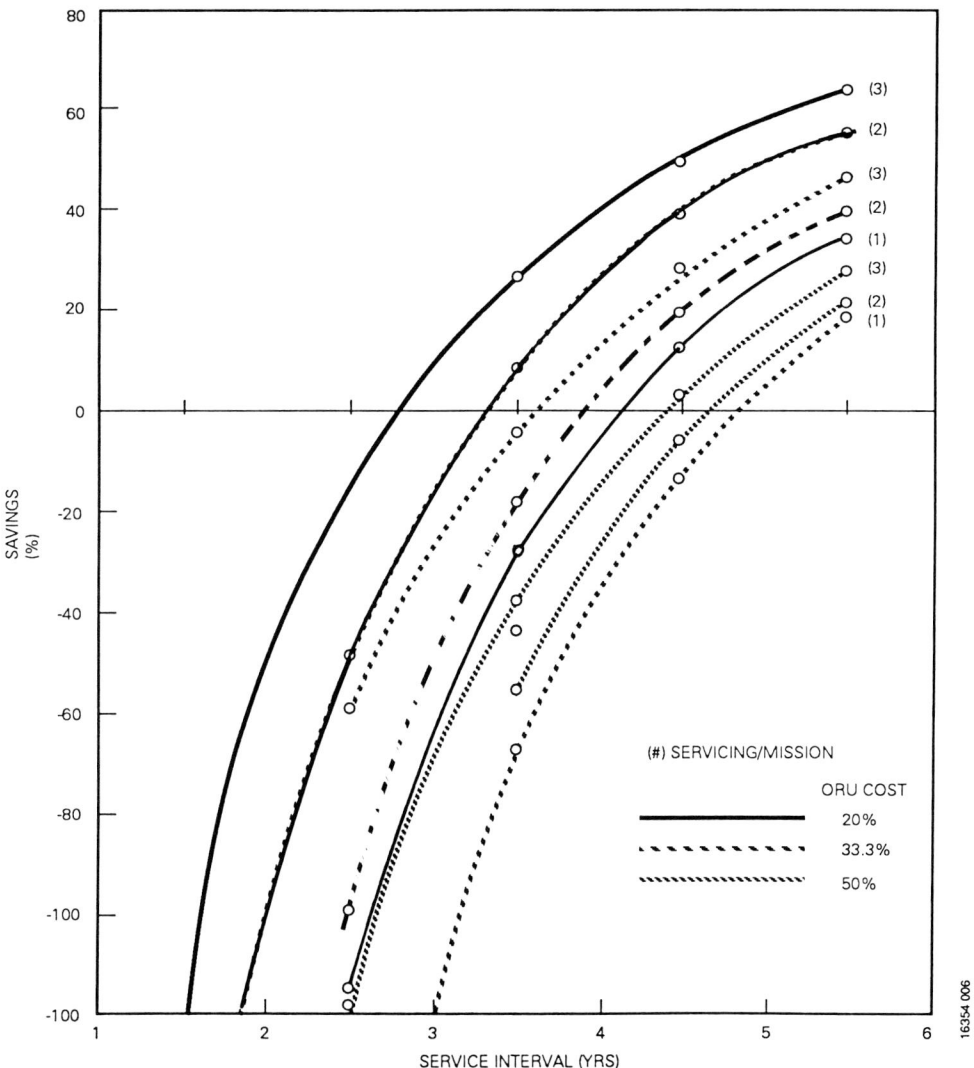

Figure 2.35 Savings versus service interval, ORU percent cost, and servicings per mission. *Courtesy of TRW.*

The operations issues that have a programmatic content are:

1. Transportation—Space Shuttle flight rate and use of expendable launch vehicles
2. Remote operating systems—operator interfaces
3. Operations of space-based vehicles (OMV, OTV)
4. NASA/DoD interagency decision systems
5. Test-bed operations, schedules, and costs

Many of the above issues were taken from Reference 7, *Book of Satellite Servicing Key Issues*. The NASA Johnson Space Center Space Assembly and Servicing Working Group (SASWG) was the organization that sponsored the assembly and publishing of this reference book. The SASWG consists of NASA, DoD, industry, university, and international participants. The purpose of the SASWG is to better focus technology thrusts and management techniques on satellite servicing hardware and operations. SASWG meetings provide an open forum for the exchange of information and ideas associated with on-orbit servicing of space systems.

References

1. *NASA Commercial Programs, A Progress Report 1988*. Published by NASA's Office of Commercial Programs, Public Affairs Officer, NASA Headquarters, Washington, DC, 1989.
2. *Space Assembly, Maintenance, and Servicing Study (SAMS)* Final Report, Volume I, Executive Summary, TRW No. SAMSS-196, Volume II, System Analysis, TRW No. SAMSS-195, Volume III, Design Concepts, TRW No. SAMSS-197, Volume IV, Concept Development Plan, TRW No. SAMSS-198, Volume V, Neutral Buoyancy Simulation, TRW No. SAMSS 199. TRW, Redondo Beach, CA, July 6, 1988.

   and

   Space Assembly, Maintenance, and Servicing Study (SAMS) Final Report, Volume I, Executive Summary, Volume II, System Analysis, Volume III, Design Concepts, Volume IV, Concept Development Plan, Volume V, Simulation Report. Lockheed Missiles and Space Company, Inc., Sunnyvale, CA, July 6, 1988.
3. *Space Transportation Architecture Study*. Contract performed by four

aerospace contractors under sponsorship by both NASA and Air Force Space Division. Study conducted from March 1986 through June 1987.
4. *NASA Civil Needs Data Base*. Compilation of NASA programs for the next ten years indicating those programs requiring servicing operations. NASA, 1986.
5. *Space Station Freedom Media Handbook*. Technical & Administration Services Corporation, Washington, DC, April 1989. Publication sponsored by Public Affairs Office, Office of Space Station, NASA Headquarters.
6. *Satellite Servicing—A NASA Report to Congress*. NASA Office of Space Flight, Washington, DC, March 1, 1988.
7. *Book of Satellite Servicing Key Issues*. Space Assembly and Servicing Working Group. Published by the Engineering Directorate, Flight Projects Engineering Office, NASA Johnson Space Center, August 1987.

Chapter 3

Mission Operations

By the year 2000, the world's governments and industries may be spending about $100 billion per year on spacecraft, space stations, planetary bases, launch vehicles, and the Earth facilities to operate them.

Barring unforeseen political, economic, or international events, the decade of the 1990s will see significant expansion in many areas of space activity. Communications/tracking satellite launches are increasing as are thrusts into space of remote sensing/Earth resources spacecraft, meteorological spacecraft, surveillance/early warning space-based systems, earth environmental surveys and planetary exploration spacecraft, experimental orbiting laboratories, navigational satellites, and large astronomy and scientific platforms. The permanently manned International Space Station Freedom is planned for assembly in space in the late 1990s. The United States, Europe, Japan, and Canada are currently developing their elements of the Station under NASA leadership. The 13 nation European Space Agency is building a space infrastructure through the Ariane, Columbus, and Hermes developments and the construction of an impressive list of scientific and applications spacecraft.

The role of on-orbit servicing operations to support this plethora of space activity is under indepth study by civil and military cognoscenti worldwide. This chapter presents a survey of those ground and flight activities required to conduct space assembly, maintenance, and servicing (SAMS) missions. Space servicing architecture, work in space operations, transportation systems, logistics operations and space assembly are discussed. The time period examined is 1985 through 2010, divided into two parts. Part 1 is 1985 through 1996, the early years of on-orbit servicing operations. Part 2 is 1997 through 2010, the years that will feature expansion of servicing flights to higher energy orbits and the blending of manned and robotic operations.

Mission operations encompasses:

- Year by year mission models of candidate NASA and DoD servicing flights and the resources required to complete the flight program
- The time-phased servicing infrastructure and architecture and related logistics planning and implementation to safely support servicing missions
- Launch and orbit-to-orbit space transportation vehicles
- Operational scenarios of how and when specific classes of SAMS missions should be conducted, accounting for planned and unplanned operations, optimal timelines and orbital maneuvers, the effective use of merged EVA and robotic techniques, and crew training programs
- Mission command, control, and communications technologies and applications
- The existing, enhancing, and enabling technologies required to perform the mission model flights

Figures 2.14 and 2.18 in the last chapter depict a possible mission data base for on-orbit satellite servicing. While NASA has indeed considered SAMS functions for several years, DoD and SDIO are just now at the point where SAMS functions are being evaluated in the space systems design process. SAMS requirements for DoD and SDI systems are evolving in a dynamic environment of rigorous technical analysis and life cycle cost studies led and sponsored by the Plans and Advanced Programs Organization in the Air Force System Command Headquarters, Space System Division.

The spacecraft population of potential SAMS candidates is quite large. It is estimated that by 1995 the on-orbit servicing events per year, for all users, will be about 15 to 20; with the number growing to around 100 per year by 2010. This estimate of 100 per year is predicated on the deployment and servicing of some form of an SDI operational program. A viable SAMS capability will serve to foster a robust U.S. and international space force servicing structure consisting of transportation, servicing tools/equipment, space-based servicing stations and/or facilities, communications/control elements, ground processing facilities, and trained crews ready to operate anywhere in the infrastructure.

Space Servicing Architecture

The ability to perform on-orbit satellite servicing assumes the manned and/or robotic operations to assemble, maintain, repair, resupply, upgrade, deploy, retrieve, or return various spacecraft and/or facilities at many orbital locations. Space servicing requires some form of a space assist support system (SASS) with appropriate infrastructure to implement the concept. This SASS, when fully deployed, will include one or more items in each of these 16 elements:

- Manned space station with a servicing facility
- Manned or man-tended space-based assembly and repair facility
- Space-based warehouse
- Space-based unmanned or man-tended servicing support platform (SBSP)
- Space-based cryo facility and fuel farm
- Communication satellites
- Global positioning satellite system
- High energy upper stage
- Ground or space-based reusable orbital transfer vehicle
- Space-based unmanned, and then manned, orbital maneuvering vehicle
- Space Shuttle II fleet
- Medium and heavy lift launch vehicles as suggested in the Advanced Launch System Program
- Ground and space-based hardware/tools and equipment, man operated (EVA/IVA) and machine intelligent robotic devices such as the satellite servicer system (SSS) to perform SAMS functions
- Ground and space-based mission control centers to conduct SAMS missions
- Ground launch site facilities to accommodate SAMS missions
- Ground-based supply depots, ORU repair facilities, equipment manufacturers, and training centers

A concept of how the total architecture/infrastructure will look in the year 2010 to support NASA, SDIO, DoD and commercial user needs is illustrated on Figure 3.1 [1].

As a national space assembly, maintenance, and servicing system program continues to evolve, the elements of an integrated SASS and ground architecture will be identified and designed. The ground systems elements must be developed in parallel with the space elements so that an optimal integrated architecture can evolve at minimum cost.

The near term (1985–1996) will include the present Space Shuttle and expendable launch vehicles (Delta, Atlas-Centaur, Titan, and Europe's Ariane), Space Station Freedom, and possibly an OMV with a satellite servicer system (SSS).

The far term (1997–2010) will see the addition of an upgraded Space Station Freedom and advanced Shuttle, SAMS support hardware matched to standard interfaces, a reusable orbital transfer vehicle, spaced-based support platforms, storage and warehouses facilities, a manned OMV with a second generation SSS, an assembly and overhaul station, a propellent carrier, man-tended polar and geosynchronous orbit servicing work platforms, a high energy upper stage coupled with a heavy lift launch vehicle, and possibly, the National Aero-Space Plane (NASP). These hardware items were identified as a result of examination of the scenarios supporting about 10 to 12 design reference missions (DRMs) in the 1986–87 USAF/SDIO/NASA sponsored Space Assembly, Maintenance, and Servicing Study [1] performed by TRW's Space and Technology Group and by Lockheed's Missiles and Space Company.

Work in Space Operations

There are five major servicing categories associated with the life of a system: activation, maintenance, repair, retrofit, and deactivation. In many cases, the functions and general actions are the same, but this classification provides a convenient method to ensure inclusion of all different types of servicing activities associated with a system. Figure 3.2 expands these five servicing categories.

Figure 3.3 lists the 20 events, in the sequence they should be performed, that are common to any mission to conduct on-orbit assembly, maintenance, or servicing to a space system.

Close-in maneuvering starts at the point where (1) the main interorbit propulsion mode is shut down, (2) the spacecraft to be serviced is within range of the command and control system sensors to be used for proximity operations, and (3) the control decision has been made to change to the close-in maneuver in propulsion mode. For an OMV, this would be at the point where the main hydrazine or bipropellant propulsion is secured and the cold gas propulsion system is activated. Clean, low impulse propulsion systems should be used on any OMV for proximity operations to avoid any contamination problems and to reduce the risk of collision with the spacecraft to be serviced.

Inspection refers to a local or remote visual examination of the spacecraft to assess its status and to locate any obvious or previously undetected problems. If there has been some type of damage to the spacecraft, this inspection may be the first opportunity for assessment of the damage. A decision could be made at this time that the spacecraft is too badly damaged to proceed with the servicing mission.

Approach and docking involve the location, alignment, and attachment of the servicing vehicle to the target spacecraft. The location and alignment functions require sensor systems such as servicing spacecraft radar, direct pilot's vision, or some other type of range/rate system on a remotely operated spacecraft. Soft docking or capture is the initial contact with the target spacecraft, and is followed by final alignment and hard docking. In the case of the Orbiter, the RMS provides capture, soft docking, and alignment by acquiring the spacecraft and translating it to the payload retention device in the payload bay. This can also be achieved by manual EVA. Hard docking is achieved when the space-

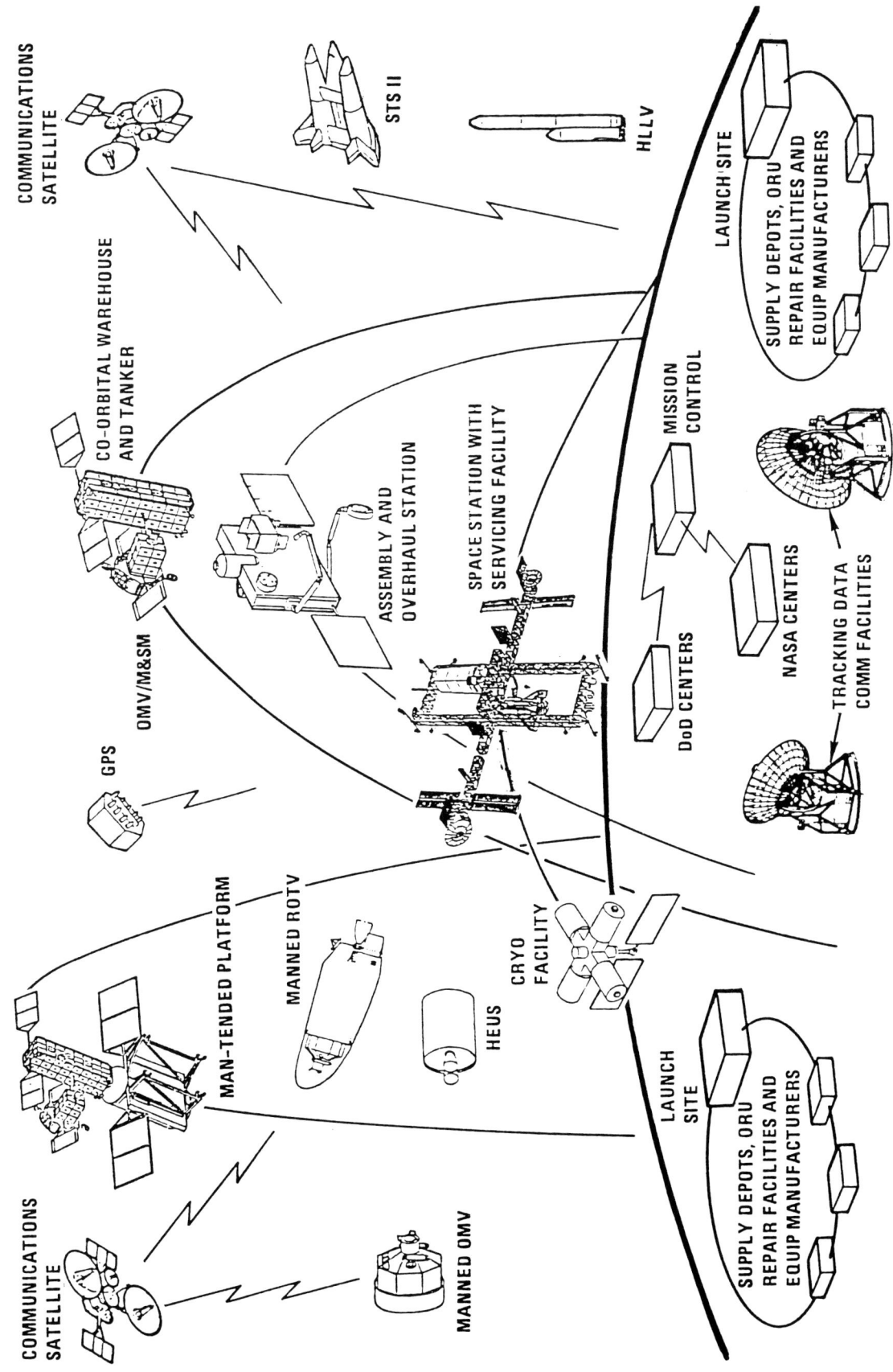

Figure 3.1 Integrated ground and space-based satellite servicing architecture. *Courtesy of TRW.*

SERVICING CATEGORY	CHARACTERISTIC	GENERAL ACTION		FUNCTIONS	
ACTIVATION	PLANNED AND SCHEDULED	• ACTIVATE • PREPARE • INITIATE		• ASSEMBLE • DEPLOY • LOAD	• INITIAL FUELING • TEST/CHECKOUT
MAINTENANCE	PLANNED AND SCHEDULED	• MAINTAIN • SUSTAIN • CONTINUE	• PRESERVE • PREVENT • CHANGEOUT	• MODULE REPLACEMENT • REFUEL • OBSERVE/INSPECT • INST REPLACEMENT	• RESUPPLY • TEST/CHECKOUT • CLEAN • REMOVE CONTAMINATION
REPAIR	UNPLANNED AND UNSCHEDULED	• RESTORE • FIX • REMEDY	• CURE • RECTIFY	• CORRECTION OF ANY UNPLANNED FAILURE OR DEGRADATION, THEN VERIFY PERFORMANCE	
RETROFIT	PLANNED AND UNSCHEDULED	• EXTEND • UPGRADE • IMPROVE	• MISSION CHANGE • REFURBISH	• MODULE REPLACEMENT • FRONT END KIT INSTALLATION • S/C MATE TO OMV	
DEACTIVATION	PLANNED AND UNSCHEDULED	• DEACTIVATE • TERMINATE • REMOVE		• TURN-OFF AND SAFE • RETRIEVE • STORE FOR RETURN TO EARTH • STORE ON ORBIT IN DORMANT MODE	

Figure 3.2 On-orbit servicing task categories.

craft is firmly attached to the payload retention device. A remote system should possess separate soft docking and hard docking systems, with built-in alignment aids. Position recognition or initialization is normally used by a remote system, which would need a reference point on the spacecraft to initialize the "map" in the servicer computer memory.

After the spacecraft has been de-activated and declared safe, the umbilical connections are made next, including electrical power (if required), command and control, and data. This step must be completed before continuing. Inspection of spacecraft connector pins before connection is desirable, but not necessary if positive alignment connectors are used.

Systems interrogation and fault analysis are necessary to determine the status of the spacecraft before any actual servicing functions are performed. This interrogation and fault analysis is for corroboration of status previously identified by ground, NSTS, or Space Station Freedom based communications. It could also be for primary fault analysis if the spacecraft was not communicating, thus precluding prior fault analysis, but was stable and approachable. The fault analysis is accomplished through ground communication and external analysis, or through Built-In Test Equipment (BITE) or Automatic Test Equipment (ATE) on board the servicing vehicle. This choice will depend on spacecraft size and complexity, servicing vehicle capacity and capability, and extent of the spacecraft BITE. It is of course desirable to determine spacecraft status before commencing a servicing mission, in order to identify the required ORU complement for the servicer. In some cases, such as lost communications or mission-of-opportunity, this may not be possible.

By docking and connecting the required power and signal lines into the spacecraft, it may be possible, with the help of ground or other service base communications, to isolate faults and develop fixes or workarounds, or even remove modules for subsequent repair and reinstallation on a future mission. As during the inspection stage, a decision must be made at this point to discontinue the servicing mission if there are any conditions identified during interrogation which would prevent successful servicing.

The integration of servicing craft and target spacecraft control systems may be required because of the relative mass of the two vehicles. Either the control system of the servicer craft or the target spacecraft will be required to assume a primary role in stabilization. This is a major area for analysis and development, particularly for very large spacecraft, where the stability system of the servicer may not have adequate control authority to stabilize both servicer and spacecraft. It may be necessary to develop a servicer vehicle control system with variable thrust, momentum CMGs, and frequency control, in order to accommodate target spacecraft of various sizes, configurations, and mass. The other option is to reconfigure the servicer spacecraft stability and control system for each different target spacecraft and mission.

Worksite preparations include debris removal, access hatch removal, multilayer insulation removal, disarming of certain systems, or other required preparation before the actual servicing function can begin.

evolving support concepts and architectures. Technologies required to implement these concepts are being studied and evaluated to ensure that the support concepts evolved can be implemented [2].

On-Orbit Support—Future

A clear requirement for an on-orbit support capability has been established and documented [2]. Many studies have concluded that on-orbit support is feasible with current or near-term technology. In addition, it will result in economic savings and provide flexibility and capabilities never before possible. The question now is, How do we develop this capability? To answer this question, let us look at two emerging programs.

Satellite Servicer System Flight Demonstration

The Satellite Servicer System Flight Demonstration (SSS/FD) program is a result of two requirements. The U.S. Congress H. R. 2782 directed NASA to a develop an on-orbit servicing capability. The previously mentioned SAMS study recommended that DoD also develop this capability. As a result, NASA, the USAF, and SDIO have formed a partnership for a combined effort to initiate a program that will develop and demonstrate satellite servicing capabilities using current and near-term technologies. SSS/FD will integrate the products of several ongoing programs, particularly the OMV and the flight telerobotic servicer (FTS), into a system capable of performing on-orbit support and demonstrate this capability through a series of flight demonstrations. The first of these demonstrations will demonstrate autonomous rendezvous and docking and Space Station Freedom proximity operations. The second will be an in-cargo-bay demonstration of ground and on-orbit supervised autonomous control for ORU exchange and fluid transfer. Then a free-flight demonstration will be conducted using autonomous rendezvous and docking and supervised autonomous control for ORU exchange and fluid transfer. TRW and Martin Marietta were selected by NASA to design servicer concepts for these flight demonstrations, but because of NASA budgetary problems the scope and schedule for this effort are undecided.

SASS Program Office

McDonnell Douglas Space Systems Company, under subcontract to General Electric, the SDI systems engineering and integration contractor, has prepared a Space Asset Support System (SASS) implementation plan. This plan defines the required capabilities, assesses the technologies (existing, in development, and planned), and describes the associated spacecraft design considerations to implement an on-orbit support concept for the SDS space assets. The plan indicates that new technology development in support of a SASS application is not required. What is required is a government decision on whether or not to establish a SASS capability and how to develop that program. The implementation plan recommends that once a decision is made to develop an on-orbit support capability, an SDI SASS System Program Office (SPO) be established to manage the development and acquisition of the SASS.

COSEMS: A Dynamic Simulation of Space Logistics

The Comprehensive Operational Support Evaluation Model for Space (COSEMS) is a largescale, discrete event simulation of space and ground operations associated with selected logistics support concepts that will do the maintenance of the operational space-based segment of the Strategic Defense System (SDS) [3]. It was developed in the ADA programming language by Advanced Technology, Inc., as a subcontractor to General Research Corporation for the U.S. Air Force Space Systems Division, Director of SDI Logistics. The support concepts available for evaluation in COSEMS are based on several alternative approaches for implementing on-orbit support, as well as satellite replacement from the ground and/or activation of on-orbit spare satellites—the current concepts for maintaining satellite constellation availability. On-orbit support involves complex, dynamic interactions between the SDS and the support and expendable launch systems.

Background

Replacement from the ground subsequent to critical failure or in anticipation of critical failure and/or activation of on-orbit spares (depending on launch response times and the mean time between critical failure of satellites in a constellation) are the current concepts for maintaining satellite constellation availability. However, alternative concepts have been proposed for maintaining the availability of constellations consisting of satellites with replaceable modules, i.e., orbital replacement units (ORUs). An integral part of these latter concepts is on-orbit support originating from the ground or from space-based support platforms (SBSPs). One such platform could be the NASA Langley Research Center In Space Assembly/Servicing (ISAS) facility [5].

Several alternatives have been suggested for implementing on-orbit support from the ground. One involves placing an orbital maneuvering vehicle and other required resources in an orbital ring of satellites to be supported by an expendable launch vehicle subsequent to the first critical failure of a satellite in the ring. The OMV remains in the ring for further coorbital support on an as-needed basis.

On-orbit support originating from space relies on support platforms permanently based in space. Two different types of support architectures involve SBSPs. One places the SBSPs in the same orbits as the satellites they are intended to support. These platforms are referred to as "coorbital" SBSPs. The other places them at similar inclinations, but at lower or higher altitudes than the satellites they are intended

Mission Operations

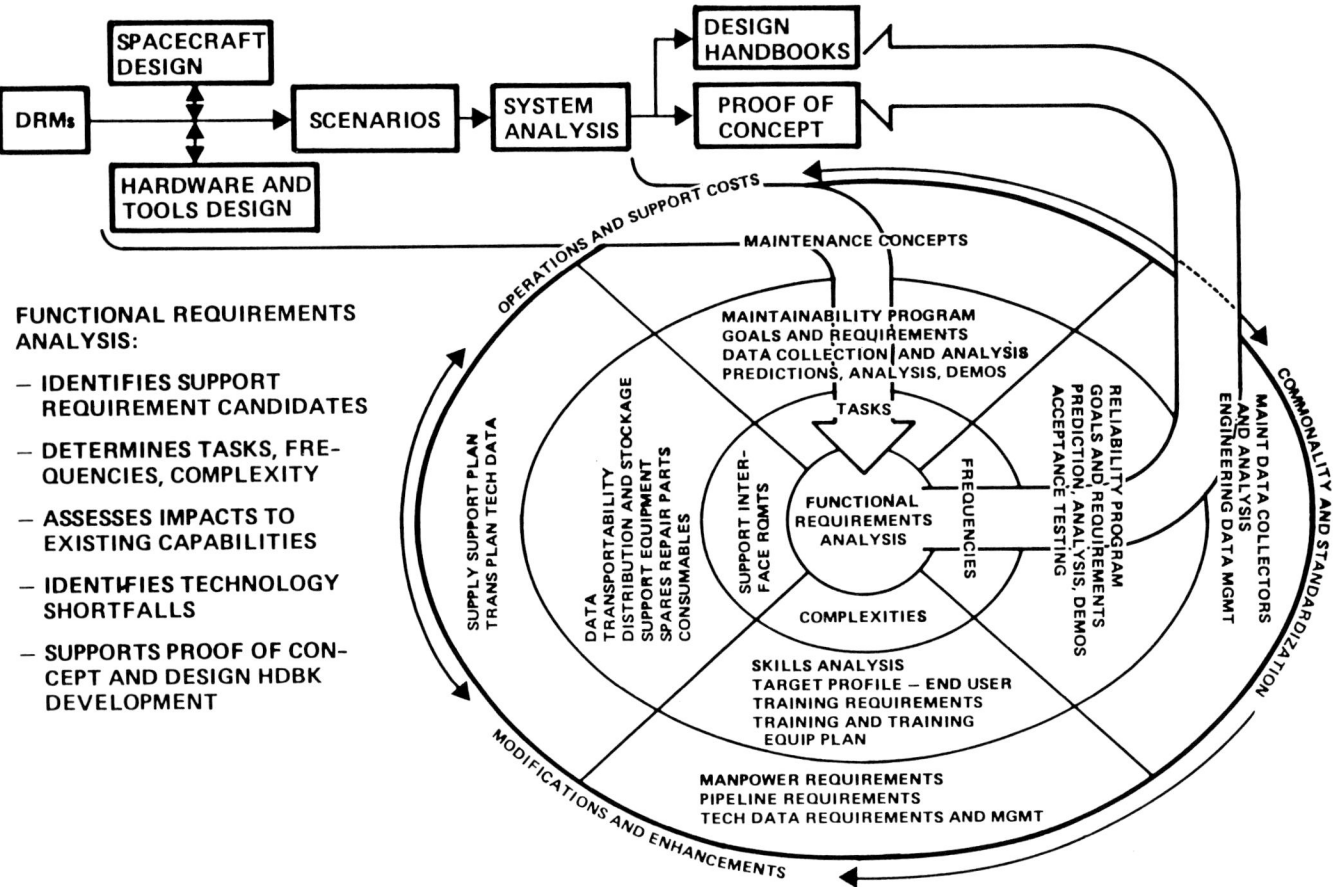

Figure 3.5 Logistics support methodology. *Courtesy of Planning Research Corporation.*

dition, the using agencies are establishing requirements for more flexibility in system operations. The capability to accomplish on-orbit maintenance and support operations becomes increasingly attractive as a means to reduce the cost of supporting these new space systems to an affordable level and also to provide user flexibility [2].

On-Orbit Support—Past

For the last three decades man has been launching and operating satellites in space. These spacecraft have had varied missions such as astrophysics research, Earth surveys, communications, navigational aids, space experiments, weather monitoring, surveillance, and military support. During this period the strategy for maintenance of these spacecraft has been reconfiguration by telemetry to correct individual element failures and abandonment of satellites after critical or final failure. Under such a concept, abandoned spacecraft are replaced by launching new satellites or activating prepositioned spare satellites. This strategy has served us well for the relatively small constellations or one of a kind satellites that have been typical during this period. With highly reliable parts and built-in redundancy, a satellite could be made to operate reliably for long periods of time, often outliving its design life. Replacement satellites, besides restoring an operational capability, were a means of introducing new technology and improved capabilities.

NASA has demonstrated the ability to deploy, retrieve, and repair satellites in space. The Strategic Defense System, with potentially hundreds of highly complex satellites, will be located in an environment more hostile than man has ever worked in. An adequate logistics base and infrastructure is required in order to sustain such complex, long term operations in space. On-orbit support with robotics is a possible solution to these supportability demands [2].

On-Orbit Support—Present

Significant strides have been made in expanding the knowledge base of on-orbit support concepts and technologies. Potential users of space systems have defined requirements that seem to be most effectively met through some form of on-orbit support. As space logisticians strive to meet these requirements, new strategies for on-orbit support and conceptual support architectures are evolving. Trade studies are being conducted using new and sophisticated analytical tools to evaluate the effectiveness and affordability of these

Assembly Tasks

1. Unloading of components or assembly elements from the Shuttle Orbiter or from Expendable Launch Vehicle(s)
2. Transfer of components to assembly site by OMV, special tug or the assembly crew with EVA propulsion such as the MMU
3. Positioning of components into assembly sequence
4. Temporary restraint/holding of components
5. Attachment of components to assembly/subassembly units by mechanical bolting/latch fastening, bonding or welding
6. Electrical power distribution systems assembly
7. Command, control, data systems assembly
8. Piping distribution systems assembly
9. Electrical power source assembly for: power units, fuel tanks, battery modules and RTG (installed by a remote system)
10. Appendage assembly/deployment. Examples are: solar array/solar dynamics, antennas, sensor packages, structural booms and instruments/payloads
11. Attachment of subsystem modules. Examples are: propulsion and propellant tankage, attitude control and communications
12. Attachment of special equipment such as: covers/protective shields, multi-layered insulation for thermal control, hand bracket's and foot holds for future servicing visits and docking adapters
13. Alignment and calibration for: structures and subassemblies, appendages, instruments and sensors
14. Initial or start-up events such as: functional testing and complete systems check-out

Maintenance Tasks

1. Observation of the target spacecraft
2. Inspection and assessment
3. Replacement of modules for instruments, payloads, sensors, and spacecraft sub-assemblies. These are generally called Orbital Replacement Units (Oru's)
4. Surface restoration and multi-layered insulation replacement
5. Realignment/recalibration of instruments, sensors, antenna and optical systems
6. Fault analysis and repair as deemed necessary
7. Contamination removal
8. Storage for Earth return
9. Storage on-orbit in dormant mode
10. Check-out and test of unit assemblies, subsystems, payloads and entire spacecraft

Servicing Tasks

1. Battery replacement
2. RTG replacement
3. Solid propellant replacement
4. Fluid replenishment: Propellants, pressurants (gases), cryogens and coolant
5. Solids replenishment such as solid cryogens
6. Life support replenishment. Examples are: food, water, respiratory gases, agents and hardware
7. Data storage media replenishment. Examples are: photographic film and magnetic tape

Figure 3.4 Example assembly, maintenance, and servicing tasks that might be performed on orbit on a serviceable spacecraft.

as a function of time, considering the near and far term. This allows identification of requirements and key technologies, commonality across DRMs, and the importance of each to mission fulfillment.

On-Orbit Support

As space systems grow larger, with increasingly complex technology and capabilities, new methods of supporting satellite constellations must be developed to optimize system availability and affordability. In the past, space systems were largely developed and operated by the using agency and were operated and supported as experimental or research and development systems. As the individual satellites in these systems experienced failures that could not be corrected by telemetry, they were replaced with a new satellite or abandoned. The cost of maintaining these large complex systems, such as those in the NASA Great Observatories program and the U.S. DoD Strategic Defense program, by the traditional method of satellite replacement is prohibitive. In ad-

SERVICING EVENTS SEQUENCE

1. Close-in maneuvering (pre-approach)
2. Inspection of the satellite to be serviced
3. Spacecraft stabilization (as requred)
4. Approach and docking maneuvers
5. Postion recognition, initialization
6. Spacecraft deactivation or safing, as required
7. Electrical power, command control, data interface connections
8. Systems interrogation, fault analysis
9. Integration of servicer craft and spacecraft control function (as required)
10. Worksite preparations, covers removed, debris collected and stored
11. Support equipment installation
12. Performance of SAMS mission tasks. See Figure 3.4 for example assembly, maintenance, and servicing tasks
13. Systems interrogation/checklist
14. Support equipment de-installation
15. Worksite return to operational status
16. Electrical power, command and control, data disconnections
17. Separation and standoff
18. Inspection of spacecraft; then powerup (including remote or ground checkout)
19. Close-in maneuvering (post standoff)
20. Debris evaluation, capture, and stowage

Figure 3.3 The events necessary to complete a typical on-orbit satellite servicing operation.

Support equipment installation involves the attachment of required aids for the servicing function. In the case of EVA servicing, this would be the portable foot restraint. Support equipment also includes hook racks, trash bags, or alignment systems.

At this point, the actual mission maintenance or servicing is performed. Figure 3.4 shows example tasks. Once servicing is complete, a series of steps is required before the serviced spacecraft returns to normal operation and the servicing spacecraft leaves the orbit.

The systems interrogation and checkout is an inspection to determine if the servicing operation was successful, or if it caused any new problems that require attention. If everything is not nominal, a decision is required as to whether to try to correct the problem or to leave the spacecraft and return at a later time. The latter could occur if the required ORUs or tools are not available.

The support equipment de-installation is simply the removal of any equipment or servicing aids that were installed prior to the servicing functions.

Worksite return to operational status is the replacement of whatever cover or insulation was removed to gain access. This could also refer to the removal of any protective coverings used during the mission.

Umbilical disconnect is the next step, breaking any electrical power, command and control, or data connections with the spacecraft.

Another function that must occur during the closeout procedure, depending on the specific mission, is debris evaluation, capture, and stowage. The timing and tool requirements for this procedure depend on the mission and type of debris produced.

Separation is the actual breaking of the mechanical connection between the servicing spacecraft and the serviced spacecraft. The servicing spacecraft will then stand off at a nominal distance to perform the next several functions.

A visual inspection follows to make sure the spacecraft appears normal. The serviced spacecraft will then be powered up and functional checks performed before the servicing vehicle departs the area. Again, if something is not right, a decision is made as to whether to return to reservice or to depart.

The servicing vehicle or serviced spacecraft is then maneuvered away, using a clean propulsion system, prior to departure using main propulsion [1].

Logistics Operations

In developing a planned integrated logistics support infrastructure, a functional requirements analysis must be performed for each mission scenario. Individual tasks, the frequency of tasks, task complexities, and support interface requirements must all be analyzed. The type of task to be performed determines the maintenance concept, task frequency yields availability and reliability goals, task complexity governs training requirements, and support interface requirements identify critical technology/functional items. Figure 3.5 illustrates the iterative methodology used for this analysis.

Servicing design reference missions (DRMs), addressed as a consolidated group to define a unified space logistics infrastructure, can account for common requirements and potential supportability overlaps in individual DRMs. One space base may be able to serve two or more DRM missions with only a minor change in operating procedures, or a common set of space maintenance tools can be developed to serve most of the DRMs. This aggregate analysis may lead to cost savings in both the development and operations of several programs.

This approach to derive SAMS logistics infrastructure entails the plotting of logistics requirements of the DRMs to support. These platforms are referred to as "nodal regression" SBSPs.

Once deployed, coorbital SBSPs are available for support on a continuous basis, provided they can be resupplied in a timely manner. Consumable replenishment and/or ORU replacement on board SDS satellites can be scheduled independently of orbital constraints. Intraorbit transit/phasing times on the order of hours are achievable by adjusting the OMV fuel consumed during the transfer. Herein lies the primary advantage of coorbital SBSPs: they have the potential for maintaining high availability for all satellites in the same orbit as the platform. A primary disadvantage for multiplane constellations is that a platform or facility needs to be deployed in each plane.

Nodal regression SBSPs are deployed, in general, at a similar inclination to the satellites they are intended to support, but at a different altitude. The rationale for deploying a platform in an orbit other than the satellite orbit is based on two considerations: (1) the oblateness of the Earth causes the longitude of the ascending node of an orbit to rotate through 360 degrees (except at inclinations of 0 and 90 degrees) at a rate that depends on orbit inclination and altitude, and (2) although limited, OMVs do have some plane change capability. These two considerations have two important implications. First, the ascending node of the orbit of an SBSP that is at a lower or higher altitude than that of a ring of SDS satellites will periodically have the same longitude as that of the satellites. When this occurs, the angle between the SBSP and satellite orbit planes will be equal to the difference in inclination between the planes. Second, as the ascending node of the SBSP orbit approaches that of the satellite orbit, there will be an interval of time (or "window of opportunity") during which the angle between the two orbit planes, as determined by the constant difference in inclination and time-varying difference in longitude, may be small enough that an OMV has sufficient fuel capacity to rendezvous with one or more satellites before returning to the SBSP.

This implies that an OMV based at a single SBSP, at a different altitude, but with an inclination similar to that of a multiplane SDS constellation, could rendezvous with all satellites in every plane of the constellation, given enough time—provided the altitude differential is not excessive. The key to this statement is "enough time." At inclinations near 90 degrees, differential rates of longitudinal change are of the order of small fractions of a degree per day. This means it could take many years before an OMV based at a single SBSP could reach all satellites in the constellation. Under these circumstances, increasing the number of SBSPs and OMVs and/or adjusting their orbital parameters could greatly reduce the time required. For example, at intermediate inclinations (between 0 and 90 degrees or between 90 and 180 degrees), differential rates of longitudinal change may increase to the order of degrees per day depending on altitude. However, the limited plane change capability of an OMV constrains to a few degrees the allowable inclination difference between the SBSPs and satellites they are intended to support. Differential rates of longitudinal change can also be adjusted by increasing the altitude separation between the SBSPs and satellites. However, the available altitude range is bounded below by atmospheric drag and above by the lift capacities of the OMVs and the launch vehicles intended for SBSP resupply.

Regardless of origin, on-orbit support involves complex dynamic interactions between the SDS, the support infrastructure, and an expendable launch system (ELS). These interactions result from SDS demands for support to maintain required constellation availabilities and attempts by the support infrastructure and ELS to respond to these demands in a timely manner. COSEMS was developed to simulate these interactions with sufficient detail to provide a realistic quantitative assessment of on-orbital support relative to satellite replacement from the ground and/or activation of on-orbit spare satellites.

Overview of COSEMS

COSEMS is a germinating discrete event simulation [4]. Multiple Monte Carlo replications of a user-specified scenario duration are executed to estimate mean values and confidence intervals for key output quantities that characterize system dynamics. The rationale for the emphasis on system dynamics is that the loads on the support and expendable launch systems are not uniformly distributed over time. Therefore, a support system and expendable launch system that are sized for time-averaged SDS loads may not always be able to maintain required SDS operational availabilities.

The three major systems represented in COSEMS are the SDS, a space asset support system (SASS), and an ELS. These systems are modeled with varying degrees of detail. Because system level issues associated with conducting on-orbit support in space are not well understood, these operations are modeled in more detail than those associated with the expendable launch system. The orbital positions of all satellites in an SDS constellation are monitored as a function of time. The status of each subsystem and the modules that make up each subsystem for each satellite in an SDS constellation are also monitored as a function of time. If the support concept under investigation calls for on-orbit support from SBSPs, the orbital positions of these platforms, such as an ISAS, and the onboard inventories of satellite ORUs and consumables are also monitored as a function of time. The timelines, incremental velocities, and fuel consumption associated with all OMV orbital transfers between the SBSPs, at which the OMVs are based, and the SDS satellites as well as intraorbit transfers within each orbital ring of the SDS constellations, are determined as well. The timelines associated with prelaunch processing for each

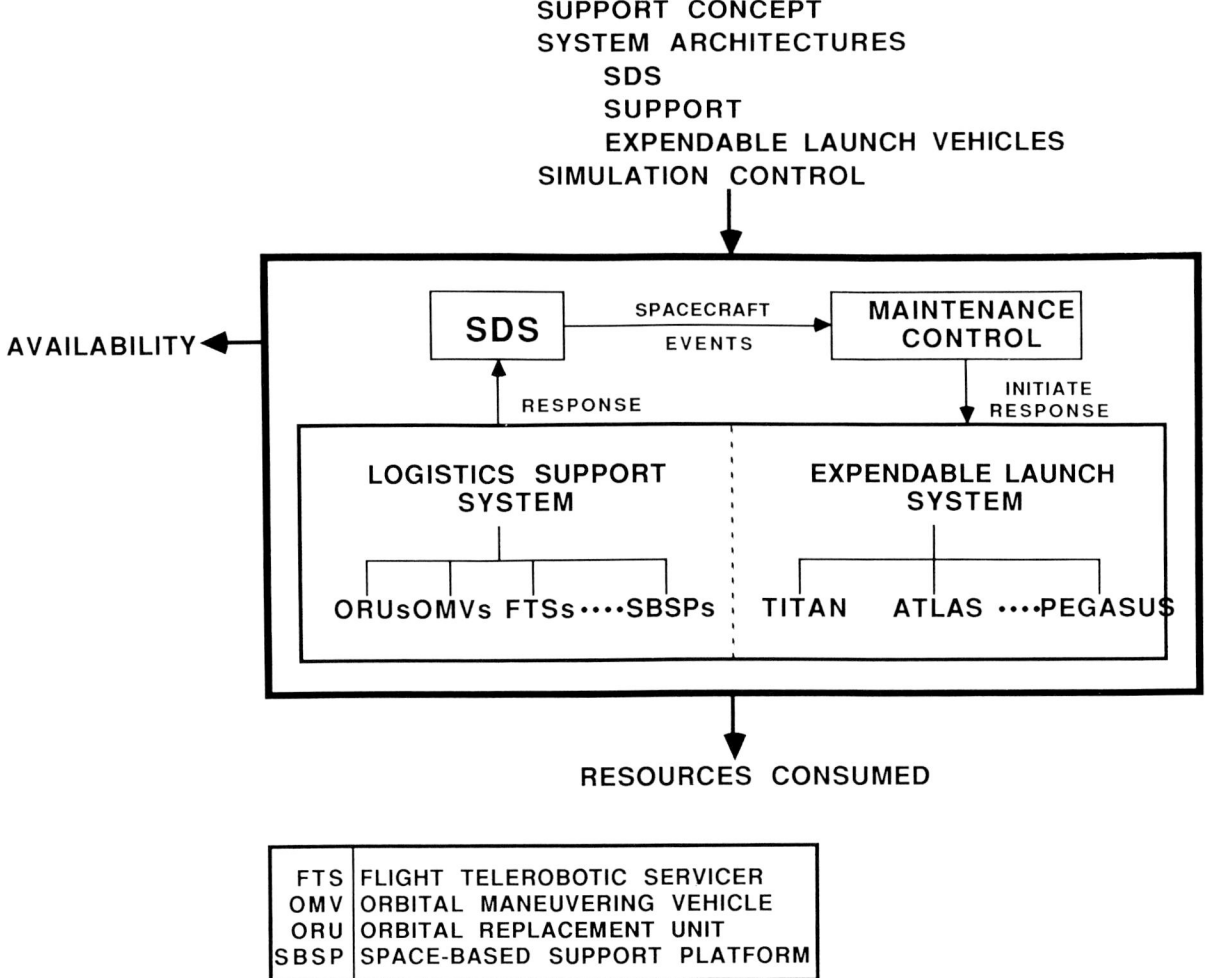

Figure 3.6 COSEMS design. *Courtesy of Planning Research Corporation.*

launch vehicle in the ELS are determined from nominal launch response times for each vehicle type at each launch site represented in the model, and the delay (if any) caused by congestion at a launch site leading to temporary unavailability of a launch pad.

As indicated in Figure 3.6, the primary inputs to COSEMS fall into three categories, delineated by the parameters required to define: (1) the support concept under investigation; (2) the architectures for the SDS, SASS, and ELS; and (3) control of the simulation (scenario duration, number of replications, selection of random number seeds, level of statistical confidence in key output quantities). The primary outputs of COSEMS fall into two categories: (1) resources consumed (e.g., payloads deployed, resupplied, and lost due to launch failures; number of launches by vehicle type, etc.) and (2) SDS status. The current version of COSEMS characterizes SDS status by time series and statistical summary portrayals of operational availability at the constellation, orbital ring, and payload level, as well as instantaneous satellite status (operational, degraded, at risk, undergoing support, failed).

The top level design of COSEMS is also illustrated in Figure 3.6. The primary events that cause simulation time to advance are random failures of spacecraft modules, scheduled planning for the predictable depletion of onboard consumables, and the implementation of responses from the support infrastructure. The maintenance control function in COSEMS implements the decision logic that ground control personnel exercise subsequent to critical satellite failures or prior to anticipated critical failures. It is in this area of COSEMS that the specific resources required to implement the support concept under investigation are integrated into a mission plan. These resources will normally include elements of an ELS. They will also include one or more elements of the space-based segment of a logistics support system (i.e., the SASS) when the support concept calls for on-orbit support.

The four primary support concept categories that are currently available for investigation in COSEMS are as follows: on-orbit support from space (nodal regression SBSPs or coorbital SBSPs); on-orbit support from the ground; satellite replacement (or abandon and replace); and no satellite

Mission Operations

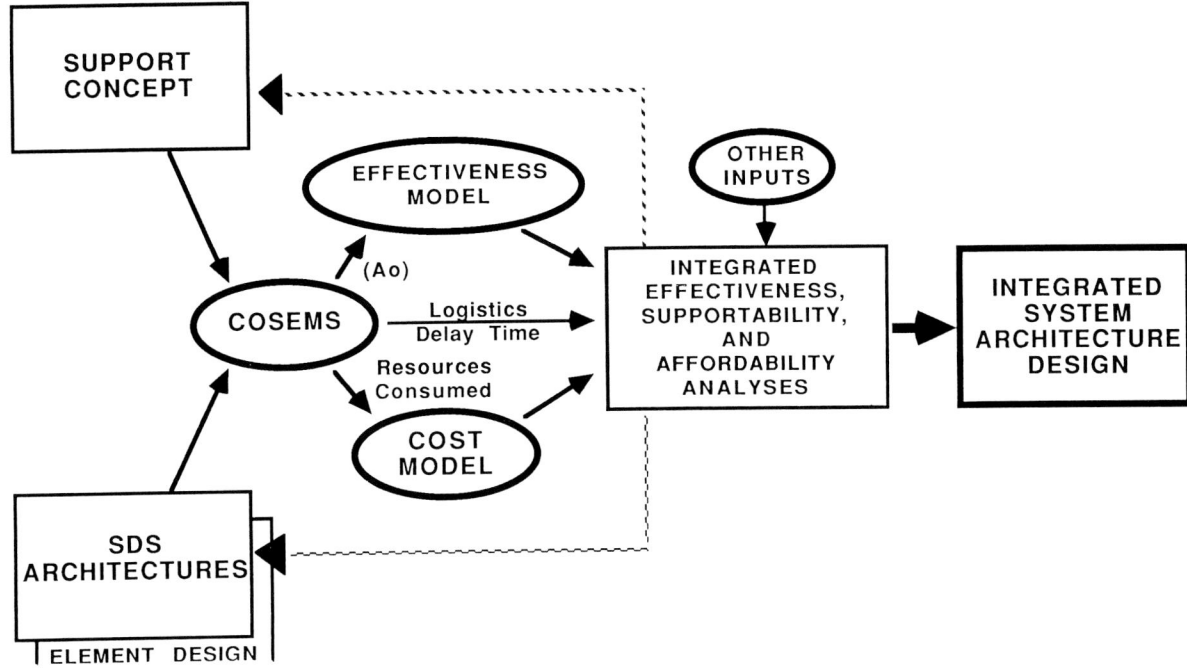

Figure 3.7 COSEMS application to support planning. *Courtesy of Planning Research Corporation.*

replacement or on-orbit support other than telemetric switching to redundant modules.

As implied in Figure 3.7, COSEMS does not determine mission effectiveness or the cost of implementing a specific support concept. However, it does determine the time varying utilization of major resources that influence cost, and the time varying operational availabilities (A_o) of SDS satellite constellations—major factors in assessing the potential of the space-based segment of the SDS to perform its intended missions. When used under stated conditions a system will operate satisfactorily at any time. A_o can be expressed by the following:

$$A_0 = \frac{OT + ST}{OT + ST + TCM + TPM + TST + (A+LDT)}$$

where

OT = Total operating time during a specific interval
ST = Total standby time during a specified interval
TCM = Total corrective maintenance time during the same specified interval
TST = Total service time that precludes other activities from occurring due to safety reasons
TPM = Total preventive maintenance time during the same specified interval
$A+LDT$ = Total administrative and logistics downtime during the specified interval

COSEMS is designed to interface with a separate cost model being developed for the government by Tecolote Research, Incorporated. During the spring of 1989, COSEMS was installed on the national test bed, where it will be available to interface with the effectiveness models already resident there. The combination of COSEMS, a cost model, and an effectiveness model will provide an integrated modeling capability for analyses of effectiveness, supportability, and affordability. In addition, COSEMS itself will continue to be used for independent studies of the influence of system reliability, maintainability, and supportability of availability.

Space Transportation Operations

Except for those operations where a spacecraft proceeds on its own propulsion to a space-based facility, all on-orbit servicing missions will require some form of Earth-to-orbit and/or orbit-to-orbit transportation. Figure 3.8 is a summary of U.S. and international space transportation systems.

When all the potential near and far term servicing mission scenarios are examined, it is evident that a wide variety of manned and unmanned ground launch vehicles and orbital transportation systems will be required. Space systems servicing requirements and support logistics strategies must be considered in the design and operational protocols of new or modified manned and unmanned launch vehicles. Once payloads are in orbit, transfer vehicles for short range (OMVs) and extended range (ROTVs) operations are required to support servicing missions. These vehicles should provide orbit-to-orbit transportation as well as serve as platforms from which servicing operations can be performed. Manned transportation capability must be considered as part of the ROTV design.

The design of spacecraft and servicing systems for SAMS will be driven, to a major extent, by the interface and payload

Vehicle	Country	Payload Delivery Performance (Lb)	Est. Launch Price (1991$)
Space Shuttle	United States	LEO 51K (due East), 21.9K (polar)	270M
Shuttle Derived HLLV	United States	LEO 310K	
NASP (X-30)	United States		
Delta	United States	LEO 8.6K (due East)	50M
Delta II	United States	LEO 11.1K (due East), 8.4K (polar)	60M
Titan 34D/IUS	United States	LEO 33.6 (due East, NUS) GEO 4.1K	140M
Titan IV/IUS	United States	LEO 39.3 (due East, NUS), GEO 5.1K	180M
Atlas K/Centaur	United States	LEO 14.5K (due East), GEO 3.4K	85M
Adv. Launch System	United States	LEO 300K (due East), 100K polar	
Ariane 4	Europe/France	LEO 10K (due East), GTO 8.2K	100M
Ariane 5	Europe/France	LEO 39.7K (i=28.5 degrees), GTO 15K	
Proton SL-13	Soviet Union	LEO 43K (i=51.6 degrees), GEO TBD	
Energia	Soviet Union	LEO 300K (i=51.6 degrees), GEO TBD	
Long March	China	LEO 7.3K, GTO 5.5K	
H-2	Japan	LEO 19.8K	
Hotol	United Kingdom	LEO 15.5K	
Hermes	France	LEO 9.9K	
Sanger II	West Germany	LEO 8.8K	

Figure 3.8 Summary of U.S. and international space transportation launch vehicles. Payload weight to orbit and estimated launch cost are shown for several systems.

capabilities of existing and future space transportation systems. Space Shuttle volume and payload to low and high inclination low earth orbits, single-stage-to-orbit rockets, expendable launch vehicle capacities to low and high altitude and inclination orbits (including upper stage vehicle capability), and the future Advanced Space Shuttle, the National Aero-Space Plane, and advanced launch systems (ALS) capabilities of the National Launch System will all determine the maximum size and weight limits of spacecraft and spacecraft modules. Current limits and projections of future limits are shown in Figure 3.9. One of the functions of the SAMS study [1] was to assist in developing requirements for these transportation systems, in particular as they apply to SAMS mission capabilities and limitations. An example is the mission analysis and review of several SDI spacecraft concepts, which may be limited in size, or constrained in geometry, by the then available space transportation systems and on-orbit assembly capability.

The NSTS and the Space Station Freedom are two manned facilities from which activities will be conducted to support on-orbit servicing. All servicing in the early period will require the direct support of humans in space and will be Shuttle-based. Because of the propulsive energy re-

Mission Operations

Vehicle	Country	Payload Envelope Dia. (Ft)	Max Length (Ft)
Space Shuttle	United States	15.0	60.0
Shuttle Derived HLLV	United States		
NASP (X-30)	United States		
Delta	United States	7.0	14.0
Delta II	United States	9.2	20.0
Titan 34 D/IUS	United States	9.3	30.0
Titan IV/IUS	United States	15.0	30.0
Atlas K/Centaur	United States	12.0	28.0
Adv. Launch System	United States	25.0	90.0
Ariane 4	Europe/France	12.0	35.0
Ariane 5	Europe/France	15.0	34.0
Proton SL-13	Soviet Union		
Energia	Soviet Union		
Long March	China	7.6	12.7
H-2	Japan	12.0	32.8
Hotol	United Kingdom	15.0	24.6
Hermes	France	9.8	
Sanger II	West Germany	14.8	

Figure 3.9 Current and future transportation launch vehicle payload envelope data.

quired to change the orbital inclination of Earth orbiting vehicles, and Shuttle operational capability, only free-flying satellites in orbits with inclinations in ranges 28 degrees to 57 degrees and altitudes up to about 593 km (320 nmi) can be serviced. This range can be extended up to 2,594 km (1,400 nmi) with plane changes of up to 7.5 degrees with an orbital maneuvering vehicle.

Should the Vandenberg STS launch capability eventually be activated, high inclination (polar) satellite orbits could also be accessed from an STS base. Otherwise, high inclination satellites will have to rely on remote, robotic, or telerobotic satellite servicer systems (SSS) carried to these orbits by expendable launch vehicles (ELVs). The SSS has yet to be developed and demonstrated. Orbital regions defined in the SAMS study [1] are: low altitude/low inclination, low altitude/mid-inclination, low altitude/polar orbit, high altitude/mid-inclination, and geosynchronous. The potential means of access to each of these five orbital regions is indicated in Figure 2.19.

NASA has developed an extensive array of ground support equipment, flight support equipment, and Shuttle-based servicing hardware and capabilities. Additional ground-based and space-based elements are currently being developed to support planned NASA on-orbit servicing activities, such as the maintenance and upgrade of the Hubble Space Telescope and the on-orbit refueling of the Gamma Ray Observatory. A comprehensive servicing facility, including a protective enclosure, external storage facilities, workbenches, and a dedicated remote manipulator system, has also been defined as part of the future Space Station Freedom's capability. This facility will not be implemented as part of the Phase 1 Station but may be incrementally developed, as part of a continuing Station evolution.

In the near term, the Space Shuttle will be the only manned facility available to support onorbit servicing. Although more limited than Space Station Freedom, in terms of the servicing materials that can be aggregated on orbit to support a specific servicing mission and the amount of EVA and IVA time available, the Shuttle has three advantages: flexibility of access to low to mid-altitude, and mid-latitude, orbiting satellites; ability to rendezvous with a satellite; and a small crew of trained people who can perform a wide variety of servicing tasks.

The baseline Space Station (Phase 1) will have the capability to service and upgrade its own facilities as well as the internal and external attached payloads. There may also be a capability to provide, on a case-by-case basis, some servicing of free-flyers in the vicinity of the Station. Servicing elements could include the flight telerobotics servicer (FTS); the Phase 1 Canadian mobile servicing center (MSC) which includes the RMS; a mobile transporter; and standard equipment such as servicing tools, manned work stations, and crew/cargo translation aids. These elements, in combination with the permanent presence of astronauts in an environment more controlled than that of the Space Shuttle, ensure a far more extensive EVA servicing capability than previously available [6].

NASA future plans include servicing the European Space Agency provided man-tended free-flyer at the Space Station. Additional free-flyers that may be serviced at or from the Space Station include commercial platforms, the Hubble Space Telescope, and the other space systems in the Great Observatories science program. These systems are the Gamma Ray Observatory, the Advanced X-Ray Astronomy Facility, and the Space Infrared Telescope Facility. As the Station evolves, a more extensive servicing capability will be added to it. These capabilities could include an enclosed full-sized servicing facility, orbital maneuvering vehicles, enhancements to the MSC, upgraded servicing control centers, and an additional RMS.

The use of expendable launch vehicles (ELVs) for on-orbit servicing of satellites is still in a preliminary stage of study and requires the development of expanded vehicle capabilities, such as automated docking and servicing capabilities. ELV-based servicing has the potential advantage of access to spacecraft at all orbital inclinations and altitudes and the disadvantage that the servicing payload cannot be returned to Earth after the servicing event. Possible solutions to this dilemma, currently under study by NASA and the DoD, include off-loading the robotic servicing devices from the ELV onto the spacecraft or platform and using them to support "self servicing" throughout the operational life of the spacecraft. No NASA mission has yet baselined the use of such on-orbit servicing, although the EOS mission is still considering this approach in its tradeoff studies, together with the use of nonserviceable (expendable) spacecraft.

Role of the OMV and the OTV

A key capability expansion of Shuttle-based and Station-based servicing is an orbital maneuvering vehicle (OMV) which can also be used for satellite placement, retrieval, and reboost. It is envisioned as a reusable, remotely controlled, free-flying system capable of performing a wide range of on-orbit services in support of orbital payloads.

In 1986 the TRW Space and Technology Group, Redondo Beach, California, won a NASA competitive procurement to design, build, and deliver an OMV. Authority to proceed from the NASA Marshall Space Flight Center (MSFC) in October 1986 triggered the start of OMV hardware development. Three years of OMV conceptual studies sponsored by NASA and performed by TRW and other contractors preceded OMV development.

On June 7, 1990, NASA canceled the TRW OMV program, citing budgetary pressure and few near-term requirements for the vehicle. TRW and MSFC had completed the program Preliminary Requirements Review (PRR) in 1987 and the Preliminary Design Review (PDR) in 1988. The Critical Design Review (CDR) was planned for early 1991, with launch scheduled in the mid-1990s.

It is the author's opinion that some form of a versatile, remotely controlled orbital maneuvering vehicle remains a vital ingredient in the NASA and DoD national capability to perform on-orbit servicing and related space missions. The satellite servicing infrastructure needs an OMV to:

1. Stretch the Space Shuttle's sphere of servicing access and capability
2. Complement the Space Station Freedom's servicing and logistics tasks
3. Support the robotics technology development and precursor hardware tests and demonstrations associated with the U.S. civil space leadership initiative directed at future lunar bases and Mars landings.

Multiple propulsion systems and on-board avionics will enable a future OMV to deliver and retrieve satellites in orbits not otherwise achievable from the Shuttle. Precision maneuvering for proximity operations (including docking with an orbiting satellite) could be accomplished by man-in-the-loop control via an OMV control station. Eventually, both OMVs and OTVs might also be equipped with "smart front ends" (i.e., with robotic or telerobotic servicers) so that in situ satellite servicing could be conducted. When used with the Shuttle, an OMV could be carried into orbit as part of the servicing payload. For Station-based servicing, an OMV could be stored at the Station and used for multiple servicing events before being returned to Earth. Thus, an OMV could be both a servicing tool and an element of the transportation infrastructure in the Space Station era.

The orbital transfer vehicle (OTV) [6] is planned as an advanced upper stage that will carry cargo, and perhaps humans, from low Earth orbit to geosynchronous orbit and beyond. The exact capabilities and configuration of an OTV will be defined in the early 1990s in parallel with NASA's new initiative definitions. In past studies of OTV concepts,

Mission Operations

several configuration options have been the subject of tradeoff studies. These include: the use of cryogenic or storable propellants; a reusable or expendable design; ground-basing or space-basing; an all-propulsive vehicle or the use of an aerobrake for return flights; and delivery to low Earth orbit on the Shuttle or on a cargo vehicle. In most concepts, an OTV has the flexibility to evolve to meet increased mission requirements. For example, the initial OTV could be designed to deliver payloads to geosynchronous orbit, while a later version could be used to ferry humans between low Earth orbit and the lunar surface. The initial OTV will most likely be a cryogenic vehicle with the capability to deliver at least 4,536 kg (10,000 lb) of cargo to geosynchronous orbit or propel equivalent payloads to the Earth escape velocities required for advanced planetary exploration.

Evolution of Servicing Operations

As stated earlier in this book, on-orbit servicing is a capability which is still in an early stage of evolution. From the servicer side, it depends on the successful evolution of the current Shuttlebased capability and facilities of the Space Station Freedom. Once manned servicing support and facilities are available at the Station, further development of telerobotic and robotic servicing capabilities will take place. Such capabilities will benefit all servicing users, since they will decrease dependence on EVA. They will also benefit potential servicing candidates operating in orbits inaccessible to the Shuttle and Space Station.

With the advent of an OMV, many forms of spacecraft services will become possible. An OMV or OTV can be used to return spacecraft to the STS or the Space Station for maintenance and/or resupply. In addition, direct spacecraft servicing may be performed by adapting special purpose mission kits or servicer systems to an OMV or OTV which are capable of performing remote or in situ maintenance of spacecraft, thus eliminating the need to return payloads. Fuel efficiency of telerobotic or robotic OTV servicing missions to spacecraft in geostationary orbits could be further enhanced by the use of aerobraking to assist in the return flight from geostationary locations.

The remote services may take two generalized forms: remote module exchange of failed spacecraft elements and replenishment of onboard expendables. Both may be desirable on the same mission. These functions, in effect, extend man's ability to perform maintenance and other mission support operations remote from the NSTS through the adaptation of specialized "effectors" or servicing and refueling kits.

It has long been recognized that a need exists to capture unstable and/or inactive spacecraft and certain classes of artificial space debris [7]. Studies are being directed toward the definition and development of systems to expand an OMV capability for debris capture. This interest addresses the economic value of some classes of spacecraft which have failed prior to the end of their useful life and threaten collision with other orbiting systems. Such spacecraft could be recovered and reflown as was demonstrated in the retrieval, repair, and redeployment of the Solar Maximum Mission spacecraft.

Where costs of on-orbit servicing are associated with NASA's servicing support, the agency will continue its efforts to develop more efficient and reliable support equipment and facilities. NASA will also provide potential users with guidance on how best to make use of on-orbit servicing. This advice will be drawn from the NASA user's own experience in using on-orbit servicing to assemble structures, to maintain or repair systems, to resupply consumables, to upgrade payloads, and to retrieve, deploy or return spacecraft and payloads [6].

Since the integrated logistics support required for a few-of-a-kind missions will continue to be a significant cost driver, NASA will continue to develop and promote the use of common systems and subsystems, as well as common interfaces, support equipment, and tools. This will reduce the cost of on-orbit servicing.

NASA and DoD are performing studies to determine the feasibility of employing ELV-based robotic servicing in polar orbits. The two concepts under study utilize a service carrier launched on an ELV. After rendezvous and docking, one scenario features a resident robot that exchanges ORUs and payloads which use standard interface connectors (SIC). The service carrier is then discarded. In the second scenario, the service carrier remains attached and functions as an extension of the original platform. NASA is considering generating guidelines to payload designers so they may incorporate generic design features to facilitate possible future robotic servicing.

Finally, NASA and the DoD continue to look at the requirements and benefits of on-orbit servicing capabilities on a program by program basis, and will balance the life cycle costs of such an approach with the scientific, commercial, or programmatic needs and benefits [6].

Design Reference Missions

The Air Force and NASA defined five basic design reference mission (DRM) classes or categories, shown in Figure 2.11, to be examined by TRW and Lockheed Missiles and Space Company (LMSC) in the SAMS study [1]. A data base developed for the study included (a) system profiles on most current and future DoD/SDIO/NASA/commercial satellites and (b) specific and generic missions to define the SAMS performance requirements envelope.

The SAMS requirements for defense systems are still very much in development, with fundamental questions of applicability and cost effectiveness still in doubt in the minds of

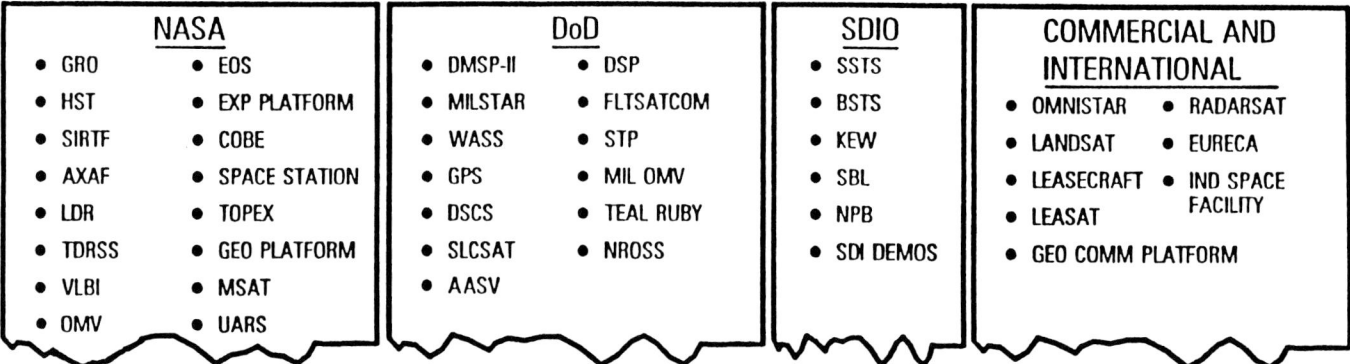

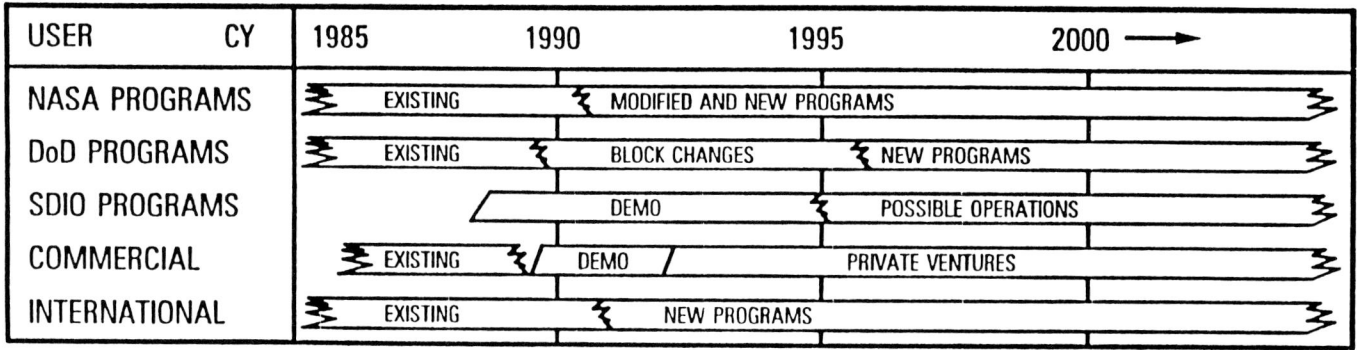

Figure 3.10 SAMS mission model showing some potential user programs. *Courtesy of TRW.*

the military decision makers. Figure 3.10 illustrates the broad structure of the mission model considered.

In developing the SAMS event mission model, two basic data sources were used:

1. The civil needs data base prepared by GRC/NASA for the National Space Transportation and Support Study (March 4, 1986)
2. The DoD transportation mission requirements definiition prepared by The Aerospace Corporation [8]

Civil needs consist of specified missions for servicing, configuration changeout, and combined servicing and configuration change. The data shown in Figure 3.11 reflects the NASA core mission scenario. The civil needs data base software was incorporated into TRW's and LMSC's software for application to the SAMS study. The DoD service missions also shown in Figure 3.11 reflect the DoD Space Transportation Architecture Study (STAS) [9] level of activity which specifically identified servicing missions in the forecast. The service mission mass forecast depicted in Figure 3.12 corresponds to the mission model shown in Figure 3.11. These mass projections can be used to derive a preliminary estimate of total transportation requirements and are useful in developing servicing mission infrastructure concepts and in logistics support planning.

In order to properly analyze the entire "performance envelope" for the SAMS system, it was necessary to expand the five basic missions defined by the Air Force in Figure 2.11 into 11 more specific servicing missions. The details of these 11 DRMs are given in Appendix B, SAMS Study Summary.

In the SAMS study the data base was analyzed to define design reference missions (DRMs) which accurately represent the population of real NASA, DoD and SDIO missions in each orbital regime. This was done by determining the orbit region characteristics for the space systems that constitute the model for both the early (1985–1996) and late (1997–2010) epochs. From this spectrum of characteristics, orbit classes were defined which contained all major elements of the data base. The data base was then searched and sorted by each class, and an assessment made of their collective needs for space assembly, maintenance, and servicing (SAMS). The data base was also searched for overall mission parameters such as spacecraft mass, size, hazards, and any other special mission characteristics. It was assumed that the identified mission operators could be-

Mission Operations

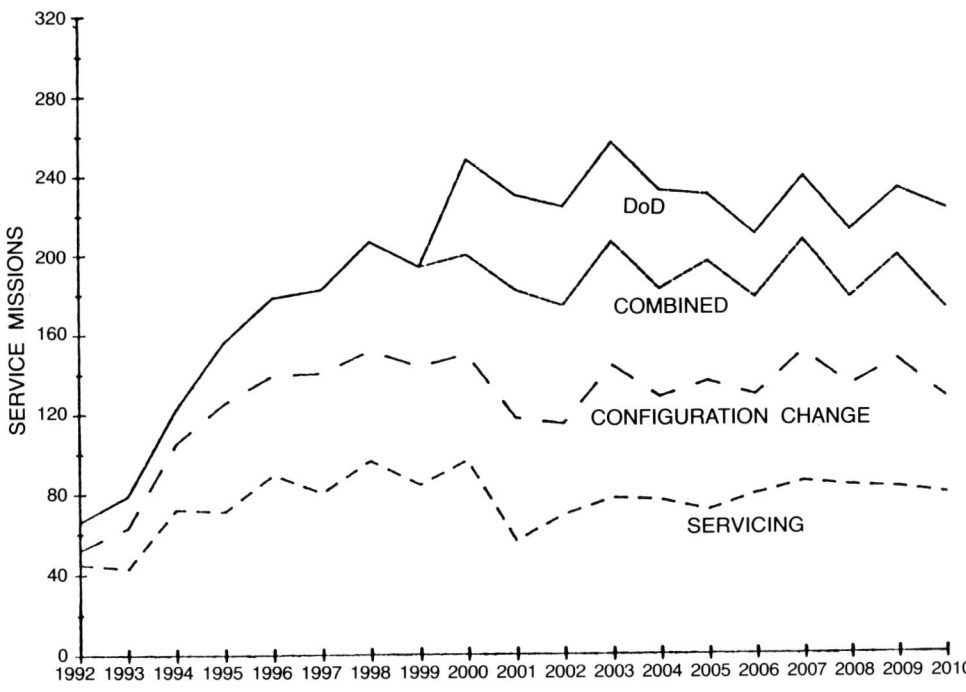

Figure 3.11 Mission model showing estimated number of NASA and DoD servicing missions from 1992 through 2010. CONFIGURATION CHANGE denotes those servicing missions where the space systems structural arrangement or subsystem modules are changed or modified. COMBINED denotes those missions where both configuration changes and servicing functions (like refueling) are performed on the space system. *Courtesy of TRW.*

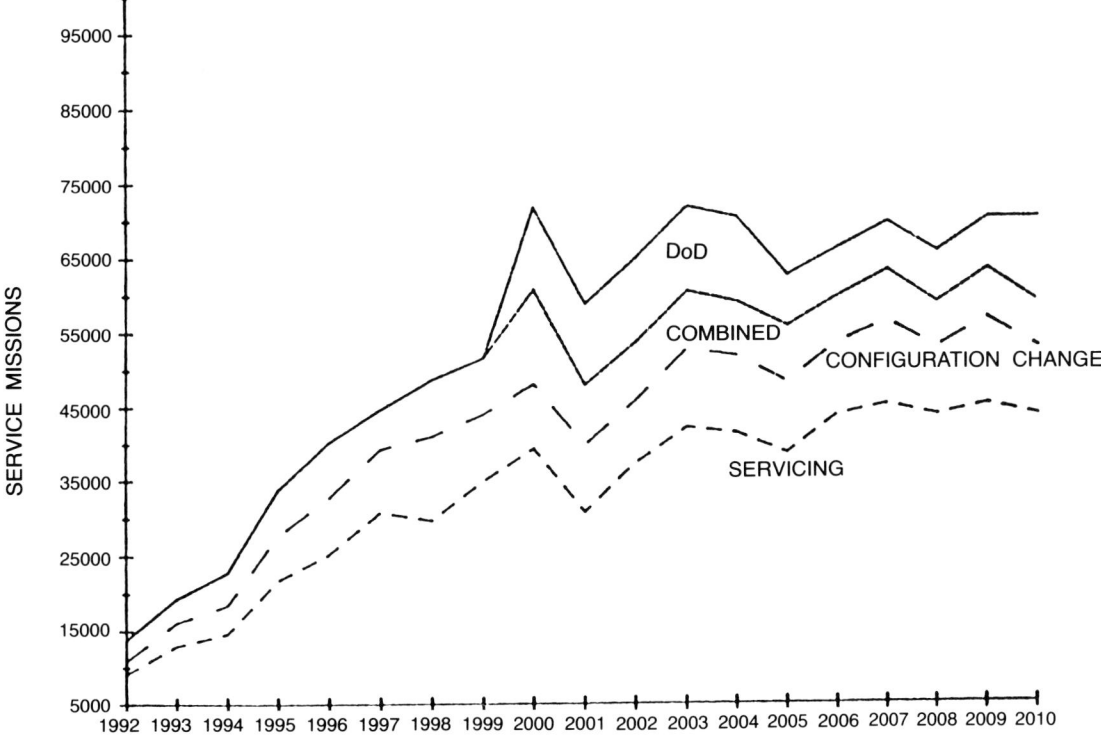

Figure 3.12 Mission mass forecast. See figure 3.11 caption for definition of terms. *Courtesy of TRW.*

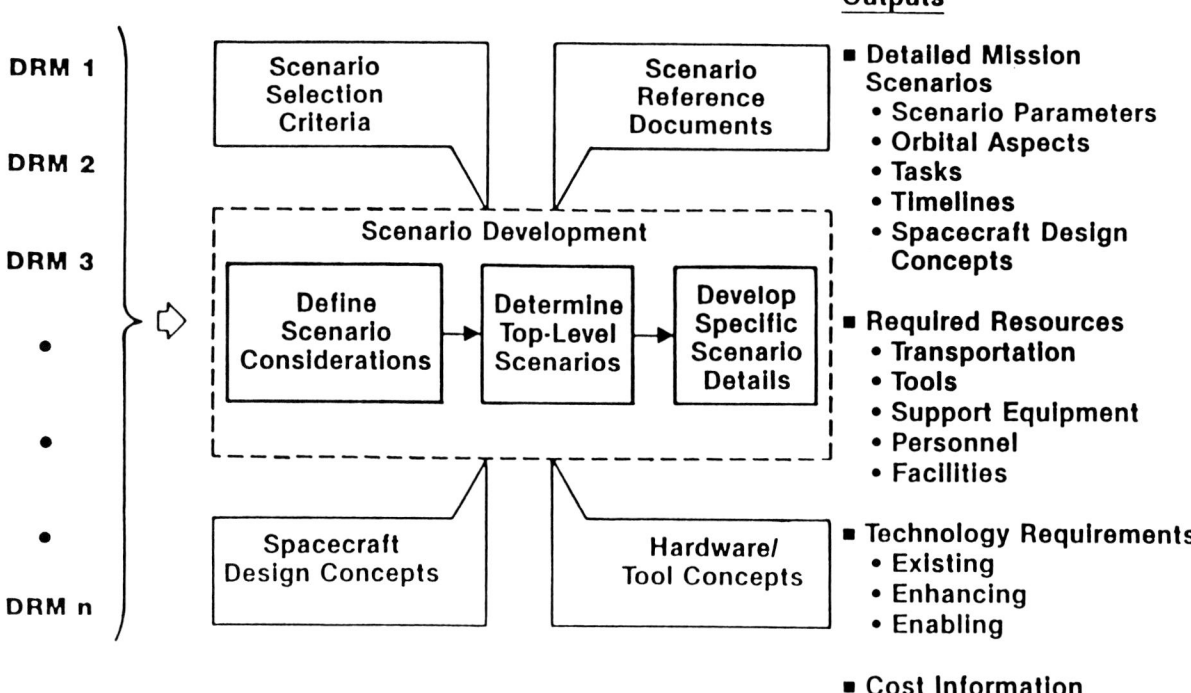

Figure 3.13 Scenario development process for DRM analysis. *Courtesy of McDonnell Douglas Space Servicing Company.*

come SAMS system users and that their spacecraft incorporated servicing capability and servicer tool interfaces.

Mission Scenarios

In order to determine the necessary resource requirements for performing the specific operations identified by the Design Reference Missions (DRMs), operational scenarios are developed. An operational scenario is defined as an imagined end-to-end sequencing of events required to establish a projected course of action. In the case of the SAMS study, the scenarios were used to define the on-orbit operations, the ground operations, and the logistic resources necessary for conducting the various mission phases for the specific operation described by the individual DRM's.

The scenario development process employed is illustrated in Figure 3.13. In this process each DRM was further defined to include more specific and refined operational considerations which then formed the basis for the specific scenario. The detailed descriptions for the DRMs were assessed relative to the spacecraft design considerations and the hardware and tool concepts which identified the details required to perform the space mission. The scenario development process then addressed the resource requirements necessary to carrying out that particular mission.

The various mission phases described by the scenario include the pre-mission activities, the on-orbit mission activities, and the post-mission activities. The pre-mission activities highlight the resource elements required for preparing the SAMS equipment for conducting the intended mission. Activities typically associated with this phase of the operation include spares/equipment activation, equipment integration, and SAMS equipment. The on-orbit mission activities encompass those tasks identified for performing the actual SAMS functions. They include the deployment/rendezvous/docking operations, performance of the required SAMS operations, and the spacecraft/SAMS equipment separation activities. The post-mission activities identify the events needed to complete the specific mission described by the DRM. These activities include equipment deintegration operations, SAMS equipment stowage operations, and where required, equipment refurbishment operations. The summary of these phases in total results in the identification of the overall resource requirements needed to support the specific DRM. Scenario selection criteria are developed to focus the efforts associated with the synthesis of the representative SAMS scenarios. These criteria are, in essence, a filter which ensure that the scenarios being developed for each DRM are suitable candidates for further analysis.

The selection criteria developed for application to the SAMS study scenario analysis were:

- Assess the firmness of the anticipated SAMS operation. That is, what is the likelihood that the mission will in fact be performed? Is it realistic and feasible?

Mission Operations

Selection Criteria / DRMs	Firmness		Urgency			Enh Perf	Success Probability			Cost Info	Combine OPS	Unique Mission
	Real	Feas	Avail	Crit	Redund		Tech Cap	OPS Cncpt	Risk			
1	5	4	4	4	2	3	4	4	5	4	5	5
2	4	4	4	3	4	3	5	4	5	4	5	5
3	5	4	2	2	1	4	4	4	4	4	1	5
4	4	4	5	5	4	4	4	4	4	4	2	4
5	4	3	5	4	4	4	2	4	4	3	4	5
6	4	4	5	4	4	4	4	4	4	4	3	5
7	4	3	4	4	3	3	3	4	4	3	4	4
8	4	3	4	4	4	4	3	4	4	4	4	4
9	4	3	4	5	4	4	3	4	3	3	2	4
10	4	4	4	4	4	4	3	4	3	3	4	3
11	3	2	2	2	1	4	2	3	3	3	1	5

Ranking 1 to 5 (with 5 being the highest)

Figure 3.14 Application of scenario selection criteria. *Courtesy of McDonnell Douglas Space Servicing Company.*

- Assess the urgency of performing the SAMS operations. That is, must the mission be accomplished right now versus can it be deferred? Here we must take into account system availability, criticality, redundancy, and enhanced system performance.
- Assess the success probability of completing the mission at its required time. That is, we can't do a GEO mission before we have the technological capability to access that particular orbital regime. Now we must consider time-phasing of technological capabilities, orbital operation concepts, and risk (safety/hazard) considerations.
- Is costing information for the system being evaluated available to allow assessment of cost benefits (or not) of performing required SAMS operation? If system cost is not available, can this system be equated to another comparable system where we do have cost information?
- Assess the opportunity to combine more than one SAMS function into a single operation. This would highlight benefits of reduced number of launches to take advantage of the repair crew while they are on orbit.
- Does the requirement highlight a different/unique operation or technology that warrants consideration? This will eliminate any duplication of effort.

Based on the above criteria, an assessment was made to evaluate each of the individual DRMs. This assessment weighed the relative merit of each DRM to the specific selection criteria considerations. The DRMs were ranked from 1 to 5 (with 5 being the highest) to determine the applicability of the DRMs to meet the established criteria. The results of the assessment in the SAMS study are shown in Figure 3.14 [1].

An evaluation of this data shows that DRMs 1–10 (Appendix B) appear to be worthy of performing further analyses. In some cases the DRM was ranked low in regard to certain criteria; however, in total they displayed respectable degrees of merit. DRM 11, large spacecraft assembly in GEO, was the lowest ranked mission in the set. However, it was determined [1] to develop this mission primarily because it represents a unique servicing operation and therefore could result in a significant impact to the SAMS supporting infrastructure. It can also be seen that even though most of the DRMs were ranked quite high, there is a progression of lower probability of success due to technological capabilities in the far-term, more complex missions.

This indicates that the technology requirements identified in these missions must be addressed to ensure that these capabilities can be realized in a timely manner.

DRM Conclusions

Analysis of design reference missions is an excellent technique for driving out the requirements and concepts for performing the satellite servicing functions for specific programs [10].

The 11 DRMs constructed during the SAMS study started with a set of relatively simple servicing tasks on a single LEO satellite (DRM 1) and progressed to more complicated servicing on satellite constellations at GEO on later DRMs. Major attention was paid to that mission segment of each DRM where the actual servicing functions were performed—either by manned EVA events or by totally robotic techniques or a blended combination of both.

Each DRM was costed [1]. In general, savings due to space repair, as opposed to satellite replacement, were in the 10 to 35 percent range.

The characteristic within the operations category which favored service and repair missions most was the capability to visit more than one satellite per mission. A huge economy of scale can be exploited for such missions assuming a sufficient number of satellites (three was the threshold) are available at a given time for such missions. Because of fuel limitations, such missions are limited to co-planar satellites. This is not a prohibitive restriction for GEO satellites, since they are all close to being co-planar because of the unique features of this orbit. Several MEO satellite systems also have multiple satellites per plane which could benefit from the multivisit technique. LEO satellites, however, probably cannot benefit from such a technique because there are a limited number of assets in LEO and they are usually not co-planar.

Blended EVA and Robotics

Once the representative scenario is developed for each DRM, an assessment can be made to determine at each phase of the mission (i.e., pre-launch, on-orbit mission operations, post-mission) the respective roles for the man and machine. Since the development of the scenario is based on a specific operational mode which meets the performance requirements and the mission needs, it was necessary to determine the human's role in accomplishing the various operational activities.

For the purpose of categorizing the various man/machine roles the TRW study team [1] adopted the definitions developed by MSFC/MDSSC in The Human Role In Space (THURIS) Study [11]. These categories describe the domains along the man/machine continuum ranging from a purely manual accomplished task up through an independent operation where all functions are mechanized and humans involvement is not required. For clarity the following definitions of the man/machine categories are provided:

- *Manual/Augmented:* Human directly performs an IVA/EVA task either unaided, with the use of simple hand tools, or through the use of powered tools/equipment which amplify human sensory/motor capabilities.
- *Teleoperated:* Use of remotely controlled sensors and actuators allowing the human presence to be removed from the work site (remote manipulator systems and teleoperators).
- *Supervised:* Replacement of direct manual control of system operation with computer-directed functions although maintaining humans in supervisory control.
- *Independent:* Basically self-actuating, self-healing independent operations minimizing requirement for direct human intervention (dependent on automation and artificial intelligence).

As an example of manned/remote task analysis we will focus on a mission scenario which involves LEO remote maintenance/servicing operations. This late 1990s mission reflects the robotic repair/servicing of a single satellite located in a polar orbit. The mission phases associated with the operational performance are as follows:

- *Pre-Launch:* The collective activities associated with the acquisition, preparation, checkout, and integration of the SAMS supporting equipment required to perform the intended mission. These activities are performed at the ground launch facility (in this case the Western Test Range).
- *On-Orbit:* These are the activities involved with conducting the actual SAMS operations. In the case of the example scenario these tasks encompass the operations from STS launch through STS de-orbit.
- *Post-Landing:* The activities performed during this phase of the mission include the safing, deintegration, defueling, and checkout of the returning SAMS equipment. These activities are also performed at the ground launch facility.

The manned involvement associated with the various tasks in each of the mission phase is summarized in Figure 3.15.

Based on the identified manned involvement, an evaluation establishes the priority of the manned performance in each of the mission phases. In many cases, such as pre-launch, the human is involved in varying levels of capacity. Therefore, it must be determined whether the manned involvement is of primary or secondary importance in accomplishing the tasks. In this example, it is decided that for the pre-launch activities the manual/augmented and teleoperated categories are of equal primary importance and the

Mission Operations

	MAN/MACHINE LEVEL			
	MANUAL/AUGMENTED	TELEOPERATED	SUPERVISED	INDEPENDENT
Pre-Launch	Support Equipment C/O – Physical Inspections Cargo Flt Prep-Build Up Payload Integration	Support Equipment Fueling Operations Payload Integration	Support Equipment C/O – Elect/Software Control Pre-Launch C/O	Pre-Launch C/O
On-Orbit	STS Deorbit	Deploy MSM Deploy OMV Dock OMV to MSM Orbital Transfer Ascent Profile Rendezvous/Dock with Spacecraft Spacecraft System Reconfiguration Deploy/Separate from Spacecraft Rendezvous with STS Retrieve OMV/MSM Berth OMV/MSM	STS Launch Orbital Transfer Ascent Profile Spacecraft System C/O Refuel Propulsion System Replenish Gas System Maintenance Exchange Payload Exchange Orbital Transfer Descent Profile STS Deorbit	STS Launch Anti-Jamming Comm Channel Switching Spacecraft Telemetry Data Statusing OMV/MSM Telemetry Data Statusing STS Telemetry Data Statusing
Post-Landing	STS Safing – Equipment Hockup Cargo Deintegration Support Equipment C/O – Physical Inspections	STS Safing – Gas Evacuation – System Purging Cargo Deintegration Support Equipment Defueling Operations	STS Safing – Checkout/Verification Support Equipment C/O – Electrical/Software Control	

Figure 3.15 LEO remote maintenance and servicing: Manned/remote operations.
Courtesy of McDonnell Douglas Space Servicing Company.

supervised category is of secondary importance. This is based upon the amount of work and time required in accomplishing the identified tasks. In the on-orbit or mission activities, the teleoperated and supervised categories are both of primary importance in accomplishing the specific SAMS operations. However, if this were an Air Force mission conducted in an escalated enemy threat environment, the activities associated with the independent category in Figure 3.15 would have an impact on the performance of this mission and thus would have a higher established priority of importance. Though independent of the particular mission, the threat considerations do or could impact the performance and direct manned involvement in accomplishment of the mission activities. In the post-landing activities, the teleoperated mode is established as the primary mode of operation and the manual/augmented and supervised categories are of second and third levels of importance respectively.

This analysis was performed on each of the mission phases associated with each of the scenarios developed for the SAMS study DRMs [1].

SAMS study results indicated man is heavily involved in all the ground operations (both pre-launch and post-mission) in both the manual/augmented and teleoperated modes of operation. The only other direct manned involvement identified is the primary application in the performance of some SAMS missions. By referring to the scenario assessment summaries [1], manned involvement was found to be required in accomplishing spacecraft assembly operations and complex depot level of maintenance/servicing activities. Other areas of human involvement through a teleoperational means of performance were seen to be significant in the pre-mission activities performed at a space-based facility. These mission activities include the integration of mission equipment with space-based SAMS equipment and the fueling, checkout, and deployment of the SAMS equipment to perform the mission objectives. Additionally, some teleoperational control of vehicles, such as the OMV, are required during the mission phase of the scenario.

The supervised man/machine category is involved extensively in the ground operations (both pre-launch and post-mission) but from a secondary level of importance. This is due to the extensive requirement for equipment checkout/verification procedures which in many cases is a computercontrolled activity which only requires the operator to initiate and monitor the task performance. It was interesting to note that this same sort of involvement is also required for the pre-mission and post-mission activities performed at a space-based facility. However, the level of importance was noticeably higher for the pre-mission tasks than it was for the post-mission activities. Investigation into this showed that the importance to accomplishing the SAMS objectives was more dependent upon the pre-mission equipment verification/checkout procedures than it was for the post-mission operations. This demonstrates the degree of criticality and its impact on task priority relative to meeting mission needs. In the case of performing tasks in the on-orbit mission phase of the scenarios, the supervised category was used extensively. The reason for this is the application of robotic systems in accomplishing SAMS operations. These robotically performed operations will be, for the most part, autonomous, however, man will be involved in monitoring the progression of events throughout the operation.

Tasks identified for the independent man/machine category follow the same line of reasoning. These functions are of secondary importance in support of the robotic implementation in accomplishing the SAMS mission operations. One point worth noting, relative to the independent operations, was that in all cases where it was called out, the operations dealt with GEO or higher altitudes. This indicates that the higher the altitude, the less important the manned involvement is when dealing with the application of robotic SAMS equipment/operations.

It has been shown in studies such as THURUS that in the near-term timeframe the demand on the manned element to meet the servicing needs of SAMS type operations will be quite high. Man has a high involvement in accomplishing the SAMS activities initially, 1985 to 1996, but as robotic technological advancements are incorporated, "hands-on" involvement will become less.

Scenario Commonality Assessment

The scenarios developed for each of the SAMS study DRMs highlighted operational approaches selected in support of the individual missions. These in essence are "micro scenarios" which focus the support requirements in all potential orbital regimes, SAMS functions, and timeframes. One feature which was noted during the scenario development process was that there was a significant amount of overlap or commonality existing between the various mission scenarios. As a result of DRM scenario assessment [1] the areas of commonality were identified as transportation elements and orbital staging bases.

Transportation Elements

An important operational consideration for SAMS missions is the effectiveness of achieving mission objectives through the use of alternate launch vehicles. Both the Air Force and NASA have recently stressed a need for the development of a fleet of expendable launch vehicles based on mission requirements, the high costs of Shuttle transportation, and the difficulty of obtaining launch dates and then meeting launch schedules. Based on the new National Launch Systems mixed fleet policy, operational decisions may be driven by (1) the use of ELVs on missions where the Shuttle's unique capabilities are not required and (2) use of the Shuttle for those missions where unique man-machine capabilities combined with the Shuttle's capabilities are required. By adopting these decisions, flight planning and manifesting problems, as currently conceived due to the backlog of missions, can be alleviated.

The Shuttle has proven itself to be a valuable asset for space maintenance operations, but despite past successes the Shuttle's capability to support SAMS missions will be inhibited by its limitations, constraints, and availability.

The major Shuttle limitations as seen for SAMS operations are the stay time on orbit, delta V capability, and delivery-to-orbit payload size and weight constraints.

Stay Time On Orbit

The present configuration for the Shuttle allows for a nominal 7 to 10 day mission duration. Several options would increase orbital operations time on the Shuttle. One is to increase the flight schedule to 12 to 16 flights per year. Another is to modify the Orbiter itself by, for example, the incorporation of the extended duration Orbiter (EDO) kit. The EDO kit gives the Orbiter the capability to support a 16 day mission. This additional stay time offers more opportunities to combine SAMS type missions along with other planned missions.

Delta V Capability

There is no doubt that the Orbiter's limited delta V capability will restrict its use for SAMS missions. An OMV can be used to relieve this limitation, but NSTS utilization for SAMS activities should be limited to those in close proximity/accessible to the Orbiter. When an OMV becomes operational it could be used to help solve the Orbiter delta V problem. An OMV will have its own limitations but will provide the capability to deliver and/or retrieve satellites to and from higher altitudes and in different orbital planes. An early OMV may not be space-based and must therefore be considered as occupying Orbiter payload bay space for each mission requiring its use. Later when OMVs are based at the Space Station, there will be a coupling of the Orbiter as a transportation mode and an operations platform for SAMS missions. Even with the OMV coupled with the Orbiter,

Mission Operations

SAMS capability will be limited to orbital altitudes up to approximately 2,594 km (1,400 nmi)

Size and Weight Constraints

The NSTS has exhibited limitations in its capabilities to transport payloads/cargo to orbit. This has an impact on SAMS missions since during the scenario development process, in many cases the NSTS is considered a primary transportation source. The size of payloads are constrained to fit within the cargo envelope of the Shuttle. This has driven programs such as the Space Station to require extensive on-orbit assembly operations. Additionally, with regard to weight, the Shuttle has the capability to launch a maximum of 24,640 kg (54,000 lb) into low inclination orbits from the Eastern Launch Facility (NASA/KSC) and a 14,515 kg (32,000 lb) payload into high inclination orbits out of the Western Test Range, Vandenberg Air Force Base, California.

These are operational constraints which must be dealt with. Once again, selecting the transportation option that best meets the needs for the mission will result in effective resource utilization.

Orbital Staging Bases

Every SAMS mission requires ground processing activities but missions where the SAMS elements are space-based will have less impact on ground-to-orbit transportation considerations. Many benefits arise from the application of orbital staging bases to support SAMS operations. First of all it is cost-effective to store SAMS equipment on orbit rather than to incur ground processing and launch from Earth for every SAMS mission. By maintaining the SAMS infrastructure elements on orbit the only hardware requiring ground-to-orbit delivery is the mission specific material. Second, space basing of SAMS equipment at orbital facilities such as the Space Station or an in-space assembly/servicing facility gives servicing missions the ability for quick response to high priority, emergency type missions. Based on this operational approach, two activity areas are identified: pre-mission activities and post-mission activities.

The pre-mission activities involve those tasks associated with orbital preparation of the servicing equipment and transportation elements required for conducting the specific mission. The initial requirement is to deliver the mission specific hardware to the orbital base, if it hasn't previously been stored there. As discussed under transportation elements, cargo elements delivered from the ground can be provided by either the NSTS or an ELV depending on the mission requirements. If the NSTS is employed it will rendezvous and dock with the orbital base itself. However, if an ELV is used, a subsequent OMV flight may be required to retrieve the ELV cargo complement and deliver it to the orbital base.

Once the cargo has been delivered to the orbital base, preparation for the mission begins. This involves the activation/checkout/preparation of the SAMS equipment (i.e., transportation vehicle, robotic servicers, tools, the integration of the SAMS equipment and the mission specific hardware) and the checkout of the aggregate system required for the particular SAMS mission. These tasks are accomplished in part by the personnel residing at the orbital base along with monitoring and some control of functions performed through ground control. Upon completion of these activities, the SAMS system is deployed for its intended mission.

When the SAMS mission is completed, the returning SAMS system again requires the services of the orbital base for the post-mission activities. The tasks associated with the post-mission activities include the deintegration of the SAMS system, the checkout and possible repair of the SAMS equipment, and the stowage of the various components. Upon return of the SAMS servicer system from its mission it is docked and prepared for deintegration. This includes primarily the defueling of consumables and the deintegration of each component of the SAMS system. The SAMS equipment is then checked out to verify its functionality, repaired as required, and either prepared for a subsequent mission or placed in storage at the orbital base. The returning mission specific hardware is stored for return to Earth on the next available NSTS flight.

The incorporation of the orbital bases appears to be very beneficial to the overall operational concept of SAMS missions. They minimize the most costly segment which is the ground-to-orbit transportation, while at the same time they are flexible enough to meet the needs of multiple SAMS missions. It is noted, however, that the optimal location of the LEO staging nodes is dependent on the location of the spacecraft requiring the SAMS mission. In other words the currently conceived International Space Station Freedom, in a LEO 28.5 degree inclined orbit, is not suitable for high inclination operational needs. The approach taken in the scenario development process was to employ the orbital staging base as near as possible to the desired inclination. However, when evaluating orbital staging bases from the "macro scenario" point of view the locations of these nodes should be optimized for meeting the needs of all potential missions. The optimization of these nodes would require more detailed analyses but it does appear that a high inclination orbital base along with the Space Station would probably meet the needs for SAMS missions.

Conclusions on Commonality

There is a great deal of commonality when evaluating the SAMS mission scenarios from a "macro" viewpoint. A variety of SAMS operational missions, though different in nature, require common operational approaches/activities.

Focussing on these common elements will, in essence, help establish standardized approaches for accomplishing SAMS operations.

The approach to SAMS should take maximum advantage of common/standard elements which are flexible to support a variety of mission needs. These will include ground-to-orbit and orbit-to-orbit transportation vehicles, orbital facilities for staging and performing SAMS operations, and SAMS equipment elements which enable operations to be performed through direct manned involvement or robotic operations or a combination of the two. Further development and refinement of these supporting elements along with spacecraft designs for standardization maintainability/serviceability will result in an attractive and cost-effective approach to maintaining U.S. national assets on orbit which is the goal of SAMS.

Assembly-in-Space Operations

The goals of the U.S. civil and military space programs demand the development of a new level of orbital assembly, maintenance, and repair of large structures. While remarkable technical achievements have been made in space during the past three decades, they have been primarily related to intermittent missions of comparatively brief duration, and most have been unmanned. The United States has not yet implemented any permanent, periodic, or even long-term human presence in space.

This presence is the next logical step, but it will demand enormous technological progress, primarily in our ability to build large, flexible, integrated orbiting structures. All of our presently deployed space systems have been put into Earth orbit preassembled. Now, we envision structures, space stations, payload platforms, large antennas and sensors, and space-based support facilities which will require more challenging stages of in situ assembly before they can perform their intended functions.

The establishment of lunar and planetary bases will also challenge our space capabilities. Within the next decade, we have to develop concepts and methodologies for large-scale space construction and the techniques and hardware necessary to build a permanent, manned station in Earth orbit. We also have to develop the in-space construction expertise and the facilities necessary to fashion spacecraft for voyages to the Moon and Mars. Finally, we have to learn how to design and build lunar and planetary bases, maybe using some extraterrestrial materials.

In the past, up to 2 years have gone into the second-by-second planning of a weeklong Shuttle mission. Obviously, this kind of preparation cannot be made for missions that may be years in duration. The questions then are: Will we be able to address the demanding problems associated with long-range orbital logistics and scheduling of materials to be delivered to space stations, vehicles, and extraterrestrial bases under construction? What about the tedious but critical tasks of transporting and manipulating large masses, both solids and tanks containing liquids, precision positioning of components, and methods of permanent and dependable joining of those components?

Researchers will not only have to define the best ways to accomplish thousands of space logistics tasks; they will also have to figure out who—or what—can best perform them. The high costs of placing and supporting humans in space will limit the number of astronauts who are available for operating stations, vehicles, and bases. U.S. and Russian experience also indicates possible physiological and other physical problems associated with long-term human presence in space.

The use of robots will alleviate some of these problems. But what is the optimum mix of humans and robots needed to communicate and interact effectively in a complex space construction environment? To find out, researchers must investigate and evaluate robotics and bionic devices used by astronauts and determine how robots can best assist astronauts.

Some universities are beginning to work on this. The University of Colorado, Boulder's Center for Space Construction (CSC) has an ambitious and far-reaching program that will address all of these questions and more. CSC's research and educational endeavors will be both inter- and cross-disciplinary in nature, and will provide an integrated systems approach into such areas as structural design, dynamics and control, construction automation, and the optimization of human/machine assembly of the intricate space structures of the future. The research will be principally based on the university's existing strengths in the areas of space structures, dynamics, and controls; robotics and artificial intelligence; and space operations.

Framework for Space Assembly

As defined in Chapter 2, space assembly is the on-orbit fabrication, joining, construction, or deployment of spacecraft, space systems, or structures. It includes the extension of solar arrays, antennas, and appendages into their operational configurations; installation of special mission kits to OMVs and OTVs; mating of spacecraft to propulsion stages; and spacecraft launch and or retrieval from the Space Shuttle or Space Station Freedom. It occurs before a space system becomes operational. It relates very well to the jargon of Earth construction in that terms such as materials, transportation, job site, tools, logistics, labor force, work schedules, safety, erection devices, and completion stages have similar meanings.

Around the world, structures of various sizes, shapes, and materials stand as tributes to human ingenuity in design and engineering. Long after a way of life passes, architecture remains as testimony to the culture and values of the builders. Grand structures are the hallmarks of advanced civilizations.

Some magnificent structures erected by physical strength, such as ancient pyramids, castles, and cathedrals, still stand. With advances in technology, the craft of building has become a science, and machines augment human strength in construction efforts. The towers, dams, and bridges of modern architecture reflect the influence of new technology.

As civilizations spread across the globe, builders adapt materials and techniques to design structures for new, and sometimes difficult, environments. Engineering and technological solutions are always devised to overcome obstacles and lay the framework for continued expansion. As a result, bridges suspended from steel cables arch over waterways and chasms, and towering skyscrapers create more space for living and working in a crowded world. Around skeletal frames, we have designed the structures, varied in form and function, for today's needs.

Now the United States, with its European, Japanese, and Canadian partners, is meeting an unprecedented challenge: establishment of a manned permanent presence in space. Although NASA has launched an assortment of spacecraft into orbit—including one temporary work station, Skylab—there has never been an attempt to build a permanent habitat in space.

The Space Station Freedom represents the most advanced project, to date, in terms of on-orbit assembly. It requires novel feats of design and engineering.

Assembly Demonstrations

The first assembly of a spacecraft in low Earth orbit was the mating of a manned Gemini capsule to a modified Lockheed Agena stage in the mid-1960s. The largest structure assembled in space in the 1960s was the 9.1 meter (30 foot) diameter Advanced Technology System antenna. In 1984 a 30.5 meter (100 foot) diameter erectable solar array was deployed from the Shuttle.

In preparation for permanent space construction, the NASA Marshall Space Flight Center managed and developed two space assembly demonstrations, EASE and ACCESS, conducted as part of a Shuttle mission in 1985.

EASE/ACCESS Space Assembly

The EASE/ACCESS mission was manifested on STS-61B, Atlantis, launched November 26, 1985. The seven-person crew included Brewster H. Shaw (commander), Bryan D. O'Connor (pilot), Sherwood C. Spring, Jerry L. Ross, Mary L. Cleave (mission specialists), Rodolfo Neri Vela, and Charles Walker (payload specialists). This 6 day, 108 orbit flight was the second night launch of a Shuttle. The flight featured the successful deployment of three communications satellites and the completion of the two space construction experiments: EASE, Experimental Assembly of Structures in Extravehicular Activity, and ACCESS, Assembly Concept for Construction of Erectable Space Structures. In both experiments, astronaut crew members Jerry Ross and Sherwood Spring assembled two small components to form larger structures, just as might be done to build the International Space Station Freedom. Baseline construction efforts were completed during the first scheduled EVA space walk, and some supplementary experiments, which included assemblies using the remote manipulator system and simulating Space Station maintenance, were completed during a second EVA. Work was geared to specific construction tasks—modification, repair, repositioning, cabling,—associated with assembly of large structures in space.

The ability to build and service large structures in space is essential to future NASA projects. The EASE/ACCESS mission was the first flight demonstration of microgravity construction techniques. One of the objectives of this important mission was to correlate actual space construction with simulated microgravity ground assemblies. Actual construction in space by astronauts Ross and Spring provided valuable on-orbit experience. Data collected during this mission in the forms of video, photographs, and crew reporting will help researchers identify ways to improve the assembly, strength, and safety of erectable space structures. A further benefit of the mission was the rehearsal of Space Station maintenance procedures using both EASE and ACCESS hardware.

During past satellite repair missions, the presence of human ingenuity proved invaluable. These previous missions demonstrated the importance of the ability to make timely, on-the-scene judgments and to provide instant feedback. These capabilities were also central to the EASE/ACCESS demonstration of construction techniques on a manned spaceflight. Ross and Spring served as structural assembly experts during two 6-hour periods of EVA. They built the structures, working from and around a platform in the Shuttle Orbiter payload bay. During the second EVA, a third mission specialist, Mary L. Cleave, inside the Shuttle operated the remote manipulator system to position one crewmember at desired locations while the structures were being assembled and Space Station repair scenarios practiced. Another crewmember oversaw operations, controlled the video cameras, and took photographs through the aft flight deck windows.

During the EASE/ACCESS mission, three different structural assembly methods were demonstrated. ACCESS was constructed with the crewmembers anchored in fixed positions or work stations attached to the experiment support structure. For the EASE assemblies, crewmembers worked unrestrained, moving about more freely to various locations around the work area. The supplementary experiments were performed with one crewmember working from the manipulator arm work station, while the other assisted from a work station near the experiment platform. While participating in these EVA tasks, the crewmembers (except when working on the RMS) were tethered to slide wires located along the sides of the Shuttle payload bay.

Careful observation of crew activities was needed to understand the human factors elements of space construction; aspects such as learning, productivity, and fatigue are important as well as the biomedical effects of working during EVAs. This first flight demonstration of space construction provided insight into a number of key questions. During orbital assemblies, is it better to move unrestrained about the structure or remain fixed in one position? Is it more efficient to work moving freely or while being moved by the manipulator arm? Are structural assemblies easily manipulated in space where microgravity allows small forces to move large objects? Is it possible to quantify the relationship between working in the space environment and working underwater, where much of the training for EVA missions takes place?

One way investigators answered these questions was by creating a computer model of humans working in space. Video cameras located in the Orbiter payload bay and on the remote manipulator arm were used to record images of each crewmember at work. In addition, film cameras mounted in the aft flight deck window were synchronized to create a stereo film. This film was used to reconstruct three-dimensional images of the astronauts at work during the EASE experiment. Body positions, equipment locations, and any difficulty in completing a task were derived from the film. Planners used this data to construct computer simulations that revealed how long it might take to complete similar EVA tasks. The experiences from this mission will prove helpful in establishing on-the-ground training methods used to prepare crews for large space construction projects.

For this first on-site assembly, no tools were used during construction. The crew members "snapped" together the prefabricated components to form each structure. Both the larger EASE beams and the smaller ACCESS struts were joined by nodes, clusters of three or four sockets which were locked into place by sleeves on the end of a beam or strut. The ACCESS nodes and struts were located in canisters mounted on the sides and top of the support structure near astronaut work stations; the EASE beams were clamped to the front surface of the support platform and the nodes attached to the top of the platform. The EASE/ACCESS support structure and other experiment equipment occupied approximately one-fourth of the Shuttle's payload bay.

ACCESS

The baseline ACCESS experiment was conducted during the first half of the first EVA period on STS-61-B, Figure 3.16. The primary objective of this experiment was to test the ACCESS structural assembly concept for suitability as the framework for larger space structures and to identify ways to improve the productivity of space construction. This was the first flight demonstration of the ACCESS hardware, but it was designed to be used again as required to develop space structure designs and better building methods.

Jerry Ross and Sherwood Spring entered the payload bay and stationed themselves at the work platform. One stood in foot restraints at the support structure base and the other worked in similar restraints on top of the support structure.

The two crewmembers first unstowed a mastlike, 1.35 meter (4.5 foot) assembly fixture, raised it to a vertical position, and unfolded the fixture's three guiderails in an umbrella-like fashion. They then removed ACCESS struts and nodal joints from three stowage canisters located on the support structure. Ninety-three tubular aluminum struts 1 inch in diameter and 33 nodal joints were used to complete the entire structure. Nine struts were connected on the assembly fixture by nodal joints to form one cell, or bay. After one bay was assembled, it was slid upward along the guide rails to the top of the assembly fixture, leaving room at the bottom for assembly of the next cell. In this manner, the two astronauts constructed 10 identical bays to complete the 13.5 meter (45 foot) high ACCESS tower.

Disassembly was the reverse of assembly; beginning with the bottom bay, struts and nodes were disconnected and stowed. Each remaining bay was slid down the assembly fixture guide rails and the disassembly process repeated. When the ACCESS structure was fully disassembled, the core fixture was folded up and clamped for stowage.

A crewmember in the aft flight deck recorded the progress of ACCESS construction. Video and audio recordings, still photographs, biomedical data, and live crew reporting were taken to evaluate the on-orbit effectiveness of this assembly technique.

EASE

Upon completion of the baseline ACCESS experiment, the EVA crew immediately began the baseline EASE experiment, Figure 3.17. The primary objective of this experiment was to study how people work during space construction and compare the methods used on orbit to those used in underwater training in neutral buoyancy facilities. To enhance this comparison, the flight hardware was designed to be almost identical in size and weight to the underwater training hardware.

When assembled, EASE was a tetrahedral cell that looked like an inverted pyramid. Six aluminum beams, each 3.6 meters (12 feet) long and weighing 28.8 kg (64 lb), and four nodal joints were connected to form the EASE structure. These prefabricated components allowed speedy, safe, and uncluttered on-site assembly of a geometric framework.

Ross and Spring together unstowed the beams clamped to the front surface of the support structure and the nodes attached to the top of the work platform. One crewmember

Mission Operations

The ACCESS assembly fixture is unstowed and raised to a vertical position.

Two EVA astronauts work together, assembling each ACCESS bay and then sliding it up along the assembly fixture to make room for the next bay.

ACCESS: Assembly Concept for Construction of Erectable Space Structures

During the ACCESS baseline construction, both astronauts are positioned in fixed work stations attached to the equipment support structure.

During the supplementary experiments, one astronaut uses the Manipulator Foot Restraint attached to the Remote Manipulator System. A special component carrier (left) holds ACCESS struts and nodes.

Figure 3.16 ACCESS experiment in Space Shuttle bay. *Courtesy of NASA.*

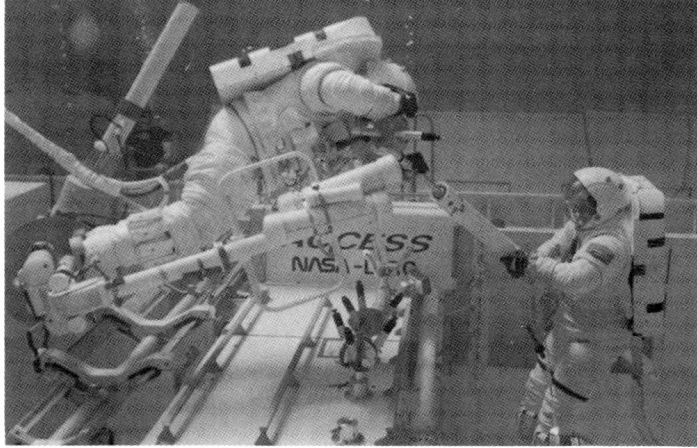

Figure 3.17 EASE experiment in Space Shuttle bay. *Courtesy of NASA.*

was positioned in a foot restraint located on the side of the work platform. The other was connected by tether to the Shuttle but floated to work at the top of the EASE structure. The astronaut located at the base of the stowage rack on the work platform handed components to his partner, who connected the ends of the beams to the nodes and slid the sleeves over the nodes to lock them securely in place.

For the remainder of the EVA, the crewmembers repeatedly assembled and disassembled the EASE structure. Two key parameters analyzed were learning (increased skill with repetition of assembly) and productivity (the crew's performance during a single assembly). During neutral buoyancy assemblies, the astronauts used a variety of scenarios to construct the EASE structure, In flight, they were given the freedom to alter the assembly method as necessary so that their ability to adapt in an innovative manner to the space environment could be studied.

Ross and Spring were photographed and timed as they assembled EASE. Video cameras located in the Orbiter payload bay and on the remote manipulator arm were used to survey overall operations. Film cameras mounted in the aft flight deck windows recorded higher resolution images of each crewmember. Intricate analysis of body motion revealed data on the amount of time and force used to complete specific tasks. Thus, the EASE experiments will help in determining which techniques and equipment are most conducive to human space construction.

Supplementary Experiments

To enhance the value of the STS-61B mission for use in Space Station planning, several tasks were added to both the EASE and ACCESS baseline experiments. Additional experiments included manipulating the large space structures, simulating Space Station maintenance operations, installing flexible cable, and using the Remote Manipulator System to assist the astronauts during structural assembly. These tasks explored alternate methods of construction and repair using the manipulator foot restraint, a portable work station attached to the end of the remote manipulator system, which can be moved around the payload bay. These extra experiments were performed during a second EVA scheduled later in the mission.

Permanent Presence

Though orbital construction is a new venture, working in space is not. For the past 25 years, people have explored the nearby space environment. We have been able not only to survive but also to work productively in this new frontier. Scientists think that low Earth orbit may be an excellent place for producing better materials, such as crystals, pharmaceuticals, and metals, or for launching voyages to distant planets. It is also an ideal location for well-equipped observatories to study Earth and the Universe.

Before these and other beneficial efforts can be sustained, a home and workplace must be established in space. The EASE/ACCESS mission was an important step toward gaining the knowledge needed to build a permanent space habitat. Essential engineering and human factors data gleaned from this mission will be used in planning other construction projects in space. The EASE/ACCESS mission began to shape the framework for the future.

Assembly of Space Station Freedom

In the mid to late 1990s, a four-member crew will climb out of a Space Shuttle, pass through an airlock, and board the International Space Station Freedom. Safely housed more than 370 km (200 nmi) above Earth in a pressurized environment approximating that on Earth, the astronauts will then watch as their transport vehicle is piloted away from the Freedom station for reentry [12].

For the next 30 years or more, Space Station Freedom will be permanently occupied. Slipping through the harsh vacuum of space, it will circle the globe in a low-inclination orbit of 28.5 degrees to the equator. Over time, the drag of thin, remnant atmosphere will pull Freedom toward Earth, but periodic bursts from onboard thrusters will rod the facility back into operational altitude.

Inside Freedom's pressurized modules, every available area will be in use, with ceiling, wall, and floor space all designated for experiments, storage, control panels, and electrical power connections. Of course, without the persistent pull of a one-gravity force, there is no up or down in space. Freedom's designers, therefore, will use lighting, computer displays, controls, and even airflow to recreate some of the up-and-down feeling humans are used to on Earth.

NASA, and its contractor team, is taking an evolutionary approach to building Space Station Freedom. The baseline program includes the U.S. elements, international components, and two unmanned platforms. The next phase will bring in additional capabilities. Observers on the ground will witness the construction of Freedom as it grows from a small to a brilliant dot crossing the night sky [13].

The heart of Freedom's manned base will be a horizontal boom structure, 145 meters (476 feet) long open truss, Figure 3.18, with solar panels extending to either side at both ends [14]. Four special purpose modules—two U.S., one European, and one Japanese—will subsequently be attached to the boom at midpoint. Each module will have an atmosphere nearly identical to Earth's: 80 percent nitrogen and 20 percent oxygen kept at sea-level pressure. In these modules, eight men and women will superintend the space complex, perform experiments, maintain equipment, handle repairs, eat, sleep, and relax.

Space Station Freedom weighs approximately 227 metric tons (about half a million pounds) and is too large and heavy to be placed into orbit by one launch vehicle [14]. Based upon the Shuttle's performance and payload bay physical limitations, the baseline planning has called for approximately 16 to 20 Shuttle flights starting in 1996, to get all of the elements, systems, and support equipment to low Earth orbit. This assembly process will take about 4 years. The sequence in which these flights occur and the packaging of selected parts are dependent on many factors. Early planning of the assembly sequence will be based on various criteria such as Space Station utilization, manning, safety, power, dependence on the Space Shuttle, and overall NASA future space exploration goals. Before final decisions are made by NASA as to SSF assembly in orbit, the agency may opt to delay the Station to about the year 2000 and launch its elements on a new unmanned heavy launch vehicle to avoid dependency on the Space Shuttle.

The following from Reference 13 is a brief description of the baseline sequence for Space Station Freedom's major assembly milestones. The manned base assembly sequence of 20 Space Shuttle flights, has four major milestones. These events are planned to be accomplished close to the completion of the 1st, 4th, 13th and 20th Shuttle flights, Figure 3.19. The launch and deployment of the U.S. and ESA polar orbiting platforms by ELVs are major assembly milestones independent of the manned base assembly sequence.

First Element Launch

The first cargo will consist of a set of integrated Space Station Freedom components to provide a fully functional spacecraft as the "cornerstone" for the fully assembled manned base. This "cornerstone" will be the starboard end of the manned base and includes a power module with solar panels and radiators, S-Band communications pallet with antenna, a reaction control system and tank farm, the mobile transporter, an assembly work platform, and associated truss and alpha joint structures. This integrated assembly will provide its own power and heat rejection, adequate orbital life, communications with the ground, and the capability to

Figure 3.18 Space Station Freedom horizontal boom structure. *Courtesy of NASA*.

rendezvous and dock with the Space Shuttle for the subsequent assembly flights.

Man-Tended Capability

Upon completion of the fourth assembly flight, added structure will extend to the approximate center of the manned base and include the star-board thermal control system, the flight telerobotic servicer and shelter, a control moment gyro pallet, propulsion thrusters, power and fluid management distribution pallets, a TDRS antenna, the aft starboard node, the first phase mobile servicing center, a pressurized docking module, module support structure, and the U.S. laboratory module outfitted to accommodate experiments. These added components and elements will provide the manned base with an early man-tended capability. The conduct of science and technology development in the U.S. lab module will be tended by astronauts on succeeding Shuttle assembly flights.

Permanently Manned Capability (PMC)

Upon completion of perhaps the 11th assembly flight, the complementing port side structure and components will have been added including the inboard power module with solar panels and radiator, tank farm, reaction control module, logistics modules, attached payloads and equipment from the extended duration orbiter flights, aft and forward nodes with cupolas, airlocks, and the habitation module. Since PMC is achieved on flight 13, this assembly plan is designated by NASA as "20/13."

Initial crew size will be at least four persons and will grow to eight when the international modules are attached. The crew will have the capability to live and work comfortably and safely in pressurized volumes indefinitely and will be able to perform full Space Station-based EVAs. PMC will mark the beginning of full scale manned base operations on a day-to-day basis.

Assembly Complete

Upon complete assembly of the manned base, the Space Station Freedom Program will be a fully international space operation. The Japanese experiment module and the ESA laboratory module will be attached, and the Canadian mobile servicing center will be in full operation. The crew will be fully integrated and composed of members from the four partners. Day-to-day operations will be centrally planned and coordinated through the Space Station Control Center

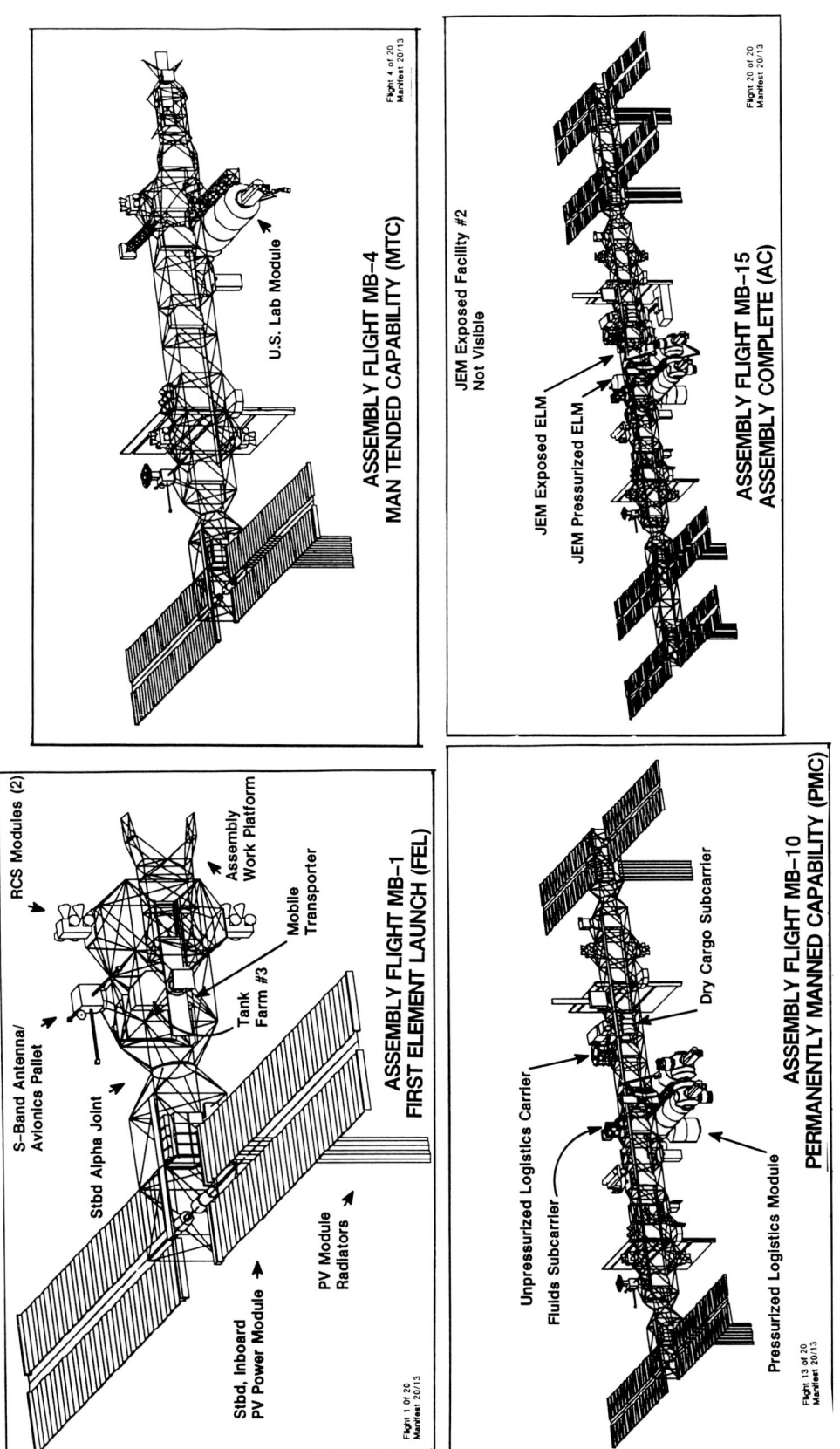

Figure 3.19 Space Station Freedom assembly sequence. *Courtesy of McDonnell Douglas Space Servicing Company.*

Before Restructuring	SSF Item	Proposed New Restructuring
479 ft. assembled on-orbit	Truss length	275 Ft. pre-integrated and assembled on grd.
122	Assembly elements	17
44 Ft. outfitted on-orbit	lab/Hab modules	27 Ft. outfitted on grd.
4	nodes	2
2	cupolas	1
all	international elements	all
18	assembly flights	16
8 people	crew size	4 people
75 kw	power	56 kw
300 mbps	comm downlink	50 mbps
14.7 psi	ECLS pressure	10.2 psi
Advanced SSF Suit considered	EVA suit	STS suit selected
20,000 MHR/Yr	average total crew hours	10,000 MHR/Yr
4 per year	utilization/logistics flights	4 per year
MTC June 96	schedule	IOC December 96

Figure 3.20 Space Station Freedom restructuring considerations. *Courtesy of NASA.*

and Payload Operations Integration Center, but execution of such activities will be initiated and controlled from remote partner payload operation centers as well as from on board the manned base.

Fifteen of the 20 SSF construction spaceflights are dedicated to the actual assembly process. The remaining five flights are for logistics support and crew transport. Up to 24 hours of planned EVA and 6 hours of contingency EVA are ground rule constraints for each of the 15 assembly STS missions [15]. Therefore, the maximum EVA available for SSF complete assembly, over the 3 years of this activity, is 360 hours if no contingency is needed and 450 hours if all the contingency is used. Flight-by-flight assembly tasks, functional flaws, EVA times, STS cargo element manifesting, hardware mass, and computer software configuration are contained in Reference 15. This document is under update.

Redesign of Space Station Freedom

Even as this text is being published, the $37 billion Space Station Freedom is undergoing redesign that will significantly change its configuration in orbit and simplify how it will be launched and assembled by Space Shuttle astronauts [14, 16, 17]. Figure 3.20 shows some of the major features of this redesign currently under review by NASA [17]. The main objective of the new concept is to solve serious engineering problems in the earlier Station design—not to redesign the facility to specifically correct budget problems or to address congressional concerns, although these factors have been considered.

NASA has issued an 11-point directive to field centers, contractors and international partners on how the program will change. Ground rules include no more than four Space Shuttle missions a year to assemble the Station, a phased, "buy it by the yard" approach to construction, and an annual budget cap of $2.6 billion—not the $4 billion peak once anticipated [16]. While top NASA officials insist that they will "consider anything," many knowledgeable engineers expect the result of the overhaul to resemble the redesign options being proposed by the Space Station Assembly Planning Group—a team of NASA, Grumman, and McDonnell Douglas engineers from the Johnson Space Center, the Lewis Research Center, and the Station program office in Reston, Virginia.

Johnson Space Center engineers have said the current Station concept, a design that has evolved since 1984, is flawed in many respects and would have required a significant reworking no matter what the program's budget. In the current design, the U.S., European, and Japanese modules are supported in the center of a (400 foot)-long truss that extends to the left and right of the modules. Solar arrays then extend upward and downward from the ends of the truss.

Under one new option, the module cluster would be retained. But instead of the modules being slung under the center of primary truss, the massive truss would be deleted, leaving the modules as a rectangular, free-flying group. In this proposed configuration, a single, smaller box beam would extend out the longitudinal axis to the aft of the module cluster. The large solar arrays and thermal radiators on the massive truss of the current design would be moved to this trailing beam. Four thermal radiators to reject heat would hang from the bottom. A set of four poles would extend perpendicular out both sides of the box beam. At the end of these poles would be two cassettes each, containing a solar array blanket that would unfurl outward, perpendicular to the box beam. Sixteen solar array blankets—eight on each side—would unfurl from the cassettes.

Viewed in total, this new Station configuration option appears as a module cluster towing its solar arrays. Other options being assessed include an integrated pallet/truss concept [16].

On November 2, 1990, Space Station director Richard H. Kohrs issued the 11-point directive outlining the "ground rules" for restructuring the program and overhauling the design. The key is to develop a new "phased approach" in which goals such as man-tended capability, full electrical power at 75 kW, and addition of permanent crew are advanced as independently as possible and in increments. Also, the directive does the following:

- Sets new annual budget ceilings—For each fiscal year, the new marks, in billions of dollars, with the old caps (in parentheses) are: 1991—2.1 (2.9); 1993—2.3 (3.0); 1994—2.5 (3.6); 1995—2.6 (3.8); 1996—2.6 (4.0).
- Reduces the initial crew from eight to four astronauts.
- Establishes life sciences and materials sciences as the highest priority work to be performed on the Station.
- Tries to hold a March 1996 target for initiating assembly of the Station in orbit.
- Limits Shuttle flights for Station assembly to four per year. Including missions to use and resupply the base, Station managers are to count on only six to seven Shuttle flights, not eight, as earlier planned.
- Seeks to minimize the redesign's impact on the international partners. NASA does not want to repeat the mistakes it made the last time it reworked the Space Station program when it failed to involve the European Space Agency, Japan, and Canada early enough.
- Places off limits any suggestions to alter the management structure of the program. The setup remains controversial among astronauts and workers at the field centers.

Other modifications to the Station's design, schedule, or program that NASA is unlikely to reconsider are:

- Free-flying platforms—Many materials scientists, concerned over vibrations in the Station that would be caused by nearby astronauts, continue to urge NASA to consider removing their projects to orbiting laboratories that would only be visited by astronauts. But the stretchout of the Station assembly schedule would offer 3 to 4 years of such microgravity projects before the base is staffed by astronauts.
- Expendable launch vehicles—The space agency especially does not want to wait for the development of a new booster to place Station components in orbit.

In addition, the redirection is significant in not offering a target date for completing assembly of the Station. NASA hopes to "get away from the concept of final completion" in favor of an "evolutionary" approach to Station expansion.

A particularly thorny issue is how the revamped Station might dovetail with the Space Exploration Initiative. The 11-point directive seeks to identify—and eliminate or defer—"all development that does not directly support the baseline configuration." However, no steps taken so far would preclude the eventual use of the Station as a staging point for expeditions to the Moon or Mars.

Johnson Space Center engineers and astronauts believe the no-truss configuration would:

- Reduce assembly risk. A more incremental approach to Station design would reduce the need to assemble complex structures in orbit. It would allow more integration of Station elements on the ground before launch and reduce dependence on both robotics and astronaut EVA for assembly.
- Reduce Shuttle Orbiter impact. NASA engineers believe the new concept would enable much simpler and more efficient packaging of station elements in the Orbiter cargo bay for launch.
- Reduce Shuttle flights to complete Station. Shuttle missions to complete assembly of the Space Station could be reduced to about 18.
- Reduce EVA maintenance. There would be fewer orbital replacement units on the new design and much less complex structure requiring human inspection and intervention.
- Improve orbital stability. The new concept's more compact configuration and center of mass characteristics would enable it to fly in a gravity gradient attitude without thruster firings to hold its attitude—something the massive current design cannot do. This could also benefit potential station users, such as the

microgravity processing community which has been concerned about thruster firings and other Station motions caused by the massive truss configuration.

In addition, under the redesigned concept the Space Shuttle would not dock directly with the Station. Instead it would be piloted to within 15 meters (50 feet) of the Station, where the Orbiter would be halted. The Orbiter's manipulator arm would then reach out and grapple the Station, then pull itself to a docking fixture. This berthing rather than docking concept has sparked controversy, especially over the ability of the Station to remain stable in a free drift control mode when the arm slowly brings the Shuttle to connect with the facility.

NASA engineers believe the more flexible growth plan that is part of the no-truss concept would enable it to return to the current configuration if necessary or the dual keel concept desirable for lunar/Mars operations. However, little work has been done to verify this [16].

Assembly of Space Station Freedom in orbit will be a challenge of enormous proportions; the task will blaze the trail for ambitious missions in the future which will require on-orbit assembly, test, checkout, and operation [12]. Where possible, flight elements will be fit checked on the ground before they are carried into orbit aboard the Shuttle. High fidelity mockups and electronic simulators will be used to ensure the compatibility of elements that cannot be tested on the ground.

During the early stages of assembly, a U.S. made flight telerobotic system will aid in the construction of the manned base. A Canadian built mobile servicing system (MSS) equipped with a manipulator arm will also be installed to assist in Freedom's assembly. The MSS is based on expertise acquired in creating the Canadian remote manipulator arm, the versatile robot arm used extensively in U.S. Space Shuttle missions. The MSS's robot arm will perform jobs controlled from work stations situated both inside and outside Freedom's pressurized modules, reducing the need for EVA space walks. It will eventually help deploy, dock, and redeploy a visiting Shuttle orbiter; assemble, retrieve, and transport payloads around the Freedom station; and position astronauts for access to its various parts. Mounted atop a NASA built flatcar, to be developed in a future phase of the Freedom program, the MSS will be able to move along Freedom's extensive truss structure.

Two of the pressurized modules that will be attached to the horizontal boom are being supplied by the United States [13]. One will serve as a laboratory, the other as a living area. Each module will be taken up individually into space inside the Shuttle's cargo bay. Astronauts, assisted by the manipulator arm, will remove the modules from the Shuttle and secure them in their proper locations on the Station's truss work.

Each U.S. module is nearly 13.6 meters (45 feet) long and about 4.5 meters (15 feet) in diameter. Early in the assembly sequence, the laboratory module will be positioned on Freedom's supporting frame. With this first module in place, astronauts can start equipping the laboratory to carry out experiments while the Shuttle is docked, periods which could extend from 2 weeks to a month. A Space Shuttle will deposit the habitation module at the orbiting construction site several flights later. With the addition of the forward nodes and logistics module, Freedom will be ready for permanent occupation.

The Japanese and European modules, slightly smaller than their American counterparts, are scheduled to arrive, respectively, after Freedom is permanently staffed. The Japanese experimental module (JEM) will accommodate scientific and technological development research, including microgravity experiments. A space exposed deck will hold a variety of experiments that can be reached by a manipulator arm. The JEM will include a detachable, experiment logistics module that will hold consumable goods, experimental specimens, and various kinds of gases for the JEM. This logistics canister can be hauled into orbit aboard the Shuttle or by a Japanese expendable launch vehicle which will be operational by the mid-1990s.

ESA's design for a permanently attached laboratory module is based on Spacelab, its contribution to the Space Shuttle program. Spacelab has successfully flown several times in the cargo bay of a Space Shuttle. Derived from that design, the ESA module will consist of four cylindrical segments forming a pressurized module. Once it is permanently attached, the ESA laboratory module will become a research site for crews to perform experiments in the physics of fluids, life sciences, and materials research. At this point, the basic foundation for living, working, and studying aboard Space Station Freedom will be in place.

ESA is also developing, as part of its contribution to Space Station Freedom, a man-tended free flyer (MTFF) to be made up of two Spacelab segments and a resource module that holds supplies. The MTFF will function independently of the manned base.

As a self-contained automatic laboratory circling Earth, this spacecraft could produce space-grown crystals and other specialized materials in an undisturbed environment. Exchanging the harvest of MTFF produced products with raw stock would be handled by Freedom's crew, Space Shuttle astronauts, or the crew of the European space plane Hermes, now under development.

Freedom's crewmembers will move between the various modules through four interconnecting resource nodes, or sets of pressurized cylinders. The nodes are spacious and outfitted with command-stations, control-work stations, and other hardware. The forward nodes will contain the primary and backup docking ports for Shuttle Orbiters.

Two nodes are fitted with airlocks through which astronauts can leave for work outside Freedom's pressurized modules. The forward nodes are sure to be a favorite lookout point. With two windowed cupolas, one looking toward Earth and the other looking outward to space, they offer on-top-of-the-world sightseeing at its best. The panoramic view of all space above and below the manned base will permit astronauts to monitor an incoming Shuttle, guide robot arms performing external payload assembly or maintenance tasks, conduct scientific observations, and observe fellow crewmembers on EVAs.

Electric power for Freedom's manned base will come from arrays of solar cells which are deployed from the power modules located on the ends of the horizontal boom. The four outstretched solar arrays at each end of the boom contain a total of about a half-acre of solar cells. The solar arrays provide all the power needs during the sunlit portions of an orbit. In addition, they provide electric power to charge batteries which then provide all the power needs during the dark portions of each orbit. Together, the solar arrays and batteries will be able to provide 75,000 watts of electricity, enough to power 25 all-electric homes on Earth. The largest electric power level required in space to date, this will support all the housekeeping loads as well as the power needs of scientific equipment, computers, communications equipment, and equipment such as a materials-processing furnace.

Regular visits to Freedom by the Space Shuttle will bring new crewmembers, visiting scientists, and a logistics module loaded with new supplies to exchange with the logistics module already in place. To further sustain the Freedom station with new supplies and experiments, unmanned rockets, possibly including Japanese or European boosters, may also be commissioned [13].

Recently, with their space laboratory Mir, the C.I.S. has shown that humans can live and work in space for almost a year without the benefit of artificial gravity. NASA plans to have the crewmembers serve tours of duty on Freedom that will gradually expand from 90 to 180 days as more is learned about the physiological effects of prolonged weightlessness.

Space Station Freedom, which NASA and its partners will assemble in space, will be much more than a collection of Earth-circling girders and cylinders housing hard-working astronauts and scientists. It is also a commitment to the bold pursuit of scientific knowledge, technological prowess, and new commerce—all ingredients essential to meeting the demands and opportunities of the 21st century.

References

1. *Space Assembly, Maintenance, and Servicing Study (SAMS) Final Report*, Volume I, Executive Summary, TRW No. SAMSS-196, Volume II, System Analysis, TRW No. SAMSS-195, Volume III, Design Concepts, TRW No. SAMSS-197, Volume IV, Concept Development Plan, TRW No. SAMSS-197, Volume IV, Concept Development Plan, TRW No. SAMSS-198, Volume V, Neutral Buoyancy Simulation, TRW No. SAMSS-199. TRW, Redondo Beach, CA, July 6, 1988.
   and
   Space Assembly, Maintenance, and Servicing Study (SAMS) Final Report, Volume I, Executive Summary, Volume II, System Analysis, Volume III, Design Concepts, Volume IV, Concept Development Plan, Volume V, Simulation Report. Lockheed Missiles and Space Company, Inc., Sunnyvale, CA, July 6, 1988.
2. Burger, John H. *On-Orbit Support: Past, Present, and Future*, Planning Research Corp., AIAA/SOLE Second Space Logistics Conference, Costa Mesa, CA, October 3, 1988.
3. Janz, Ron. Planning Research Corp., and Neal Ely, Air Force Space Systems Division. *COSEMS; A Dynamic Simulation of Space Logistics*, AIAA/SOLE Second Space Logistics Conference, Costa Mesa, CA, October 3, 1988.
4. Borrego, J., F. Cheng, and R. Janz, Planning Research Corp., *A Space Logistics Simulation Implementation in Ada*, 1988 Winter Simulation Conference, San Diego, CA, SCS 761–764.
5. Cockrell, Charles E. *In-Space Assembly/Servicing Facility Workshop*, NASA-Langley Research Center, July 24–26, 1991.
6. NASA. *Satellite Servicing, A NASA Report to Congress*. NASA Office of Space Flight, Washington, DC, March 1, 1988.
7. Johnson, Nicholas L., and Darren S. McKnight. *Artificial Space Debris*. Krieger Publishing Company, Melbourne, FL, 1987.
8. *The DoD Space Transportation Mission Requirements Definition*. The Aerospace Corporation, Report No. TOR-0086 (6960-01)-1, October 25, 1985.
9. *Space Transportation Architecture Study*. Sponsored by Air Force Space Division and NASA, performed by several contractors, 1985–1987.
10. Chucker, S. M. *Space Asset Support System Implementation Plan*. McDonnell Douglas Space Systems Company, Huntington Beach, CA, Contract No. SDI 0084-88-C-0020, October 1989.
11. Johnson, G. A., and S. M. Chucker. *The Human Role in Space (THURIS) Study*, McDonnell Douglas Space Systems Company, Huntington Beach, CA, Study performed for NASA/MSFC, 1984–1986.
12. David, Leonard. *Space Station Freedom, A Foothold on the Future*, NASA Office of Space Station, NP-107-10-88.
13. Hess, Mark. *Space Station Freedom Media Handbook*. Office of Space Station, NASA Headquarters, Washington, DC, April 1989.
14. Morata, L. P., and F. D. Riel. *Space Station Freedom Assembly and Operations*. Paper presented to Twenty-Eight Space Congress, Cocoa Beach, FL, McDonnell Douglas Space Systems Company. MDCH7024, April 1991.
15. *Space Station Freedom Assembly Sequence*, NASA SSF Level II System Engineering and Integration, Doc. No. SSE-E-R20, September 23, 1988.
16. Aviation Week and Space Technology Article, *NASA Concept Deletes Station Truss. Changes Assembly to Solve Design Flaws*. Page 26, November 12, 1990 written by Craig Covault and James R. Asker.
17. Robinson, Robert L. NASA/JJC Space Station Freedom Program Office. *Space Station Freedom Restructure*. Briefing to Southwestern Aerospace Professional Representatives Association, February 12, 1991.

Chapter 4

Orbital Maneuvers

The influences of orbital mechanics on space systems design and operations are complex. A spacecraft is not simply placed in orbit. A number of attitude, propulsion, and flight control strategies must be considered in determining the orbit the spacecraft will fly. The reality of Keplerian motion must be reconciled with mission needs, orbit control technology, and costs.

Satellite servicing missions will eventually occur in all the useful orbits, including orbits termed: circular, elliptical, geosynchronous, Sun-synchronous, repeating, polar, high eccentricity, and planetary.

This chapter addresses the astrodynamics, guidance, navigation, and propulsion technologies as related to satellite servicing missions. Orbital mechanics, as an academic subject, will not be discussed as other references [1 through 6] have thoroughly presented the definition and mathematics of this subject. Instead three kinds of orbital maneuvers will illustrate mission operations important in the planning of space servicing missions. These examples are:

1. Servicing of multiple satellites
2. Cost-effective propellant expenditure for orbital transfer maneuvers required to reach or retrieve satellites for servicing
3. Satellite proximity operations near the Space Shuttle or the Space Station

Servicing of Multiple Satellites

The information for this example was derived from Reference 7. A problem of principal concern in polar orbiting satellite servicing and resupply is to define cost-effective rendezvous and docking scenarios of the servicing vehicle for a constellation of satellites. Alternative scenarios include single or multiple space-based transfer vehicles and expendable transfer vehicles. If several satellites are to be serviced on a cyclical basis by one servicing vehicle, such as an orbital maneuvering vehicle (OMV), at a given servicing interval of, say, 36 months, the time between revisits to the various platforms may be sufficiently long to allow using orbit transfer modes that reduce the propellant required for orbital change maneuvers. For example, a less costly in-plane altitude change maneuver can be employed to produce nodal position changes due to the resulting change in nodal drive rate, however, at the expense of extra time.

The four polar orbiting platforms (POP) of the proposed NASA Earth Observing System (EOS) program are used to demonstrate the parameters associated with servicing multiple satellites and the magnitude of altitude change maneuvers, i.e., the associated propellant cost, and the time elapsed for achieving the desired nodal change. Other maneuver modes, such as combined altitude and inclination changes to accelerate the nodal change, are also presented. These parameters generally depend on propulsion performance characteristics of the servicing vehicle and also on ELV resupply capabilities and launch schedules. Trades of propellant cost versus time expended for orbital transfer are subject to revisit time constraints within the satellite constellation and the time allowed for completing each servicing cycle.

OMV Orbit-to-Orbit Maneuvering Cost

The polar orbit servicing scenario calls for multiple orbit-to-orbit transfer maneuvers to be performed by the OMV in visiting user satellites in orbits with different nodal positions. In addition to performing servicing tasks on satellites in these orbits, the OMV also may be called on to transfer a spacecraft from one orbit to another to replace one that has failed. Retrieval of satellites from polar orbit is not an objective as long as the Shuttle Orbiter is not projected to operate from the Western Test Range, i.e., at orbit inclinations required to reach satellites in polar orbit.

Plane Change by Differential Nodal Drift

Purely propulsive transfers between orbits with large differences in the ascending nodes are ruled out because of the unacceptably large out-of-plane ΔV maneuvers that would be required. Even a plane change of only one degree requires about 128 meters/sec (420 ft/sec). Instead, a change in nodal position can be accomplished with the aid of the natural nodal drive due to the gravitational force initiating

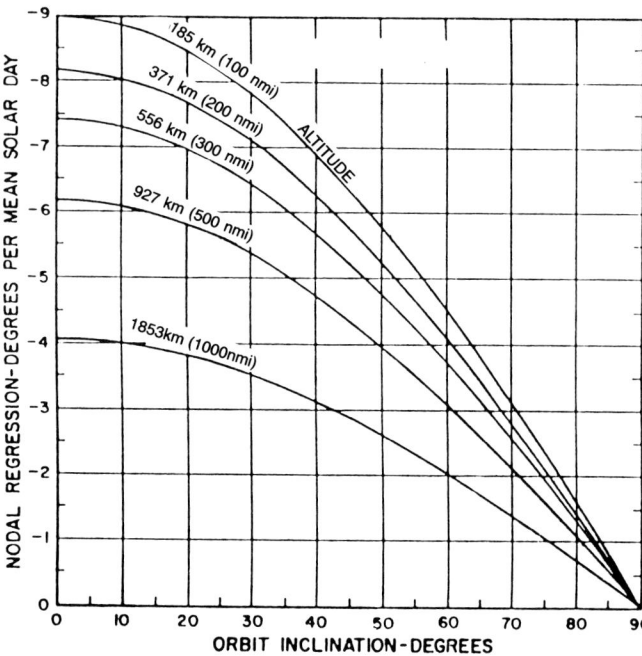

Figure 4.1 Nodal regression rates per day versus orbit inclination with altitude as a parameter. *Courtesy of TRW.*

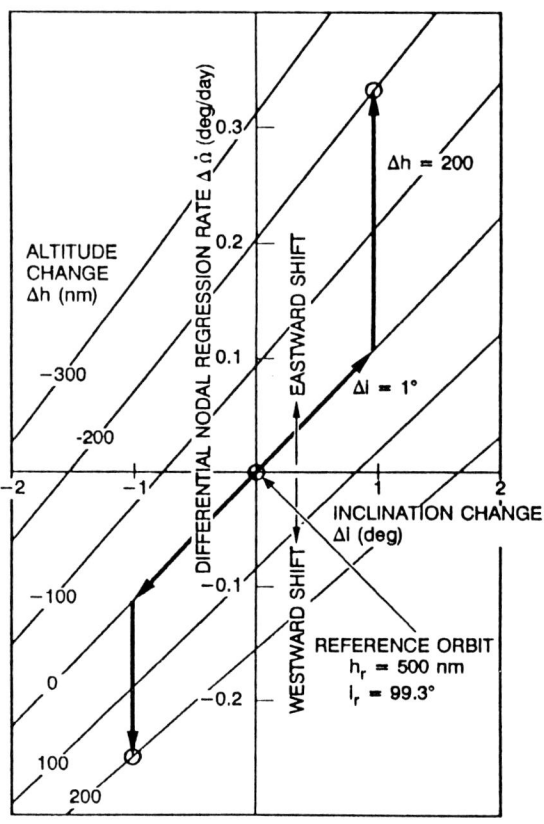

Figure 4.2 Differential nodal regression resulting from combined altitude and inclination changes. *Courtesy of TRW.*

this drift by an in-plane altitude change maneuver. Figure 4.1 shows the nodal regression rate $\dot{\Omega}$ as a function of orbit inclination with orbit altitude as parameter. At 30 degree inclination and 555 km (300 nmi) altitude, a differential rate of nodal regression as large as 0.6 degree per day is produced by a 185 km (100 nmi) altitude change maneuver at the cost of 107 meters/sec (350 ft/sec). Allowing a waiting period of 50 days, a 30 degree shift of the orbit node from its original position is attained. At this time, the altitude maneuver must be reversed to stop the differential nodal drift. A total maneuver expenditure of about 214 meters/sec (700 ft/sec) accomplishes a nodal shift that otherwise would require a 15 degree propulsive plane change maneuver at the antinode at a cost of 1,920 meters/sec (6,300 ft/sec).

The 30 degree nodal change could be further increased as much as desired if there is no time constraint. On the other hand, the nodal shift can be accelerated by a larger altitude change. The trade is between time and the amount of in-plane ΔV maneuver expenditure. The ΔV expenditure is inversely proportional to the transfer time for a given amount of required nodal shift.

Differential nodal regression rates at near-polar orbit inclinations used for Sun-synchronous Earth observation platforms, typically in the 95 to 100 degree range, are much smaller than the rates at low inclination for a given altitude change as shown in Figure 4.1. However, even these small differential regression rates may suffice to achieve the desired nodal shift if enough time is allowed. The servicing scenario and the length of the revisit period for a constellation of user spacecraft dictate the allowable drift time.

Note that the regression rates for 95 to 100 degree inclinations correspond to those for 80 to 85 degree inclinations in Figure 4.1 except for a sign reversal. As an example, a 185 km (100 nmi) altitude change from a reference Sun-synchronous orbit at 925 km (500 nmi) altitude and 99.3 degree inclination produces only about 0.15 degree per day of differential nodal regression, and the waiting time for completing a sizable nodal shift may become excessively long.

Enhanced Differential Nodal Regression

To increase the differential regression rate achievable by a given ΔV maneuver magnitude, it is advantageous to add a small inclination change to the altitude change. Figure 4.2 shows the benefit of combined altitude and inclination changes for near polar orbits using an enlarged section of Figure 4.1 in that region. By combining the in-plane and out-of-plane maneuver components vectorially, a total ΔV is obtained at a lower propellant expenditure than if the two maneuvers were performed separately. One half of the plane change is combined with the perigee maneuver and the other half with the apogee maneuver at the respective times of nodal crossing.

The most effective combination of ΔV components for altitude and inclination change can be determined analytically by an optimization procedure, but is also readily ob-

Orbital Maneuvers

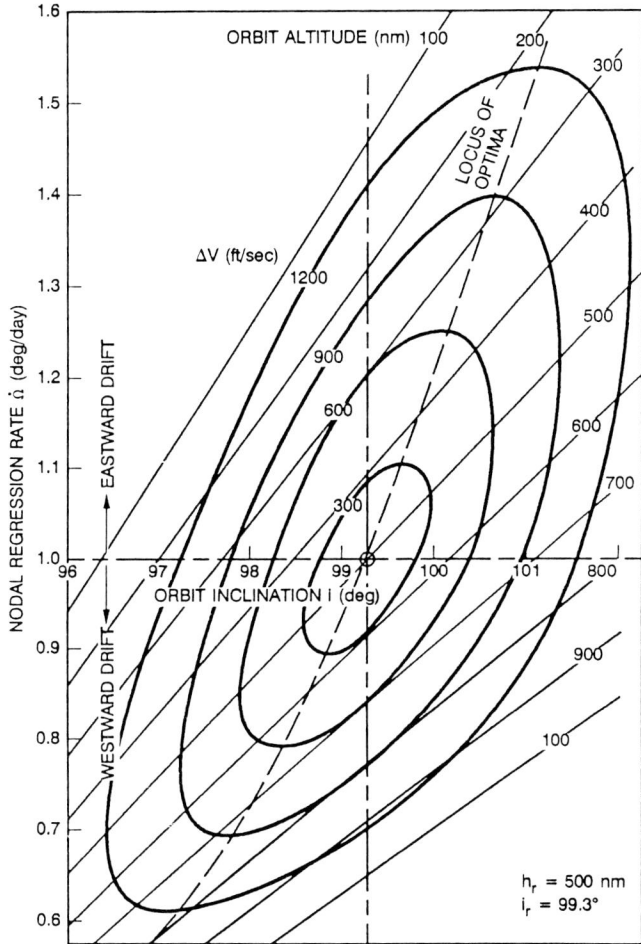

Figure 4.3 Contours of constant maneuver velocities and resulting nodal regression rates. *Courtesy of TRW.*

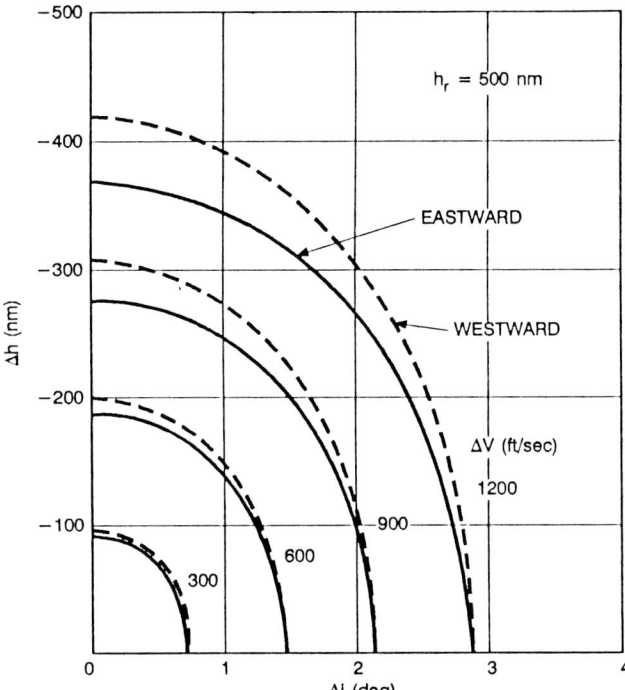

Figure 4.4 Contours of constant maneuver velocities for combined altitude and inclination changes. *Courtesy of TRW.*

tained from a plot of constant ΔV contours superimposed on an $\dot{\Omega}$ versus inclination and altitude diagram, such as Figure 4.2, centered on the reference altitude and inclination point in that diagram. Figure 4.3 shows such contours around the reference point of 925 km (500 nmi) altitude and 99.3 degree inclination for ΔV maneuvers ranging from 91 to 366 meters/sec (300 to 1,200 ft/sec). The contours are obtained by mapping the corresponding curves of Δh versus Δi for the same ΔV values, Figure 4.4, onto the $\dot{\Omega}$ versus inclination and altitude diagram, Figure 4.2. Similar contours can be readily plotted for orbits other than the reference orbit shown in Figure 4.3.

The maximum and minimum regression rates on each contour designate the most effective combination of altitude and inclination change for the highest eastward or westward differential regression rates. The dashed line drawn through these points is the locus of optimum combined maneuvers. The data in Figure 4.3 shows a 35 percent increase in differential regression rate by the combined maneuver above that obtainable from a change in altitude alone at the same maneuver velocity, $\Delta V = 366$ meters/sec (1,200 ft/sec). At lower velocities the percentage is smaller. The maximum

eastward nodal drift rate (obtained by altitude reduction) is about 25 percent greater than the maximum westward drift rate (achieved by the corresponding altitude increase) at the same ΔV expenditure. This is explained by the spreading of the lines of constant altitude with increased orbit inclination to the right of the reference point in Figure 4.3.

Figures 4.5 and 4.6 show optimal nodal regression rates, eastward and westward, and the time required for a full 360 degree nodal change, as functions of maneuver velocity, accounting for both the start and stop maneuver involved in accomplishing the nodal change. The 360 degree change in Figure 4.6 is of interest in scenarios where a return to the first of several user spacecraft within a specified time interval is required. Note that the maximum eastward drift rate in Figure 4.6 is limited by the lowest orbital altitude to which the transfer vehicle can be transferred without incurring excessive atmospheric drag effects.

Figure 4.7 shows contours of constant ΔV maneuvers similar to those in Figure 4.3, but for a reference orbit of 55 degree inclination and 925 km (500 nmi) altitude. It is apparent that because of the smaller tilt angle of the altitude parameter lines in the diagram there is little, if any, gain in regression rates to be achieved by combining inclination changes with altitude changes. The combined maneuver concept is most useful only in near-polar orbit regions.

Maneuver Strategy

In a multiple spacecraft servicing scenario not all transfer maneuvers need to be performed at the highest drift rate. If

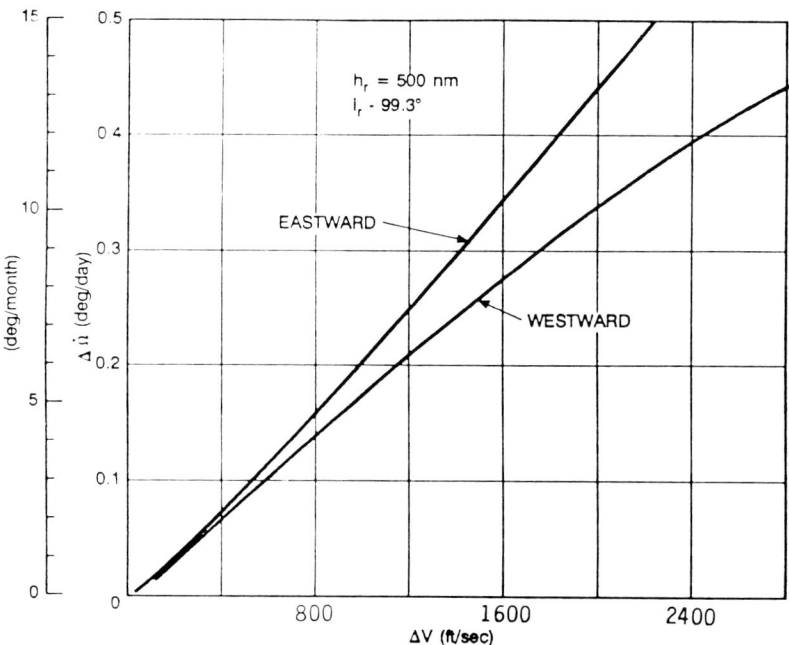

Figure 4.5 Nodal drift rate versus maneuver velocity. *Courtesy of TRW.*

the nodal difference between two adjacent user satellites is small, a lower rate can be used to conserve ΔV expenditure without much impact on the total time expended for completing the servicing cycle. Transfers between satellites with the largest nodal difference in turn should be performed at the highest rate.

The best drift rate combination depends on the relative distribution of the satellite orbit nodes and can be determined by an optimization approach, which is carried out below for a three-satellite servicing scenario. Of the three transfer times, T_1, T_2, and T_3, within a specified total cycle time, two are independent variables. The optimization consists of minimizing the sum of ΔV expenditures, subject to the time constraint T = constant. The nodal differences D_1, D_2, D_3, are specified with $\varepsilon D_1 = 360$ degrees.

To minimize the total ΔV expenditure requires minimizing the sum of the three transfer rates, $S = \varepsilon/R_1 = \varepsilon/D_1/T_3$. Thus, we set the partial derivatives $\partial S/\partial T_1$ and $\partial S/\partial T_3$ equal to zero, as

$$\frac{\partial S}{\partial T_1} = -\frac{D_1}{T_1^2} + \frac{D_2}{(T - T_1 - T_3)^2} = 0,$$

$$\frac{\partial S}{\partial T_3} = -\frac{D_3}{T_3^2} + \frac{D_2}{(T - T_1 - T_3)^2} = 0,$$

This leads to two equations

$$\frac{T - T_3}{T_1} = 1 + \sqrt{D_2/D_1} = Q_1$$

and

$$\frac{T - T_1}{T_3} = 1 + \sqrt{D_2/D_3} = Q_3$$

from which the optimal values T_1 and T_3 can be found. They are

$$T_1 = T\frac{1 - Q_3}{1 - Q_1 Q_3} \quad \text{and} \quad T_3 = T\frac{1 - Q_1}{1 - Q_1 Q_3}$$

Results obtained for three nodal point distributions are listed in Figure 4.8. The reduction in ΔV expenditure is proportional to the reduction in the sum of transfer rates achieved by the rate optimization. As shown in the three cases investigated, it amounts to 9.9, 14.5, and 6.8 percent,

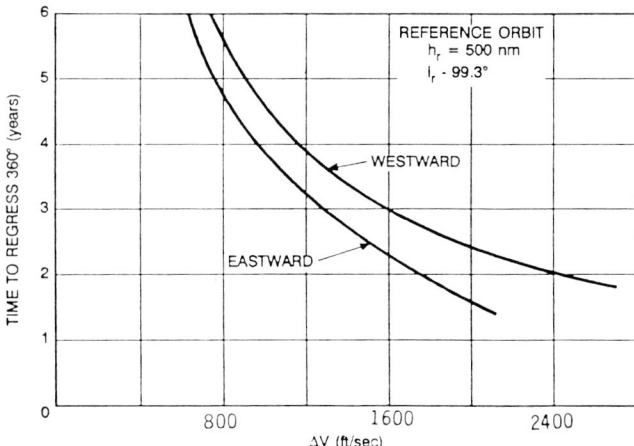

Figure 4.6 Time to complete 360 degrees of nodal regression versus maneuver velocity.

Orbital Maneuvers

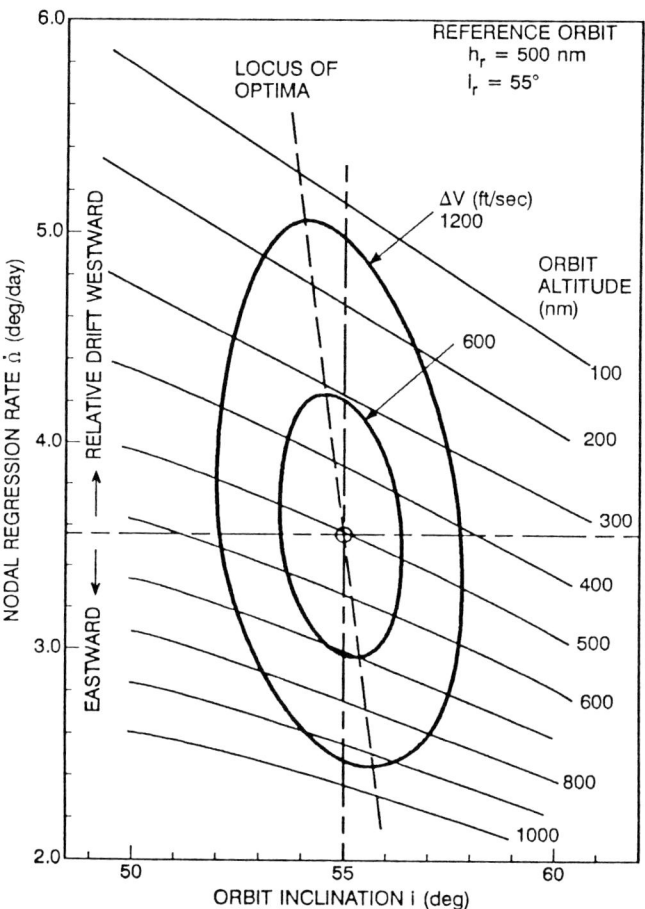

Figure 4.7 Nodal regression rates resulting from combined altitude and inclination changes for inclination of 55 degrees. *Courtesy of TRW.*

respectively. It is interesting to note that the above equations for T_1 and T_3 yield the result $T_1 = T_2 = T_3$ if the nodal points are uniformly spaced, i.e., if both terms D_1/D_3 and D_2/D_3 in these equations are equal to 1. Therefore, the same drift rate is appropriate for all mission phases.

The above analysis applies to node distributions that call for unidirectional transfers, i.e., a scenario where a reversal of the transfer direction after the last satellite visit in the servicing cycle would be less efficient than continuing in the original eastward direction in order to return to the starting point. The exact point, D_r, at which transfer direction reversal becomes less efficient, depends on the equivalent eastward and westward nodal drift rates, R_e and R_w, and is given by

$$D_r = 360°/(1 + R_e/R_w)$$

As previously discussed, $R_e > R_w$ and, therefore, D_r is less than 180 degrees. For the condition where $R_e = 1.25\, R_w$ for the same amount of ΔV expenditure, the nodal position in question is 160 degrees.

Typical Scenarios for Servicing Four Polar Platforms

Three alternative scenarios are considered for servicing a constellation of four Earth Observing Systems (EOS) polar platforms in their respective orbits. The constellation includes two NASA platforms (NPOP-1 and 2), one ESA platform (EPOP), and one extra platform (XPOP). The platforms are located in Sun-synchronous orbits at 824 km (445 nmi) altitude and 98.7 degree orbit inclination, but with different ascending nodes: both NPOP-1 and 2 cross the equator at 11:30 p.m. local time, EPOP crosses at 9:30 p.m.,

ORBIT NODE POSITIONS D_1,D_2,D_3 (deg)	RATIO D_2/D_1	D_2/D_3	OPTIMAL TRANSFER TIMES AND RATES T_1 (R_1)	T_2 (R_2)	T_3 (R_3) (mo) (deg/mo)	MANEUVER COST INDEX ΣR_i (deg/mo)	MANEUVER COST REDUCTION** (%)
45, 90, 225	2	0.40	7.74 (5.81)	10.95 (8.22)	17.31 (13.0)	(27.03)	9.9
30, 60, 270	2	0.111	5.41 (5.55)	13.29 (4.52)	17.31 (15.60)	(25.66)	14.5
60, 90, 210	1.5	0.429	8.79 (6.83)	10.77 (8.36)	16.44 (12.77)	(27.96)	6.8

NOTES:
* 30-MONTH SERVICING CYCLE ASSUMED, WITH ORBIT TRANSFER IN ONE DIRECTION ONLY
** NOMINAL ΣR_i = 30 deg/month COST REDUCTION DETERMINED BY RATIO OF ΣR_i TO NOMINAL ΣR_i

Figure 4.8 Optimal transfer times and rates for several sample POP constellations. *Courtesy of TRW.*

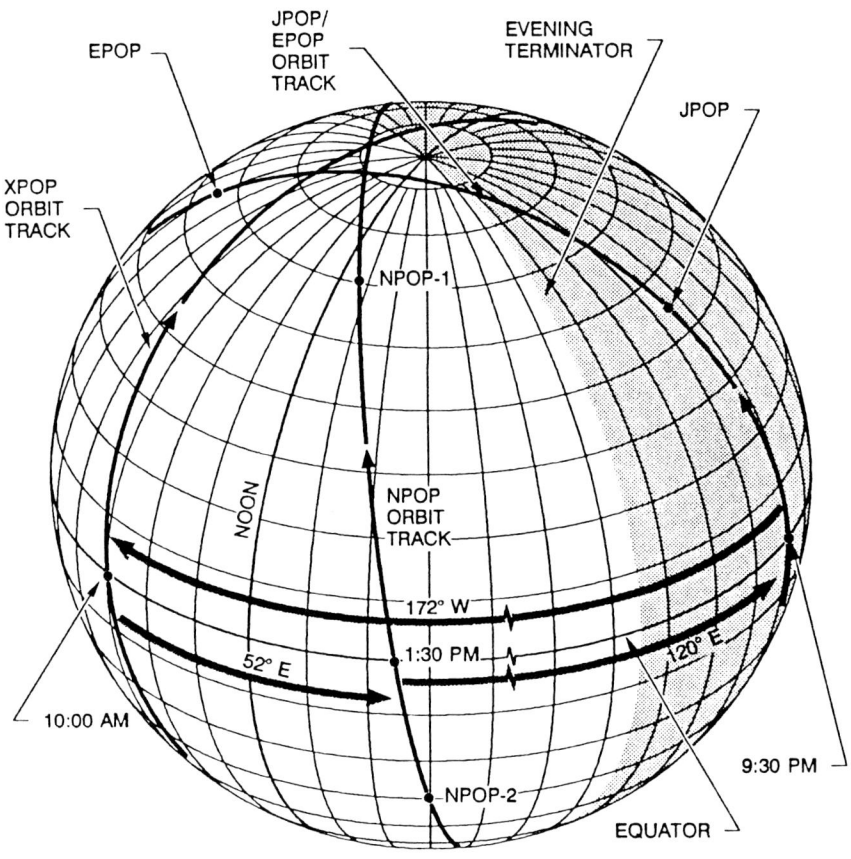

Figure 4.9 Ground tracks of a typical POP constellation. *Courtesy of TRW.*

and XPOP at 10:00 a.m. Figure 4.9 shows the ground tracks.

Actually, the EOS system constellation includes the Japanese platform, JPOP, with a 10:00 p.m. ascending node, rather than the arbitrarily selected XPOP platform considered in this discussion. The 10:00 a.m. XPOP platform requires OMV short range vehicle (SRV) nodal traverses substantially larger than those to be performed if JPOP, with nearly the same node as EPOP, had been part of the scenario. The constellation assumed here illustrates greater SRV eastward/westward orbit transfer capabilities than those required by the actual NPOP/EPOP/JPOP constellation.

A common OMV-derived servicing vehicle is assumed to revisit these POPs periodically to exchange orbit replacement units (ORUs) that contain subsystem elements and scientific experiments. The servicing interval for each platform is 3 years, such that with the planned four successive servicing calls, each platform is anticipated to achieve a 15-year orbital life. The servicing or rendezvous/docking vehicle (RDV) remains space based and uses the 3-year servicing cycle to traverse the differences in nodal positions between the three platform orbits by techniques discussed in the preceding sections, thus minimizing propellant consumption. As shown in Figure 4.9, it must negotiate a nodal change of 120 degrees east to go from the NPOP orbit to the EPOP orbit and 52 degree west to go from NPOP to XPOP. The arrows in Figure 4.9 indicate the preferred sequence of nodal point changes.

The three scenarios are summarized in Figure 4.10. They differ primarily in their use of either a Delta II (Scenario 1) or a Titan IV (Scenarios 2 and 3) launch vehicle to deliver the ORUs and science experiments, as well as the requisite resupply propellant for the RDV, all contained on a payload carrier that will be picked up by the RDV from the launch vehicle after it reaches a rendezvous-compatible orbit. In contrast to servicing low-inclination satellites with Shuttle-launched replacement units, the polar platforms require expendable launch vehicles (ELVs) such as the Delta or Titan-class rockets assumed in these scenarios. Titan IV can deliver servicing supplies for all four platforms during the assumed 3-year cycle, whereas the lower payload capacity of Delta II demands more frequent launches, one dedicated to each POP. In either case, the RDV must perform the transfer maneuvers required to proceed from one POP orbit to the next within an allocated time interval.

Scenarios 2 and 3, both of which are based on using Titan IV for resupply, differ primarily in their orbit transfer sequences as outlined in Figure 4.10. In Scenario 2, the RDV

Orbital Maneuvers

SCENARIO	REPEATED EVERY 3 YEARS ELV REQUIREMENTS	ELV CAPABILITY	REPEATED ONCE EVERY 3 YEARS RENDEZVOUS & DOCKING VEHICLE	RDV CAPABILITY	ARCHITECTURE ELEMENTS (ASSUME 15-YEAR POP MISSION LIFE)
#1	• 4 ELV LAUNCHES, IN 3-YEAR PERIOD. ONE TO EACH POP • TOTAL 16 ELV LAUNCHES	• DELTA-II CLASS • LIFTS 7,000 lb TO POP ORBIT	• 1 SPACE-BASED RDV (SRV-D) SHUTTLES FROM POP #1 TO #4 • AT EACH POP LOCATION – RENDEZVOUS WITH ELV – TRANSFER EXCHANGE CARRIER TO POP	• SRV-D CLASS (OMV DERIVATIVE) • REFUEL CAPABILITY	• 16 ELV – DELTA-II CLASS • SINGLE SPACE-BASED SRV-D (OMV DERIVATIVE VEHICLE) • 16 CARRIERS – DELTA COMPATIBLE
#2	• 1 ELV LAUNCH TO POP #1 • ELV LOADED WITH 4-MODULE CARRIER/PAYLOAD STACK. ONE FOR EACH POP • TOTAL 4 ELV LAUNCHES	• TITAN-IV CLASS • LIFTS 35,600 lb TO POP TRANSFER ORBIT	• 1 SPACE-BASED RDV (SRV-T) RENDEZVOUS WITH ELV NEAR POP #1 ORBIT • SHUTTLES FROM POP #1 TO #4 • DISTRIBUTES CARRIER/PAYLOAD, ONE TO EACH POP	• SRV-T CLASS (OMV DERIVATIVE) • REFUEL CAPABILITY	• 4 ELV – TITAN-IV CLASS • SINGLE SPACE-BASED SRV-T (OMV DERIVATIVE VEHICLE) • 16 CARRIERS – TITAN COMPATIBLE
#3	• SAME AS SCENARIO #2 • TOTAL 4 ELV LAUNCHES	• SAME AS SCENARIO #2	• 1 SPACE-BASED RDV (SRV-T) RENDEZVOUS WITH ELV NEAR POP #1 ORBIT • POP #1 USED AS STORAGE DEPOT FOR CARRIER/PAYLOAD STACK • RDV DISTRIBUTES ONE MODULE AT A TIME TO EACH OF THE POP • SHUTTLES FROM POP #1 TO #2; POP #1 TO #3; POP #1 TO #4	• SAME AS SCENARIO #2	• SAME AS SCENARIO #2 • NPOP-1 DESIGNED TO ACCOMMODATE STACK OF 4 MODULAR CARRIERS FOR STORAGE (ORBITAL DEPOT)

Figure 4.10 Servicing scenarios for four POP platforms. *Courtesy of TRW.*

carries all supplies delivered by the Titan IV except those left at each of the POPs it visits during the servicing cycle. This includes replacement ORUs and experiments and the carrier rack which is attached to and remains at the POP. In Scenario 3, the RDV deposits most of the supplies at one platform, the NPOP-1, and returns to the depot between trips to other POPs to pick up the supplies required for the next sortie. This scenario, therefore, requires additional stop-and-go maneuvers but has the advantage of using less propellant per sortie because of less dry weight. Mission analysis results presented below show that Scenario 3 requires somewhat less total propellant than Scenario 2.

Scenario 1 Characteristics

Figure 4.11 presents performance parameters and architecture elements used in Scenario 1. The space-based OMV derivative servicing vehicle, resupplied by the Delta II ELV, is designated as SRV-D. The term SRV is taken from the short-range vehicle of the MSFC/TRW OMV program which does not carry a bipropellant propulsion module. The SRV-D uses hydrazine monopropellant only, with added tankage that is located centrally in the space not utilized for bipropellant tanks.

The payload carrier delivered to the SRV-D by the Delta II ELV is transferred to the POP and remains attached to it at the docking location. The old ORUs and experiments, after being replaced by the new ones, are stored in the carrier to avoid release of orbital debris. A telerobotic servicing arm stationed on the POP is used to perform ORU and experiment exchange manipulations. As new payload carriers are attached to the POP on each servicing call, four such carriers will be stacked at the POP at the end of 15 years. The servicing manipulator arm is designed for the reach requirements imposed by the stacked carriers.

The SRV-D and its payload carrier have been defined in preliminary design studies by TRW [8] and General Electric.

Figure 4.12 shows the ΔV expenditures and propellant consumption versus time for Scenario 1; Figure 4.13 summarizes the performance data, along with that of Scenarios 2 and 3. Four Delta II launches are required in every 3-year servicing cycle to carry the 3,175 kg (7,000 lb) payload for equipment replacement on each POP, as well as the propellant required by the SRV-D to perform the respective orbit transfer maneuvers. The delivered weight includes 454 kg (1,000 lb) for the carrier, 1,878 kg (4,140 lb) of ORU/experiment equipment and 839 kg (1,850 lb) of propellant.

The payload carrier includes all necessary attach mechanisms and interface hardware (electrical, mechanical, and

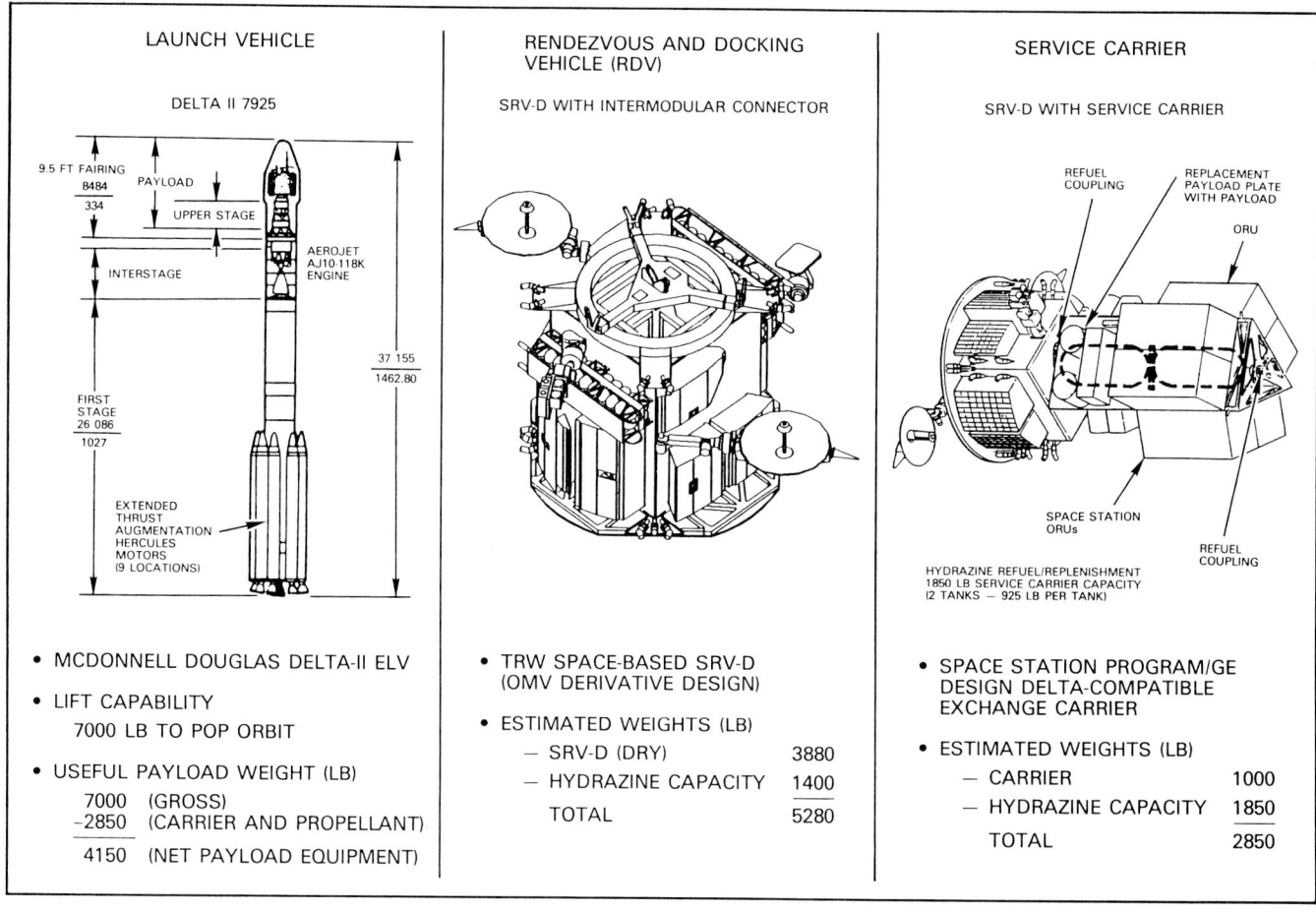

Figure 4.11 Performance parameters and architecture components used in scenario 1. *Courtesy of TRW.*

fluid coupling) used in mating with the SRV-D, the POP, and for retention and release of ORUs and experiment modules.

A disposable payload carrier, designed for controlled re-entry into the atmosphere, would be a viable alternative in this and the other servicing scenarios. In that option, the carrier would be equipped with a solid rocket motor. The SRV would be used to spin up and point the carrier for its de-orbit maneuver.

The SRV-D, designed by TRW in 1989 as an OMV-derivative vehicle, has a 635 kg (1,400 lb) hydrazine storage capacity and provides for remotely controlled or automated refueling. The internal 635 kg (1,400 lb) of hydrazine is intended as a contingency propellant reserve. The 839 kg (1,850 lb) of additional propellant, transferred to the SRV from the payload carrier's storage tanks, is used for the maneuvers required to achieve the nodal point traverse in moving from one POP orbit to the next.

Based on the results of transfer maneuver optimization, a slight advantage would be gained by continuing the orbit transfer from the EPOP to the XPOP platform in an eastward instead of westward direction because of the greater nodal regression rate achievable. However, the 172 degree nodal point separation between the two platforms is close enough to the break-even point (i.e., 160 degree nodal point separation) to make these options approximately equivalent.

Scenario 2 Characteristics

Figure 4.14 presents performance parameters and architecture elements used in Scenario 2. The space-based OMV derivative servicing vehicle, resupplied by the Titan IV ELV, is designated as SRV-T. This vehicle, its payload carrier, and the ELV to be used are depicted in Figure 4.12. A single Titan IV launch carries 16,148 kg (35,600 lb) of payload into an NPOP-compatible transfer ellipse 185 × 825 km (100 × 445 nmi). Prior to the ELV launch, the SRV-T servicing vehicle transfers into an elliptical orbit of the same dimensions to be ready for rendezvous with the ELV and the payload stack attached to it. The best timing for this procedure occurs during the eastward nodal transfer phase of the SRV-T after it has completed XPOP servicing and approaches the NPOP orbit with its nodal point 52 degrees east of the XPOP node. During this transfer phase, the SRV-T is in a 370 km (200 nmi) circular orbit from where it performs the descent to 185 km (100 nmi) perigee and the sub-

Orbital Maneuvers

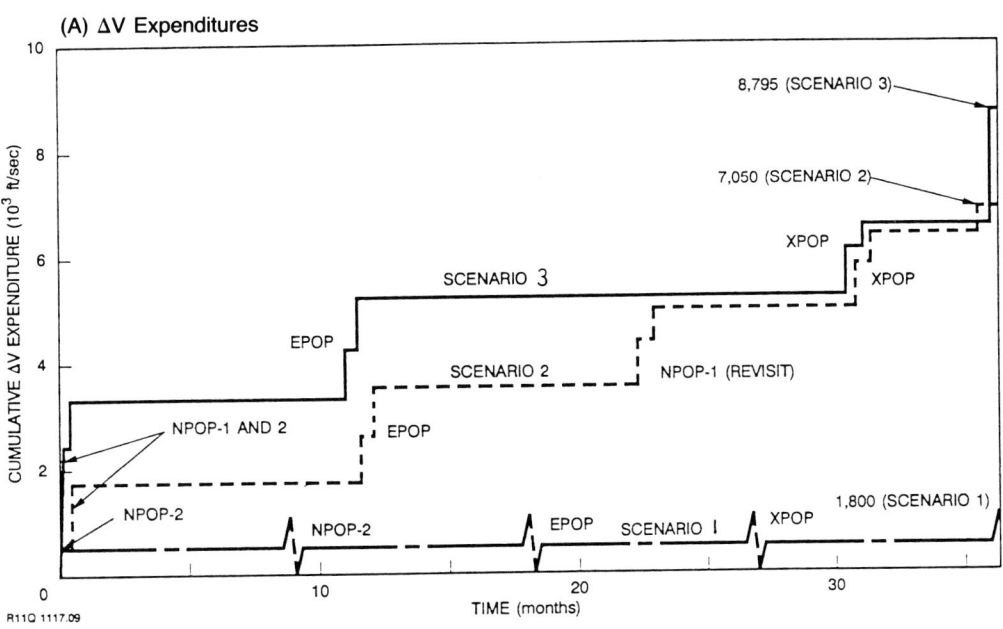

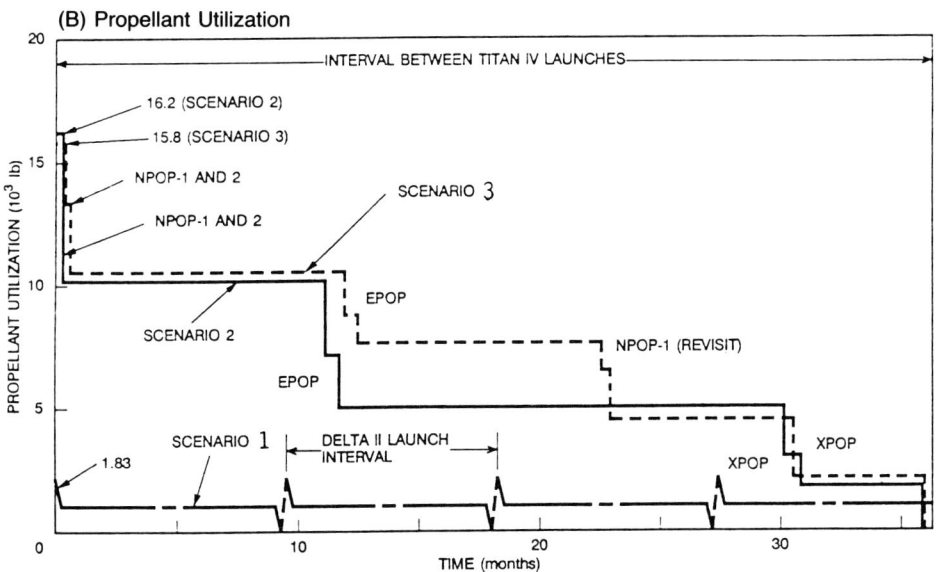

Figure 4.12 Delta V expenditures and propellant utilization versus time of three servicing scenarios. *Courtesy of TRW.*

sequent orbit change to the 185 × 825 km (100 × 445 nmi) orbit dimensions compatible with Titan IV rendezvous. Only a minor increase in the total maneuver ΔV requirements for return to the NPOP orbital altitude occurs in this resupply scenario. The small out-of-plane component used to enhance the nodal drive rate, as previously explained, will be reversed in nearly equal increments during the perigee and apogee maneuvers required for orbit transfer back to POP altitude.

The Titan IV payload, 16,148 kg (35,600 lb), consists of a stack of four payload carriers that are interconnected by design elements developed under the NASA Space Station program. Each payload carrier is loaded with 1,837 kg (4,050 lb) of hydrazine and 1,860 kg (4,100 lb) of equipment for ORU and experiment changeout at the POP for which it is intended. The total weight delivered to each POP is 2,200 kg (4,850 lb).

Like the carriers used on the SRV-D, the SRV-T carrier contains all necessary attach mechanisms and interface hardware. The SRV-T itself has a 2,041 kg (4,500 lb) hydrazine storage capacity and refueling capability to receive up to 1,588 kg (3,500 lb) of propellant contained in the respective carrier tanks. The payload carriers delivered to the POPs at each servicing visit are designed to remain at-

CHARACTERISTICS		SCENARIO 1	SCENARIO 2	SCENARIO 3
ELV USED FOR RESUPPLY		DELTA II	TITAN IV	TITAN IV
NUMBER OF ELVs LAUNCHED PER 3-YEAR SERVICING CYCLE		4	1	1
SRV TYPE		SRV-D	SRV-T	SRV-T
SRV GROSS WEIGHT	(lb)	5,280	6,935	6,935
GROSS PAYLOAD WEIGHT PER LAUNCH	(lb)	7,000	35,600	35,600
– CARRIER WEIGHT PER LAUNCH	(lb)	1,000	3,000 = (4 × 750)	3,000 = (4 × 750)
– PROPELLANT WEIGHT PER LAUNCH	(lb)	1,850	16,200	15,800
– USABLE PAYLOAD WEIGHT PER CARRIER	(lb)	4,150	4,100	4,200
USABLE PAYLOAD WEIGHT DIFFERENCE, PER PLATFORM*	(lb)	–	–50	+50
USABLE PAYLOAD DIFFERENCE PER LAUNCH*	(lb)	–	–200	+200

* COMPARED TO DELTA II LAUNCH, SCENARIO 1

Figure 4.13 Servicing scenario performance summary. *Courtesy of TRW.*

tached, as in the SRV-D scenario, so that ultimately four carriers are stacked at each POP during the last years of their 15-year mission life. The alternative of deorbiting the payload carriers with the defunct modules removed during payload change-out also has been investigated but is not part of this scenario.

Figure 4.12 shows the ΔV expenditures and propellant consumption versus time, and performance data is summarized in Figure 4.13. The mission is designed to be repeated in 3-year cycles to meet POP equipment replacement requirements, relying on the large payload delivery capability of Titan IV. The analysis shows that this can be achieved with some margin in cycle completion time.

It is apparent from the results that the orbit transfer performance gain derived through a small inclination change component (see above) is essential in meeting the multiple POP servicing objectives with the specified payload weight capability. Otherwise, the propellant requirements would have to be increased at the expense of delivered payload weight. A 30 percent increase in propellant requirement would be at the expense of a comparable percentage of replacement module weight per servicing call.

Scenario 3 Characteristics

Performance parameters and architecture elements used in Scenario 3 are comparable to those described in Figure 4.14 for Scenario 2. The main difference is the delivery sequence to the respective POPs, which in Scenario 3 involves a temporary storage of payload carriers at one platform, assumed here to be NPOP-1. This permits the SRV-T to execute its transfer maneuvers with much less dry weight than in Scenario 2 and, thereby, saving propellant weight. The rendezvous and docking procedure with the Titan IV, after delivering the last payload ensemble to XPOP at the end of each servicing cycle, is the same as that described in Scenario 2.

Figure 4.12 shows the resulting ΔV expenditure and propellant consumption versus time in comparison with those of Scenarios 1 and 2, and performance data is summarized in Figure 4.13.

Scenario 3 permits a total propellant saving of 181 kg (400 lb) compared to Scenario 2 while delivering 45 kg (100 lb) more useful payload weight to each platform. This weight saving is achieved even though the total ΔV expenditure is significantly greater 532 meters/sec (1,745 ft/sec), due to the more frequent stop-and-go maneuvers in this scenario.

Summary and Conclusions

The data presented in this multiple satellite servicing example is concerned primarily with techniques for reducing orbit-to-orbit transfer maneuver requirements and costs associated with servicing of multiple satellites or platforms in Sun-synchronous polar orbits with different ascending nodes. Combining the in-plane maneuver for altitude change with an out-of-plane maneuver component for simultaneous

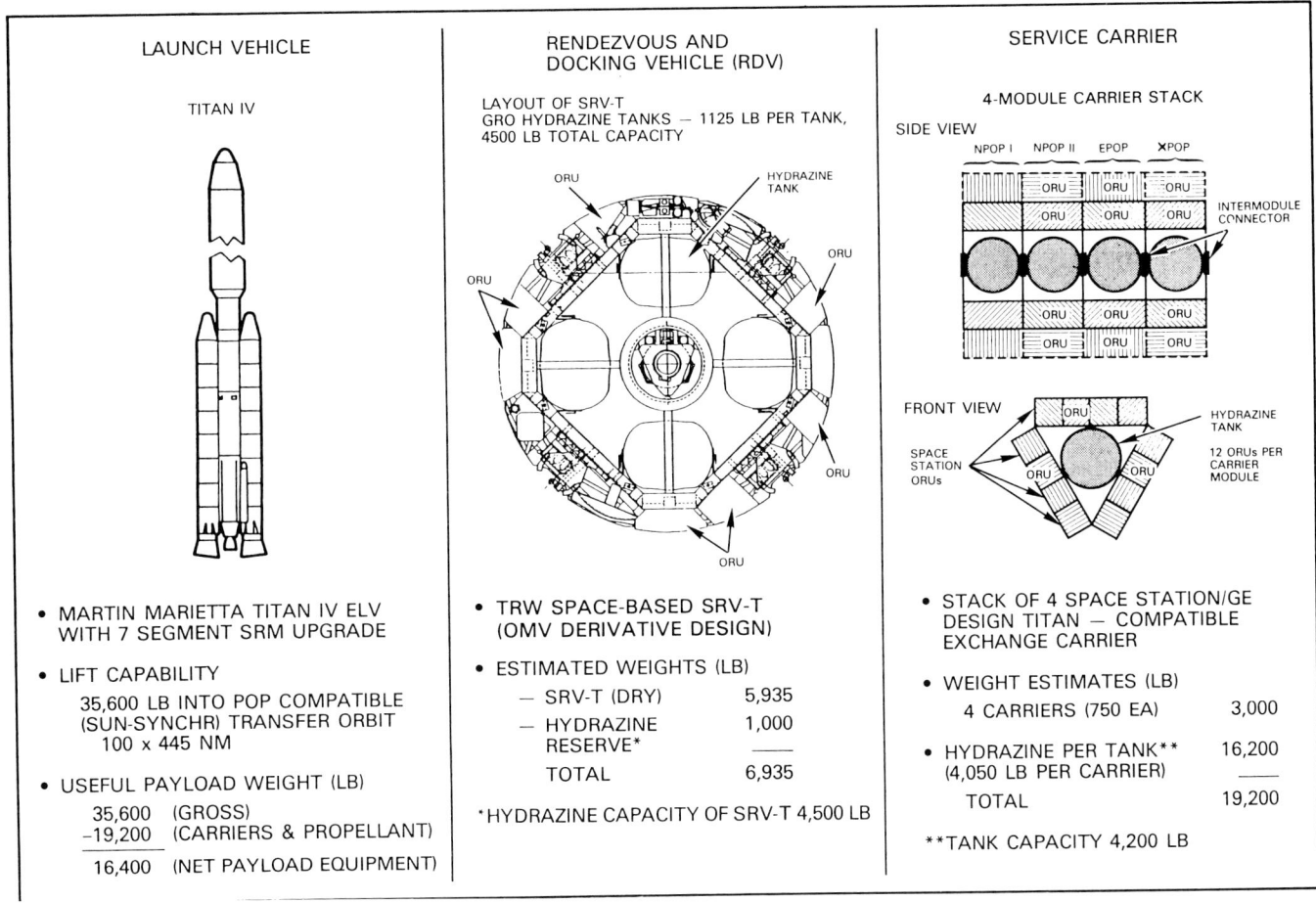

Figure 4.14 Performance parameters and architecture components used in scenario 2. *Courtesy of TRW.*

small inclination change can increase the nodal drift rate by as much as 35 percent when performing a transfer in an eastward direction. Westward transfers gain less performance enhancement by a comparable plane change but still benefit from this technique.

These results were applied to three alternate scenarios for using a space-based servicing vehicle to carry replacement modules (i.e., ORUs and science experiments) to four POPs deployed in three polar orbits. ELVs of the Delta II and Titan IV class were assumed as launch vehicles used to resupply the POP constellation in periodic three-year servicing cycle with a total of 16 versus 4 launches, respectively, in the course of the 15-year POP mission life. It was found that both launch scenarios are feasible, although the Delta-based Scenario 1 (assuming one ELV launch dedicated to each POP servicing call) does not provide as much usable payload equipment as the Titan-based alternative servicing Scenario 3.

The resulting usable payload weight delivered to each platform is 1,832, 1,860, and 1,905 kg (4,150, 4,100, and 4,200 lb), respectively, in these scenarios. Scenarios 2 and 3 have a distinct cost advantage over Scenario 1 to the degree that it is less costly to launch a single Titan IV than to launch four separate Delta II ELVs. However, the Delta II ELVs used in Scenario 1 offer greater flexibility in terms of the lead time for equipment to be used in POP resupply. In all scenarios, whether Delta or Titan based, the servicing intervals between the various POPs are constrained by the SRV traversal time of the SRV between the platform orbits. However, adaptation of the planned servicing sequence to changes in POP servicing schedules is more difficult for Titan ELV resupply, unless a payload storage depot is available. Scenario 3, with its payload storage depot, alleviates scheduling problems and also offers the best and most cost-effective performance for servicing the POP complex.

As a general comment, note that accelerating the servicing schedule would increase propellant needs in direct proportion and consequently reduce the amount of equipment deliverable to each platform. The carrier weight is largely independent of these variations; therefore, the usable (net) payload weight delivered by the SRV is highly sensitive to any reduction in the time allocated for platform-to-platform transfer.

In deriving the data presented here [7], no attempt was

- EARTH'S OBLATENESS CAUSES NODE TO MOVE ALONG EQUATOR.
- RATE DEPENDS ON ALTITUDE AND INCLINATION.
- FREE-FLYER REGRESSES AT DIFFERENT RATE FROM BASE.
- PLANE CHANGE GREATLY INCREASES PROPELLANT REQUIRED FOR TRANSFER.

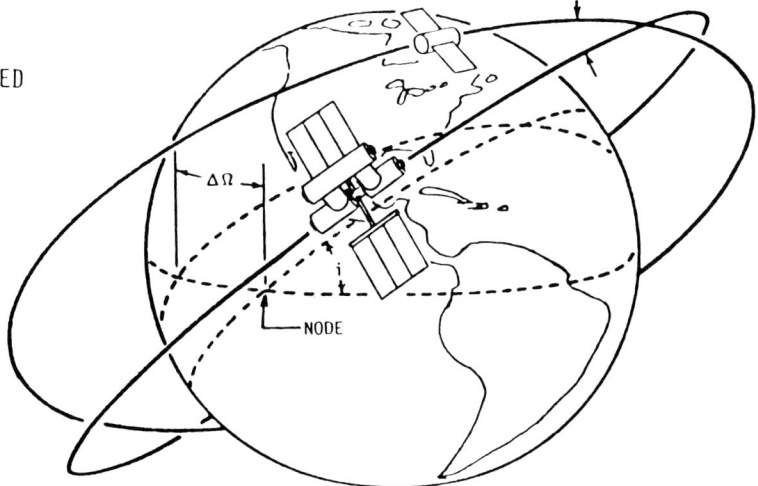

Figure 4.15 Differential nodal regression model. *Courtesy of TRW.*

made to define an optimum vehicle design or mission operation sequence. The objective was to demonstrate the advantages to be gained by using cost-effective orbit transfer modes and by alternative approaches to resupplying the servicing vehicle with replacement modules required by the POP platforms, and with maneuver propellant.

Cost-Effective Orbit Transfer Modes

The data in this example, taken from Reference 9, pertains to orbital transfer maneuvers required to reach satellites for servicing. It presents approaches that may be used to perform these maneuvers most economically, that is, with as little propellant expenditure as possible.

Given the high cost per unit cargo mass carried into low Earth orbit by the Space Shuttle, the propellant cost savings achievable by selecting inexpensive orbit transfer modes can be a significant factor in making satellite servicing economically attractive. For example, by saving 10,000 kg (22,050 lb) of orbit transfer propellant in the course of 1 year, transportation cost savings of at least $50 million can be realized. The exact amount depends on whether the satellites or space platforms to be serviced are in low inclination or polar orbits.

This topic addresses:

1. Access, for the purpose of servicing, from the Space Station to satellites flying at similar or different orbital altitudes
2. Limitations in accessibility due to differential nodal regression between the orbits in question; avoidance of high ΔV expenditures for plane change maneuvers orbit
3. The orbit transfer alternatives of retrieving a satellite from higher orbit for servicing at the Station or Shuttle versus in situ servicing at the satellite orbit altitude by using an orbital maneuvering vehicle carrying a telerobotic servicer and the necessary replacement parts

Satellite Orbit Accessibility

Free-flying platforms or satellites are accessible from the Space Station for retrieval or in situ servicing by an OMV at acceptable ΔV expenditure only if the respective orbit planes are nearly co-planar.

In general, conditions of co-planar alignment will occur infrequently due to small differential nodal regression rates of the Station and target satellite orbit planes, unless the orbital inclinations and altitudes are identical, Figure 4.15. For 28.5 degrees inclination of both orbits, the difference in nodal regression rates between the Station at 360 km (194 nmi) and a target satellite at 560 km (302 nmi) altitude is about 0.66 degrees day, Figure 4.16. The time period between alignments of the ascending nodes thus is 550 days. The length, ΔT, of this cycle depends on the altitude difference, ΔH, between the station and the target satellite and is approximately $\Delta T \simeq 300/\Delta H$ years, where ΔH is in kilometers, Figure 4.17. The interval is 2 years for a differential altitude of 150 km (80 nmi) and 6 years for 50 km (27 nmi). The principal concern is with satellites in orbits of small differential altitude.

A one degree plane change at the Space Station orbital

Orbital Maneuvers

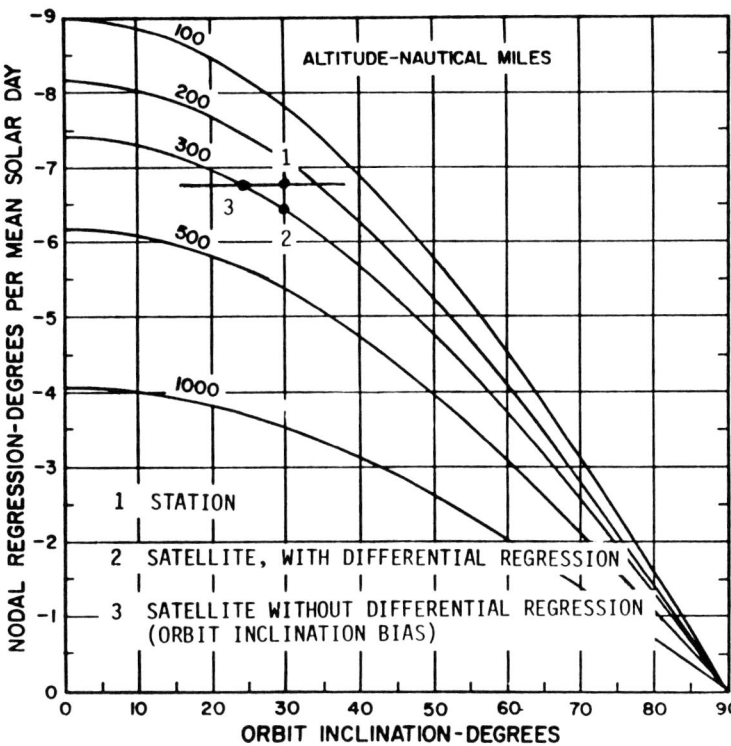

Figure 4.16 Nodal regression rate versus orbit inclination for several orbit altitudes. *Courtesy of TRW.*

altitude requires a velocity increment of about 130 meters per second (426 ft/sec). Even when combined with altitude change maneuvers to rendezvous with the target satellite, a plane change of more than 3 to 4 degrees significantly increases the ΔV required for the orbit change maneuvers at perigee and apogee. This extra amount must be expended twice, i.e., going to the satellite and returning again to the Space Station. At 30 degrees orbit inclination, 1 degree of nodal difference between the two orbit planes corresponds to 0.48 degree of plane change. For the differential nodal regression rate given in the above example, a maximum, plane change of about 3.2 degrees is required at both ends of a 20 day time interval, 10 days before and after the date of nodal alignment. "Windows" of satellite accessibility, defined by what would be allowable as an acceptable plane change ΔV penalty, are shown in Figure 4.18.

Alternative Modes of Satellite Access for Servicing

The following alternatives should be considered:

1. Accept the potentially long intervals between accessibility windows.
2. If mission objectives permit, select large differential altitudes ΔH to keep ΔT acceptably small. However, this increases the in-plane transfer maneuver cost.
3. Use dedicated Space Shuttle missions to service satellites that would remain inaccessible from the Space Station for long time periods.
4. Select favorable orbital characteristics and perform periodic orbit corrections to provide near co-planar conditions and provide frequent access, e.g., several times per year.
5. Operate the satellite at the specified altitude and at the same 28.5 degree inclination as the Station. At the time of requiring servicing access, the satellite first

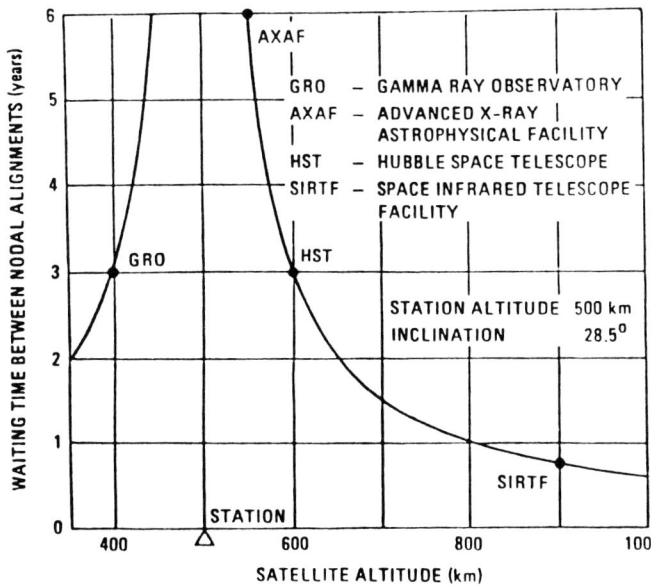

Figure 4.17 Waiting period versus altitude. Courtesy of TRW.

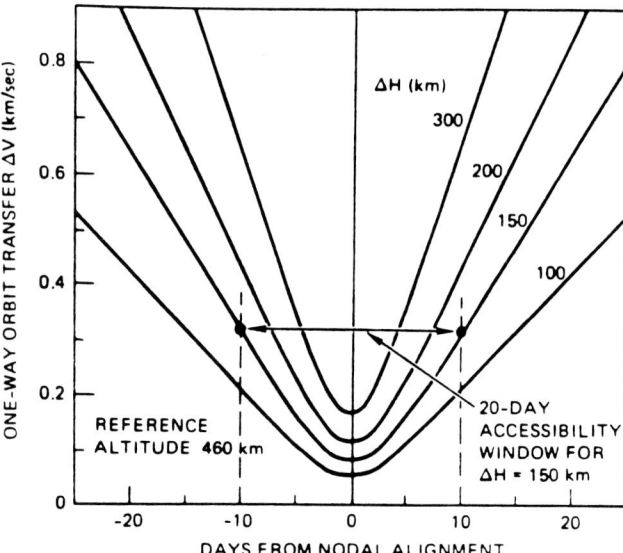

Figure 4.18 Satellite accessibility window for several differential altitudes. *Courtesy of TRW.*

performs an in-plane maneuver to higher altitude to increase the differential nodal regression and thus reduce the waiting time. Then, at the time of nodal alignment, the satellite descends to the Station altitude for rendezvous.

One suboption of alternative 4 above is the near-coaltitude formation flying mode, Figure 4.19. The co-orbiting satellite experiences a slow orbital decay that depends on its ballistic coefficient (W/C_DA) while the Station is assumed to be maintained at a constant altitude. The two relative trajectories shown correspond to two values of W/C_DA producing different orbit decay rates. The drop in altitude produces a change from the initial eastward to a westward nodal drift, with crossover at the time of coaltitude passage occurring at the tip of the relative motion pattern. Periodic reboots restore the satellite to the relative altitude and nodal position it had at the beginning of the cycle, 32 or 82 days earlier.

The second suboption of alternative 4 involves operating the satellite at the specified altitude, but with a small orbit inclination bias, such that its average nodal regression rate equals that of the Space Station, Figure 4.20. This permits access for unplanned servicing practically at any time. However, it demands major extra ΔV expenditures in deployment and retrieval of the satellite or in performing in situ servicing. For example, the extra maneuver expenditure to achieve the inclination bias of 2.8 degrees required to maintain continuous nodal alignment with the Space Station at a 50 km (27 nmi) altitude difference is 360 meters per second (1,181 ft/sec) each way.

Due to different orbit decay rates of the Space Station and the target satellite, however, a buildup of differential nodal rates from the initial zero difference will occur, Figure 4.21. In order to maintain near co-planar conditions continually, a sequence of Station and satellite reboost maneuvers will be required. Figure 4.22 shows the small periodic nodal variations occurring between reboost events, starting with a slow initial westward drift followed by a reversal to a slow eastward drift of the satellite's orbit node relative to that of the Station. This pattern is comparable to those presented in References 10 and 11 for near-coaltitude station keeping. Figure 4.21 explains this reversal by the relative altitude decay rates assumed for the Space Station and the satellite. Note that in every cycle there are two minimum ΔV transfer opportunities which occur at the zero-crossings of the nodal difference, $\Delta \Omega$.

Alternative 5 is operationally simpler than Alternative 4 as it does not require frequent reboosts. Altitude change maneuvers will be needed only at times when servicing access is required to reduce the waiting time. For example, with a 50 km (27 nmi) initial altitude difference, i.e., 6-year intervals between access windows, the remaining waiting time to the next window can be reduced by 80 percent through a 220 km (118 nmi) in-plane maneuver, at a cost of 116 meters per second (380 ft/sec) each way; see Figure 4.23. Obviously, this alternative is more economical than the out-of-plane maneuver options discussed before, but is restricted to satellites having self-contained propulsion capability. The same restriction does not apply in the two modes of alternative 4.

Comparison of Satellite Access Alternatives

Figure 4.24 summarizes advantages and disadvantages of the principal satellite accessibility options considered. The preferred approach depends on the altitude difference between the Space Station and satellite/platform orbits. Three differential altitude regions are distinguished:

1. Large differential altitudes between SSF and target satellite/platform >300 km (162 nmi) correspond to acceptably short intervals (1 year or less) between accessibility windows. Generally there will be no need to modify this pattern.

2. Moderate differential altitudes, 50 to 100 km, (27 to 54 nmi) correspond to excessively long intervals, 3 to 6 years, between accessibility windows for servicing calls if the orbit inclinations are identical. The intervals can be reduced to fractions of a year or the satellite can be made continuously accessible by applying an inclination bias, with the target satellite/platform at 1 to 3 degrees lower inclination than the Space Station. The extra plane change expenditure appears acceptable in exchange for the accessibility advantage.

3. Near-coaltitude formation flying permits unconstrained access for servicing at any time. To avoid clustering of several satellites in overlapping flight

Orbital Maneuvers

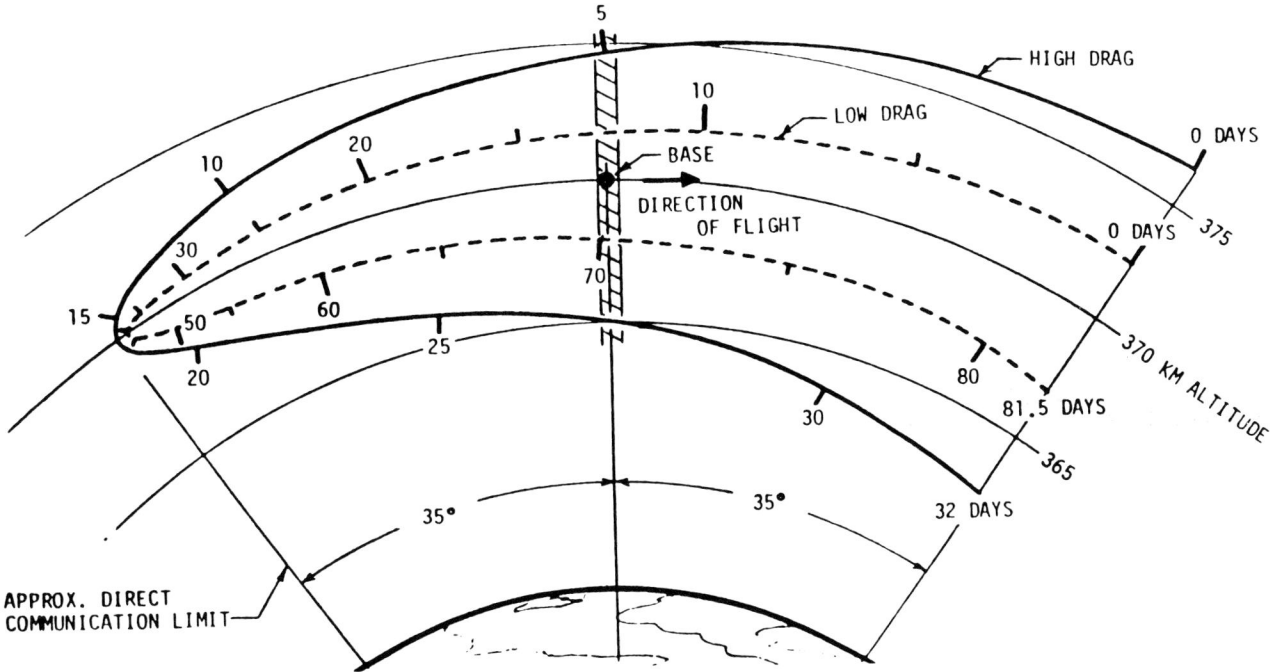

Figure 4.19 Formation flying near the space station. *Courtesy of TRW.*

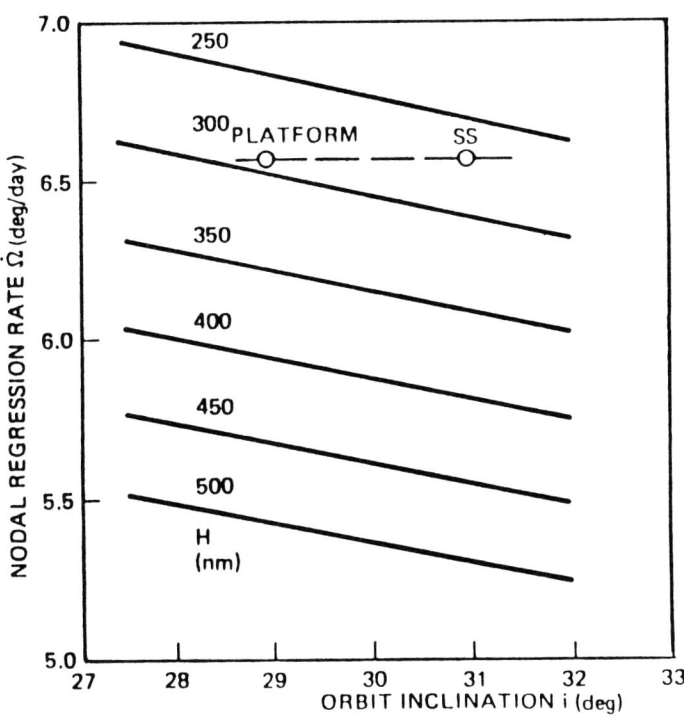

Figure 4.20 Orbital inclination bias effect on platform and Space Station nodal regression — without orbit decay. *Courtesy of TRW.*

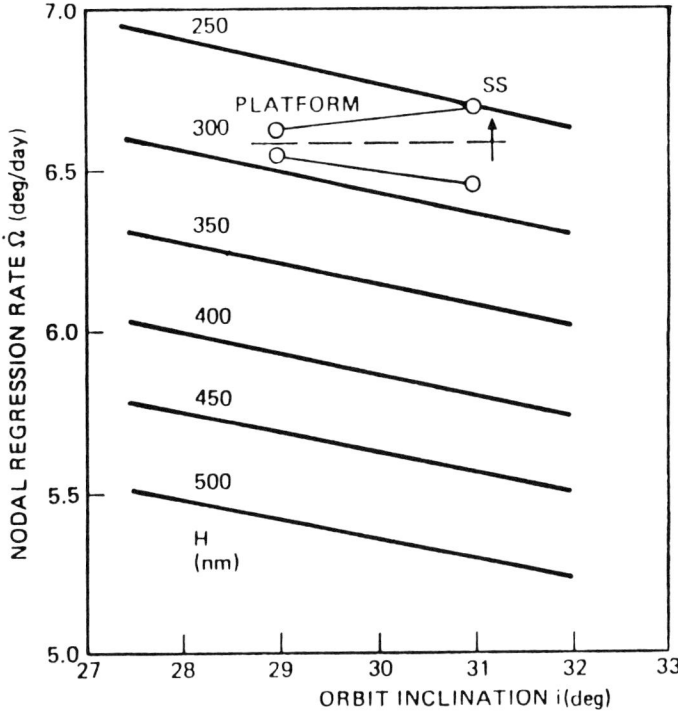

Figure 4.21 Orbital inclination bias effect on platform and Space Station nodal regression — with orbit decay. *Courtesy of TRW.*

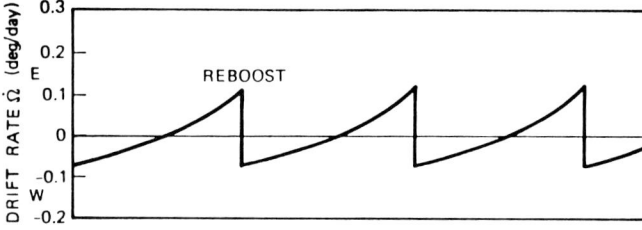

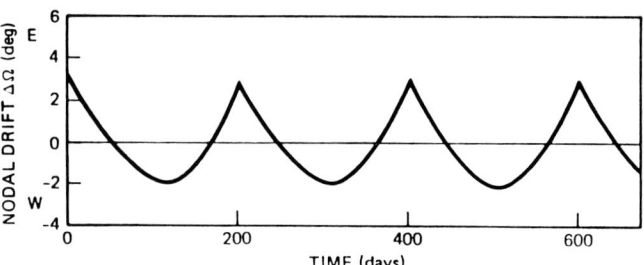

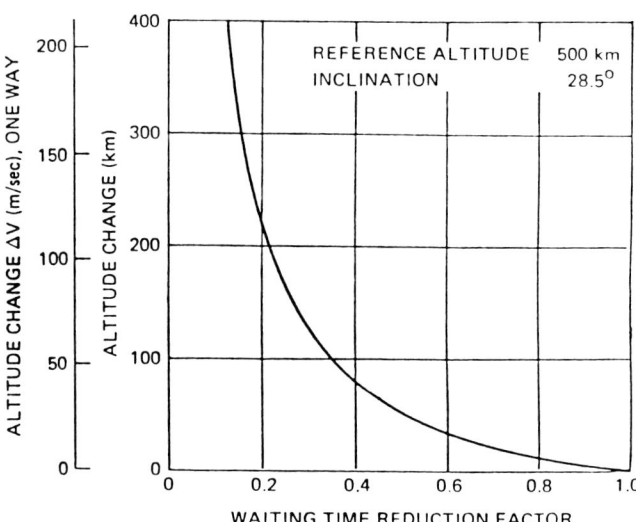

Figure 4.22 Differential nodal drift of platform relative to Space Station. *Courtesy of TRW.*

Figure 4.23 Waiting time reduction through altitude change. *Courtesy of TRW.*

profiles, the periodic descent/reboost patterns of these satellites can be separated with their centers shifted to different orbit locations relative to the Space Station.

The choice of any of the alternatives depends on the required access frequency and the preferred satellite/platform operating altitudes that may be dictated by payload functional constraints.

The foregoing discussion shows the greater flexibility and cost-effectiveness of using alternatives 4 and 5 in achieving satellite accessibility. These options avoid the costly alter-

INFREQUENT ACCESS OF WINDOWS AT PERIODIC ALIGNMENT OF ORBIT PLANES (UNCONTROLLED NODAL DRIFT)	FREQUENT ACCESS THROUGH MAINTENANCE OF NEAR-COPLANAR CONDITIONS (CONTROLLED NODAL DRIFT)
SUBOPTIONS • SERVICING PLANNED AT NODAL ALIGNMENT INTERVALS • RECOURSE TO SHUTTLE-BASED SERVICING IF NEEDED WHEN INACCESSIBLE FROM SS	SUBOPTIONS • NEAR-COALTITUDE ORBITS ($\Delta H < 3$ nm) • MODERATE DIFFERENTIAL ALTITUDE ($\Delta H < 50$ nm) ORBIT INCLINATION BIAS
ADVANTAGES • LEAST ΔV REQUIREMENT (WITHIN ACCESSIBILITY WINDOW) • MINIMUM ORBIT CONTROL (DRAG MAKEUP) REQUIREMENTS ON SS AND FREE-FLYERS	ADVANTAGES • MANY ACCESS OPPORTUNITIES PER YEAR • SMALL OUT-OF-PLANE ΔV PENALTY IF TRANSFER DATE NONOPTIMAL
DISADVANTAGES • POTENTIALLY LONG (> 3-year) CYCLES OF INACCESSIBILITY FOR LOW DIFFERENTIAL ALTITUDES • TRANSFER ΔV INCREASES RAPIDLY ON BOTH SIDES OF OPTIMUM ACCESSIBILITY DATES	DISADVANTAGES • STATION KEEPING IS COMPLEX • FREQUENT SPACE STATION DRAG MAKEUP MANEUVERS REQUIRED (THIS DOES NOT MEAN INCREASE IN DRAG MAKEUP PROPELLANT WEIGHT)

Figure 4.24 Satellite orbital access from Space Station. *Courtesy of TRW.*

Orbital Maneuvers

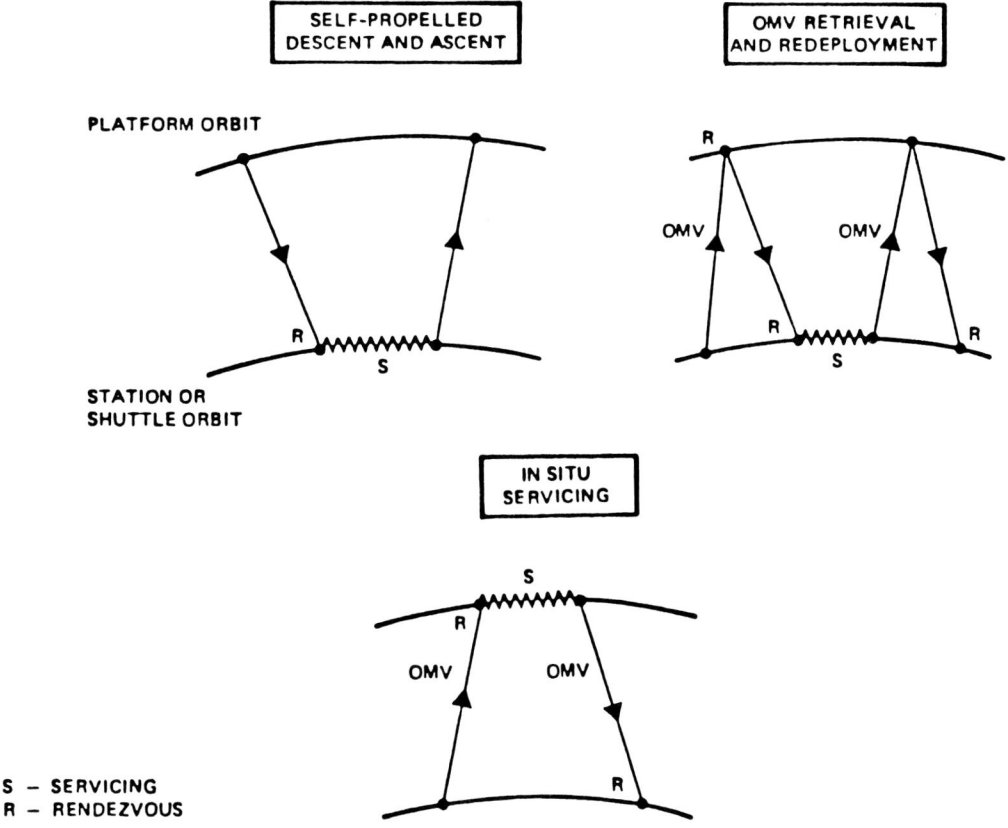

Figure 4.25 Servicing scenarios. *Courtesy of TRW.*

native of retrieving satellites by the Shuttle, alternative 3, or the necessity of waiting for several years after a satellite failure, alternative 1, before retrieval by an OMV is possible.

Polar Platform Servicing Modes

Periodic servicing of the polar platform for maintenance, repair, resupply, and/or payload equipment changeout can be performed either on board the Space Shuttle or in situ at the platform's orbital altitude. The principal advantage of the latter servicing approach is the significantly lower propellant expenditure required to transfer the servicing module, an OMV equipped with a "smart front end," to and from the platform orbit rather than transferring the much more massive platform to and from the Shuttle. However, the in situ servicing mode, while it saves several thousand kilograms of propellant each time, requires more technologically advanced teleoperated or fully automated servicing equipment as well as special design features of the platform, its subsystems and payloads, to permit effective remote servicing. This involves a trade between the lower initial development cost but higher recurring costs of the first mode versus higher initial cost for advanced technology and lower recurring costs of the second mode.

NASA is placing major emphasis on the further development of automation and robotics as applied to space system operations, in general, and thus provides an incentive for the evolution of the more cost-effective in situ servicing approach.

A quantitative comparison of the propellant requirements of three orbital transfer modes, illustrated in Figure 4.25, is of interest.

- In Mode 1, the platform uses its own propulsion system in transferring to and from the Shuttle for onboard servicing. One round trip is involved.
- In Mode 2, as in Mode 1, servicing is performed on board the Shuttle but an OMV is used in retrieving and redeploying the platform. This involves two OMV round trips.
- In Mode 3, servicing is performed in situ, at the platform altitude, by an OMV equipped with a servicer kit. It involves one OMV round trip.

In calculating the transfer propellant requirements of each of these modes, the following weight characteristics are assumed:

- Platform: 10,800 and 16,200 kg (23,814 and 35,721 lb)
- OMV: 2,700 kg (5,954 lb)
- Servicer plus supplies: 810 kg (1,786 lb)

SERVICING MODE		PROPELLANT WEIGHT (kg)	
NO.	DESCRIPTION	10,800 kg PLATFORM	16,200 kg PLATFORM
1	SELF-PROPELLED PLATFORM	2,367	3,550
2	TWO OMV ROUND-TRIPS	3,551	5,030
3	IN SITU SERVICING BY OMV	592	

Figure 4.26 Propellant weight comparison of three servicing modes. *Courtesy of TRW.*

- Resupply propellant carried by OMV: 540 kg (1,191 lb)

Other assumptions are:

- Platform altitude: 740 km (400 nmi)
- Shuttle altitude: 260 km (140 nmi)
- Transfer velocity (one way): 283 meters per second (928 ft/sec)

The calculation results are shown in Figure 4.26.

The results show the high to very high propellant weight penalty associated with Modes 1 and 2 in comparison with Mode 3. Mode 2 requires 35 to 50 percent of the maximum cargo capacity of a single Shuttle launch into polar parking orbit, while Mode 3 requires only 6 percent of that capacity.

The differential nodal shift due to altitude differences between the Shuttle and platform orbits will probably be less than 1 degree even during extended on-orbit servicing activities. By planning the servicing mission profile with an initial small nodal position bias of the Shuttle orbit such that its total eastward drift will straddle the platform nodal position during the servicing time interval, the out-of-plane maneuver penalty to compensate for the differential drift can be made negligibly small. By combining the small plane change velocity in equal parts with the respective perigee and apogee maneuvers required for the orbital transfer between the Shuttle and platform altitudes, the out-of-plane maneuver penalty is reduced to less than 2 percent of the total.

Changes in Nodal Position of a Polar Platform

Observation objectives of different platform users may call for different nodal positions, and the question of a possible change in nodal position after changing payload instruments is of interest.

Changing the nodal position in an eastern or western direction by 1 hour requires a plane change of about 15 degrees. Since a propulsive plane change requires a ΔV of about 130 meters per second per degree, a direct nodal position change of even 1 hour by a propulsive maneuver would be prohibitively expensive. However, by exploiting the differential nodal progression or regression relative to the Sun line obtained through a change in orbital altitude, major changes in nodal position may be obtained by a much more affordable ΔV expenditure. This indirect mode requires two pairs of comparatively inexpensive in-plane altitude change maneuvers to initiate and terminate the nodal drift phase. A westward drift is accomplished by raising the altitude, and an eastward drift by lowering the altitude from that required to maintain Sun synchronism. Figure 4.27 shows the trade between the magnitude of the altitude change maneuver and the time required to achieve the desired nodal shift [12].

Note that altitude changes and differential drift rates are not strictly proportional. Thus, the eastward drift due to a given altitude change downward tends to be faster than the corresponding westward drift due to an upward change of the same magnitude.

A change from Sun-synchronous to non-synchronous conditions with continued variation of the ascending node may be desirable in missions with the objective of monitoring atmospheric and oceanographic phenomena occurring at different times, day or night [12]. The mission versatility of the polar platform provided by the execution of relatively small in-plane maneuvers thus becomes an important asset in accommodating different classes of users.

As a matter of practical interest, the orbit maneuvering vehicle involved in performing periodic platform servicing and payload instrument changeouts also may be used to assist the platform in the altitude change maneuvers required to achieve a desired nodal repositioning.

As an alternative, an OMV may provide enough propellant to enable the platform to perform the desired altitude changes on its own. As another alternative, the platform may perform the desired altitude changes in connection with redeployment to its Sun-synchronous altitude after a Shuttle revisit when payload instruments have been exchanged.

Conclusions

The cost benefits achievable by using properly selected, favorable orbit transfer modes in satellite retrieval and servicing missions can be substantial. They include:

1. The cumulative saving of large amounts of transfer propellant in orbit change maneuvers performed either by the satellites themselves or by an OMV serving as transfer vehicle
2. Avoidance of excessive waiting periods between infrequent windows of accessibility and the associated loss of satellite utilization that would occur during that time
3. Timely repair of malfunctions that might cause the loss of the satellite if delayed too long

Orbital Maneuvers

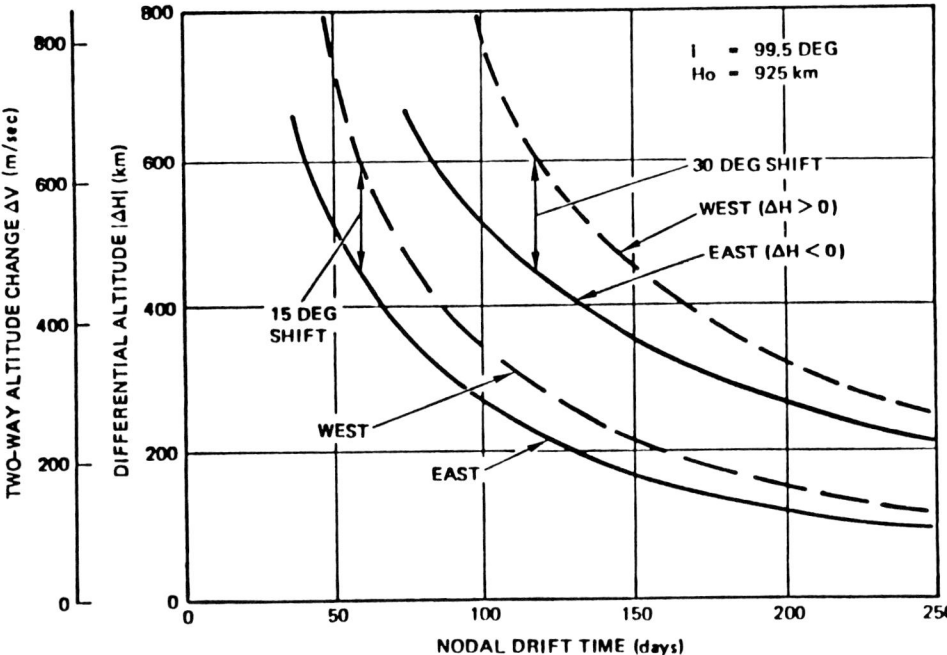

Figure 4.27 Trade between time and altitude change maneuver for 15 and 30 degree nodal changes. *Courtesy of TRW.*

4. Accommodation of multiple users by one polar platform without requiring retrieval and relaunch, e.g., through changes of the platform's nodal crossing time
5. The potentially most significant benefit of eliminating the need for some dedicated Shuttle launches to reach satellites at times when they are inaccessible from the Space Station

Some particularly interesting examples of cost savings achievable by using orbit transfer modes are:

- Propellant weight savings of 10,000 kg (22,050 lb) per year can represent savings of charges to Shuttle users of about $50 million for low inclination orbits and nearly twice as much for polar orbits.
- Reduction in waiting time resulting from more frequent or continuous orbital access can mean tens of millions of dollars saved in avoiding loss of space system utilization or even of the satellite due to deterioration during prolonged repair delays.

Many servicing and retrieval missions can be accomplished without depending on Shuttle support other than periodic delivery to the Space Station of equipment, spare parts, tools, and consumables. In fact, using the Shuttle in this way is part of a well-planned logistics and supply scenario. The application of cost-effective orbit transfer modes will enhance the utility of the Space Station as the future center of on-orbit servicing activities. To derive maximum benefits, the mission design should make effective use of OMV capabilities either as a transfer vehicle or as a telerobotic servicer. Relying on the Space Station Freedom, supported by OMV sorties, is an economical means of accomplishing on-orbit servicing. Appropriate selection of satellite orbit parameters, mission profiles, and maneuvering modes is essential to employing the Space Station and an OMV effectively in on-orbit satellite servicing missions.

Satellite Proximity Operations

Extensive use was made of References 13 and 14 in the preparation of this example of orbital maneuvers important to satellite servicing missions.

The evolution of Earth orbital satellite servicing operations in the era of the Space Shuttle and future space stations will rely on the development of accurate, efficient, and risk-free techniques for controlling spacecraft orbiting in close proximity to one another. This includes the mission phases of:

1. Orbital departure and rendezvous
2. Satellite retrieval
3. Formation flying and station keeping
4. Satellite visits for inspection, resupply, maintenance, repair, and rescue

The first Earth orbital rendezvous maneuvers in the U.S. space program were performed during the Gemini missions in the early 1960s. They were followed in the late 60s and early 70s by the application and further development of the Gemini maneuver techniques in the Apollo lunar exploration missions which depended critically on successful lunar orbital rendezvous; in the joint U.S.-U.S.S.R. Apollo-Soyuz mission; and in the repeated visits by an Apollo-type command and service module carrying men and supplies to Skylab, the forerunner of manned space stations. U.S. Space Shuttle missions have routinely demonstrated proximity operations with the deployment and/or retrieval of a number of satellites.

The U.S.S.R. space program, too, has developed and used orbital rendezvous and docking in frequent revisits to the Salyut Space Station by manned spacecraft and unmanned supply vehicles such as the Progress Tanker.

In all of these missions the required proximity operations were performed without difficulties thanks to extensive ground testing and flight simulation. However, further development of routinely performed, automated sequences will be essential in order to save crew time and effort, improve maneuver efficiency, and reduce potential risks.

Reference 13 gives a set of linearized orbital equations which provides a good approximation of the relative motion of one satellite with respect to another in an adjacent circular orbit. These rendezvous equations serve as a useful analysis tool. They were incorporated in modified form in the rendezvous guidance computer used in the 1962 Gemini mission and still provide a basis of Space Shuttle short-range maneuver computations where speed is critical.

The range of validity of the linear approximations used in obtaining these relative motion equations has been extended in the more recent literature by introducing curvilinear, especially coelliptic, coordinates to replace the original rectangular coordinates rotating with the orbital motion of the spacecraft.

Reference 13 refers to, and reviews, the extensive published literature on orbital proximity operations, but also includes the Reference 13 author's new results concerning characteristics of zero velocity departure and approach, guidance error propagation, and some simplifying algebraic techniques in trajectory analysis.

Proximity operation problems include: rendezvous under specified approach conditions, avoidance of retrorocket exhaust plume impingement, and zero-velocity (soft) rendezvous. A step-by-step maneuver sequence can be adopted which permits successive refinements of guidance accuracy while extending the remaining time-to-go.

Analysis of guidance error propagation indicates that the trajectory dispersion of the target point only depends on the transfer time interval (or angle), regardless of the type of relative trajectory flown and its approach direction to the target.

The laser range and angle sensor promises terminal guidance accuracy improvements by one to two orders of magnitude. Together with software developments for rendezvous control and guidance, this accuracy improvement constitutes a major step toward establishing the technology for automated rendezvous operation to be used by the Shuttle, by teleoperators and orbital transfer vehicles, and by future manned space stations.

Autonomous and Remotely Controlled Rendezvous and Docking

Remotely controlled rendezvous/docking sequences are subject to stringent constraints on target detection and tracking conditions and two-way orbit-to-ground communication characteristics in the feedback control loop. Relay link delays and operator perception lag affect remote control performance [14]. Fully autonomous control avoids these constraints and provides uninterrupted rendezvous/docking opportunities. In practice, remote monitoring and supervisory control of unmanned satellite rendezvous and docking sequences by a ground-based operator may still be desirable, at least as a transitional step before full autonomy is introduced.

This topic outlines the sensing, viewing, and communication and control requirements that apply at various levels of rendezvous/docking autonomy, and presents results of comprehensive analyses of the underlying orbital geometry, orbital mechanics, proximity operations, and control system characteristics. Of particular interest are the implications on mission timelines and permissible daily rendezvous/docking windows.

For this discussion, a rendezvous and docking control scenario assumes automated rendezvous based on GPS navigation and radar data, to a nominal target distance of 305 meters (1,000 ft). From here, the terminal approach and docking phase is controlled remotely from a ground control console (GCC) by commands transmitted by a human operator via TDRSS link. The operator receives feedback data from the OMV docking camera(s) and from telemetry of range, range rate, angle, and angle rate information, at least during the major part of the terminal approach, as long as commensurate with radar performance. During the final approach, nominally within about 15.4 meters (50 ft) of the target, the operator relies entirely on the data presented on the video monitor that indicates range and angular alignment.

Addition of auto-docking capability permits docking without the need for the communication link and human operator control. Therefore:

- The adverse effect of communication time delay, nominally 3 seconds, in the control loop is avoided.
- Automated control responds instantly to multiple channel error signals.

- Rendezvous and docking can be completed faster and generally at lower propellant consumption.
- The work load for GCC operators is greatly reduced.

If necessary, the control capability of the initially assumed manual docking system is still available to provide for manual override of the automated approach. To provide this capability, the navigation and control channels of the "chaser" vehicle are modified to incorporate an onboard docking sensor and to accommodate its output data. Sensor characteristics must meet auto-docking accuracy requirements and operational constraints.

Remote Rendezvous/Docking Control via TDRSS Relay

Relay communication via the Tracking and Data Relay Satellite System (TDRSS) is an essential link in remote rendezvous/docking control from the ground, but also is needed for supervised autonomous rendezvous and docking. In addition to the signal path via the TDRS satellite, relay communication, via a domestic comsat, between the ground control console and the TDRSS ground terminal may be required. This increases the number of signal round trips to and from synchronous orbit altitude from two to four. Full control autonomy avoids the signal delay due to relay transmission, and its potential degradation of the feedback control system performance. It also eliminates the human operator's perception lag and his slower response to control errors in 6 degrees of freedom, compared with the autonomous simultaneous system response in all control channels.

Time Delay Effect on Remote Human Control Process

A time delay in the feedback control loop is introduced by the relay link between ground control and chaser (about 3 seconds of round trip delay) and the human control operator's perception lag (estimated to be 2 to 5 seconds). The perception lag decreases with target distance, since the target's image (and its relative motion as displayed on the video screen) becomes larger.

The phase lag in control response due to pure time delay tends to destabilize the control system behavior. In the limit, for a given time delay and a sufficiently high frequency content of the control input, the effective phase lag could be as large as 190 degrees and cause instability. In actual remote control simulations with time delay, an experienced operator slows down his input commands to an equivalent driving frequency of at most 0.05 to 0.10 Hz, i.e., he allows 10 to 20 seconds between command pulses to avoid destabilizing effects.

Predictive display techniques will ease the control task and permit faster completion of the docking approach and are worth exploring. Control simulations performed at MIT and JPL with time delays between 3 and 10 seconds showed a marked improvement of control accuracy, speed, and repeatability.

Candidate Docking Sensors

Two types of range and angle sensing techniques suitable for application to automated docking are currently under development. Candidate sensors are:

1. Imaging systems such as a docking camera on an orbital maneuvering vehicle/satellite servicer system, with appropriate image signal processing that provides relative target range and angle information.
2. Precision range and angle tracking systems such as the Laser Docking Sensor (LDS); an engineering prototype is currently being developed under NASA/JSC contract by McDonnell Douglas Space Systems Company (MDSSC).

Both types of docking sensors require active or passive target augmentation to simplify target detection, and range and angle data extraction. Auto-docking is achievable more readily with cooperative targets that support the process by carrying appropriately mounted and configured retroreflector arrays. The chaser will preferably be equipped with illuminating devices such as laser diodes to enhance the return signal and permit discrimination between true and false target points in the sensor field of view. Flashing lights arranged in a suitable pattern on the target will further enhance detection and signal extraction capabilities by the docking sensor, and will also enhance discrimination against false targets.

The laser docking sensor (LDS) is based on an engineering demonstration model designed and built at NASA/JSC in the early 1980s and incorporates improved target acquisition and tracking characteristics. The sensor provides range and bearing angle data as well as attitude measurements from the reflected beams. Output signals include not only range and angle data but also their rates of change obtained as derived quantities.

The geometry of range and angle determination by image data extraction from the video image of a T-shaped three-point target array carried by the target satellite is illustrated schematically in Figure 4.28. Range is determined on the basis of stadiametric measurement, using the subtended angle of the outer two reflecting targets 1 and 3 having a known separation ℓ. Bearing information is provided by the location of the centroid of targets 1 and 3 relative to the optical axis of the camera, which is assumed to be accurately aligned with the chaser body axis (x-axis). The off-axis view, on the right, illustrates the principle by which the misalignment angle σ can be detected, as a result of the apparent parallax of the tip of the central target pin (2) from the centroid of the outer two targets. In this view the angle deviation is shown in the vertical plane. Note that this parallax effect is the same as that being used in the manually

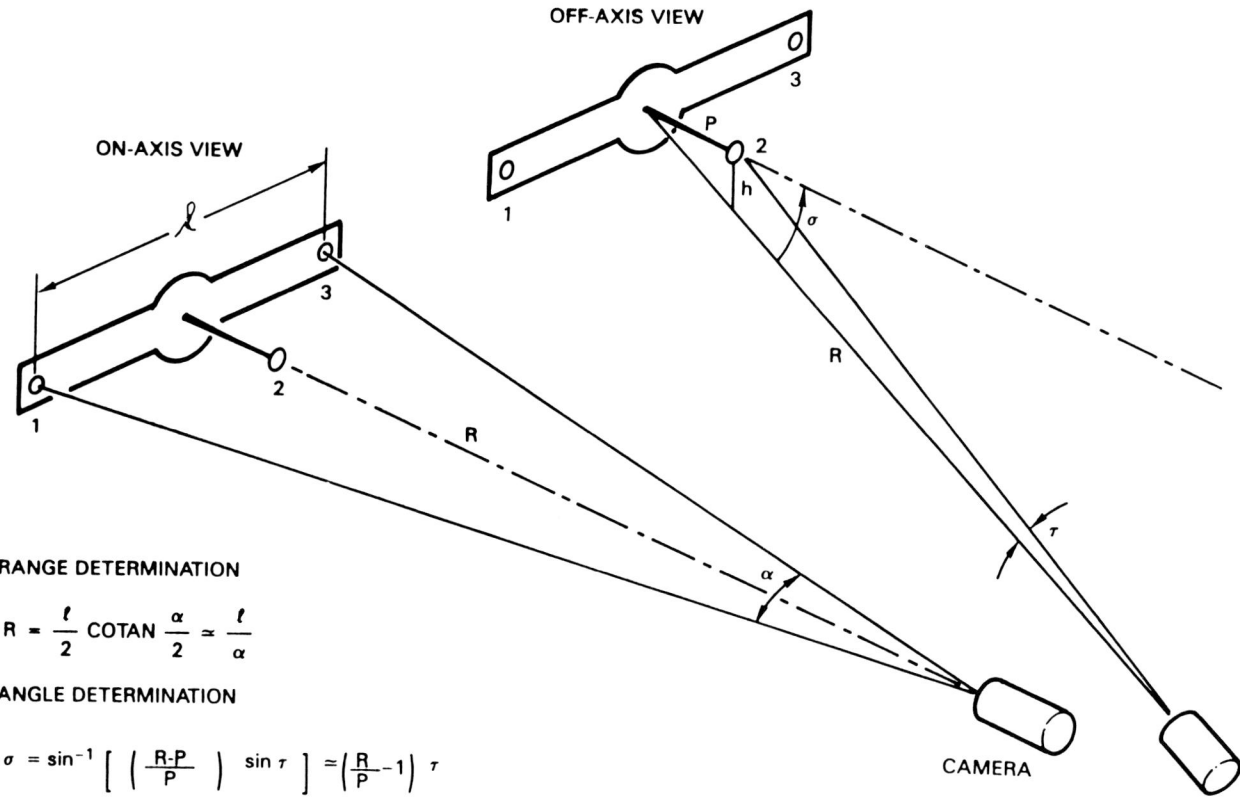

Figure 4.28 Range and angle determination from TV image. *Courtesy of TRW.*

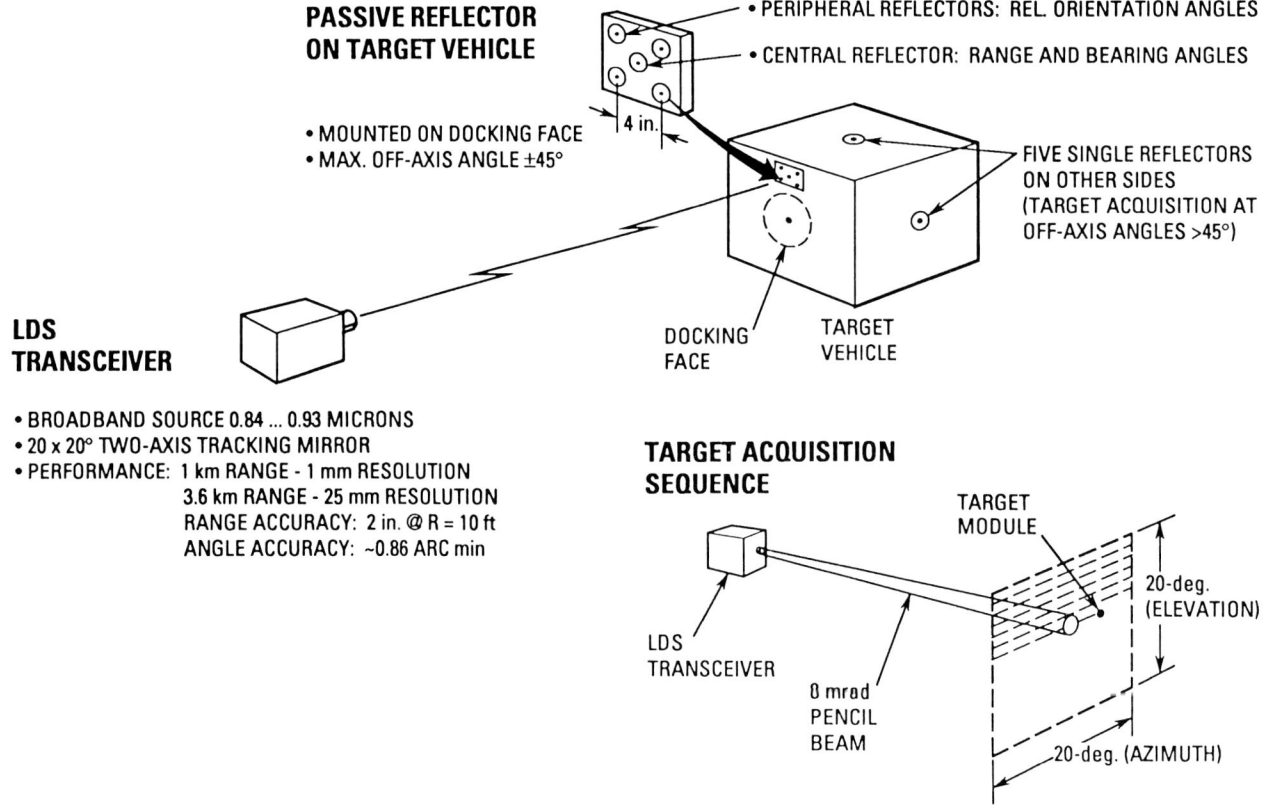

Figure 4.29 Laser docking sensor. *Courtesy of TRW.*

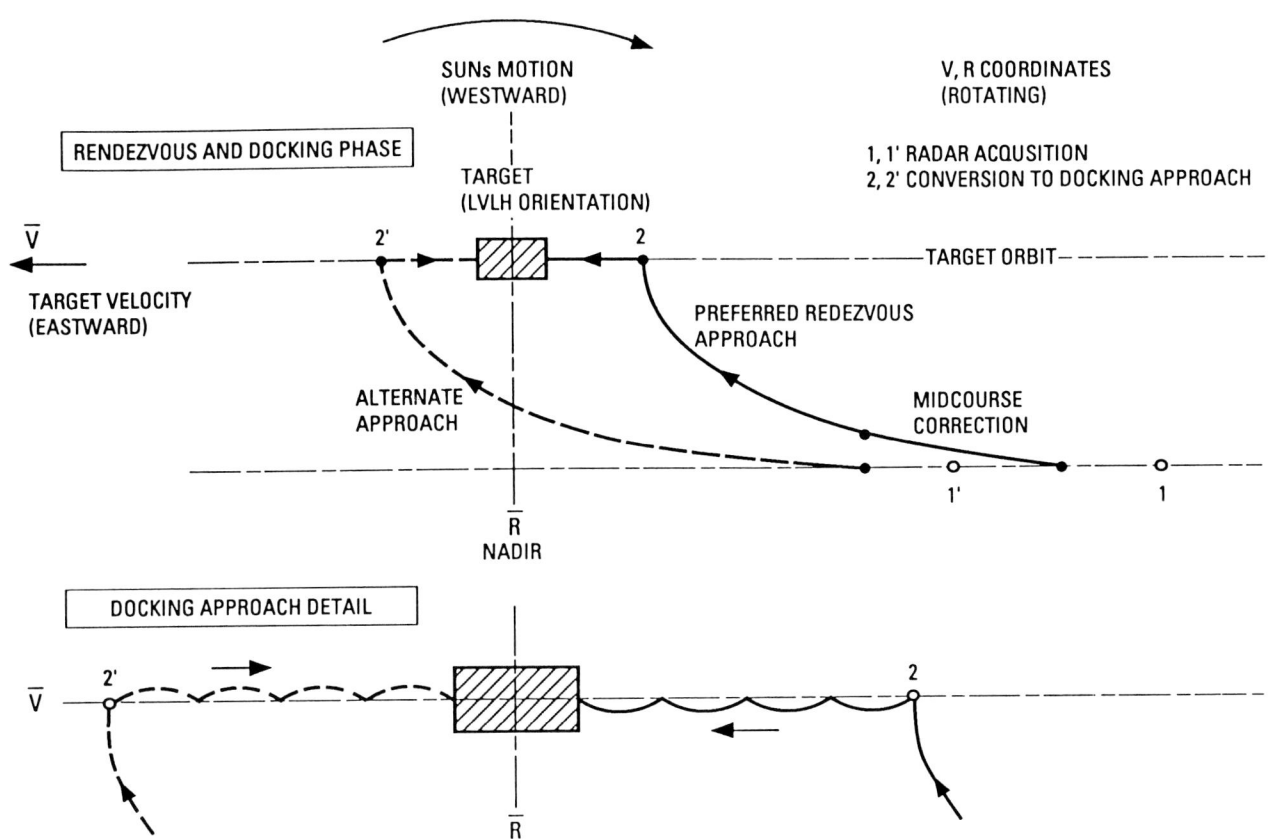

Figure 4.30 Remotely controlled rendezvous and docking profile with target in local vertical/horizontal attitude. *Courtesy of TRW.*

controlled docking mode to detect the angular misalignment of the camera axis (and hence, the chaser body axis) with the docking face of the target satellite.

The pin parallax is a first order effect governed by the sine of the misalignment angle. It is a sensitive parameter at small excursions, but the sensitivity decreases as the angle approaches 90 degrees.

Stadiametric ranging based on the known distance ℓ is affected by the target body angle which reduces the apparent distance between the outer target reflectors, 1 and 3. This is a second order effect which only becomes significant for large off-axis angles. It can be corrected by taking the off-axis angle into account.

The MDSSC laser docking sensor (LDS) system is a potentially simpler and highly accurate alternative for use in auto-rendezvous and docking. Figure 4.29 describes the LDS operating principle and some of the projected performance characteristics. The maximum acquisition range originally specified as about 1.8 km (1.0 nmi) being increased to about 185 km (100 nmi). A five-point retroreflector array attached to the docking face of the target vehicle permits determination of relative range and bearing angles at large distances, and relative orientation angles at closer range, to permit 6 DOF control of the close approach and docking phase. Sensor signal processing yields the rates of change of all six relative motion coordinates. Additional single-point retroreflectors mounted on other sides of the target vehicle facilitate acquisition by the LDS, if the docking face is not initially visible, and may even indicate the required maneuver direction by the chaser.

Remotely Controlled Rendezvous and Docking Profile with Target in Local Vertical/Horizontal Attitude

Figure 4.30 depicts two nominal rendezvous and docking profiles where the docking sequence is controlled from the ground via TDRSS command and feedback communication links. The rendezvous phase starting at Point 1 with radar acquisition of the target is performed autonomously, supported by the rendezvous radar with GPS navigation data, with a midcourse correction occurring somewhere between Points 1 and 2 if necessary. The final approach phase to docking under manual control is initiated at Point 2, supported by radar data, video image, and other telemetry data transmission to the ground.

Of the two approach profiles shown, the one approaching the target satellite from behind is generally preferred, since

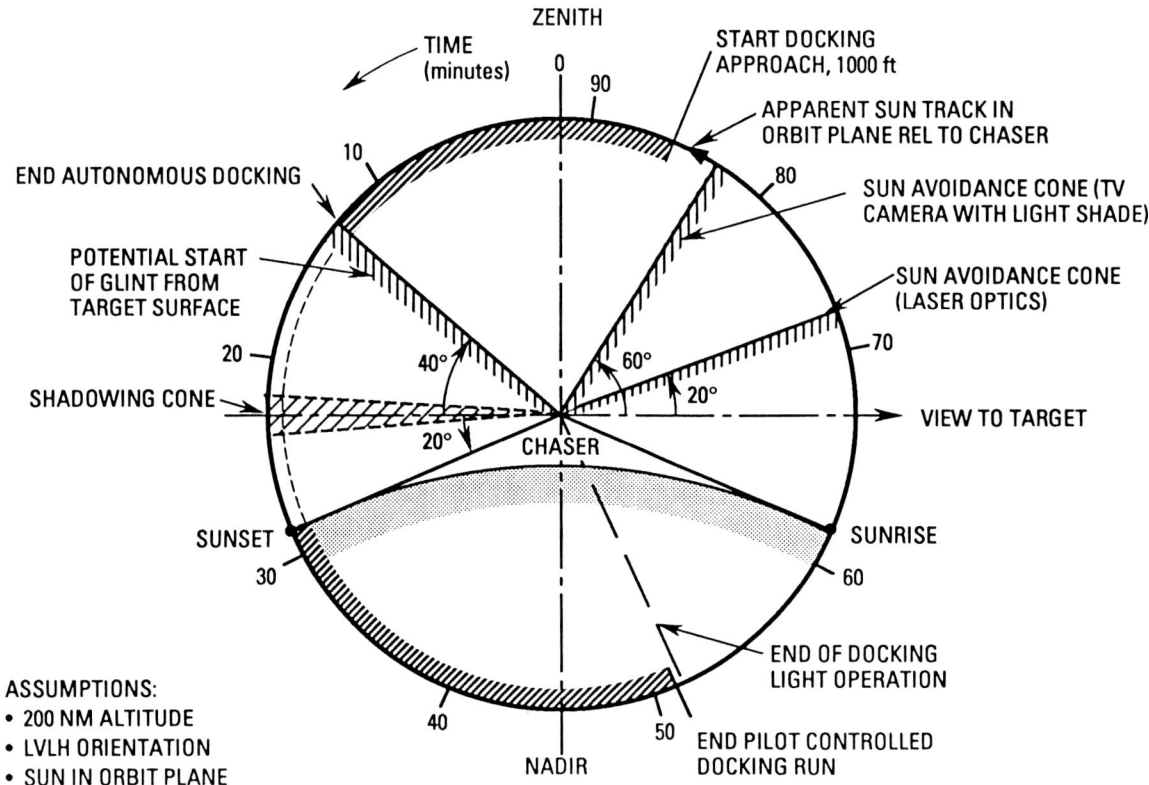

Figure 4.31 Potential sunlight interference with docking phase. *Courtesy of TRW.*

in this case, with both spacecraft flying in an easterly direction, the Sun will illuminate the docking face of the target rather than being in the field of view of the camera during the critical close approach phase. That phase typically occurs in the later part of the daylight portion of the orbit, i.e., during the final 15 to 20 minutes before sunset.

In the auto-docking mode similar rendezvous and docking approach profiles will be flown if the target has a horizontal orientation, such as the one shown on Figure 4.30. If the video camera is used as the docking sensor, the same restriction on avoiding sunlight in the camera's field of view applies as in manually controlled docking. If a laser docking sensor is used instead, that restriction is not as binding. However, to permit manual control intervention or backup, based on using video information, the Sun orientation constraint still must be observed. For target orientations other than those shown on Figure 4.30, the approach profile, whether manual or automated, must be modified appropriately.

Potential Sunlight Interference with Docking Phase

Potential sunlight interference with TV camera or laser docking sensor operation must be prevented by placing appropriate constraints on the docking approach timeline. Figure 4.31 shows the apparent Sun motion as seen by the chaser during its orbital revolution, assuming a "local vertical/local horizontal" (LVLH) body orientation. An approach from behind, rather than from in front of the target vehicle is selected, such that the setting Sun is in the aft hemisphere of the chaser vehicle toward the end of the docking approach.

To avoid direct sunlight entry into the sensor optics the Sun's orientation must remain outside a 60 to 90 degree (half angle) avoidance cone of the TV camera's field of view, and outside a 20 degree cone of that of the laser sensor after docking phase initiation. Therefore, the start of the terminal approach is deferred until the Sun is within about 30 degrees of zenith. Auto-docking is estimated to be completed within 15 minutes, i.e., before the Sun line enters a 40 degree cone in the aft hemisphere, where reflection from the target's docking face may cause glint in the TV camera's field of view.

In a human-operator controlled rendezvous/docking sequence, or in a supervised autonomous mission mode, the final docking phase, at distances of less than 30 meters (100 ft), is deferred until after sunset and completed under docking light illumination. The nominal duration for using docking lights in darkness is 25 to 30 minutes, to keep power consumption within acceptable limits.

Note that due to seasonal changes the Sun can be as much as 52 degrees out of the orbit plane (for a 28.5 degree orbit inclination), in which case the Sun interference restrictions would be less severe.

Requirements and Constraints Summary

Reference 14 lists functional requirements and constraints in the important rendezvous and docking access of: navigation, guidance, control, sensors, relay communication from and to the ground station, target approach strategy, and safety. Some of the key issues and related requirements in Reference 14 distinguish between remote control and autonomous control (with or without ground-based supervision). Future full autonomy implies technology advances, needed in rendezvous and docking in planetary orbit as in the Lunar/Mars Space Exploration Initiative program where direct control or supervision from an Earth station is precluded.

As an intermediate, evolutionary step, supervised autonomous control will be used in Earth orbital application, building up confidence in system performance capability and dependability, permitting man's involvement, if necessary, to assure successful mission completion.

Conclusions

The data presented in this discussion shows the transition from an initially assumed, remotely controlled rendezvous and docking capability to one that would use fully autonomous techniques, with the rendezvous/docking sensor replacing the human control operator's response. Advantages include the avoidance of TDRSS relay communication with its unavoidable inaccessibility periods during passages through the "zone of exclusion"; the need for handover of coverage from one relay satellite to the other interrupting the control process; and the time delay effect on control performance. Supervised autonomy is an interim stage in this transition, still depending on the TDRSS relay, but otherwise speeding up the completion of the rendezvous and docking phase of a satellite servicing mission.

References

1. Woodcock, Gordon R. *Space Stations and Platforms.* Krieger Publishing Company, Melbourne, FL, 1986.
2. Damon, Thomas D. *Introduction to Space.* Krieger Publishing Company, Melbourne, FL, 1989.
3. Madonna, Richard. *Orbital Mechanics.* Krieger Publishing Company, Melbourne, FL, in prep.
4. Wertz, J. R., Microcosm, and W. J. Larson, U.S. Air Force Academy. *Space Mission Analysis & Design.* 1991.
5. Clohessy, W. H., and R. S. Wiltshire. *Terminal Guidance System for Satellite Rendezvous.* J. Aerospace Sciences, Vol. 27, No. 9, September 1960.
6. Wolverton, R. W., ed. *Flight Performance Handbook for Orbital Operations.* John Wiley and Sons, New York, 1961.
7. Graves, Carl D., Hans F. Meissinger, and Alan Rosen. *Servicing of Multiple Satellites Using an OMV-Derived Transfer Vehicle.* TRW Federal Systems Division, Redondo Beach, CA, presented at AAS/GSFC International Symposium on Orbital Mechanics and Mission Design, NASA Goddard Space Flight Center, AAS Paper 89-184, April 24–27, 1989.
8. Graves, Carl D. *Orbital Servicing from an Expendable Launch Vehicle.* TRW Federal Systems Division, Redondo Beach, CA, April 1989.
9. Meissinger, Hans F. *Cost-Effective Orbit Transfer Modes for Satellite Retrieval and Servicing.* TRW Federal Systems Division, Redondo Beach, CA, presented at ESA/DGLR In-Orbit Operations Technology Symposium, Darmstadt, West Germany, September 7–9, 1987, and at the 23rd Satellite Services System Working Group Meeting, NASA Johnson Space Center, Houston, TX, March 14, 1990.
10. Livingston, L. E. *Co-Orbiting Mechanics.* Proc. Satellite Services Workshop, NASA Johnson Space Center, Houston, TX, June 22–24, 1982, 166–181.
11. Snoddy, W. C., W. E. Galloway, and A. C. Young. *Use of the Orbital Maneuvering Vehicle (OMV) for Placement and Retrieval of Spacecraft and Platforms.* Paper AAS 86-041, Annual AAS Guidance and Control Conference, Keystone, CO, February 1–5, 1986.
12. Meissinger, Hans F., A. Rosen, and P. C. Wheeler. *Polar Space Platform Orbit Selection Considerations.* Paper AAS 85-433, AAS/AIAA Astrodynamics Specialist Conference, Vail, CO, August 12–15, 1985.
13. Meissinger, Hans F. *Satellite Proximity Operations Near a Space Shuttle or a Future Space Station.* Presented at the 1983 Annual Conference of Hermann-Oberth-Gesellschaft, Koblenz, Germany, September 16–17, 1983.
14. Meissinger, Hans F. *Operational Requirements and Constraints in Autonomous and Remotely Controlled Rendezvous/Docking Missions.* Presented at NASA/JSC Autonomous Rendezvous and Docking Conference, Houston, TX, August 15–16, 1990.

Chapter 5
Steps to Spacecraft Design

This chapter addresses fundamental considerations in the design and support of satellites which can be assembled and/or serviced on orbit. Spacecraft design steps are suggested.

On-orbit satellite servicing should be evaluated on the basis of requirements, technology, and costs. In turn, these items are dependent on a government (NASA and DoD) integrated policy for designing, planning, and executing space servicing missions. Among the most important aspects of this policy are infrastructure interface standards, commonality of ground processing and launch support systems, crew training facilities, technology sharing, development of mission common servicing support hardware and tools, and the level of servicing appropriate for each mission situation.

The military concept of aircraft maintenance echelons on the ground is applicable to satellite servicing in space. In the military, the first echelon maintenance, the least complex level, involves elements designed for repair-by-replacement. These tasks may be accomplished in space by EVA, remotely operated devices, or automation. They typically pertain to removal and reinstallation of units equipped with quick-disconnect features, such as the GSFC multimission spacecraft type modules. Second echelon maintenance relates to elements that are repairable or replaceable but are not necessarily designed for servicing. This task can be done in space by EVA and is of intermediate complexity. Servicing of the main electronics box on the NASA Solar Max mission exemplifies the second echelon of maintenance. In the military, the first and second echelon activities are performed at operational and training sites. The third and fourth echelons of spacecraft maintenance today occur on the ground, but by the late 1990s will occur in space. In the military, these activities are performed at a large maintenance and repair depot rather than in the field. In level 3 maintenance, black boxes within systems are replaced, while in level 4, elements inside a box, such as switches and circuits, are repaired or replaced. The fourth echelon, servicing at the individual component or piece part level, is the most complex.

The SAMS study [1] indicated that it will be exorbitantly expensive to bring observatory-class instruments to the ground for level 3 and 4 servicing. Cost estimates for full refurbishment range as high as 50 percent of the original development cost of observatory systems. Furthermore, the delicate telescopes would be subjected to g-loads, vibration, and contamination during return and landing. As the military strategy is to keep vehicles and systems in service with minimal downtime, so the strategy for NASA's operational science missions is to keep its assets in space and keep them working there. While return of smaller, less complex spacecraft may be feasible, the economically practical approach to large-scale servicing is to *do it in space*.

Figure 5.1 summarizes the above levels or echelons of maintenance and servicing.

Spacecraft Design "Must Do" Considerations

Serviceable spacecraft incorporates design features that make the spacecraft "service-friendly," safe, and rugged. "Service-friendly" implies the capability of the spacecraft to accommodate both the manned and automated, and a blending of both, modes of servicing with a minimum of crew training, special equipment, and mission operations support. Flexibility to permit manual servicing by the crew or remote/automatic manipulation is necessary to meet a variety of servicing scenarios for servicing at a manned/mantended servicing base in space or in situ, using a remote servicer, such as an OMV with a satellite servicer system (SSS) attached. Crew and mission safety will always be a design driver, especially for operations in the EVA mode. Rugged construction is needed to reduce the potential for spacecraft damage as it is subjected to the events of docking, manipulation, and handling.

Figure 5.2 summarizes the design feature of serviceable spacecraft while Figure 5.3 lists key design issues of serviceable spacecraft and principal interfaces with the servicing facility, with reference to the major on-orbit servicing functions. Also listed are the likely locations, either on a suitable servicing base or in situ. A blending of EVA crew involvement and remotely operated servicing could be performed for most of the functions. In many instances this

LEVEL	FEATURES
FIRST ECHELON "FLIGHT LINE" OR "FIELD" SERVICE	· LEAST COMPLEX · REPAIR BY REPLACEMENT · SPACE ENVIRONMENT · TASKS BY EVA OR AUTOMATION · QUICK DISCONNECT ORUs
SECOND ECHELON "HANGAR" OR "FIELD" SERVICE	· INTERMEDIATE COMPLEXITY · REPAIRABLE OR REPLACEABLE ELEMENTS--NOT NECESSARILY · DESIGNED FOR SERVICING · SPACE ENVIRONMENT, BUT SHELTERED FACILITIES · TASKS BY EVA OR AUTOMATION, SOME A1 SUPPORT
THIRD ECHELON "DEPOT" OR SPACE FACILITY OR SPACE STATION SERVICE	· HIGH LEVEL OF COMPLEXITY · BLACK BOX REPLACEMENT · SHELTERED, PROTECTED SHIRT-SLEEVE ENVIRONMENT · MAJOR MAINTENANCE TASKS
FOURTH ECHELON "DEPOT" OR ADVANCED SPACE FACILITY SERVICE	· MOST COMPLEX · ELEMENTS INSIDE BLACK BOX REPAIRED OR REPLACED · COMPLETE OVERHAUL OF SPACECRAFT, · REFURBISHMENT TASKS · SHELTERED, PROTECTED SHIRT- SLEEVE ENVIRONMENT

Figure 5.1 Maintenance and servicing levels as described by military and commercial organizations. *Courtesy of TRW.*

desired flexibility will drive the spacecraft design as well as the facility design.

The objective of identifying features and issues at a general level, not specific to any particular spacecraft or servicing mission, is to provide the spacecraft or servicing system designer with a summary and checklist of servicing design requirements. They are an aid to identifying all the factors to consider in developing designs for spacecraft and servicers as well as in developing servicing equipment and tools for specific spacecraft and servicing missions. These requirements are also a starting point for developing more specific design requirements for a particular spacecraft and its related servicing and servicing support equipment.

Although designing a spacecraft for on-orbit assembly, maintenance, and servicing is a complex matter, program managers, systems engineers, and subsystems designers will find it useful to integrate these nine "must do" steps into their planning:

1. Generate a program servicing strategy, then develop a set of integrated requirements for the design of the spacecraft and its servicing support equipment.
2. Incorporate EVA crew systems safety specifications and regulations into the strategy and requirements.
3. Based on cost models and reliability failure mode analysis, determine the amount of modularity, ORU levels, and number of non-ORU but replaceable units to be designed into the spacecraft.
4. Make maximum use of company, government, and technical society (AIAA, ASME, IEEE) standards and standardization techniques on the internal and external spacecraft interfaces.
5. Strive for simplicity of design and hardware integration.
6. Analyze the access (orbital rendezvous and docking) and the EVA or servicer system access to the spacecraft to be serviced. Account for docking ports, lighting, crew hand holds, foot restraints, and color codes when synthesizing the initial spacecraft design.
7. Provide for sufficient ground tests and simulations in the validation and demonstration of spacecraft serviceability.
8. Integrate maintenance and supportability logistics into the early planning of spacecraft servicing operations. Develop the logistics plan in parallel with the design evolution.
9. Install a thorough spacecraft and servicing support equipment configuration management system into the design process.

Servicing Strategy and Spacecraft Design Requirements

Once on-orbit servicing has been established as desirable for a program, studies must be undertaken to determine factors such as: systems architecture; level of modularity of spacecraft and payload; degree of reliability; frequency of servicing needed during lifetime of spacecraft; and level of

```
┌─────────────────────────────────────────────┐
│ GENERAL   - "SERVICING-FRIENDLY"            │
│           - "AUTOMATION-FRIENDLY"           │
│           - SAFE TO SERVICE                 │
│           - RUGGED                          │
└─────────────────────────────────────────────┘
```

1. **LAYOUT**
   - MODULAR DESIGN (INSIDE OUT APPROACH)
   - STANDARDIZED ORUs
   - STANDARIZED, SIMPLE ORU/SPACECRAFT INTERFACES
   - MUTUAL NON-INTERFERENCE IN ATTACHMENT/REMOVAL

2. **SERVICING MANIPULATION FEATURES**
   - CONVENIENT HANDHOLDS, FOOT RESTRAINT SOCKETS (CREW SERVICING)
   - VISUAL ALIGNMENT AIDS (MANUAL OR REMOTE SERVICING)
   - GRAPPLE FIXTURE(S) FOR REMOTE MANIPULATOR/SERVICER/END EFFECTOR
   - ALIGNMENT GUIDES
   - SIMPLE ONE-POINT/ONE TURN ATTACHMENT MECHANISM
   - TOOLS (STANDARDIZED, SIMPLE)

3. **INTERFACE DESIGN**
   - MECHANICAL/ELECTRICAL/FLUID CONNECTORS
     DESIGNED FOR SAFE, RELIABLE ENGAGEMENT/DISENGAGEMENT
   - INTERFACE
     - SIMPLICITY
     - DESIGN COMMONALITY/STANDARDIZATION
     - SELF-CHECK FEATURES TO VERIFY MATING (MICROSWITCHES, LEAK TEST, ETC.)
     - LOW ENGAGEMENT FORCE BY CREW, END EFFECTOR

4. **CREW OPERATION COMPATIBLE**
   - TRAINING/SIMULATION OF SERVICING MODES
     CREW SAFETY ISSUES (NASA SAFETY POLICY NHB 1700.7A, OTHERS)
     - HAZARD AVOIDANCE
     - NO SHARP EDGES, PROTRUSIONS
     - SURFACE TEMPERATURES
     - FLUIDS, GASES
     - ELECTRIC POWER

5. **AUTOMATIC/REMOTE MANIPULATION COMPATIBLE**
   - MANIPULATOR ACCESS, "ELBOW ROOM"
   - ALIGNMENT PROVISIONS
   - PROGRAMMABLE MANIPULATION SEQUENCE
   - INTERIM HOLD/STOWAGE PROVISIONS

6. **CHECKOUT, CALIBRATION, VERIFICATION FEATURES**
   - BITE (BUILT IN TEST EQUIPMENT)
   - DIAGNOSTIC SEQUENCES (COMMAND, AUTOMATIC)
   - FINE ALIGNMENT AND BIAS CORRECTION
   - CALIBRATION SEQUENCES
   - CHECKOUT SOFTWARE AND AI SUPPORT
   - COMPUTER-AIDED TEST PROGRAMS (E.G., SPACE-BASED ISATs)
   - GROUND-BASED CHECKOUT PROCEDURE

7. **LOGISTICS**
   - SPARES KEEPING (ON GROUND, IN SPACE)
   - TRANSPORTATION PALLETS
   - SIZE, WEIGHT CONSTRAINTS ON ORUs
   - TOOL STOWAGE, TOOL CADDY
   - MANIPULATOR BASING, TRANSFER, STOWAGE
   - INTERCHANGEABLE MODULES (BETWEEN PROGRAMS). MMS TYPE
   - TRANSFER VEHICLES (OMV, SERVICER, OTV, MMUs)
   - CONFIGURATION MANAGEMENT
   - ELECTRONIC DOCUMENTATION

Figure 5.2 Design features of serviceable spacecraft. *Courtesy of TRW.*

FUNCTION	KEY ISSUES	PRINCIPAL INTERFACES	SERVICING AT BASE*	LOCATION IN SITU
1. REPAIR · Equipment · Structure · Circuits	- Readily accessible units required for servicing - Delicate repairs require work in pressurized workshop	- Crew support system - Power, data system - Test equipment	X	(X)
2. ASSEMBLE, MATE, DEMATE · Construct · Reconfigure · Satellite segments · Satellite/upper stage · Install, attach P/L	- Standardized interface designs and operating routines essential to effective task handling	- Mechanical, electrical and fluid interfaces between segments and with SS support equipment	X	
3. RESUPPLY CONSUMABLES · Propellants · Coolants · Pressurants · Raw materials	- All fluids, gases handling potentially hazardous and may cause contamination - Emphasis on safety issues	- Crew support system - Fluid handling system - Test equipment - Monitoring equipment	X	X
4. MODULE REPLACEMENT · ORU's · P/L instruments · Harvest products	- Standardized interfaces essential to effective operation especially if automated		X	X
5. DEPLOY/RETRACT · Appendages · Mate/demate umbilicals · Prepare for Earth return	- Early operations will depend on hands-on EVA mode - Evolution to more automated modes desired	- Crew support systems - Specialized control interfaces	X	(X)
6. REFURBISH, CLEAN · Optical surfaces · Solar panels · Thermal radiators	- Refurbishment/cleaning may cause additional particle flux, must control contamination effects	- Automated, mechanical system/crew interfaces - Observation and monitoring equipment	X	
7. TEST AND CHECKOUT · State of health · Verification · Operation readiness	- Depends on test data readout, processing, evaluation and/or diagnosis - Usually requires umbilical connection. Otherwise RF link	- Servicing control center - Facility data system or dedicated computer - Test and checkout I/F equipment - Crew support system - Comm. links	X	X
8. MONITORING · Inspect · Observe · Determine status · Give warning	- Requires sensors and other monitoring equipment - Data readout, processing, evaluation and/or diagnosis - On-board or remote	- Servicing control center - Facility data system or dedicated computer - Monitoring I/F equipment - Comm. links - Crew support system	X	X

* Servicing Base: 1. Shuttle Orbiter, 2. Space Station, 3. Other, man-tended bases

Figure 5.3 Serviceable spacecraft design issues. *Courtesy of TRW.*

commonality of systems or subsystems internal and external to the spacecraft's mission. Inherent in the design decisions associated with serviceability are those which make the planned on-orbit servicing activities safe and efficient for human servicers and feasible for the less dexterous robots.

For example, a means must be provided to make the spacecraft safely approachable. If the spacecraft is to be transported to another location before servicing commences, its structure must be capable of surviving the loads (i.e., shock, acceleration, and vibration) imposed by the transporter. The spacecraft also must be protected from contamination sources during all phases of the servicing process, and the thermal environment must be maintained in a safe range throughout the operation [2]. Here are some trade studies or analyses that should be performed to determine program servicing strategy and spacecraft servicing equipment integrated requirements:

1. What to service on the spacecraft
2. Frequency of servicing and location of the servicing events
3. Servicing cost constraints (servicing cost versus spacecraft value)
4. Technology status of servicer systems
5. Concurrent engineering principles to be employed
6. Transportation to orbit and in-orbit transportation needs for both planned and contingency on-orbit servicing
7. Servicing support equipment required
8. Levels of EVA hours anticipated and crew skills
9. Design approaches to the major interfaces of:
   — Spacecraft bus to the ORUs
   — Spacecraft bus to the robotic servicer system
   — ORUs to the robotic servicer system
   — ORUs to manned (EVA) servicing techniques
   — ORUs to servicing tools, fixtures, and support equipment
   — Existing and new ground support equipment to the spacecraft
   — Spacecraft to the space and interorbit transport vehicles
   — Spacecraft to simulation and crew training facilities
   — Spacecraft refueling and other expendables replacement design concepts to the servicing tools and devices for these events

The program management philosophy should be to establish maintenance and servicing (M&S) requirements and responsibilities early in the spacecraft design schedule, and should include astronauts thinking in all design and development phases if EVA is contemplated. M&S should be incorporated in all planning and program design reviews. Determining the servicing requirements for the spacecraft and its payloads is highly dependent on the engineering reliability index which is used as the planned interval to replacement, and on the maintenance philosophy (management decision) on the level of operational system reliability. Changes in these ground rules cause all logistics for upload to be recalculated, and may influence whether or not a service vehicle is able to offer a satisfactory operational scenario.

When a satellite is successfully launched, it has a very high probability of being reliable (typically 0.99), as would be expected from a new product which has been thoroughly tested. However, as time passes, the chances of failure increase due to the increasing probability of random failure and wear-out of components. The rate at which reliability will fall to 0.9 will vary, depending on the function and design of the instrument. This rate, known as years to 0.9 reliability, may be used as an index to determine the date at which the satellite requires parts or subsystem modules or payload instruments replacement in a servicing mission.

However, a satellite is designed to operate for a number of years; this is known as the design life. It is inevitable that during the design life the probability of continued reliability will fall below 0.9. A satellite could continue to operate at full performance, but as time progresses the probability of failure will be increasingly great, and the reliability may fall to as low as 0.5 at the end of the design life. It thus becomes obvious that the design life is always a longer period than the years to 0.9 reliability index. The design life may also be used as an index to determine the date at which a satellite requires selected components replacement on a servicing mission.

Systems Safety

Figure 5.4 lists various sources of potential hazards to which the crew might be exposed on a servicing mission. It is essential to recognize such hazards and to provide the necessary means of hazard avoidance for the servicing crew by (a) installing monitoring and alerting systems; (b) providing protective equipment and design features; (c) imposing stringent design requirements that exclude hazard sources in spacecraft configurations, tools, and support equipment; and (d) establishing operating rules and restraints that exclude inadvertent exposure of the crew to hazardous conditions. All of these safety and protective features are governed by safety standards and codes issued by NASA and DoD for manned space system operations. The principal document is *Safety Policy and Requirements for Payloads Using the Space Transportation System*, NASA, NHB 1700.7.

Safety considerations with regard to serviceable spacecraft, servicers, and EVA personnel fall into several major categories [3]. The first relates to personnel safety of the EVA crew members while performing spacecraft assembly, maintenance, or servicing. There is a considerable body of information, procedures, and requirements which have been developed in connection with the NSTS EVA activities.

The second area is the safety of the satellite and related remote servicing equipment. Again, the approaches and procedures that have been developed for launch, operation, and NSTS recovery of existing satellites are generally applicable. The principal concerns are mechanical collision damage, contamination by effluents or servicing fluids, generation of particulate matter by maintenance tasks, and electrical discharge or ignition. Mechanical damage must be prevented by adequate design strength, shields, or covers to withstand EVA collision loads, and tether control of tools.

Crew Safety Considerations - Spacecraft Design

- EXPOSURE TO POTENTIALLY HAZARDOUS SYSTEMS/DEVICES/SOURCES:

  - PYROTECHNIC SYSTEMS
  - PROPULSION/PROPELLANT SYSTEMS
  - CRYOGENIC SYSTEMS
  - PRESSURIZED SYSTEMS
  - RADIOACTIVE SOURCES
  - ELECTRICAL/ELECTRONIC SYSTEMS
  - MECHANICAL SYSTEMS AND MOVING PARTS

- PROVISION OF HAZARD MONITORING AND ALERTING SYSTEMS

- PROVISION OF CREW PROTECTION EQUIPMENT AND SYSTEMS

  - EMERGENCY FIRST AID KITS
  - EMERGENCY RESCUE AND SITE EGRESS PROVISIONS (INCLUDING SAFE HAVENS)
  - EMU OVERGARMENTS AND DECONTAMINATION PROVISIONS
  - EMERGENCY TOOLS AND AIDS
  - MANUAL OVERRIDE PROVISIONS

- PROTECTION OF CONTROLS FROM INADVERTENT ACTUATION, ESPECIALLY RELATED TO:

  - PYROTECHNIC SYSTEMS
  - STORED ENERGY DEVICES

- SPECIAL PROTECTION FOR EMERGENCY/CONTINGENCY SUPPORT EQUIPMENT
- PROTECTION OF CREW FROM SHARP EDGES, CORNERS, PROTRUSIONS, AND OBSTRUCTIONS

  - LATCHES, LEVERS, SCREWS, BOLTS, FASTENERS, PINS, RINGS, CRANKS, HOOKS, CONTROLS, ETC.

- DESIGN TO PRECLUDE CREW ENTRAPMENT BY EQUIPMENT/STRUCTURES

  - UNCOVERED HOLES AND OPENINGS
  - GLOVED HAND PROTECTION

- THERMAL TOUCH HAZARD ELIMINATION
- ADEQUATE CORD, CABLE, AND UMBILICAL MANAGEMENT SYSTEMS
- PROVISION OF ADEQUATE TETHERING AND RESTRAINT SYSTEMS FOR BOTH MOBILE AND FIXED-POSITION CREWMEMBERS AND THEIR EQUIPMENT
- DESIGN TO PROVIDE ADEQUATE PHYSICAL AND VISUAL ACCESS AND ILLUMINATION
- APPLICATION OF CAUTION AND WARNING LABELS AND COLOR CODES, AS REQUIRED
- PROVISION OF NON-SLIP HANDLING FEATURES ON LARGE ITEMS REQUIRING CREW TRANSFER
- TOOL/EQUIPMENT DESIGN TO ACCOMMODATE NORMAL AND INADVERTENT CREW LOADS/ FORCES APPLICATION
- HAZARD PROTECTION PROVISIONS RELATIVE TO ELECTRICALLY OPERATED TOOLS AND AIDS

Figure 5.4 Spacecraft design crew safety considerations. *Courtesy of TRW.*

Electrical discharge due to charging of either spacecraft or servicer has not yet been adequately addressed. The sources are understood in a general way, and some test data has been collected from existing spacecraft in orbit. The general solution involves dissipating the charge on the spacecraft, or at least equalizing the charge on the spacecraft and servicer. The servicer will probably have a lower charge level due to shorter time in orbit although this might not be the case if it passed through an area of high activity before accessing the spacecraft. If the soft docking grapple is electrically insulated from the docking arm or boom, and the spacecraft docking interface is grounded to the charged surface or structure of the spacecraft, the excess potential could be bled off through a conductor attached to the grapple. The charge can be balanced by a controlled transfer current between vehicles (the present approach) or by dissipation into the surrounding plasma. For high charges in difficult environments, both approaches involve theoretical and practical problems and will need analysis, laboratory testing, and orbital testing to identify and prove a practical solution.

The third and somewhat new area is that of safety considerations associated with the specific assembly, servicing, and maintenance operations. The major areas of concern are (1) safety during servicer or servicing vehicle approach

Steps to Spacecraft Design

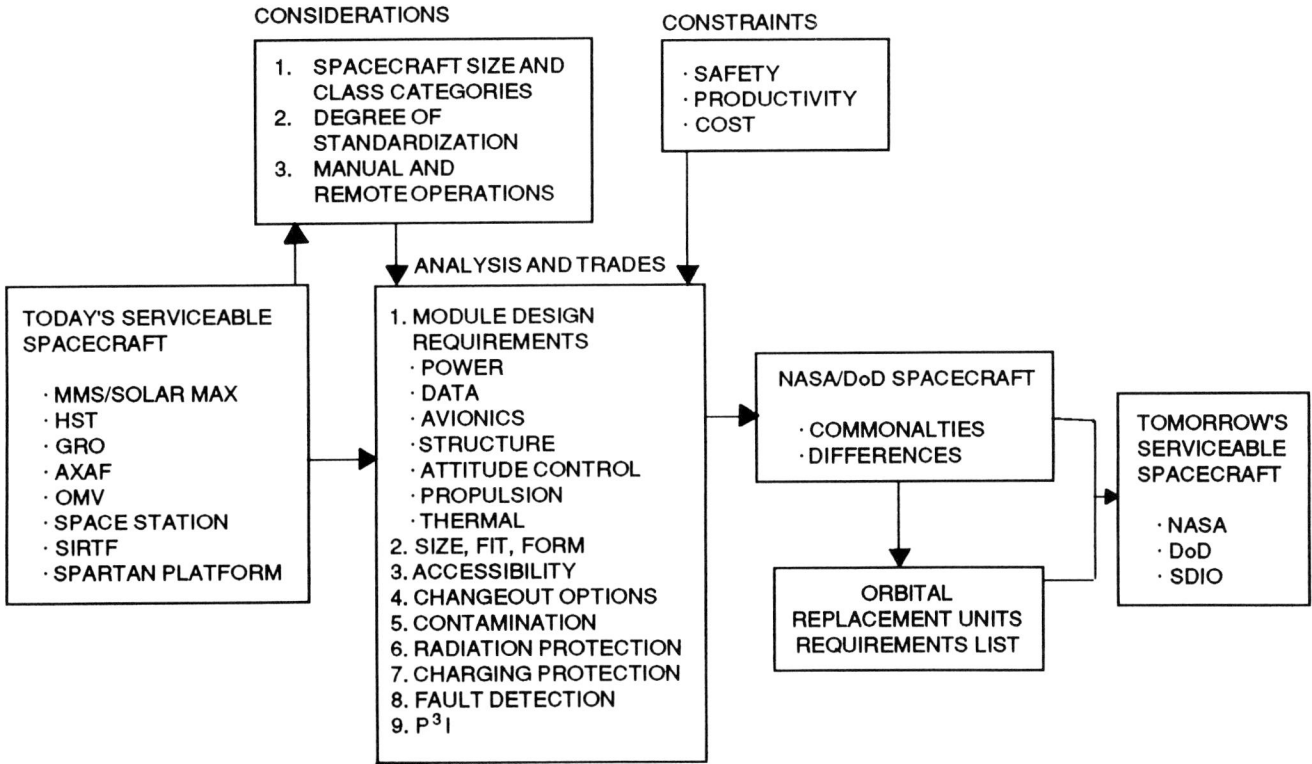

Figure 5.5 Logic flow for modularity analysis.

and docking to the subject satellite and (2) interactions between robotic or teleoperated servicing arms or other mechanisms and the serviced spacecraft as well as EVA personnel [3].

The types and sources of hazards during docking are basically the same as those presented to SAMS spacecraft by EVA servicing personnel. The level of mechanical hazard is of course much greater with a servicer, due to its greater mass. The docking operation presents the greatest physical hazard in terms of force and acceleration, but the servicing arm presents a greater likelihood of damage to the spacecraft due to its proximity, the length of exposure, and the complexity of tasks, controls, and collision avoidance measures. The intricate spacecraft interfaces, including bus, appendages, and sensors, will make the collision avoidance task complex, whether programmed sensing or viewing by an operator is involved. To prevent teleoperator error or lack of adequate information for control from resulting in mission or spacecraft critical damage, some form of backup proximity sensing and/or force limitation is desirable. The other listed hazards are similar to those for EVA servicing and will require similar control measures. Some controls, such as warning labels, will probably not be effective with telepresence limitations.

Another unique set of safety problems is introduced when EVA servicing is combined with robotic assistance or used to supplement remote servicing. The principal addition is the danger of collision between EVA personnel and the servicer spacecraft and servicing arms. The first hazard can be prevented or controlled by limiting EVA access until docking of the servicer and spacecraft is complete, and stability and control established by the servicer, Shuttle arm, or other means, assuming the spacecraft is to be shut down to allow servicing access. The second problem, of EVA collision with an active servicing arm, will be more difficult to control. Although the astronauts can be restricted from entering the servicer arm operating envelope, human error, arm malfunction, or joint EVA/robot task requirements could prevent this from being an adequate solution. In automated factories, light or other beam screens and sensors are used to shut down robots when their operating space is penetrated. This would be difficult to implement in a servicer due to spacecraft and work volume envelope complexity. An alternative might be multiple collision avoidance sensors on the servicer arm, or velocity/force sensors to limit energy input in case of a collision. This is an area of remote servicing requiring much more analysis, design, and test.

Modularity

Spacecraft modularity addresses how many modules, module size, and the content of each module. It focuses on the trade of refurbishable versus throwaway modules and entails

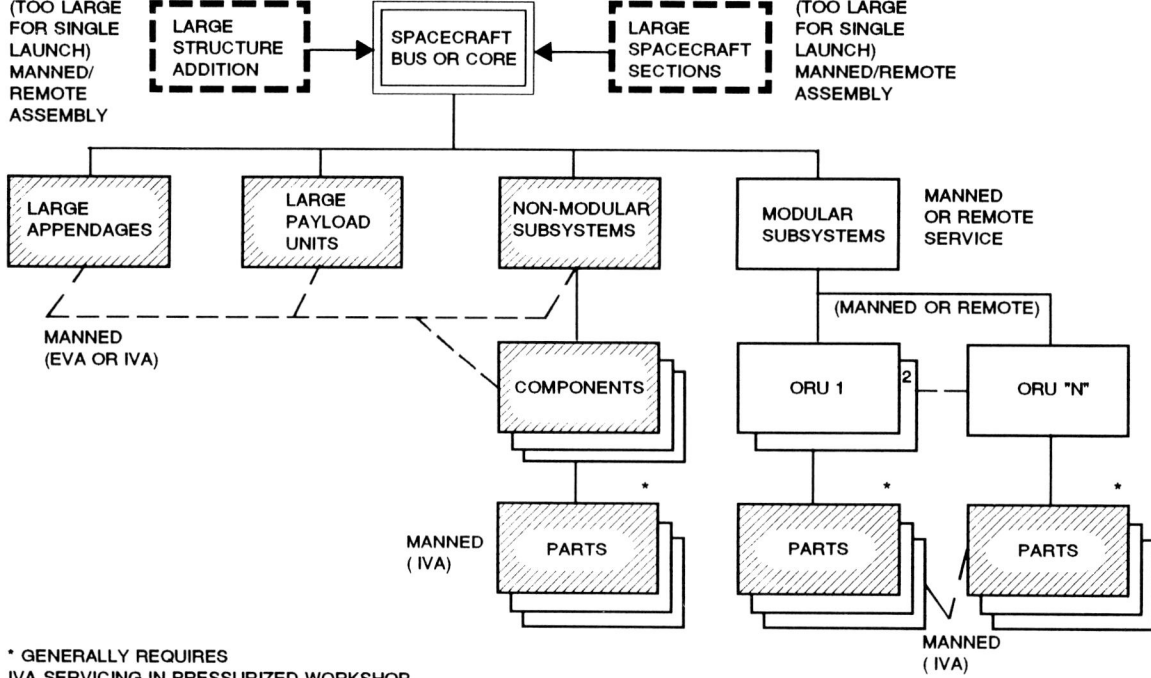

Figure 5.6 Partitioning of modular spacecraft. *Courtesy of TRW.*

reliability assessment of the critical parts in each module. The procedure is to identify all ORUs and candidate ORUs as soon as possible in the spacecraft design process, then put them, figuratively speaking, into boxes or modules. Treat any critical item with a risk of early failure as an ORU. Plan to replace parts. On-orbit repair takes too much time and effort which translates into costly space operations.

Modularity is the keynote to spacecraft serviceability. Figure 5.5 shows a methodology and flow to define the modular approach to making spacecraft receptive to SAMS functions. Note that today there are eight to ten spacecraft in operation or advanced development that are committed to accept some form of on-orbit servicing. They are all NASA spacecraft; however, a NASA modular OMV could have military applications.

Modularity in the design approach provides easy access and quick changeout to critical components but not all spacecraft components lend themselves to modular configuration. Examples are structure, electrical harness, antennas and solar arrays. These can be replaced or added to but probably on a case-by-case basis with unique requirements imposed on the design and hardware location.

The NASA serviceable OMV, that was under development by MSFC/TRW, is probably the best example of a modular serviceable spacecraft with 1985–1990 technology.

The difference between DoD and NASA spacecraft, reflected in the considerations of survivability, security, and availability needs by military missions, causes the approach to modularity to differ but does not change the concept.

Modularity can improve the timeliness of ground integration and test of new or block change space systems. This has a potential cost benefit for a space systems program.

Partitioning of Modular Spacecraft

Figure 5.6 illustrates several categories of subdividing or partitioning a modular spacecraft into elements that facilitate on-orbit assembly or servicing.

The central bus contains the irreducible core of the spacecraft to which all other parts are attached to form the entire assembled system. Large sections or structure additions that would exceed the capacity of a single (Shuttle) launch are subsequently attached by manned or telerobotic operations, (indicated by dashed units in the top line.)

Other large appendages, payload units, or nonmodular subsystems are integrated into the system, largely by manned operation, either in the EVA or IVA mode (lightly shaded blocks in the second tier). In addition, modular units are also integrated into the system by manned or remote operation.

The next tier includes typical ORUs that are the subject of routine changeout in manned or remote servicing missions. The last tier consists of parts that are too small or inaccessible for servicing/repair in the EVA mode (darker shading), and thus require IVA crew service inside a pres-

Steps to Spacecraft Design

DATA MANAGEMENT	ATTITUDE CONTROL	RCS/PROPULSION
- CENTRAL COMPUTER - CENTRAL COMPUTER SOFTWARE - DATA BUS CONTROLLER - BUS I/O INTERFACE UNIT - TELEMETRY FORMATTER/ENCODER - COMMAND DECODER - MODULE POWER INTERFACE - MODULE THERMAL CONTROLLER - MODULE STRUCTURE	- SUN SENSORS - EARTH SENSOR - STAR TRACKER(S) - GYRO PACKAGE - ACCELEROMETER PACKAGE - REACTION WHEEL PACKAGE - MODULE STRUCTURE - MODULE THERMAL CONTROLLER - MODULE PROCESSOR/SOFTWARE - BUS I/O INTERFACE UNIT - MODULE POWER INTERFACE	- TANKS - VALVES/VALVE DRIVES - VALVE DRIVE ELECTRONICS - PLUMBING - THRUSTERS - MODULE THERMAL CONTROLLER - MODULE STRUCTURE - BUS I/O INTERFACE UNIT - MODULE PROCESSOR/FIRMWARE - MODULE POWER INTERFACE
UP/DOWNLINK COMMUNICATIONS	**CORE MODULE**	**POWER**
- U/D-LINK SIGNAL PROCESSOR - U/D-LINK ELECTRONICS - TRANSEC/COMSEC - U/D-LINK ANTENNA - U/D-LINK RF ELECTRONICS - U/D-LINK ANT. PHASING CONTR. - S-BAND OMNI ANTENNA - S-BAND RF ELECTRONICS - MODULE STRUCTURE - BUS I/O INTERFACE UNIT - MODULE THERMAL CONTROLLER - MODULE POWER INTERFACE	- CORE STRUCTURE - ELECTRICAL HARNESS - MASTER DATA BUS - ANTENNA BOOMS - DEPLOYMENT MECHANISMS	- SOLAR ARRAY(S) - SOLAR ARRAY BOOM/DRIVE ASSEMBLIES - SOLAR DRIVE ELECTRONICS - SOLAR ARRAY SUN SENSOR - BATTERIES - POWER CONTROL UNIT - POWER SWITCHING UNIT - SHUNT ASSEMBLY - ORDNANCE ELECTRONICS - BUS I/O INTERFACE UNIT - MODULE THERMAL CONTROLLER - SUBSYSTEM STRUCTURE - MODULE PROCESSOR/SOFTWARE/FIRMWARE - DEPLOYMENT SEQUENCER (EIA)

Figure 5.7 Representative module partitioning.

surized workshop if they are to be handled individually, for example, in emergencies. Generally, they remain enclosed within the next larger module and are part of a subsystem that would be returned for ground refurbishment as appropriate.

Representative Module Partitioning—Engineering Subsystems

Figure 5.7 identifies the six principal engineering subsystem ORU modules carried on serviceable spacecraft. Earlier spacecraft programs now under development did not have all of these ORU modules. For example, the GSFC/TRW GRO spacecraft has a modular power subsystem as well as data management and attitude control subsystems. The propulsion subsystem is serviceable but is not an ORU.

The fewer the ORUs, the easier the on-orbit servicing event. However, ease of on-orbit servicing must be balanced against the potential cost of replacing an ORU containing both failed and functional systems. Transportation to and from space to support the servicing event must also be taken into consideration. Ideally, today's satellite ORUs should be designed for manipulation in both EVA and remote servicer modes; this includes considerations for grappling points, mass, and geometry. The ORUs should be designed for easy insertion and removal in both modes. The considerations here are for alignment aids, interface verification, and connection designs for power, thermal, fluid, and communication services between the ORU and satellite core.

Physical Characteristics of Orbital Replacement Units

Some designers advocate the development of standard size and shape ORUs for use across many spacecraft projects. In general, satellite requirements are sufficiently diverse and stringent to cause unique solutions to be incorporated in the resulting designs. Very little progress has been made in identifying common ORUs. The selection of common standard ORU geometrics should remain a long-term goal, but in the meantime, standard interfaces have a higher priority. Variable size and shape alone will not preclude standardized on-orbit servicing approaches, but they could make servicing and transportation less efficient [3].

Generally there are five categories of ORU geometry: rectangular of uniform size, rectangular of fractional sizes, rectangular but irregular size and shape, trapezoidal or pie shape, and free form. A summary of observations about each type is made in Figure 5.8. In general, ORU designs from the three categories at the bottom of Figure 5.8 are the most flexible and therefore the most useful to the satellite designer. The recommendation of one size and shape ORU or even a series of sizes and shapes as standards is not totally practical at this time, and would not be accepted by a majority of satellite designers. Efforts in this direction should be continued and encouraged.

While the concept of satellite maintenance by ORU exchange is a conventional approach that has been explored extensively in theory and somewhat in hardware, it results

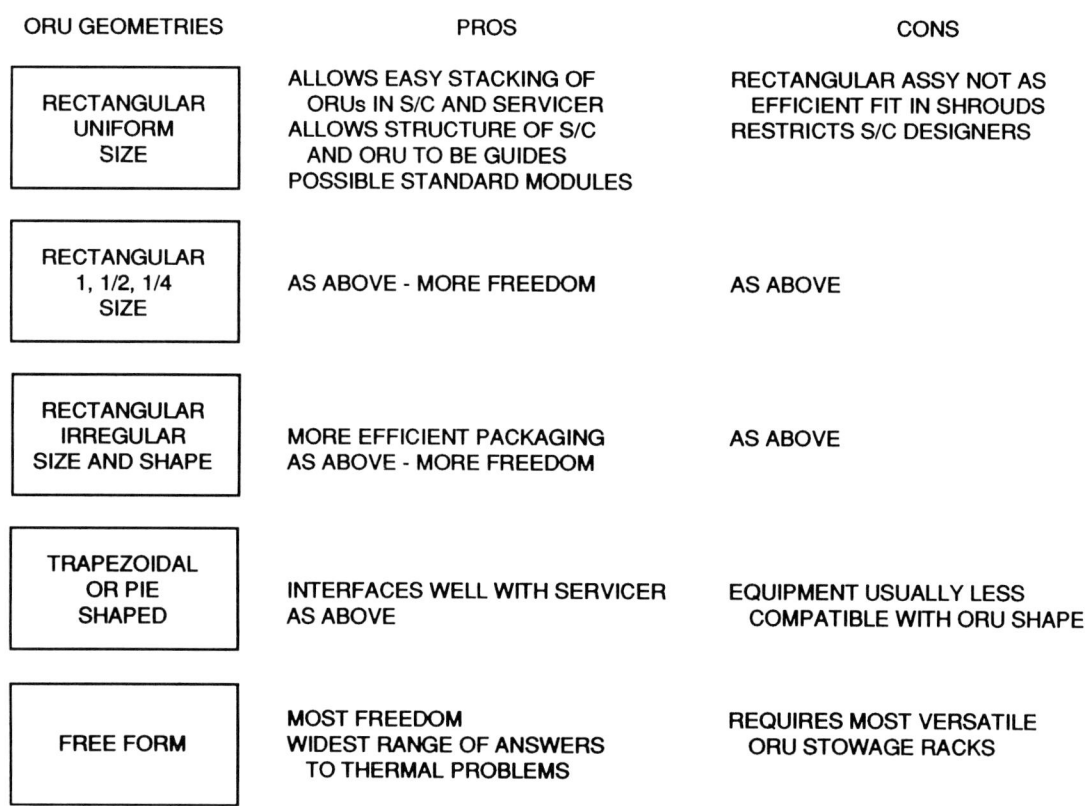

Figure 5.8 ORU geometry comparisons. *Courtesy of Lockheed Missiles & Space Co.*

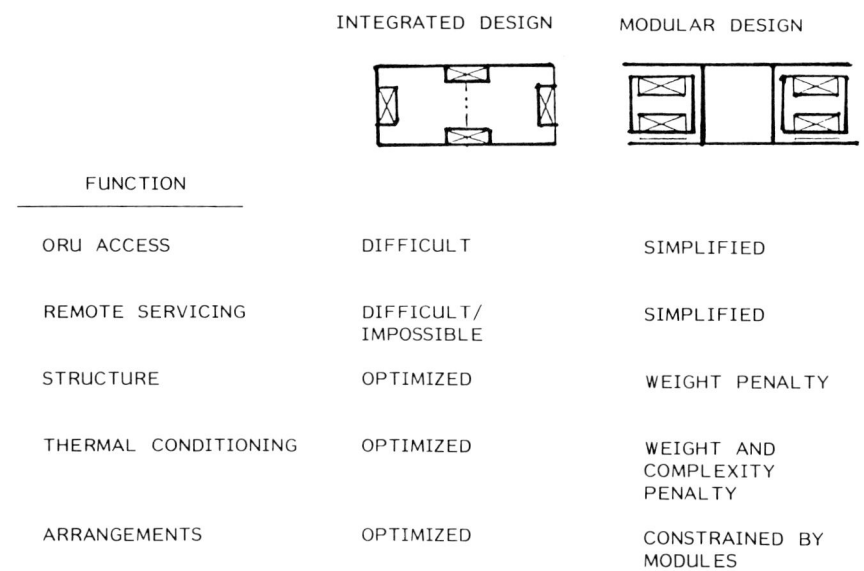

Figure 5.9 Impact of modular satellite design. *Courtesy of Lockheed Missiles & Space Co.*

in both design and cost penalties and advantages, in addition to simplifying servicing. Expendable spacecraft have an integrated design as diagrammed in Figure 5.9. Equipment shown as boxes in the figure are generally mounted to the inside surfaces of a very integrated and optimized structure. Thermal control is achieved by wrapping insulation around the entire assembly. The thermal control task is eased because some heat is transferred between equipment by radiation and conductivity through the structure.

Figure 5.9 also presents a modular design used when the satellite is partitioned to include ORUs. Because the ORUs

SATELLITE	ALTITUDE (km)	INCLINATION (deg)	MISSION
GRO	450	28	Scientific Observatory
UARS	600	57	Earth Observation
TOPEX	1300	65	Earth Observation
TRMM	350	35	Earth Observation
AXAF	600	28	Scientific Observatory
EOS A	500	98	Earth Observation
SIRTF	900	28	Scientific Observatory
ESGP	34740	0	Earth Observation
EOS B	500	98	Earth Observation
LDR	700	28	Scientific Observatory
SSTS	Mid	High	Anti-Ballistic Missile Surveillance
SBI	Low	High	Anti-Ballistic Missile Weapon
Zenith Star	LEO	Low	Anti-Ballistic Missile Weapon Test

Figure 5.10 Satellites in the ORU analysis data base. Satellite names are defined in Appendix A. *Courtesy of TRW.*

must be easily removable from the basic satellite, they must have their own independent structures separate from the backbone structure of the spacecraft. This results in duplicated or parallel structures and a weight penalty relative to an integrated design. Each ORU must be mounted to the satellite through a mechanical and structural mechanism that allows for, and eases, the exchange procedure. This interface includes alignment aids and accommodations for fastening the ORU and probably a specific interface to accommodate an ORU exchange mechanism. These features also represent a weight penalty over expendable designs. Such an interface design usually does not result in a large surface area of contact between ORU and satellite structure. As a result, it cannot be relied upon to provide a thermal conductivity path between ORU and satellite structure, unless new design and material techniques are developed.

Most ORUs removed from satellites during maintenance will be returned to Earth for test and refurbishment. For the foreseeable future, ORUs returning from space will do so in the Shuttle Orbiter payload bay. The ORUs must therefore be size compatible with a stowage rack or ORU carrier that can fit within that space. Later epoch transportation systems may relax that restraint. Eventually, some modules may be serviced at the Space Station Freedom and will have to be sized to pass through the applicable hatches.

A related subject that provides a similar limitation is the problem associated with the replacement of large deployable appendages such as antennas and solar arrays. These items are frequently much too large after deployment to be transported in reasonable sized stowage racks, or even the Orbiter cargo bay. They are launched in compact stowed packages and are deployed after reaching orbit. Mechanisms can be devised to restow most of these items, but the reliability of such devices after long exposures in space or after the deployed structure has been damaged is not assured [3].

ORU Sizing

The designs of 13 existing or planned satellites were analyzed by TRW [4] to determine their ORU characteristics as if the satellites were all designed for remote servicing. Figure 5.10 shows this satellite data base. The steps in this analysis were:

1. Data was collected from government and contractor documents.
2. Each element of each satellite which might be configured as an ORU was identified.
3. Thus, a total of 171 ORUs, for the 13 satellites, were listed.
4. The mass, volume, and shape of each potential ORU were estimated.
5. Adjustments were made to the characteristics of each element to account for the changes required to make it an ORU.
6. Types and masses of potential resupply fluids were estimated for each satellite.
7. Distributions were calculated for each physical characteristic.

The results of this study are plotted in Figures 5.11 through 5.18. Each element which might be configured as an ORU was identified. Here some judgment was required. Some items such as the EOS/HIMSS payload were judged too big and complicated (as presently designed). Other elements such as the EOS B SAR antenna and most solar arrays were also judged too large and too difficult to configure as an ORU. In the case of the SAR antenna, it was felt that

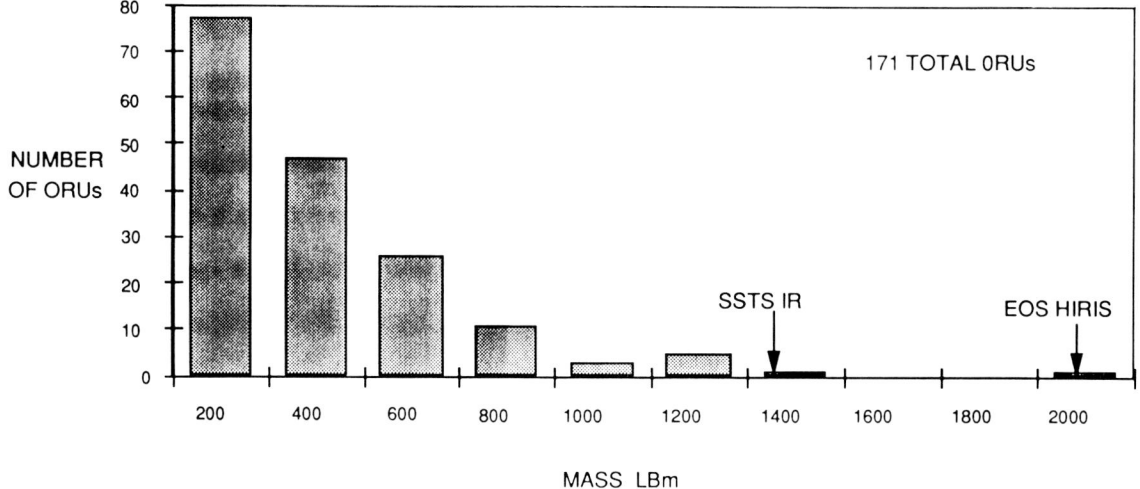

Figure 5.11 ORU mass distribution. *Courtesy of TRW.*

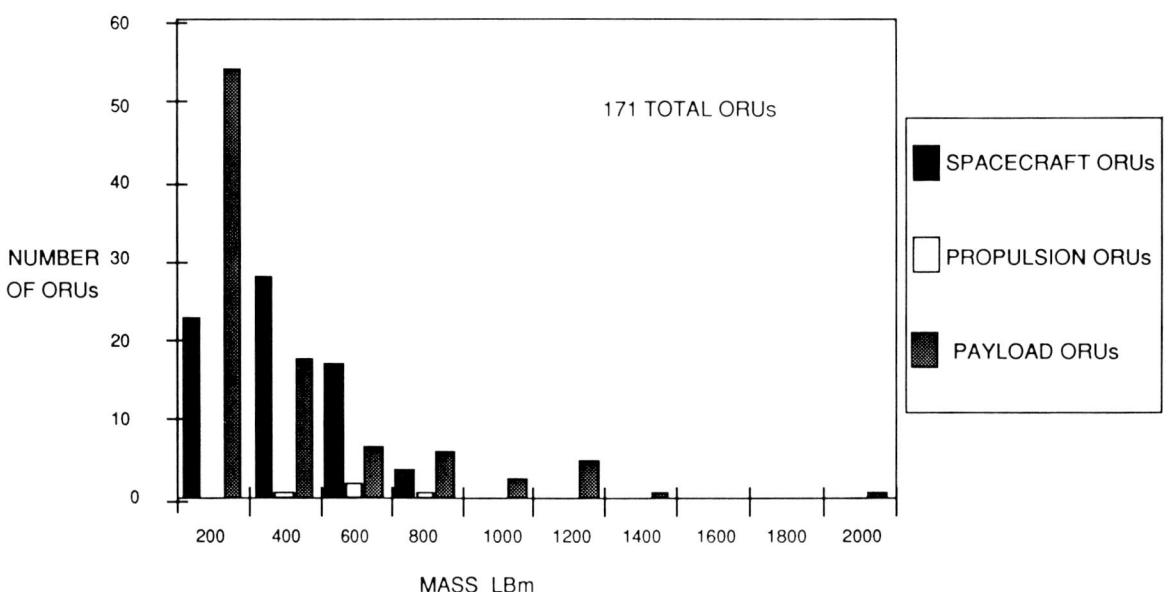

Figure 5.12 ORU mass distribution by type. *Courtesy of TRW.*

it could be built with enough redundancy to make it unnecessary to configure it as an ORU (items like the feed and antenna electronics could be an ORU but this was not considered for the study). Solar arrays can also be highly redundant (but a method of replacing the drive and joints might be useful). All told, 171 elements were identified as potential ORUs on the 13 satellites [4].

The mass, volume, and shape of each potential ORU were estimated. This was done considering its design and estimating what changes would be required to make it an ORU. For instance, standard EOS and MMS modules were not changed. The EOS characteristics entered in the data base are those currently estimated in EOS reference documents. Mass, size, and shape estimates for most other potential ORUs were modified before entering in the data base. Most mass increases were in the 10 percent range. The shapes and volumes shown in the data base are the cylindrical or rectilinear shapes that would contain the potential ORU. It is really an accommodation shape and volume which would have to be provided on the servicer. There is no duplication in the data base. An ORU was only counted once even if multiple copies existed on a satellite [4].

The engineering team on this TRW study [4] recommends the following maximums as guidelines for ORU sizing until firm user requirements are available:

- Volume: 5.66 cubic meters (200 cubic feet)
- Mass: 567 kilograms (1,250 pounds)

Steps to Spacecraft Design

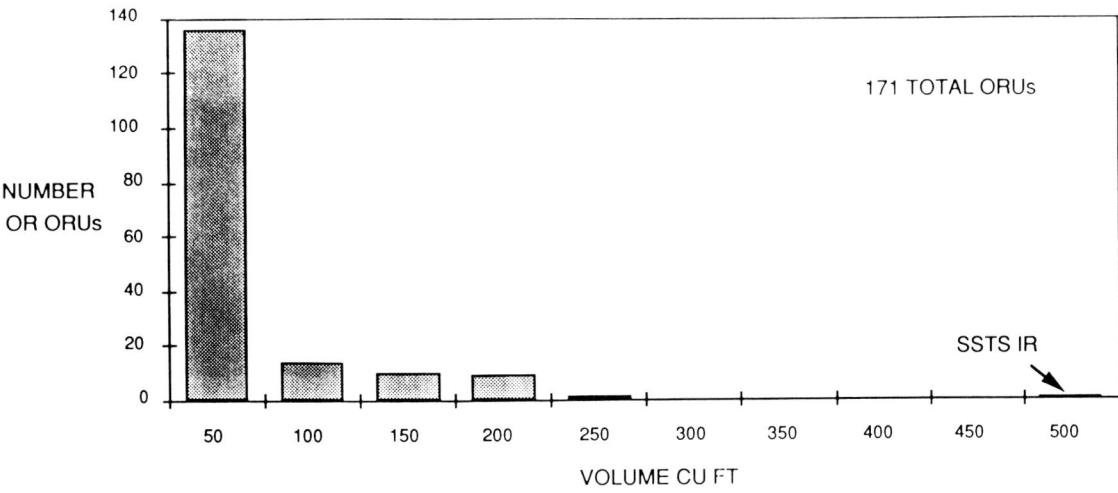

Figure 5.13 ORU volume distribution. *Courtesy of TRW.*

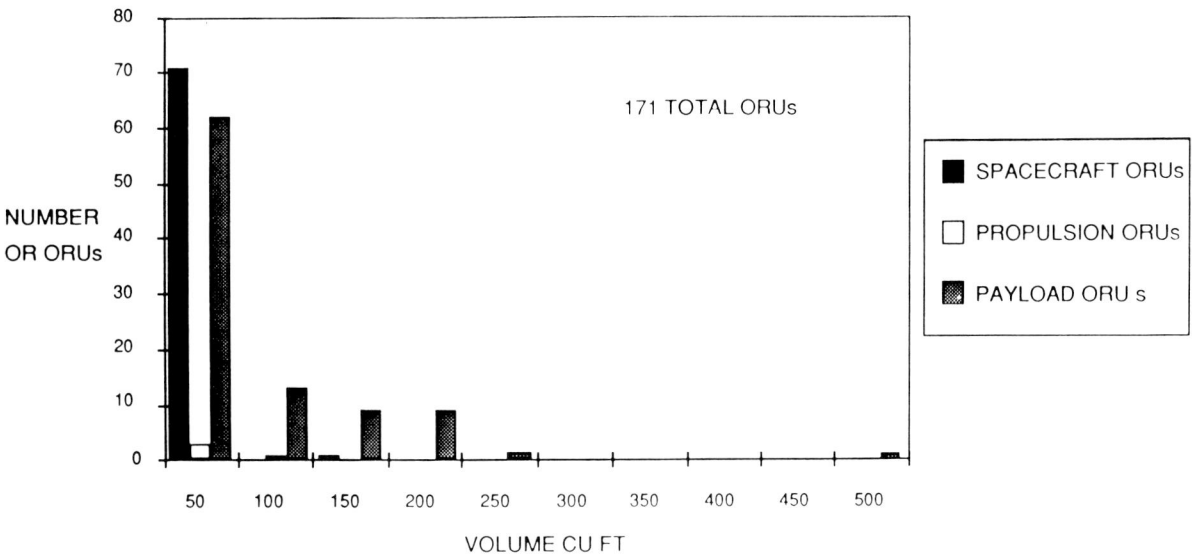

Figure 5.14 ORU volume distribution by type. *Courtesy of TRW.*

- Longest dimension: 254 centimeters (100 inches)
- Second longest dimension: 188 centimeters (75 inches)

Design of a Modular Satellite

A typical 1980s vintage military communications satellite using the NSTS/IUS launch vehicle weighs about 2,268 kg (5,000 lb). A small volume spacecraft bus can support a larger volume payload module and two antenna platforms, Figure 5.19, that stow together as an A-frame for launch.

Modularity is incorporated into the design for the next level of configured item—the subsystem or equipment group. The spacecraft and payload elements are designed as independent, stand-alone modules. The equipment groups are designed to be built on one or more standardized pallets that attach directly to the element module core structure or form part of that structure.

Each module or submodule (equipment group) is an assembly of equipment specified and verified as a complete entity, and has:

- Well-defined electrical, mechanical, data-processing and thermal functional characteristics
- Simple, standardized physical interfaces to other equipment (ideally the electrical interface is bus power and digital signals only, no RF)
- The capability to be manufactured and to undergo

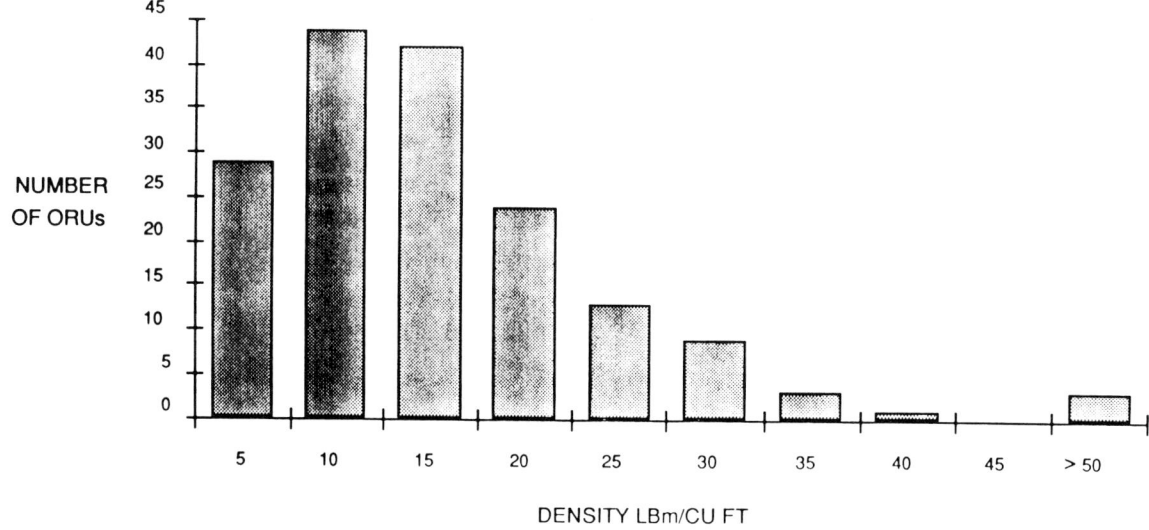

Figure 5.15 ORU density distribution. *Courtesy of TRW.*

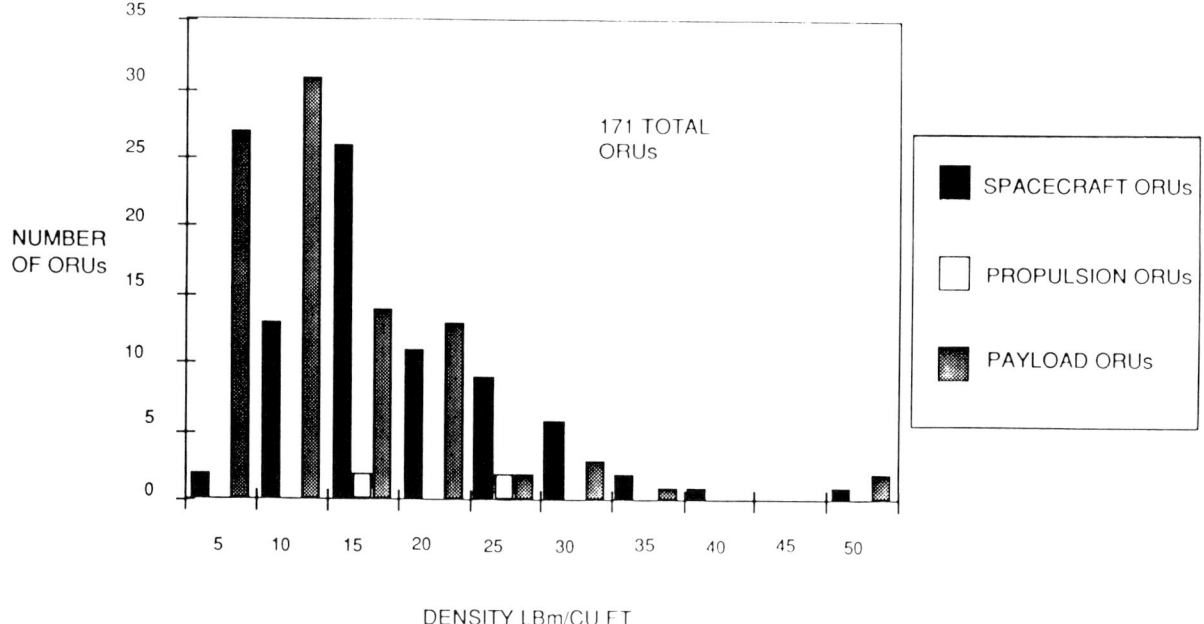

Figure 5.16 ORU density distribution by type. *Courtesy of TRW.*

functional acceptance and qualification testing as an independent entity

- Its own configured item specification, interface control documents (ICD), testable performance requirements, drawings, verification and validation program, and customer buy-off procedures

In addition, it may have a common, stand-alone structure and may also be thermally isolated if these features are weight- and cost-effective. An exploded view of the Figure 5.19 spacecraft is shown on Figure 5.20. It is characterized by a module core structure to which the pallets containing the electronic equipment groups are attached. The propulsion submodule and attitude control system (ACS) equipment submodule fit within the core structure, the two power submodules (containing both batteries and solar array) attach to its sides, and the launch adapter submodule connects the core structure to the IUS interface attach points.

The design configuration allows for each equipment group to be built, tested, and verified before being integrated into the spacecraft bus module. The same concept is used for the payload module.

Various criteria were developed to design modules that

Steps to Spacecraft Design

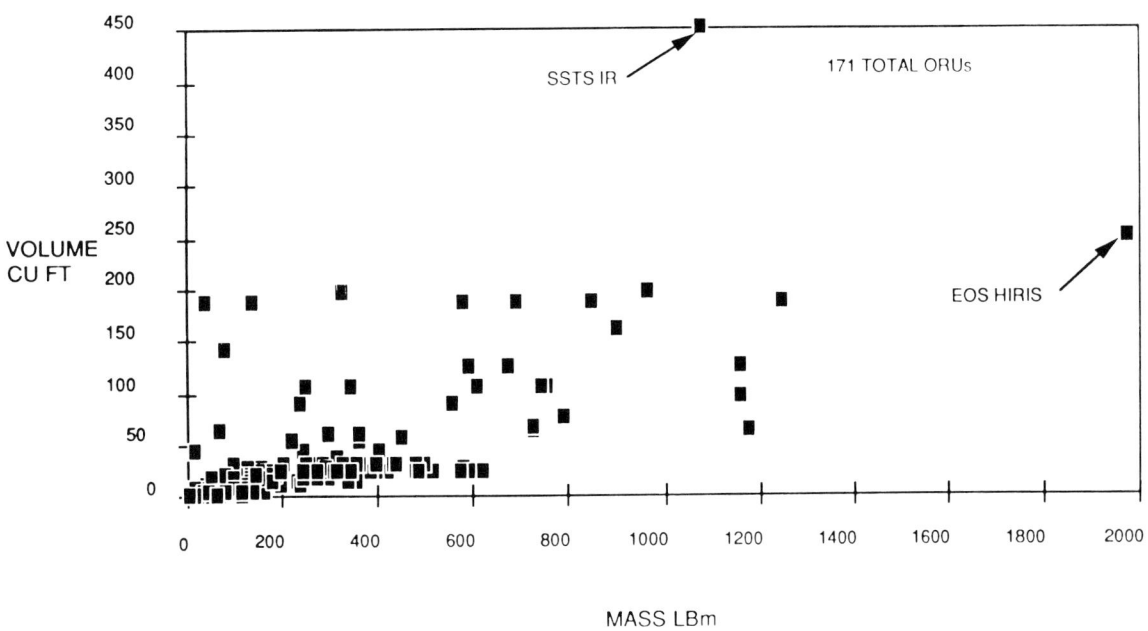

Figure 5.17 ORU volume versus mass. *Courtesy of TRW.*

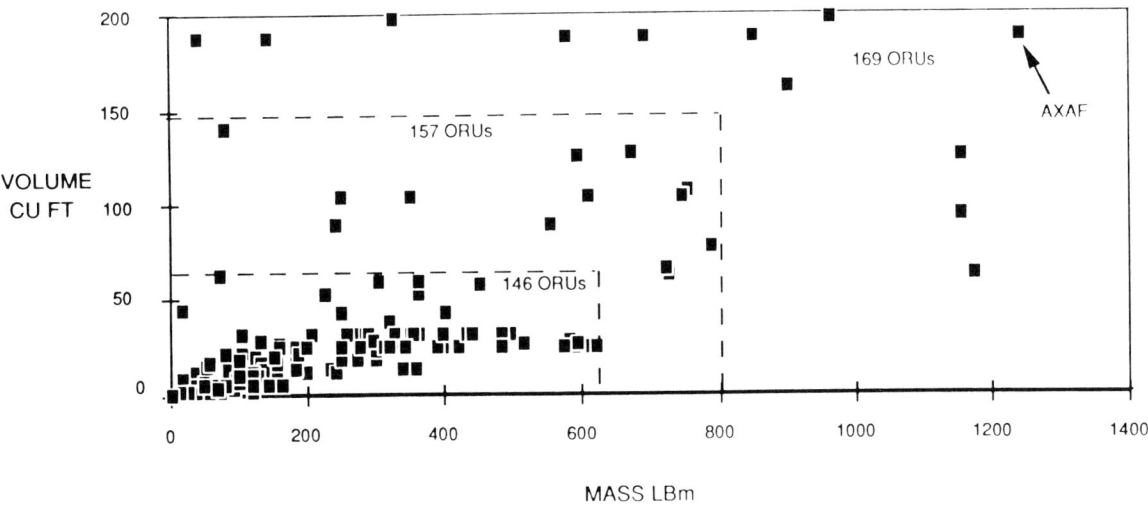

Figure 5.18 ORU volume versus mass without SSTS IR & EOS HIRIS. *Courtesy of TRW.*

minimize assembly, test, and verification time; they are shown in Figure 5.21. Use of these criteria not only saves time but reduces program risk.

The design, integration, and verification process used for this spacecraft module with modular equipment groups is as follows:

- The spacecraft contractor designs the spacecraft module, assigning functions and interfaces to the equipment groups and supporting ground support equipment (GSE) test equipment and software.
- Individual equipment groups and supporting software, GSE, and test equipment are either built by the spacecraft contractor or subcontracted.
- Each equipment group's electronic equipment is mounted on one or two (adjacent) pallets.
- Outsize equipment (e.g., reaction wheel, propellant tank) is mounted in close proximity to its equipment group pallet(s) and verified with it.
- Each equipment group is built, integrated, and verified to its own CI specification, ICDs testable performance requirements, drawings, and procedures.
- Equipment group software is validated at the pallet

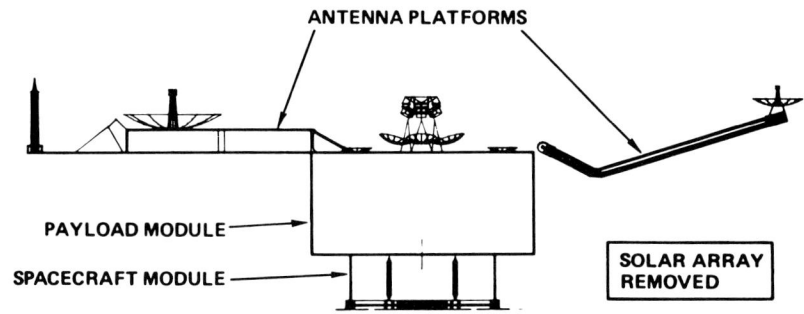

Figure 5.19 Generic communications satellite used for evaluation of modularity. *Courtesy of TRW.*

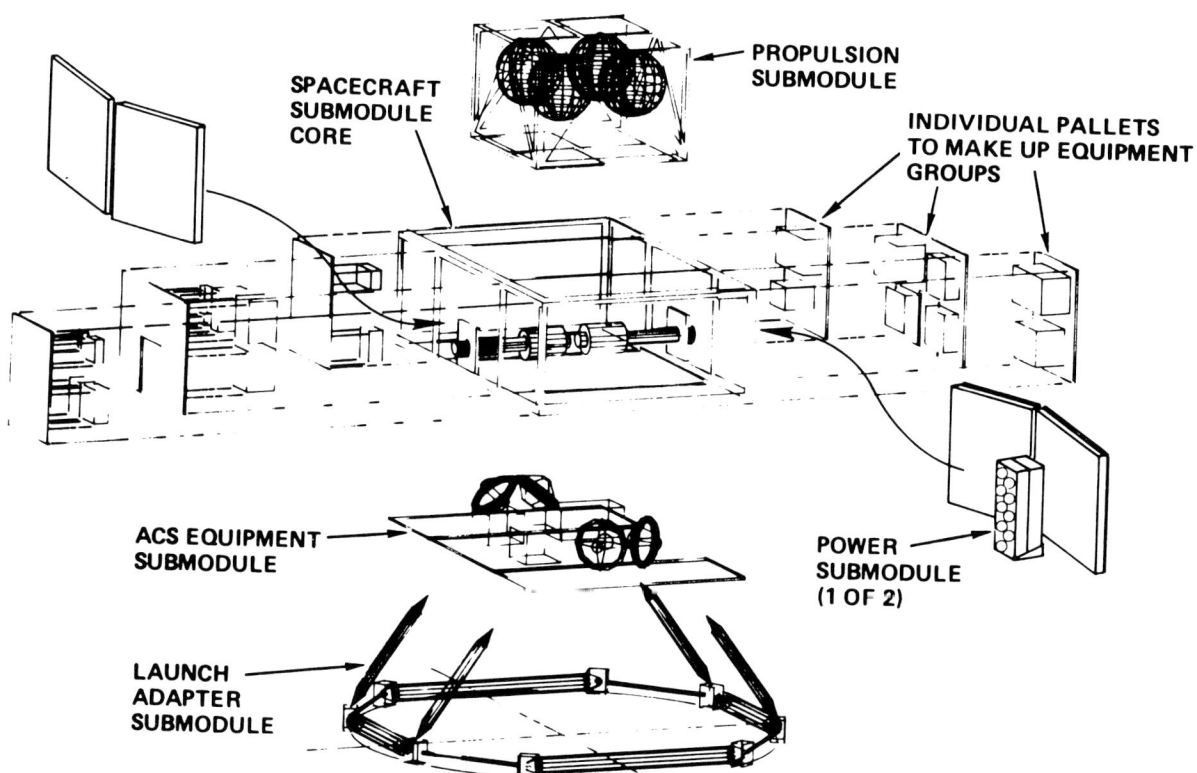

Figure 5.20 Spacecraft bus module used for evaluation. *Courtesy of TRW.*

Steps to Spacecraft Design

- ASSIGN EQUIPMENT WITHIN THE INDIVIDUAL EQUIPMENT GROUP PALLETS AND MODULES IN A WAY TO MAXIMIZE PARALLEL INTEGRATION, ASSEMBLY, AND TEST OPERATIONS. WHERE POSSIBLE:
  - SPLIT GROUPS THAT REQUIRE ESPECIALLY LONG, TIME-CONSUMING ASSEMBLY INTO INDEPENDENT GROUPS
  - COMBINE GROUPS THAT HAVE RELATIVELY SHORT ASSEMBLY TIMES
- PROVIDE STANDARD, EASILY-ACCESSIBLE ELECTRICAL INTERFACES TO EACH MODULE/PALLET
- PROVIDE SIMPLE ACCESS FOR INDIVIDUAL UNITS
- USE DUMMY UNITS/PALLETS AS SHORT-TERM REPLACEMENTS
- PROVIDE GENEROUS MARGINS FOR GROWTH IN EQUIPMENT SIZE, WEIGHT, POWER AND THERMAL DISSIPATION
- PROVIDE VISUAL ACCESS TO CONNECTORS, THRUSTERS, VALVES, AND ALIGNMENT REFERENCES
- MOUNT UNITS WITH HIGH REPLACEMENT RATES IN THE MOST ACCESSIBLE LOCATIONS
- PROVIDE SEPARATE, TESTABLE APPENDAGE MODULES
- VALIDATE SOFTWARE AT BOTH THE PALLET AND THE ELEMENT MODULE LEVELS

Figure 5.21 Pallet and module design criteria to reduce assembly, test, and verification time. *Courtesy of TRW.*

level as well as at the element module and satellite levels.
- Checked-out equipment groups are available as backup during the spacecraft module buildup period.
- The completed spacecraft module is verified by an integrated system test before integration with the payload module to form the satellite.
- Selected environmental tests are performed on the spacecraft module to reduce satellite-level testing.

To demonstrate the advantages of this modular design concept, a typical satellite integration flow is shown in Figure 5.22.

A modular satellite that is structured to facilitate efficient verification provides several ways to save both schedule time and program cost and reduce risk over the classical series integration, verification, and test flow.

First, assembly and test crew man-hours are reduced due to performing operations in parallel and testing at lower assembly levels. Figure 5.23 shows how a modular satellite assembly and test flow reduces assembly and test time and cost over a classical flow, especially as the number of flight satellites increases. The time and costs associated with submodule/pallet build-up and qualification must be included in the total assembly and test savings assessment.

Second, delays due to faulty equipment are greatly reduced because submodules are simply removed and replaced with a checked-out replacement. History indicates that there are 6 to 12 significant equipment or software failures per satellite during assembly and test. After taking account of work-arounds, an average of 7 calendar weeks are lost per satellite. With a modular satellite, the number of in-line failures will be reduced due to detailed checkout at the pallet and submodule levels; also, the time for removal and installation is much less. It is conservatively estimated that the 7 weeks of lost time can be reduced to 3 weeks per satellite.

Third, performance and interface problems can be identified early, before they become satellite schedule-pacing items. The cost of delayed software error detection could double in each phase of software development (requirements, design, code, unit test and integration, acceptance test, and operation). Discussions with spacecraft project office personnel at TRW, Lockheed Missiles and Space Company, and Martin Marietta Corporation indicate that a similar progression occurs for the phases of satellite hardware development (requirements, design, fabrication, module test and integration, satellite integration, satellite test). Verification at the pallet and equipment group levels can reduce the risk that hardware/software errors will be discovered late in the process.

Fourth, delays due to late hardware and software can be reduced by using engineering model and brassboard simulators and modules as placeholders during the early parts of payload/spacecraft module integration. Placeholders allow a significant amount of useful work to be accomplished while waiting for late hardware. Their use, however, does require that the early models or simulators of potentially late hardware and software be sufficiently like flight models to allow them to be useful as placeholders.

The net cost savings offered by modular spacecraft concepts during the satellite assembly and test phase are itemized in Figure 5.24. The impacts of lost time and other program time-related costs show that the net cost savings during the assembly and test phase can be as much as 35 percent for a four-satellite program or 42 percent for eight satellites.

There are other individual schedule and cost factors that may either increase or decrease for a modular satellite concept. These include the following:

- A modular system design requires an early, intensive engineering effort to establish and document modular performance, physical requirements, and interfaces specifications.
- A modular system design will likely have slightly greater cost for simulators and software development.
- The time and cost of submodule/pallet assembly, test, and verification must be added to the data shown in this analysis.
- The time and cost savings achieved by using logic model simulation and development and test-procedure validation for test software are significant and should be added to the modular concept savings.

It appears that these effects do not significantly modify the major payoffs identified during the assembly and test phase. The payoff of modularity is in reducing the net space-

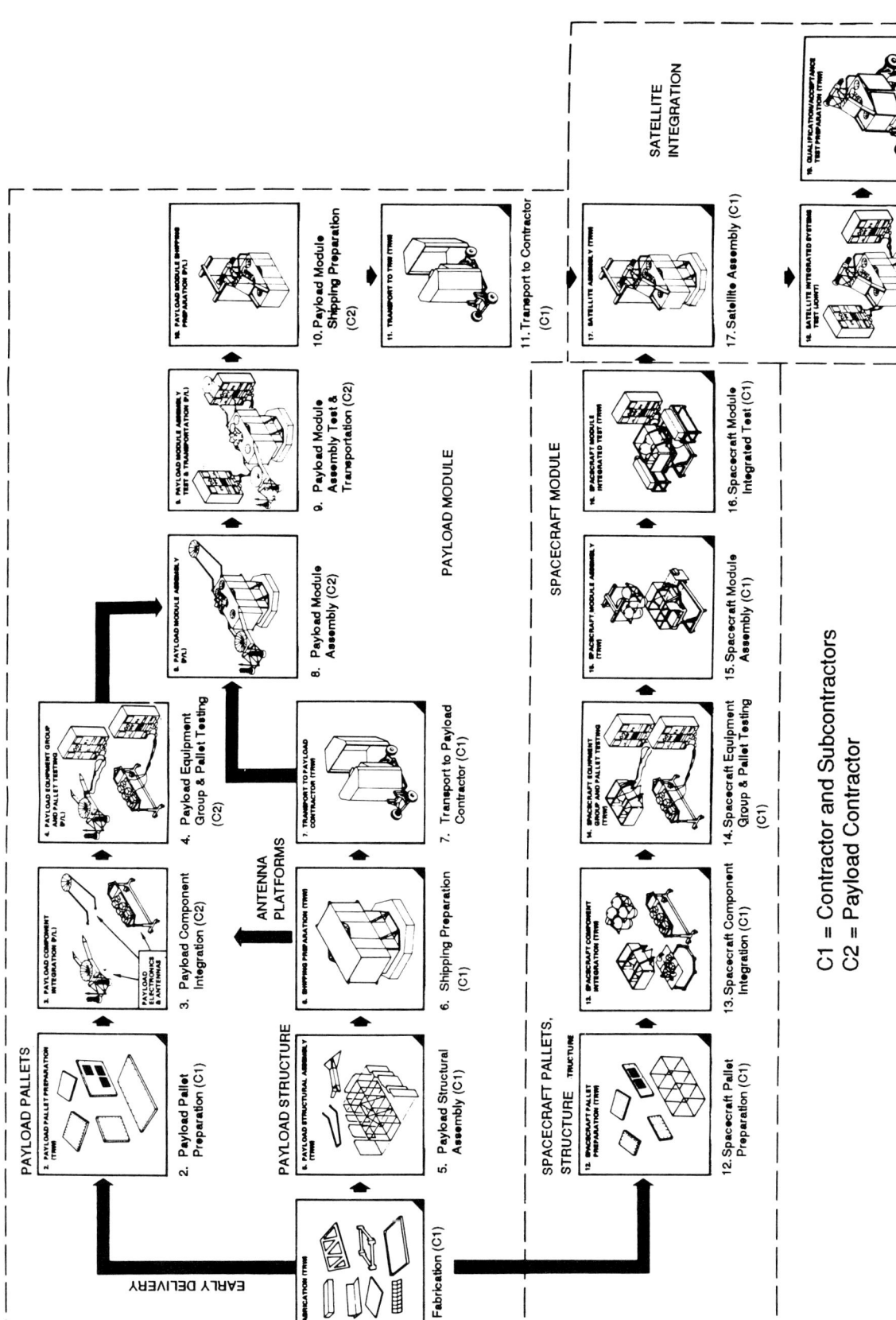

Figure 5.22 Modular satellite integration flow. *Courtesy of TRW.*

Steps to Spacecraft Design

PARAMETER	CLASSICAL FLOW	MODULAR FLOW
CALENDAR TIME FOR A 2-SHIFT, 5-DAY A&T SCHEDULE (WEEKS)	41	21
A&T CREW LABOR PER SATELLITE (MAN-MONTHS)	522	270
A&T CREW LABOR COST PER SATELLITE ($)	4.2 M	2.2 M
GROUND SUPPORT EQUIPMENT COST ($)	13.9 M	19.1 M
TOTAL A&T CREW COST — 4 FLIGHT SATELLITES ($) (SAVINGS)	30.7 M (0)	27.9 M (−9%)
8 FLIGHT SATELLITES ($) (SAVINGS)	47.5 M (0)	36.7 M (−23%)

NOTES:
(1) IUS-CLASS SATELLITE
(2) NO CONTINGENCIES
(3) 1982 DOLLARS

Figure 5.23 Relative satellite assembly and test, crew time, and cost. *Courtesy of TRW.*

		CLASSICAL FLOW		MODULAR FLOW	
NUMBER OF SATELLITES		4	8	4	8
SCHEDULE COMPARISONS					
SCHEDULED A&T TIME	(WEEKS)	2 × 41	4 × 41	2 × 21	4 × 21
AVERAGE LOST TIME DUE TO EQUIPMENT FAILURES	(WEEKS)	25	56	12	24
TOTAL A&T TIME PER PROGRAM	(WEEKS)	110	220	54	108
COST COMPARISONS					
TOTAL SCHEDULED A&T COST	($M)	30.7	47.5	27.9	36.7
A&T CREW COST FOR LOST TIME	($M)	2.8	5.6	1.2	2.4
TOTAL A&T COST PER PROGRAM	($M)	33.5	53.1	29.1	39.1
OTHER PROGRAM TIME-RELATED COSTS	($M)	44.0	88.0	21.6	43.2
TOTAL PROGRAM COST DURING A&T PHASE (SAVINGS)	($M)	77.5 (0)	141.1 (0)	50.7 (−35%)	82.3 (−42%)

Figure 5.24 Relative program schedules and costs during assembly and test phase (satellites assembled two at a time). *Courtesy of TRW.*

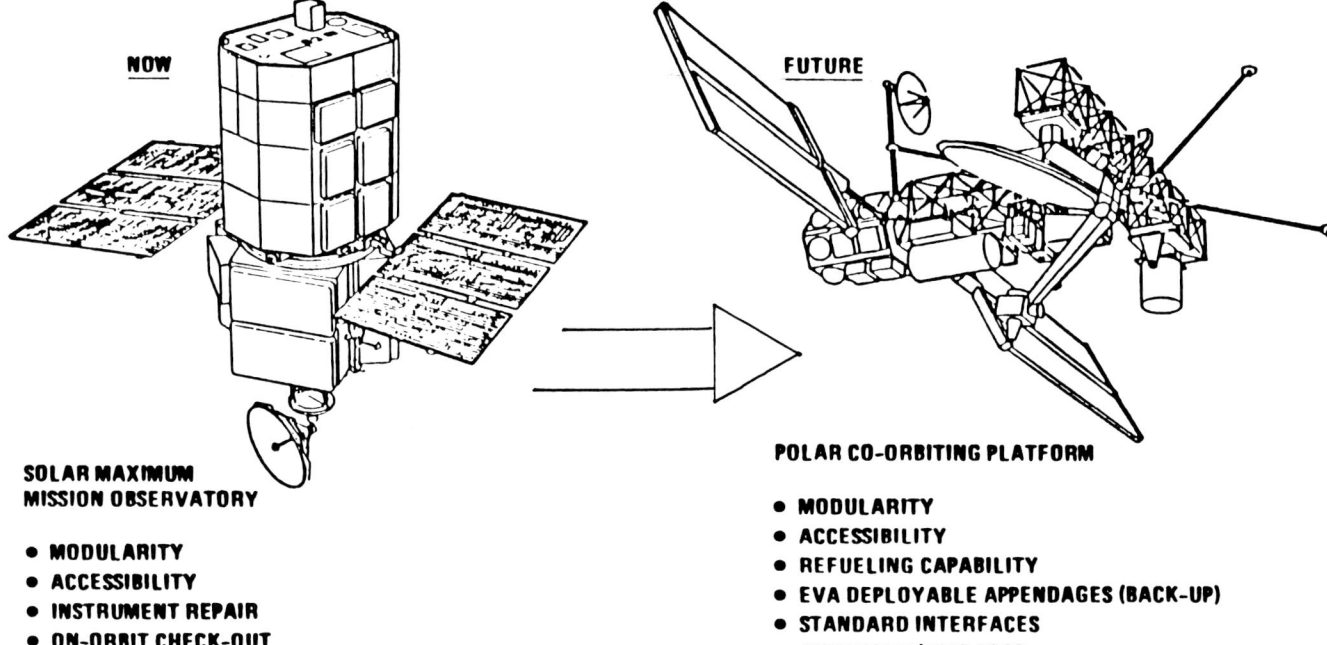

Figure 5.25 Serviceable features will evolve as spacecraft move from the Solar Maximum Mission Observatory era to the decade of the 1990's where spacecraft like the Polar Co-orbiting Platform will include new spacecraft servicing capabilities.

craft life cycle cost, not to simply reduce assembly and test costs.

Standardization

If modularity is the keynote to spacecraft serviceability, standardization is the thrust that makes it economically attractive. Standardization should focus on interuser servicing simulators, training methods, command/control functions and software, transportation modes (launch vehicles and orbit-to-orbit vehicles), NSTS ORU carriers, robotic servicer software, and servicing tools/equipment. But the two standardization drivers that influence on-orbit servicing the most are standardized spacecraft hardware and standard interfaces both internal and external to the spacecraft.

Standardized Spacecraft Hardware

Today's technology for the standardization of replaceable modules on serviceable spacecraft is embodied in the NASA/GSFC Multimission Modular Spacecraft (MMS) components and the Solar MAX Mission (SMM) Observatory.

In the evolution of the MMS, GSFC focused on standard spacecraft development to capture the broadest number of possible remote sensing, Earth and astrophysics observatory missions. Another focus was on making the MMS compatible with the Delta launch vehicle. Thus evolved today's three-sided MMS configuration. The MMS design was validated in 1975 through Earth Observation System Study contracts performed by General Electric, Grumman, and TRW. Then SMM was selected in 1977 to be the first user of the MMS standard bus. SMM and MMS were developed in parallel but in separate GSFC organizations.

The Multimission Modular Spacecraft contained these design guidelines:

- The design reference missions were: Sun, Earth and stellar pointing from low Earth orbit, and Earth pointing from GEO.
- The MMS incorporated only compatible common hardware for all design reference missions but employed full flexibility to integrate mission unique hardware such as antennas, solar arrays, and propulsion systems.
- Launch system compatibility expanded to include the Delta, Atlas, Titan, and NSTS launch systems.
- The MMS was to be serviceable through modular design and retrievable by the NSTS with no single point failure preventing NSTS repair or retrieval.

The design emphasis was on a low cost design suitable for a broad range of missions. A service spacecraft design was a subordinate guideline.

The spacecraft concept on the right side of Figure 5.25 shows an approach to design of future serviceable systems. Note the concept features: a truss structure, accessible modules, refueling capability, deployable appendages, removable payloads, safing, self diagnosis/testing autonomy, and on-

Steps to Spacecraft Design

orbit servicing accomplished interchangeably by either manned operations or robotic servicing.

Candidate ORUs and Components for Standardization

An ARINC Research Corporation's study [5] identified six spacecraft ORU modules and nine subsystems or component modules that were found to offer substantial benefits through fit, form, and function (F3) standardization. The ORU module candidates for standardization are communications and data handling, structure, attitude control, electric power, thermal, and propulsion subsystems. Component candidates for standardization are the battery, power control unit, inertial reference unit, reaction wheel, Earth sensor, Sun sensor, magnetic-torquer, thruster, and fuel and pressurant tanks.

Standard Interfaces

The NASA JSC Space Assembly and Servicing Working Group (SASWG) has undertaken the important task of preparing a set of space assembly and servicing interface standards.

Within the membership of the SASWG, an Interface Standards Committee, chaired by Robert Radtke of Tracor Applied Sciences, was chartered to: (1) identify and prioritize servicer satellite interfaces; (2) propose the same to professional standards organizations, including AIAA, AIA, IEEE, and SAE; (3) aid the preparation and approval of aerospace standards by these organizations in accordance with ANSI requirements; and, finally, (4) to gain approval by the International Standards Organization, ISO. The Interface Standards Committee was further organized according to six functional areas and chairmen were elected as follows:

- Mechanical Interface Standards, Al Thompson, Martin Marietta Space Division
- Electrical Interface Standards, Bob Davis, NASA GSFC
- Fluid Interface Standards, David Ball, Edwards Air Force Base
- Optical Interface Standards, Al Haddad, Lockheed Missiles and Space Company
- Thermal Interface Standards, Otto Ledford, Advanced Technology Incorporated
- Data Communications Interface Standards, Shlomo Dolinsky, NASA JPL

Here are some typical interface items each of above six functional areas are attacking in a concerted effort to bring standards practices to on-orbit servicing:

- Mechanical
  - Multimission Modular Spacecraft, module structure assembly form and fit
  - ORU container interface, including alignment
  - ORU, size and weight
  - Satellite grasping/berthing
  - Flight releasable grapple fixture
  - Robotic end effector exchange system
  - Robotic end effectors
  - Tool interfaces
  - Standardized tools
  - Tethering devices
- Electrical
  - Panel mount and inline connectors
  - Robocon subminiature electrical connectors
  - Low force connector pins
  - Connector mate/demate tools
  - Tool interfaces
  - Cable tie wraps
  - Satellite and servicer power buses
- Fluid
  - Automatic refueling coupling
  - Automatic umbilical connector
  - Robotic fluid coupling
  - Universal refueling interface system
  - Leak detection techniques
  - Tank gauging techniques
- Optical
  - Cameras and mounts
  - Access envelope and viewing angles
  - Lighting
  - Fiber optic connectors and cables
  - ORU status indicator
  - ORU inspection techniques
  - Labeling and color coding, NASA Std 3000 and Fed-Std 595A (combined with automatic vision identification, such as bar code labels)
- Thermal
  - Replaceable thermal insulation panels
  - Insulation thermal resistance
  - Conductive ORU to satellite interfaces
  - Convective ORU to satellite interfaces
  - Test methods with space ratings
- Data Communication
  - Communication control architecture and protocols
  - Data formats
  - Laser communication wavelengths
  - Warning/message signals
  - Servicer data buses
  - Satellite data buses

To assure the broadest scope of thinking gets into the generation and correlation of standard practices, these professional societies and industry associations have been invited to participate in the committee's work:

- Aerospace Industries Association
- American Institute of Aeronautics and Astronautics
- American Society of Mechanical Engineers

- Compressed Gas Association
- Electronic Industries Association
- Institute of Electrical and Electronics Engineers
- National Fluid Power Association
- Optical Society of America
- Robotics Industries Association
- Society of Automotive Engineers

Ultimately, the SASWG hopes that government and industry people concerned with planning, designing, manufacturing/testing, operating, and managing space systems using on-orbit servicing will have enough recognized and accepted standards that serviceable spacecraft and servicing equipment will reach new levels of safety, utilization, and affordability.

Known, standardized interfaces will be essential to allow a smooth transition from Space Shuttle-era servicing to Space Station-era servicing, while continuing to allow for the use of the Shuttle [6]. The spacecraft operator should not be forced to decide, at an early point in spacecraft development, precisely where in space the servicing will take place, due to interface restrictions or unique designs. Nor should the servicer be forced to accommodate a multitude of unique spacecraft requirements. Standardized interfaces will alleviate many of these problems by making design details on either side of the interface independent of the opposite side. The spacecraft designer is thus allowed to optimize the vehicle for its particular mission needs while still taking full advantage of the services offered at a variety of orbital locations. This also reduces the possibility of a single choke point if a particular servicing location becomes inoperative or otherwise unavailable.

Many examples of standardized interfaces exist in daily life [6]. The ability to fill an automobile with gasoline does not depend on which service station the driver enters. Nor does it depend on the type of car being driven. Similar statements can be made for a wide range of pneumatic tires and common household appliances. Thus standard interfaces are not only commonplace, but quite convenient and taken for granted once they are in existence.

It is not difficult to extrapolate this concept to satellite servicing operations. The remote manipulator system end effector and its grapple fixture, the refueling coupling developed for Gamma Ray Observatory, and the modular mission spacecraft/flight support system berthing interface are all examples of servicing interfaces currently available and assumed to be de facto standards by a wide range of users. For this reason, both the servicer and servicee can design future space systems to accommodate these interfaces. This practice has already proved its worth in an operational situation. The Solar Maximum Mission repair operations was salvaged because the bus carried a standard grapple fixture even though the spacecraft was launched more than a year before the Space Shuttle completed its first flight. Applying this same philosophy to future designs will allow a wider range of vehicles to be serviced at a wider range of facilities even as both are being developed.

Previous and current procedures for developing a servicing interface are typically in response to a specific application or need. This does not mean that, once developed, an interface could not be used in different situations. However, spacecraft and subsystem designers may not be aware that a conceptual or flight-qualified interface exists which could be utilized for their application. A consolidated source of information of this type will help accelerate the design process for serviceable spacecraft by reducing the effort spent on interface development. The aforementioned NASA sponsored Space Assembly and Services Working Group is attempting to provide the groundwork and guidelines for this source.

Simplicity

Straightforward, relatively uncomplicated hardware interfaces, operations, crew EVA functions, and robotic procedures should be incorporated into space assembly, maintenance, and servicing missions. An example of hardware interface simplicity is the grouping of electrical connectors so they mate and demate automatically when the ORU is replaced. A simple ORU to spacecraft interface or an ORU to servicer tool interface will probably be more reliable, less costly, and generally lighter than a sophisticated one from the viewpoint of successful ORU exchanges.

Access

Spacecraft designers must give plenty of room for astronaut eye and hand access to ORUs and their fasteners and connectors on EVA servicing missions. Place ORUs on the outside of the spacecraft bus or payload module to increase productivity of EVA activity and to comply with both allocated time constraints for servicing functions and the EVA safety standards. When designing servicing tool interfaces, human factors must be considered. The interface, for example, needs to be designed so that the maximum working envelope of (8 inches) wide by (18 inches) deep is established for the EMU glove. This volume allows a gloved hand to manipulate most hand-operated controls such as latches and knobs. Handles should be conformal or oval, and knobs should have knurled or other nonskid surfaces. When only tool access is required, a (1 inch) minimum clearance should be provided around a fastener or drive stud, for insertion, actuation, and removal of the drive end of the tool. Clearances of at least (3 inches) are needed between a tool handle engaged on a fastener or drive stud and the nearest piece of hardware.

Simulation and Ground Tests

Satellite servicing concepts for spacecraft, servicers, and tools/support equipment should be tested with computer-aided design analysis and three-dimensional real-time solids modeling; then moved to full-scale mockups and neutral buoyancy facilities to verify assembly, maintenance, and repair capabilities. Development simulations, combined with operational time lines, to compliment crew IVA and EVA training, is very useful.

Computer graphics programs can demonstrate clearances, showing where parts may interfere with each other or where astronauts cannot reach. Neutral buoyancy mockups built from mature designs provide higher fidelity simulations and have a positive influence on the final design. Neutral buoyancy simulations during hardware development are extremely valuable precursors to neutral buoyancy crew training activities.

An extended ground test period can uncover surprises while the spacecraft is within easy reach. Procedures can be "debugged" so problems do not appear during orbital verification and possibly endanger the spacecraft. Reverification of the spacecraft can also be simulated, thus shortening the period of orbital reverification after a maintenance mission.

Conduct a final series of maintenance demonstration tests before the spacecraft leaves the factory for launch. These tests should use the flight article and mockups, plus the assigned EVA crew assisted by engineers and technicians directly involved in servicing work. Similar to electronics "burn-in" and debugging, such tests will catch subtle problems that may have slipped through even the best management systems.

Train and test after launch. A high-fidelity electrical simulator, using ORUs where possible, should be built for software development and debugging, operator training, failure simulations, and ORU preflight checks. Simulated failures give ground crews experience in isolating them and developing "work-arounds" that may save the spacecraft in case of unanticipated problems. A mechanical simulator—possibly the neutral buoyancy mockup—is valuable for testing procedures, checking fits, and giving crews the "feel" of the hardware before a mission.

Logistics

Discussed in Chapter 3, logistics includes spares in flight article contracts. One problem may be the assurance of vendor life and supply capability during the life of the space system. An integrated logistics plan must be incorporated into the serviceable spacecraft design from the start. It has to include the servicing strategies, supportability approach and architecture, and the logistics requirements over the expected life of the space system.

Configuration Management

As designs are completed, the spacecraft elements will move into manufacture at different rates and may be subject to many revisions at varying levels of importance. Spacecraft serviceability assurance must keep pace with work on the factory floor and in the contracting offices.

Control the configuration and plan for spares once the design is set and tools are selected. Close monitoring of all changes as a spacecraft design evolves and as the hardware is built will ensure that tools and replacement units are compatible with flight hardware. The spacecraft program's Maintainability Office should be represented on the Configuration Control Board and have signature approval in its proceedings.

The importance of this control can be seen in the space servicing missions to date; the principal difficulties have arisen from differences between "as built" hardware and blueprints for tools and spares. Recognizing that exact documentation of the "as built" design is essential, Hubble Space Telescope managers instituted a stringent configuration control program that included all drawings and documents as well as still photography of the hardware and video interviews with key personnel. As a further precaution, tools to be used in servicing were physically fitted to the bolt and screw heads they must remove.

Use the Maintainability Office as a "Bureau of Standards" that maintains master copies of tool fittings and jigs and manufactures copies for use on the work floor and in training. A master copy of the footprint and bolt hole pattern for each ORU will assure identical fits for spares installed in orbit. Similar approaches can be taken for electrical connectors and other interfaces. Simulators should be built to match the layout and mountings of the flight article, then upgraded to reflect changes as the design evolves. These steps can prevent those rare cases where variations within tolerances keep parts from fitting together.

Retain all test procedures and equipment, particularly unique items, for later use with replacement hardware. Test data should be achieved and accessible for future reference.

Buy spares in the same contract that buys qualification and flight elements to reduce prices and to guarantee availability. Since qualification units must match flight standards, they can be recycled as spares. Development models often can be upgraded for less than half the cost of a new unit, with further savings in lead time. Sustaining engineering to keep spares viable through the life of a spacecraft must be considered when a contractor is selected so there is some assurance that the vendor or his skills and documentation will be available when needed.

Servicing Roles for Space Science (Astrophysics) Missions

From experience on the Skylab Program and the Solar Maximum Repair Mission, we perhaps conceive of satellite servicing as a function performed by an astronaut-mechanic equipped with an EVA tool kit and some special mobility aids and grapple devices [7]. For the Space Station era, how-

ROLE I	MAINTENANCE/SERVICING OF SCIENCE SATELLITES • MULTI-MISSION S/C	REPAIR/REPLACEMENT OF FAILED/DEGRADED PARTS • PLANNED REPLACEMENT • RANDOM REPLACEMENT TO EXTEND USEFUL S/C LIFE
ROLE II	MAINTENANCE/SERVICING OF ORBITAL OBSERVATORIES • HST • AXAF • GRO • SIRTF • SMM	ROUTINE MAINTENANCE, RESUPPLY OF CONSUMABLES, REPAIR BROKEN PARTS, AND INSTRUMENTS UPGRADE • PLANNED SERVICING • UNPLANNED SERVICING TO EXTEND USEFUL S/C LIFE AND IMPROVE DATA ACQUIRED
ROLE III	MAINTENANCE/SERVICING OF MODEST-SCALE MISSIONS • EXPLORER CLASS • PROTEUS CLASS	FLEET OF STANDARD S/C BUSES, EACH WITH 10-15 YEARS LIFE WITH SERVICING, • PLANNED PAYLOAD CHANGEOUT TO PERFORM DIFFERENT RESEARCH AGENDA
ROLE IV	DELIVERY AND RETRIEVAL OF SATELLITES • XTE • HETE • ALEGRE	SPACE BASED, REUSABLE, LOW AND HIGH ENERGY TRANSFER VEHICLES (LIKE OMV) TO EXTEND SHUTTLE • DELIVER TO EFFECTIVE ORBITS TO ENHANCE SCIENCE MEASUREMENTS

Figure 5.26 Servicing roles for space science astrophysics missions.

ever, this "Mr. Goodwrench" concept is too simplistic. Four major servicing roles for future astrophysics missions are contemplated in Figure 5.26. The Space Station Freedom can be the nexus for all of these functions.

The most readily apparent servicing function is the repair or replacement of failed or degraded parts. Failures at the system, subsystem module, component, or piece part level jeopardize every mission. The best design in the world does not guarantee against random failures or limited life cycle parts in today's complex instrumentation. Some failures and wear-outs are inevitable, so the challenges facing us as we plan astrophysics missions are to identify repairable and/or replaceable elements, develop a servicing philosophy, and design our hardware appropriately to implement that philosophy.

A second servicing role is the maintenance of the large and expensive orbital observatories with their expected operational lifetimes of 10 to 20 years or longer. This spacecraft class includes the Hubble Space Telescope (HST), the Advanced X-Ray Astrophysics Facility (AXAF), the Gamma Ray Observatory (GRO), and the Shuttle Infrared Telescope Facility (SIRTF). There is compelling justification for servicing these observatories on the basis of our large investment in their development and because of the scientific necessity for extended observation periods.

Another innovative servicing role is the support of modest-scale missions in the GSFC NASA Explorer or Proteus class space systems. A fleet of standardized free-flyer buses is envisioned, capable of carrying several different payloads with each bus having a lifetime of 10 to 15 years. The reusable bus would consist of standardized serviceable parts and a standardized instrument interface. An instrument set mated to the bus would operate for 2 to 3 years and then be removed, in orbit, for replacement by a new payload of instruments with perhaps an entirely different research agenda.

A fourth significant servicing role is the delivery and retrieval of satellites. The use of space-based, low and high energy transfer vehicles to deliver newly launched spacecraft from the Space Station to operational orbit will extend the cost benefits provided by the Shuttle as a delivery vehicle. The use of the Space Station as a transportation node will be further enhanced by efficient retrieval of malfunctioning or obsolescent spacecraft and transport to the Space Station for either maintenance, repair, or retrofit. An additional servicing benefit is the cost-effective reboost of functional spacecraft into operationally more effective orbits by reusable transfer vehicles, based on the Space Station, rather than by dedicated Shuttle missions.

As the new servicing technology evolves, so will our philosophy of mission planning. Instead of designing a spacecraft to accommodate preselected scientific objectives, we will be able to tailor science to the flexible capabilities of multimission spacecraft. We will also be able to improve our capabilities to meet future mission requirements by upgrading these serviceable spacecraft. Using the same bus, with no redesign, to do different kinds of missions, such as Sun pointing, stellar pointing, Earth pointing from low-Earth orbit, and Earth pointing from geosynchronous orbit is cost-effective. A versatile, serviceable spacecraft offers many mission opportunities. Standardization of certain spacecraft elements and designs need not limit our imaginative planning for astrophysics research in space.

Conclusion

The thrust of all servicing functions—repair, maintenance, refueling, product improvement, mission support, and orbit transfer—is the same: to use and maintain our capital assets in space. Satellite servicing is the key to gaining the most value from these facilities. Rather than accept failures that convert sophisticated spacecraft into expensive junk, we must adapt and repair our assets. Further, the focus will be on space servicing operations with ground return as a last resort. This will help to avoid subjecting the flight equipment to the hostile environment of a round trip back to space, avoid nonessential repairs, and minimize "marching armies" on the ground.

References

1. *Space Assembly, Maintenance and Servicing Study (SAMS)* Final Report, Volume I, Executive Summary, TRW No. SAMSS-196, Volume II, System Analysis, TRW No. SAMSS-195, Volume III, Design Concepts, TRW No. SAMSS-197, Volume IV, Concept Development Plan, TRW No. SAMSS-198, Volume V, Neutlral Buoyancy Simulation, TRW No. SAMSS 199. TRW, Redondo Beach, CA, July 6, 1988.

    and

    Space Assembly, Maintenance, and Servicing Study (SAMS) Final Report, Volume I, Executive Summary, Volume II, System Analysis, Volume III, Design Concepts, Volume IV, Concept Development Plan, Volume V, Simulation Report. Lockheed Missiles and Space Company, Inc., Sunnyvale, CA, July 6, 1988.

2. NASA. *Satellite Servicing, A NASA Report to Congress*, NASA Office of Space Flight, Washington, DC, March 1, 1988.

3. Chappelear, D. N., and T. R. Danielson. *Space Assembly, Maintenance and Servicing Design Concepts Handbook*. Advanced Systems, Astronautics Division, Lockheed Missiles and Space Company, Inc., July 1, 1988.

4. Mitchler, Lawrence. *Mass and Volume Requirements for Remote ORU Exchange and Fluid Resupply of Satellites*. Space Assembly and Services Working Group, Meeting 25, November 7, 1990.

5. FEDSIM. *Spacecraft Partitioning and Interface Standardization*. FEDSIM, Report No. 85603-02, USAF, Washington, DC, February 1987.

6. Rysavy, Gordon, and Stephen J. Hoffman. *The Transition from Space Shuttle to Space Station On-Orbit Servicing*. Presented at Session II, Satellite Servicing Studies and Planning, GSFC Satellite Servicing Workshop III, June 11, 1987.

7. Waltz, Donald M. *Design and Support of Serviceable Spacecraft*. Presented at Session III, Serviceable Spacecraft Designs, GSFC Satellite Servicing Workshop III, June 9–11, 1987.

Chapter 6

Serviceable Spacecraft

The design of serviceable spacecraft is now acquiring maturity and sophistication. Goddard Space Flight Center led the way for NASA with their Multimission Modular Spacecraft in the late 1970s. Now all NASA centers can claim some knowledge of, and a stake in, space systems that include on-orbit assembly, maintenance, or servicing in the implementation of their mission.

The military, especially the Air Force and the Strategic Defense System (SDS) Organization, has studied the application of on-orbit servicing for their space assets. Top-down technology and economic assessments of USAF and SDS operational requirements have determined the servicing capabilities needed to establish a future Space Asset Support System (SASS). The military is particularly interested in applying multi-component ORU changeout designs as a means of saving weight and reducing vehicle complexity and for using in-orbit refueling as a way to extend the life of their space systems. Their servicing scenarios options are similar to NASA's: telemetry-based repair and workaround, abandon and replace, on-orbit nodal regression, and on-orbit ring-dedicated. Their first priority will be to use the telemetry-based repair option. The abandon and replace option can always be considered as a last resort. Therefore, it is the last two options that address the operational aspects associated with on-orbit support and design of their spacecraft.

This chapter presents eight space systems where on-orbit servicing has been introduced at some point in the design process into the spacecraft's configuration.

1. Multimission Modular Spacecraft (MMS)
2. Hubble Space Telescope (HST)
3. Advanced X-Ray Astrophysics Facility (AXAF)
4. Gamma Ray Observatory (GRO)
5. Space Infrared Telescope Facility (SIRTF)
6. Earth Observing System (EOS)
7. Zenith Star Program
8. Orbital Maneuvering System (OMV)

Those aerospace engineers embarking on the design layout of a spacecraft considered to benefit from on-orbit repair, maintenance, and servicing will want to study these systems in detail before starting their configuration synthesis.

Details vary between these serviceable spacecraft but the overall philosophy of extending the mission lifetime and enhancing the effectiveness of the program is the common goal prompting on-orbit servicing to be incorporated into the spacecraft concept. In all cases an economic analysis favored orbital-based spacecraft servicing versus total replacement of a failed vehicle or ground-based servicing in the life cycle cost of the program.

These eight spacecraft represent the latest in spacecraft technology as applied in a practical manner to serviceable spacecraft design. Recent NASA studies and 1992–94 budget considerations may result in changes to the orbital operations of the AXAF, SIRTF, and EOS. These changes could impact the spacecraft design so as to preclude on-orbit servicing.

Multimission Modular Spacecraft

The Multimission Modular Spacecraft (MMS) was created by Frank Cepollina and others at the NASA Goddard Space Flight Center.

The MMS is a standard spacecraft, Figure 6.1 that can accommodate the requirements of a wide variety of missions. In Figure 6.1, it is shown as part of the Solar Maximum Mission Observatory. While most MMS applications reference the Delta and STS launch systems, no design feature of the MMS precludes its being utilized on either the Atlas-Agena with a 2.4 meters (8-foot) diameter fairing or the Titan III. The mechanical and environmental specifications address Delta and Shuttle launch systems since they effectively constitute the boundary or envelope requirements for the MMS design for all applicable launch systems. Capable of being used at orbital altitudes from near Earth to geosynchronous, it supports U.S. government, foreign

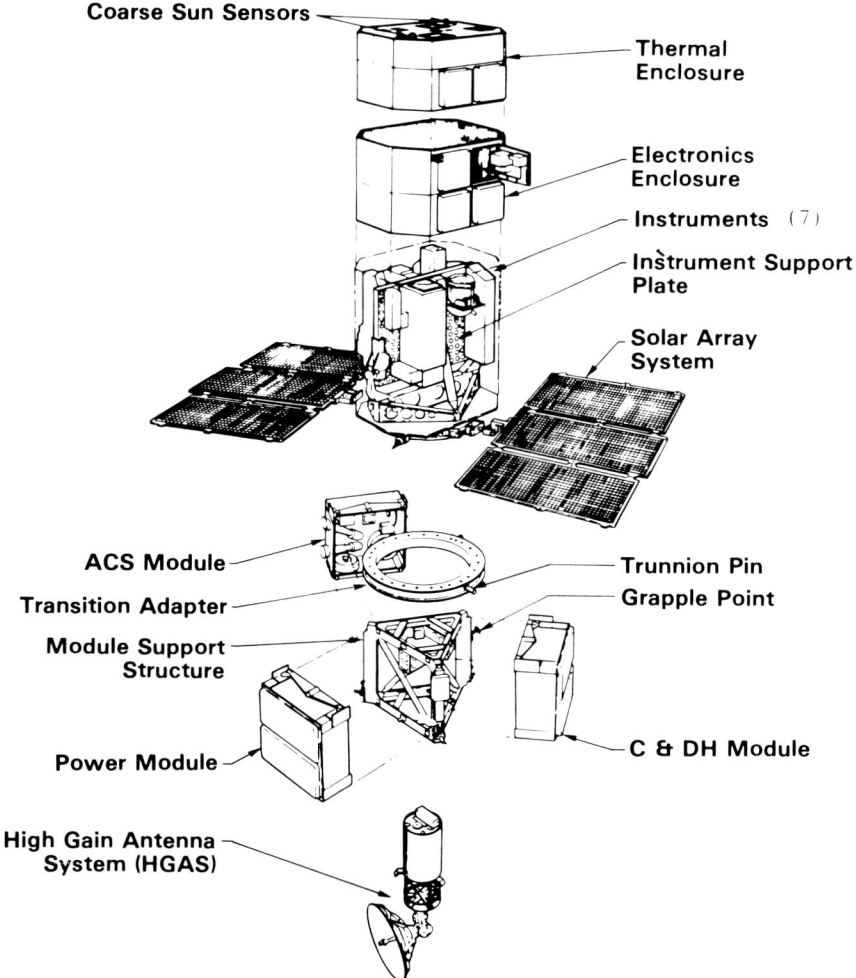

Figure 6.1 Solar Max Mission observatory exploded view showing modular design. *Courtesy of NASA.*

government, and commercial space programs. It presents a standardized telemetry and command interface both for integration and test, and for on-orbit operations with the Ground Space Tracking and Data Network (GSTDN) and the Tracking and Data Relay Satellite System (TDRSS).

The MMS spacecraft was developed as a multipurpose "spacebus" for use on a variety of low Earth orbit missions. It utilizes standardized subsystem modules to provide basic spacecraft functions such as power, attitude control, communications, data handling, and propulsion. In developing the MMS concept, the flexibility for accommodating a variety of mission operational modes and payload types was of prime importance. Various payload configurations, size, weight, power, telemetry, command, computer memory size, look angles, thermal, and other considerations had to be considered in order to maintain the feasibility of the concept.

One of the most significant features of the MMS is its modular design which facilitates on-orbit repair or refurbishment. Modules are structurally mounted by the use of only two retention bolts and all electrical connections are through self-aligning, blind mate connectors. Application of the modular principle to the instrument module on future missions will permit ready on-orbit repair or upgrading of instruments as well.

The MMS incorporates a high level of standardization, but retains a very high degree of versatility. This is possible because standard options are available to perform a wide variety of functions for a wide range of mission requirements. Such things as battery capacity, telemetry rates and format, and spacecraft orientation are selectable as standard options for MMS. Additional flexibility for communication and data handling functions is provided by allotting 0.56 square meter (6 square feet) of area in the Communications and Data Handling (C&DH) module to mount mission unique equipment. The interface for such items as tape recorders and additional computer memory has been provided in the baseline MMS.

In the 1960s and early 1970s, GSFC was responsible for numerous observatory programs. The OAO and the NIMBUS are examples. Anticipating larger observatories in the

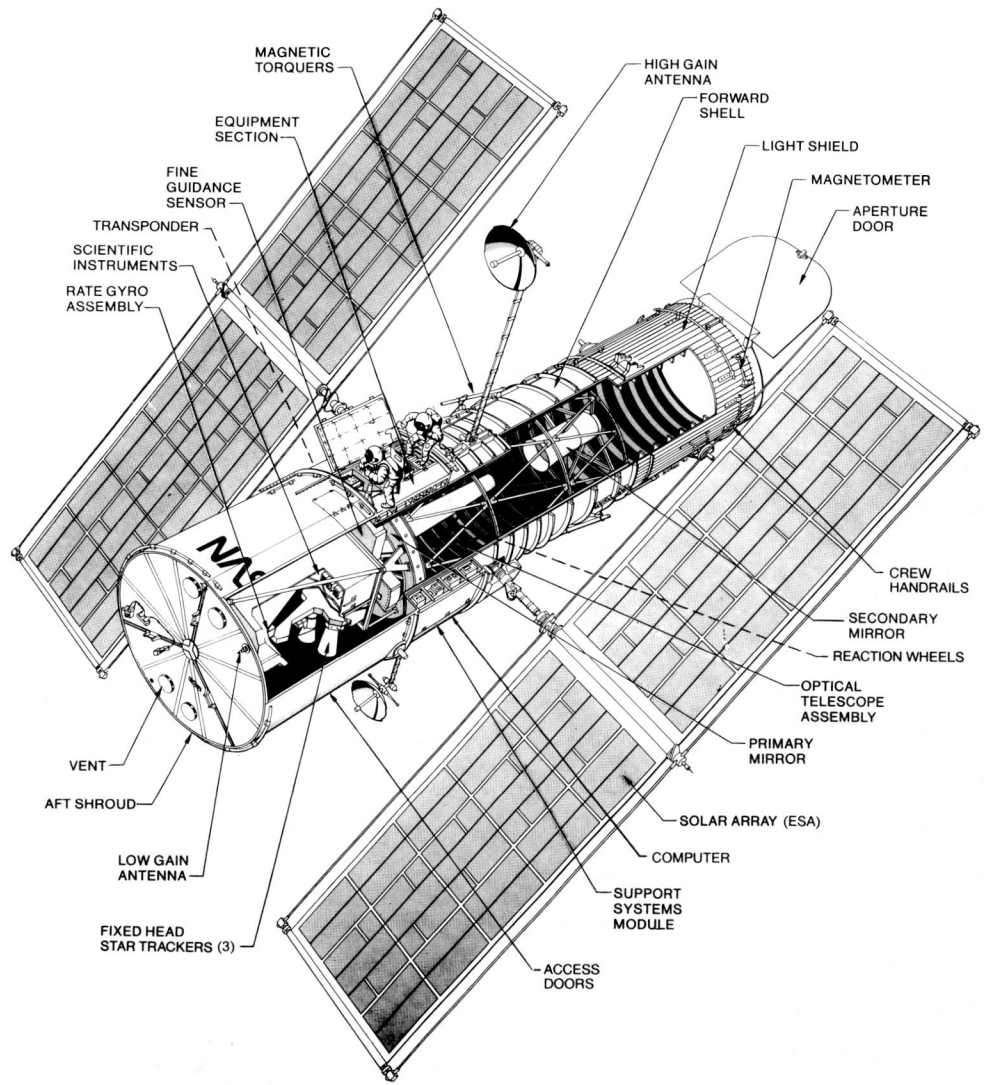

Figure 6.2 Hubble Space Telescope showing replaceable units. Note the two EVA astronauts servicing the equipment section of the HST. *Courtesy of NASA.*

future, GSFC reasoned that there was a need for a flexible configuration spacecraft bus to support large payloads [1]. With STS program go-ahead, it was decided that the bus would be serviceable by the Shuttle Orbiter. GSFC then focused on standard spacecraft development: (1) for remote sensing, Earth and astrophysics observatory missions, (2) to capture the broadest number of possible missions, and (3) to be Delta launch vehicle compatible. Thus evolved today's three-sided MMS architecture configuration, Figure 6.1.

The MMS design was validated in 1975 by phase B study contracts awarded to GE, Grumman, and TRW. GSFC won the NASA intercenter competition for low cost standard spacecraft development from the NASA Headquarters Low Cost Systems Office. The Solar Maximum Mission (SMM) Observatory was selected in 1976 to be first user of the MMS standard bus. SMM and MMS were developed in parallel but in separate project organizations at GSFC.

The MMS was first launched in February 1980, as part of the Solar Maximum Mission Observatory. The second and third MMSs were flown on Landsat 4 and 5, respectively. The MMS system includes a flight support system (FSS) for the MMS which provides a platform capable of launch, on-orbit servicing, and retrieval by the NSTS. The first on-orbit satellite repair was the Solar Max Repair Mission, which took place in April of 1984. Other missions utilizing the servicing capabilities of the MMS and the FSS are planned through the year 2000.

The MMS—with its modular, serviceable subsystems and its successful application as the spacecraft bus for the SMM, Landsats 4 and 5, and the Upper Atmosphere Research Satellite (UARS)—has provided spacecraft design and mission planners with valuable information. MMS lessons learned include the following:

1. Standard interface systems engineering is key. Technology is transparent.
2. Parts become obsolete, vendors go, but functional interfaces can be maintained.
3. Configuration control is an absolute must.
4. Prepare for the unexpected.
5. Be prepared for failures during tests. Have spares on hand for critical parts.
6. Engineers do not need to design totally for repair, just anticipate repair. The SMM instrument module was repaired because it was accessible. The GRO did neutral buoyancy water tank tests to assess access to modules.
7. LEO space is a benign environment. The returned to Earth MSS attitude control system gyros today look as good as when qualified.

Astrophysics Observatories

NASA's next generation of astrophysics observatories will be the primary facilities for conducting major astronomical research programs in the next two decades. These facilities—the Hubble Space Telescope, Advanced X-Ray Astrophysics Facility, Gamma Ray Observatory, and Space Infrared Telescope Facility—will be designed for 10- to 15-year lifetimes. Reliability analyses show on-orbit maintenance and refurbishment will be required to avoid unacceptable performance degradation. These activities will include:

- Replenishment of consumables
- Replacement of worn-out and failed equipment
- Installation of next-generation science instruments

Hubble Space Telescope

Named for the modern astronomer Edwin P. Hubble, this telescope is the most sophisticated scientific facility ever conceived for observing the Universe [2]. Operating above the obscuring atmosphere, it will enable astronomers to observe a volume of space 350 times larger than previously possible and to view objects 14 billion light years away, seven times more distant than objects viewed through the best observatories on Earth.

The Hubble Space Telescope (HST), shown in Figure 6.2, is the first observatory designed for extensive maintenance and refurbishment in orbit. It weighs about 10,886 kg (24,000 lb) and has a length of 13 meters (44 feet) and a diameter of 4.27 meters (14 feet). While a few other U.S. spacecraft have been retrieved or repaired by astronauts, none is so thoroughly designed for orbital servicing as this one. Although Skylab science instruments were serviced through planned EVA tasks, ingenuity more than design accounted for the impromptu major repair of Skylab in 1973, and only part of the Solar Maximum Mission spacecraft repaired successfully in 1984 was originally designed for servicing. A number of other spacecraft have been lost because equipment failed after launch and no provisions were made for retrieval and orbital servicing.

The Hubble Space Telescope is a major scientific resource with a price tag of about $1.5 billion. Orbital servicing is a desirable and practical means to return that investment by ensuring a long and successful period of operation.

The design of the HST reflects a commitment to on-orbit servicing based on new opportunities, prior experience, and compelling scientific, technical, and economic reasons. It is expected to revolutionize modern astronomy. For maximum benefit, however, the telescope must remain in operation for many years. If total failure occurs, replacement of the observatory is not affordable, scientifically or economically, nor is it practical to design the spacecraft with sufficient backup systems to protect against all contingencies.

Despite multiple redundancy in critical spacecraft components and a number of safe modes that can be initiated automatically or by ground command to protect the observatory, manned servicing visits are occasionally required to keep the observatory operational. The telescope will remain in operation for at least 15 years, considerably longer than the operational life of some of its components, such as batteries. Therefore, periodic servicing is required to keep it functioning properly and the data flowing for as long as possible. These servicing missions are scheduled at approximately 3-year intervals.

Furthermore, servicing is required to correct malfunctions that could jeopardize the telescope. The scientific loss incurred by an inoperative observatory is incalculable. Orbital servicing is an efficient method for correcting problems and restoring the telescope to fully operational status.

Finally, technology is advancing rapidly, and improved detectors will become available during the telescope's lifetime. The gain in knowledge merits upgrading the observatory with new technology. Therefore, orbital replacement of science instruments with new or different models is planned.

Originally, planners assumed that the Hubble Space Telescope would be retrieved and returned to the ground for major maintenance and refurbishment every 5 years. However, this plan was set aside on technical and economic grounds. Servicing on the ground could require as much as 2 1/2 years, a period without any new scientific data, and an expensive clean room and support facility with a large engineering staff. Servicing in space can be accomplished within a week, a brief interruption of normal operations, and without additional facilities and staff. The risk of damaging or contaminating the highly sensitive telescope during return and servicing on the ground is significant, and the costs of orbital servicing compare favorably with those for return, refurbishment, and relaunch. Therefore, all servicing of the Hubble Space Telescope will occur in space at either the Space Shuttle or the Space Station Freedom, when the Station has the capability and facilities to accommodate servicing of space systems.

Maintenance and refurbishment program costs are justified by an extended spacecraft operational life for optimal scientific return. In addition to designing the spacecraft for orbital servicing, the program must provide and maintain an inventory of space parts, detailed records of the actual design of all flight articles, and a ground support network of key facilities and personnel. These investments, however, serve as the insurance policy for a long-lived scientific mission. They also permit the cost of replacement hardware to be controlled.

The maintenance and refurbishment plan for the HST provides for the various conditions that warrant orbital servicing. The plan also provides for sustained servicing capabilities throughout the full 15-year life of the telescope.

HST Initial Flight Problems

Successful deployment of the Hubble Space Telescope on April 25, 1990, from the 31st NSTS flight established the observatory at 613 km (331 nmi) miles above the Earth. A series of problems ensued, principally:

- The solar arrays did not fully extend and lock.
- The telescope aperture door closed accidentally due to an error by a ground controller.
- The motion of the number two high gain antenna, used to communicate with the TDRSS, was impeded by a wire that moved out of place.
- Communications from Goddard ground control to HST via a 500 bit/sec rate failed.
- Two of the four primary attitude control gyros dropped off-line.

It took NASA and contractor ground operations control and engineering people just one month (until May 1990) to correct all of the above problems through telemetric, electronics, and command work-around transmissions between HST and the ground stations. This demonstrated one form of on-orbit servicing—namely by software, remote command/control, and communications.

But the biggest problem with the HST came to light during the first days of the Phase I Orbital Verification Plan when NASA ground controllers embarked on the calibration and focusing of the optical system. A spherical aberration in the 240 centimeter (94.5 inch) main mirror was discovered which is producing distorted images in many of the astronomy measurements. However, the images are still better than those from ground-based telescopes, despite the mirror flaw.

Present NASA thinking is to replace, by EVA, in a few years (estimate is late 1993 or early 1994) Hubble's main camera, the JPL wide field/planetary camera (WF/PC), with one compensated for the spherical aberration. Other changes are contemplated on the same repair/replace mission. In fact, repairs being planned for the HST will require astronauts to perform more EVA activities than yet attempted on one Space Shuttle mission [2]. In a three EVA sequence under consideration, the astronaut repair crew would:

1. On EVA one, replace the HST's solar arrays in an attempt to eliminate slight vibrations of the telescope. Jitters have been a persistent problem that threatens to make long-duration astronomical observations impossible.
2. On EVA two, replace the wide field/planetary camera, as mentioned above. The WF/PC is one of the spacecraft's five main science instruments.
3. On EVA three, one or more of the other four instruments might be modified or removed. If a major instrument is removed, HST's mission will be ended to make room for an assembly of corrective optics called COSTAR, shorthand for corrective Space Telescope axial replacement.

The other major HST astronomy instruments are the high speed photomaster, ESA's faint object camera, the faint object spectrograph, and GSFC's high resolution spectrograph.

Meanwhile, despite the problems, the NASA/ESA Space Telescope continues to produce valuable pictures and data. The U.S. space agency hopes a slowly growing record of scientific accomplishment, when coupled with a convincing plan to fix the spacecraft, will allow the program to cut through the wave of bad publicity that began in mid-1990 with the discovery of the gross optical flaw.

A panel of astronomers, opticians, and engineers convened by the Space Telescope Science Institute compiled a list of two dozen possible corrective measures for the Hubble, and NASA is analyzing technical benefits and risks, along with costs. The goal is to allow NASA management to decide on a mission plan by late 1992. Fixes to Hubble that have been ruled out include anything that would unduly risk disturbing the mirrors or contaminating the optics. Reboosting the telescope in orbit, a step that will have to be taken periodically throughout the life of the observatory, probably will not be necessary by the time of the first repair mission (late 1993). A likely date for a reboost is 1996. However, some additional tasks are likely to be added as the estimated late 1993 repair mission approaches. For example, if a tape recorder fails, the EVA astronauts would be called on to replace it, too.

Despite the demanding work load anticipated, astronauts, Shuttle flight directors, and top NASA managers are said to be enthusiastic about the prospects for the mission. NASA managers want to perform three EVAs on a single Shuttle mission to prepare for the EVA-intensive missions that will be required to assemble the Space Station.

Servicing Missions

Several kinds of servicing missions are planned to keep the Hubble Space Telescope operating properly for an extended lifetime. Some are scheduled, and others are anticipated responses to contingencies that may never occur.

Scheduled missions are those required for routine maintenance of the observatory. They will occur at approximately 3-year intervals to replace components having limited lifetimes, such as batteries, fine guidance sensors, and solar arrays. The specific tasks performed on each mission will vary depending on the status of the observatory's components.

Unscheduled missions are those required to provide a fast response to emergencies or malfunctions that are not part of the normal degradation of the observatory, such as loss of electrical power, command, or telemetry capabilities. System failures that jeopardize continued operation of the telescope or data qualify may prompt a decision to launch a repair mission between the scheduled 3-year service dates, for example, if malfunctions had occurred during the initial months after deployment or if the telescope's orbit later decays too much and a reboost is needed.

If a failure is critical, an emergency repair mission will be launched as soon as possible; if the failure is serious but not catastrophic, repair will be scheduled within a given period, probably 6 to 9 months. Repair of noncritical failures may be delayed until the next regularly scheduled servicing mission. If a contingency servicing mission is required, the launch date will be scheduled by Shuttle flight operations at Johnson Space Center and Kennedy Space Center. The criticality of the telescope repairs will determine whether other Shuttle flights must be delayed or whether the servicing mission can wait for the next available launch date.

Plans and provisions have been made for all probable maintenance tasks, and a cadre of trained EVA astronauts are available for crew selection. Mission-unique training would occur in the weeks prior to servicing missions. In addition, trained ground support and flight operations support personnel are available for duty in the event of any Hubble Space Telescope servicing mission.

As the Hubble Space Telescope was being designed for orbital servicing, engineers and managers considered various technical and human factors criteria to reduce the complexity of maintenance tasks. The resultant spacecraft and its maintenance and refurbishment plan reflect a design philosophy based on modularity, standardization, and accessibility, with an emphasis on crew-intensive operations that differs from the philosophy governing Skylab design in the pre-Shuttle era.

Modularity

Candidate items for servicing were identified early and designed as modular orbital replacement units (ORUs). These units include critical subsystems for spacecraft operation and science data collection, as well as candidates for future upgrading. Most of these modules are self-contained boxes that are installed or removed by simple fasteners and connectors. Rather than attempt intricate handling of individual parts, planners decided to simplify the task by designing entire component assemblies for replacement.

Selection of an item as an ORU is based on several criteria, including its likelihood of failure sometime during the telescope's planned lifetime, its importance to telescope operations, its assessibility, and the cost of servicing. Predicted component lifetimes and failure rates are determined by reliability analysis.

For the Hubble Space Telescope, 26 different components, some duplicated to make about 70 individual units, were selected as ORUs; see Figure 6.3 [3]. These include the telescope's batteries, fine guidance sensors, solar arrays, computers, reaction wheel assemblies, and other major system components, as well as the focal plane detectors and cameras. The ORUs range in size from small fuses weighing only a few ounces to 318 kg (700 lb) scientific instruments as large as a telephone booth.

During the first Hubble Space Telescope servicing mission, items that degrade the fastest will be replaced—batteries, solar arrays, and sensors. On later missions, items that degrade less quickly—computers, reaction wheels, and tape recorders—will be replaced, and some instruments may be exchanged for newer models. Selection of items to be replaced on each mission is based on predicted performance and actual operational experience. Reliability assessments and flight data are among the factors taken into account when determining the failure probability of each ORU and identifying the candidates for servicing on a given mission.

Standardization

To reduce the number of unique components and tools that must be kept in inventory and packed for a servicing mission, designers have standardized many common elements, such as bolts and connectors. Although there are deviations, the HST design features considerable commonality.

Items identified as ORUs at the outset and designed accordingly, for example, are held in place by captive bolts with 7/16-inch double-height hex heads. Rather than several

Orbital Replacement Units

Unit		No.	Size (in.)	Mass (lbs.)
Support Systems Module (SSM) Equipment Section				
Battery		6	24x8x14	145
Data Interface Unit*	DIU	3	15x16x7	35
Data Management Unit*	DMU	1	28x30x7	95
DF-224 Computer		1	24x23x18	117
Electronics Control Unit (Rate Gyro)	ECU	3	11x9x9	17
Fuse Plug		12	5x3.7 dia.	0.4
Mechanism Control Unit	MCU	1	20x12x8	48
Multiple Access Transponder*	MAT	2	10x4x2	12
Power Distribution Unit*	PDU	4	18x10x6	29
Rate Sensor Unit	RSU	3	12x10x9	24
Reaction Wheel Assembly	RWA	4	25x21 dia.	104
Science Instrument Control and Data Handling	SIC & DH	1	34x26x10	136
Single Access Transmitter*	SAT	2	10x8x2	7
Solar Array Drive Electronics*	SADE	2	14x10x8	16
Tape Recorder*	TR	3	13x10x7	22
Optical Telescope Assembly (OTA) Equipment Section				
Data Interface Unit*	DIU	1	15x16x7	35
Electrical Power/Thermal Control Electronics*	EP/TCE	1	17x9x8	29
Fine Guidance Electronics	FGE	3	23x11x12	52
Fuse Plug		2	3.7x1.7 dia.	0.24
Optical Control Electronics*	OCE	1	11x13x7	18
Focal Plane Assembly Radial Bay				
Fine Guidance Sensor	FGS	3	67x46x22	494
Wide Field/Planetary Camera†	WF/PC	1	83x31x79	610
Focal Plane Assembly Axial Bay				
Faint Object Camera†	FOC	1	36x36x87	706
Faint Object Spectrograph†	FOS	1	36x36x87	689
High Resolution Spectrograph†	HRS	1	36x36x87	685
High Speed Photometer†	HSP	1	36x36x87	587
Externally-Mounted				
Diode Box		2	5x6x34	17
Low Gain Antenna*	LGA	2	8.5x5 dia.	0.7
Solar Array	SA	2	172x27x26	318
Total ORU Mass (percent of HST launch mass)				**7,836 (33)**

*Second Generation ORU
†Science Instruments

EVA Tool List

Screwdrivers
- Shrouded flex screwdriver, 4.8- and 8.6-inch shafts
- Shrouded rigid screwdriver, 3.8- and 8.3-inch shafts
- Torque-set tip tool #10, 10.3-inch shaft

Wrenches & Sockets
- Right angle drive tool
- 5/16-inch rigid hex capture tool, 10.3-inch shaft
- 5/16-inch wobble hex capture tool, 10.3-inch shaft
- 5/16-inch non-capture wobble socket, 7.3- and 10.3 inch shaft
- 7/16- and 1/2-inch box end wrench
- 3/8-inch drop-proof tether ratchet and caddy
- 7/16-inch open-end ratcheting wrench
- 7/16-inch non-capture wobble socket, 3-inch
- 7/16-inch rigid hex capture tool, 10.3-inch
- 7/16-inch wobble hex capture tool, 10.3-inch
- 7/16-inch socket extensions, 6, 12, 18, 24-inch
- 1/2-inch box ratchet wrench
- Torque limiters, 6.5, 9.0, 35 ft-lb

Electrical Connectors
- Circular connector tool, 0 and 90-deg. jaws
- Coax connector tool, hex (with and w/o shoulders) and round
- Installation and internal and external removal tools
- Multisize pin straightener

Lighting
- HST portable work lights
- EVA flashlight

Handling and Positioning Aids
- Preload tool (High Gain Antenna/Aperture Door)
- Mechanical finger
- Shepherd's hook
- Adjustable door stays
- Portable foot restraint, plus extender and socket
- Shuttle manipulator foot restraint
- Portable EVA grapple fixture
- Portable handhold plate
- Portable ORU handles
- Tool boards
- Assorted tethers
- Caddy, with French hooks
- Standard caddy
- Jettisn handle

Transfer Gear
- Transfer aid ("clothesline" system)
- Transfer bag
- Trash bags
- Multiple transfer system
- Fuse transfer containers

Power Tools
- Power tool, high-torque/low RPM

Cutting tools
- EVA scissors
- Cable cutter

Protective Covers
- Wide field/planetary camera mirror cover
- ORU electrical connector covers
- Fixed-head star tracker delta plate cover
- Fixed-head star tracker lightshade cover
- Fine guidance sensor mirror cover

Figure 6.3 Hubble Space Telescope ORU and EVA tool lists. *Courtesy of NASA.*

tools for removing and installing these ORUs, the crew needs only a 7/16-inch socket that can be fitted to a power tool or manual wrench.

Other ORUs were selected later when reliability assessments and ground testing indicated that some additional items may need to be replaced. Because these ORUs were added to the list after their design had matured, there is more variety in the fasteners, such as non-captive 5/16-inch hex head bolts and connectors without wing tabs. These departures from standardization and accessibility add more tools to the tool kit and more complexity to the servicing task.

Larger scale interfaces also are standardized to control hardware compatibility. The telescope has several features that make it compatible with the Space Shuttle, the Space Station, and an orbital maneuvering vehicle. Common grapple and berthing features permit the spacecraft to be retrieved by any vehicle similarly fitted and attached for servicing at any compatible worksite. Standardized mechanical and electrical connections facilitate servicing.

Limiting the variety of basic hardware is logistically efficient; it reduces the size and cost of the warehouse stock and the tool kits. Standardization is also economical for crew training; astronauts have fewer elements to master.

Accessibility

To be serviced in space, an item must be seen and reached by a pressure-suited astronaut or be within range of the appropriate tool. Items deep inside the telescope cannot be repaired or replaced. In the Hubble Space Telescope, most ORUs are mounted in equipment bays around the perimeter of the spacecraft. These bays open with large doors so components can be inspected and handled.

Crew aids—handrails, foot restraint sockets, tether attachments—are important features for accessibility. An astronaut needs safe, conveniently located worksites near the components to be serviced. There are 69 meters (225 feet) of handrails and 101 meters (331 feet) of restraint receptacles at strategic locations on the HST. These aids give the crew mobility and stability during servicing tasks. Other crew aids—such as portable lights, tools, and installation guiderails—improve the astronaut's visual and physical access to serviceable components.

Another design consideration for accessibility is contamination. Sensitive optical systems must be located and handled with minimal risk of exposure to vapors, particles, or excessive force. The Hubble Space Telescope was assembled in a high-quality clean room at the Lockheed Missiles and Space Company's facility in Sunnyvale, California, where nothing was allowed that might contaminate the optical surfaces or disturb the precise alignment of the instruments. Such control is difficult to maintain in space. Although the ORU bays provide no access to the telescope itself, contaminants might enter when the aft shroud doors are open. To protect the telescope during servicing periods, crew access into the aft shroud is limited, venting of water and other materials from the Orbiter is inhibited during EVAs, and optically sensitive components are handled carefully to avoid vibrations, impacts, or contamination. Both the hardware design and the maintenance and refurbishment plan for the HST take account of contamination hazards.

Time is a constraint on orbital servicing. Current spacesuits and procedures limit extravehicular activity periods to 6 hours a day, with only two planned EVAs per Shuttle mission. Furthermore, for components that are sensitive to changes in temperature, servicing must be completed quickly and efficiently. ORUs must be readily accessible for servicing within the confines of Shuttle and EVA capabilities and limitations.

Advanced X-ray Astrophysics Facility [4].

The Advanced X-ray Astrophysics Facility (AXAF) is a large aperture (1 meter), grazing incidence, multielement telescope with a cluster of scientist-developed instruments in its focal plane. Like the HST, AXAF will assure long-term scientific excellence through replacement of selected elements of the payload at later phases of the program. AXAF will also make use of modular spacecraft systems which are common to those of the GRO or the HST. Thus the AXAF servicing program will also enjoy some of the benefits of system commonality.

AXAF is the planned successor to the 1978–80 Einstein X-ray telescope mission. Designed for a 15 year mission lifetime, it will contain a mirror four times as large in area as Einstein's, covering a spectrum in energies twice as broad.

The AXAF program is managed by NASA/Marshall Space Flight Center. The prime contractor for design and development is the Federal Systems Division of TRW's Space and Technology Group, Redondo Beach, California. TRW heads a contractor team which includes Eastman Kodak Corporation, Bell Aerospace Systems Division, Boeing Aerospace, and 11 major subcontractors. Also on this team is a MSFC specified mirror elements subcontractor, Hughes Danbury Optical Systems.

NASA plans to launch to AXAF in 1999 from the NSTS into a 649 to 741 km (350 to 400 nmi), 28.5 degree Earth orbit. X-ray astronomy as a field of study is comparable in importance to radio and optical astronomy.

The heart of the Advanced X-ray Astrophysics Facility is its flight system. The modular AXAF is comprised of the spacecraft, the X-ray telescope, and a set of six science instruments (SI) that includes four focal plane SIs and two objective transmission gratings. The initial on-orbit weight of the AXAF is estimated to be 14,515 kg (32,000 lb) [4]. The AXAF system includes a variety of major ground facil-

Serviceable Spacecraft

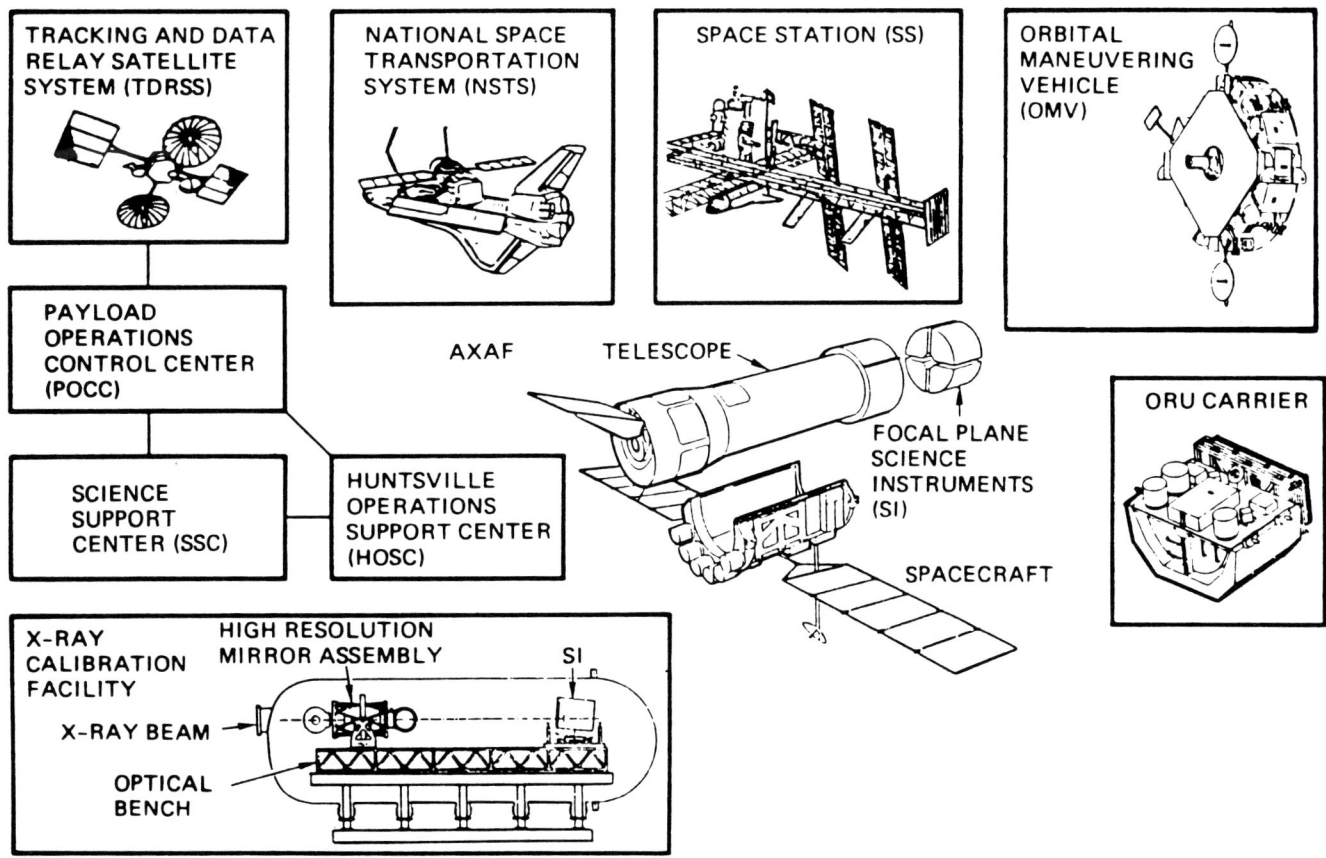

Figure 6.4 AXAF system description. Courtesy of TRW.

ities. First among these is the substantially upgraded X-ray Calibration Facility at NASA/MSFC where high-resolution mirror assembly (HRMA) and science instrumentation calibration tests will be performed. In addition, a Science Support Center (SSC), which will function as the interface between the AXAF science community and the program, and a Payload Operations Control Center (POCC), are included. The POCC will be located at MSFC. The location of the SSC is the subject of ongoing study. The Huntsville Operations Support Center (HOSC) at NASA/MSFC provides technical support during launch and orbital verification.

The AXAF system will employ the Tracking and Data Relay Satellite System (TDRSS) network as its primary means of communication, with the Ground Station Tracking and Data Network (GSTDN) available for backup. The National Space Transportation System will provide launch services as well as on-orbit servicing capabilities. An orbital maneuvering vehicle will be used to retrieve the AXAF for servicing, and to reboost and reinsert the AXAF into operational orbit. The AXAF will be serviced on-orbit at the Space Station Freedom when its facilities become available.

To complete the system, an orbital replacement unit (ORU) carrier will be used to contain and store all AXAF ORUs in the Space Shuttle during transport between the ground and in-space servicing. Figure 6.4 shows the AXAF system elements.

Servicing Philosophy

The goal of the AXAF servicing program is to provide an economic, effective means of sustaining a scientifically productive data acquisition phase for 15 years, with data analysis continuing for many years thereafter. Economy in this program is crucial. If the cost of the servicing infrastructure approaches the cost of replicating and launching another flight system, the concept of servicing loses merit. Therefore, the servicing program will be approached with cost effectiveness in mind and the proper balance between system reliability and servicing.

The strategy is to minimize any servicing conducted prior to detection of actual in-flight failures, while retaining reasonable prospects of successful repair once the decision has been made to conduct a servicing mission to replace scientific elements.

For program planning purposes, routine servicing will be scheduled at launch 1997, plus at 5 years (2002), and launch plus 10 years (2007), to replace failed subsystems and reboost the observatory if required. Instruments will be upgraded at these times, as science and technological advances dictate and programmatic considerations allow. The program plan for these missions will include changeout of those units whose replacement is virtually certain to be required. The program plan provides the flexibility to delay a routine servicing mission if the reliability of the observatory's life limited items proves to be higher than anticipated.

Unforeseen events may require an emergency servicing mission to be mounted as quickly as possible (realistically, this will take about 1 year) to assure survival of the flight system. The program plan will include all the measures required to implement such a mission.

If multiple failures occur which do not affect the survival of the observatory but do inhibit or curtail its operability, a servicing mission using the Shuttle will be launched nominally within 3 years after the NASA administrator determines that such a mission is necessary.

AXAF Servicing

The AXAF observatory complies fully with the requirement to be on-orbit serviceable. It is designed to be serviced at 5-year intervals during the mission, thereby reducing life cycle costs by more than $100 million.

The baseline approach is to service AXAF at the Space Shuttle. In the NSTS servicing activities, EVA astronauts will replace planned maintenance items, perform necessary contingency servicing, and replace consumables. The spacecraft and servicing equipment design will allow these same operations to be performed at the Space Station by astronauts using many of the same tools and procedures.

The serviceable elements of AXAF may be replaced when required to restore failed critical functions or redundancy lost by component failure, when telemetry data indicates a potential malfunction, or when units with newer designs and enhanced performance become available. Consumables such as liquid helium for the X-ray spectrometer may also be replaced during planned servicing at the Space Station or Shuttle. All serviceable unit attachment mechanisms and connections are simple and require only standard tools.

The orbital servicing space support equipment (SSE) has been designed to minimize complexity of EVA operations and take advantage of existing designs and technology. For example, the AXAF ORU carrier uses a Spacelab pallet designed for compatibility with the NSTS cargo bay and the Space Station servicing facility storage area; see Figure 6.5. It has provisions for ORU handling and temporary storage, and can accommodate two focal plane SIs and a complement of observatory ORUs.

The NASA/GSFC flight support system (FSS) described in Chapter 7 is used to berth AXAF during NSTS servicing. This universal payload adaptor will provide the same support in Space Station servicing. All elements of the AXAF except structure and cabling are designed as replaceable units. They are categorized as follows:

1. Class I ORUs include units with known wear-out mechanisms which will almost certainly have to be replaced on routine servicing missions. Science instruments are included in this category.
2. Class II ORUs are the units which may be needed on an emergency servicing mission to restore the survivability of the flight system to an acceptable level.
3. Contingency replaceable units (CRUs) are those which would be needed to restore the scientific performance of the observatory to an acceptable level.

Class I ORUs will be available in time for routine servicing missions. Class II ORUs will be available to meet the lead time for an emergency servicing mission. CRUs will be available, if required, to meet the nominal 3-year time line for missions involving restoration of scientific performance. At the conclusion of AXAF development, all residual development spares will be refurbished and stored for use in servicing.

AXAF Configuration

The AXAF spacecraft, Figure 6.6, is a highly modular system designed for parallel assembly and test and straightforward on-orbit servicing. Extensive use of existing subsystem hardware ensures a low-cost, low-risk program [3].

The pointing control and aspect determination (PCAD) subsystem points AXAF at desired science targets and provides star and fiducial information for ex post facto image reconstruction. The aspect camera utilizes 1024×1024 CCD arrays to detect the star and fiducial images. Six two-axis gyros (two active gyros required), updated by the aspect camera, provide attitude reference for pointing and slewing and image reconstruction. Six reaction wheels provide torque and momentum storage for pointing control and slews. Magnetic torques are used for momentum unloading.

Two deployable, rigid-panel arrays yield a minimum electrical power output of 7,700 watts at beginning of life and 6,190 watts after 15 years at end of life. Three internally redundant power channels, each sized to handle 50 percent of the total required load, provide a two-out-of-three channel redundancy. Nickel-hydrogen (NiH_2) batteries store energy for use during eclipse. A regulator maintains the bus voltage between 24 and 30 Vdc.

The command and data management (CDM) system acquires and formats 24 kbps science data and 8 kbps engineering data stored on two 1.2×10^9-bit tape recorders. The on-board computer (OBC) provides 256K (16-bit words) memory and 500 kbps throughput. Remote command and telemetry units (RCTUs) provide command and telemetry interfaces between the subsystems and science instruments, as well as decryption of uplink commands.

Serviceable Spacecraft

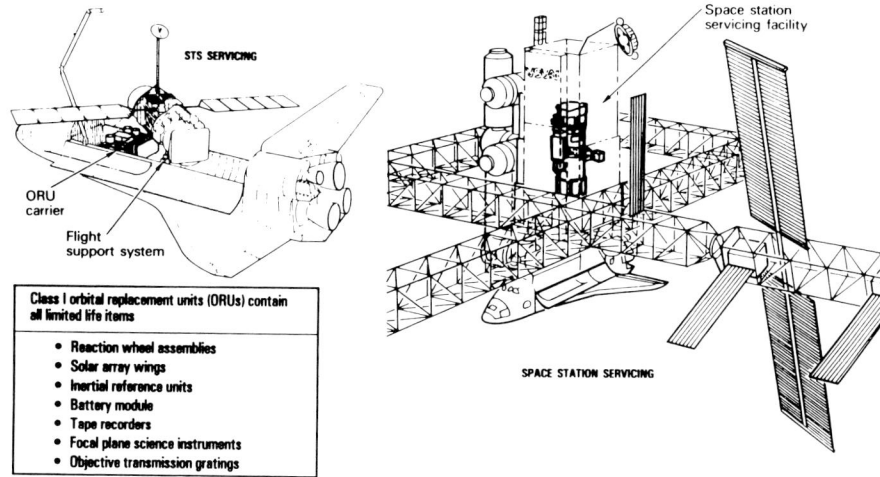

Figure 6.5 AXAF servicing. *Courtesy of TRW.*

NASA standard S-band transponders and an internally redundant high-gain antenna transmit 512 kbps data via the TDRSS S-band single access service (SAS) during tape recorder dumps and can also simultaneously transmit 32 kbps real-time data. Also, 32 kbps of data can be transmitted via the TDRSS multiple access service. Low-gain antennas provide full spherical coverage for backup command and telemetry via the TDRSS SAS or the ground network.

The spacecraft uses a welded aluminum structure. Kinematic links attach the telescope to the spacecraft structure, decoupling the spacecraft distortions from the telescope. The thermal control subsystem uses a passive design employing thermal coatings, louvers, and multilayer insulation. Heaters controlled by the on-board computer will maintain equipment temperatures. The major elements of the telescope system are the high resolution mirror assembly (HRMA), the optical bench, and compartments for four independent science instruments.

The HRMA is a temperature-controlled set of six concentric cylindrical mirror pairs ranging in diameter from 0.6 to 1.2 meters (1.9 to 3.9 feet). Each pair consists of one mirror with a parabolic curvature followed by another with a hyperbolic shape. The curvatures are very shallow, such that incoming x-rays just graze the surface of the mirrors and are directed to a focal plane 10 meters aft of the nodal point between the pair. The entrance apertures of the science instruments are located at the focal plane. The HRMA is mounted to the optical bench by means of six actuators to allow it to be pointed at each of the apertures. The aspect camera and the inertial reference units, components of the spacecraft pointing control subsystem, are located on the structure which supports the HRMA to provide a stable reference for attitude control and image reconstruction. Two objective transmission gratings are attached to the HRMA mounting structure via actuators that allow the gratings to be positioned in or out of the converging x-ray beam.

The outer surface of the telescope consists of multilayer insulation panels which are hinged to allow planned access into the interior of the telescope for on-orbit replacement of components during servicing missions. The sunshade door will be closed during launch, but when open will protect the HRMA from direct illumination by the sun. It incorporates the forward HRMA contamination cover, which also will be opened after launch and outgassing have been completed. Both the fore and aft contamination covers can be reclosed prior to servicing to prevent degradation of the mirrors by the servicing vehicles and personnel.

B. EVA servicing of AXAF focal plane instrument using Shuttle RMS

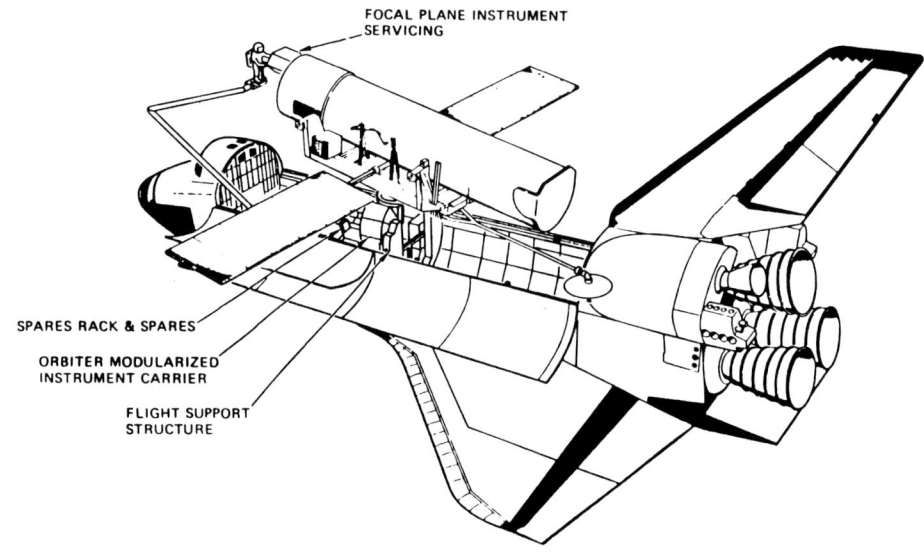

C. AXAF is designed for serviceable operations

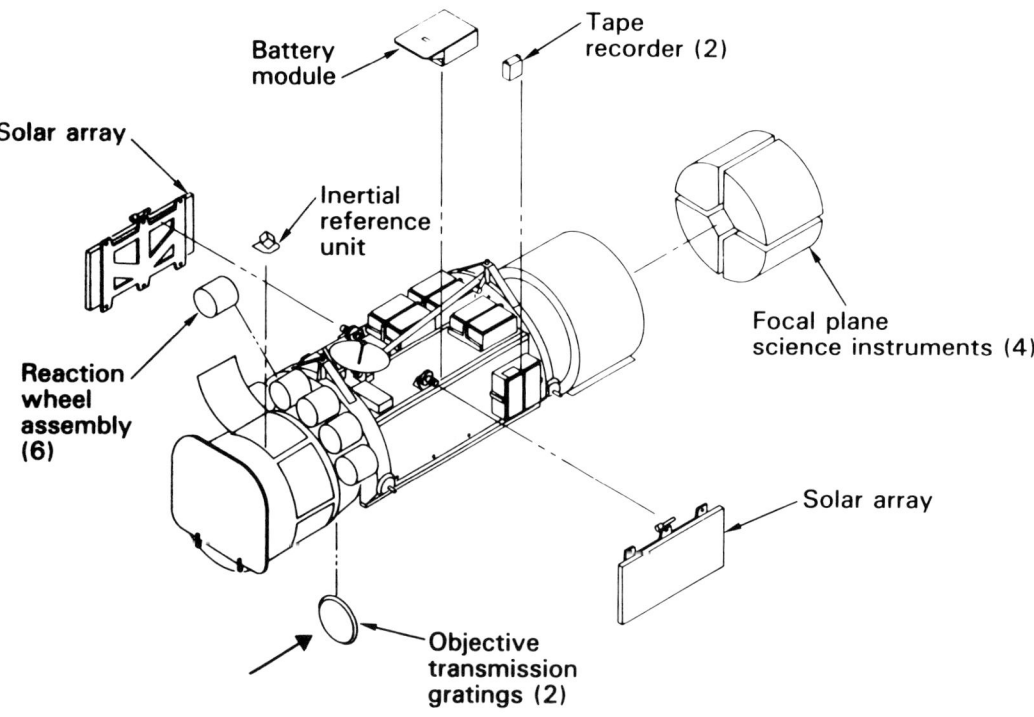

Figure 6.5 *(continued) Courtesy of TRW.*

Serviceable Spacecraft

D. AXAF on-orbit servicing sequence using an OMV and the Space Station.

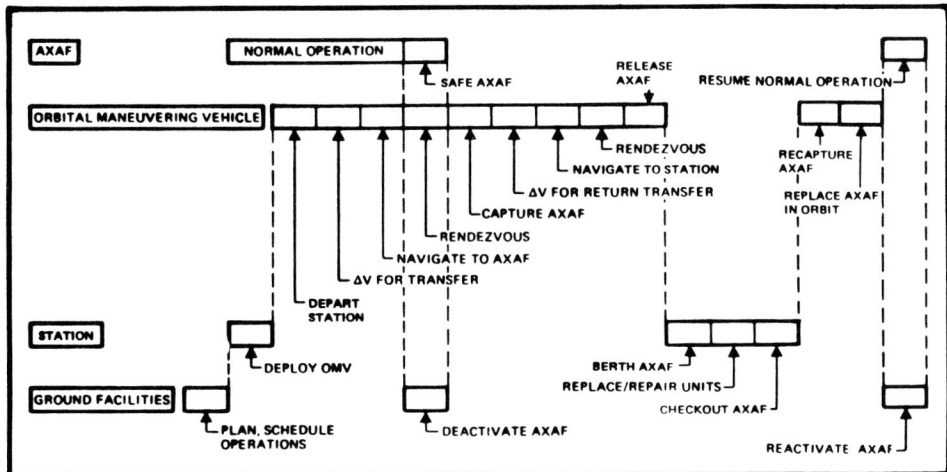

E. OMV/AXAF retrieval mission profile

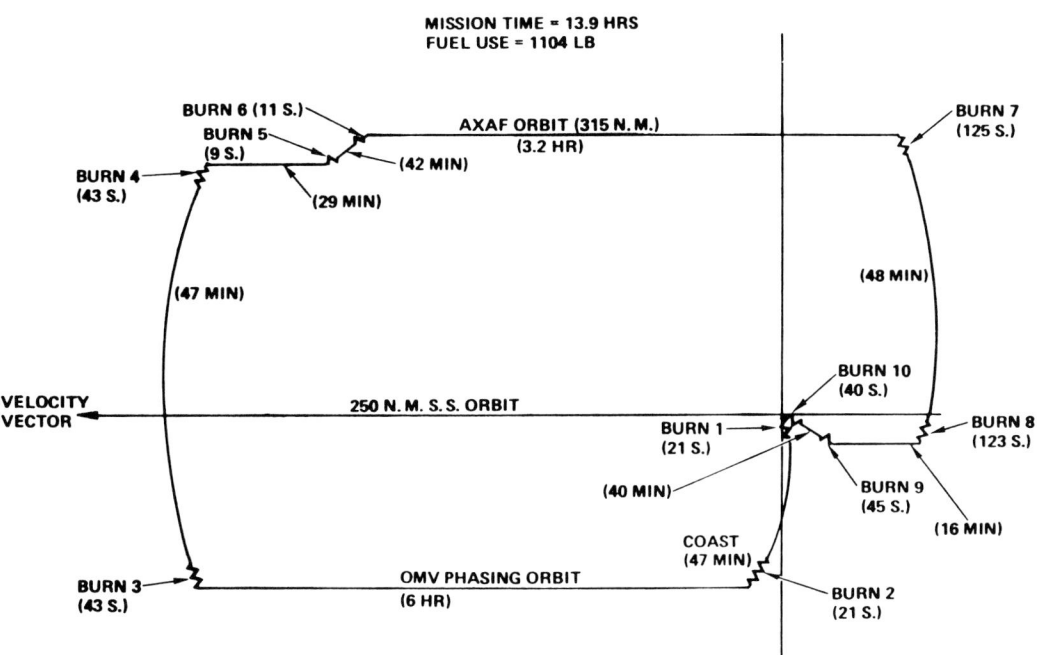

Figure 6.5 *(continued) Courtesy of TRW.*

The optical bench precisely maintains the location of the science instruments relative to the HRMA. It consists of aluminum and titanium rings, separated by graphite epoxy struts.

The science instrument compartments include latching devices which provide kinematic support to the instruments. Thermal insulation isolates each instrument from the others. The curved exterior surface provides thermal control surfaces for the science instruments.

All AXAF components except structure and cabling are candidates for on-orbit removal and replacement. The spacecraft is designed to make them as accessible as possible during EVA. Thus, the spacecraft equipment is mounted externally on MMS modules or packaged as individually replaceable units, and the telescope equipment is mounted on the optical bench behind access doors.

All new hardware will be designed for ease of access and EVA serviceability, with a minimum number of standard fas-

F. AXAF Command, Control, and communications network for on-orbit retrieval and servicing events

Figure 6.5 *(continued) Courtesy of TRW.*

teners, EVA handles, and EVA-compatible connectors. Existing hardware will be mounted in locations which provide maximum accessibility. Existing ORU II hardware will be modified to ensure serviceability, and will be verified by neutral buoyancy simulations. The serviceability of existing CRU hardware will be verified by engineering analyses and 1-g simulations.

AXAF Design for Serviceability

The AXAF design is driven by mission and science requirements and paced by technology awareness and life-cycle cost considerations. ORU component selection and allocation to ORU and CRU categories are based on reliability and mean-time-between-failure analyses. The number of ORU I types is limited, and the number of servicing missions has been reduced to two. The observatory Class I ORUs and science instruments are individually packaged to minimize the amount of hardware which must be replaced during a servicing mission. The remaining equipment units, with wear-out lives greater than 15 years, are defined as ORU IIs or CRUs but retain the capability to be changed out by extravehicular activity (EVA).

The observatory design facilitates access at the Space Station Freedom customer servicing facility and at the Space Shuttle. Placement of the flight support system (FSS) berthing interface on the top of AXAF permits optimal access to the externally mounted ORUs and CRUs. Full EVA access to internally mounted elements is provided through doors in the thermal covering.

Nearly all ORUs and CRUs are placed on the outside of the spacecraft module to increase productivity of EVA activity and comply with the time constraints for servicing. The spacecraft structure uses flat accessible faces for mounting replaceable units, and all ORU and CRU interfaces are designed for easy replacement. Flight-tested interfaces such as the Multimission Modular Spacecraft (MMS) modules are used extensively. Neutral buoyancy simulations were used to determine the best locations for EVA support equipment. EVA handholds and portable foot restraint (PFR) sockets are positioned so as to optimize crew work station configurations.

The power system modules, Figure 6.7, provide a good example of the packaging scheme selected for the AXAF spacecraft subsystems. It uses the design developed for the Multimission Modular Spacecraft which has flown successfully on the Solar Maximum Mission spacecraft, and on Landsats 4 and 5.

The 119 centimeter (47 inch) square by 46 centimeter (18 inch) deep modules are attached to the spacecraft structure with the standard MMS module retention system (MRS) hardware. The front (outboard) and side faces of the module are constructed of 1 inch thick aluminum alloy honeycomb material.

The assembly at the top of the module consists of two corner brackets, a bridge beam, and a retention system at the center of the bridge beam. The assembly at the bottom of the module consists of a retention mechanism mounted at the center of the module frame. The retention mechanisms attach to the spacecraft via two preload floating nut assemblies.

Thermal control for the modules is provided with a primarily passive design using selected thermal coatings, louvers, and multi-layer insulation (MLI) blankets. Heat flow through the sides of the modules is restricted with MLI blankets, and its flow through the outer radiator surface is con-

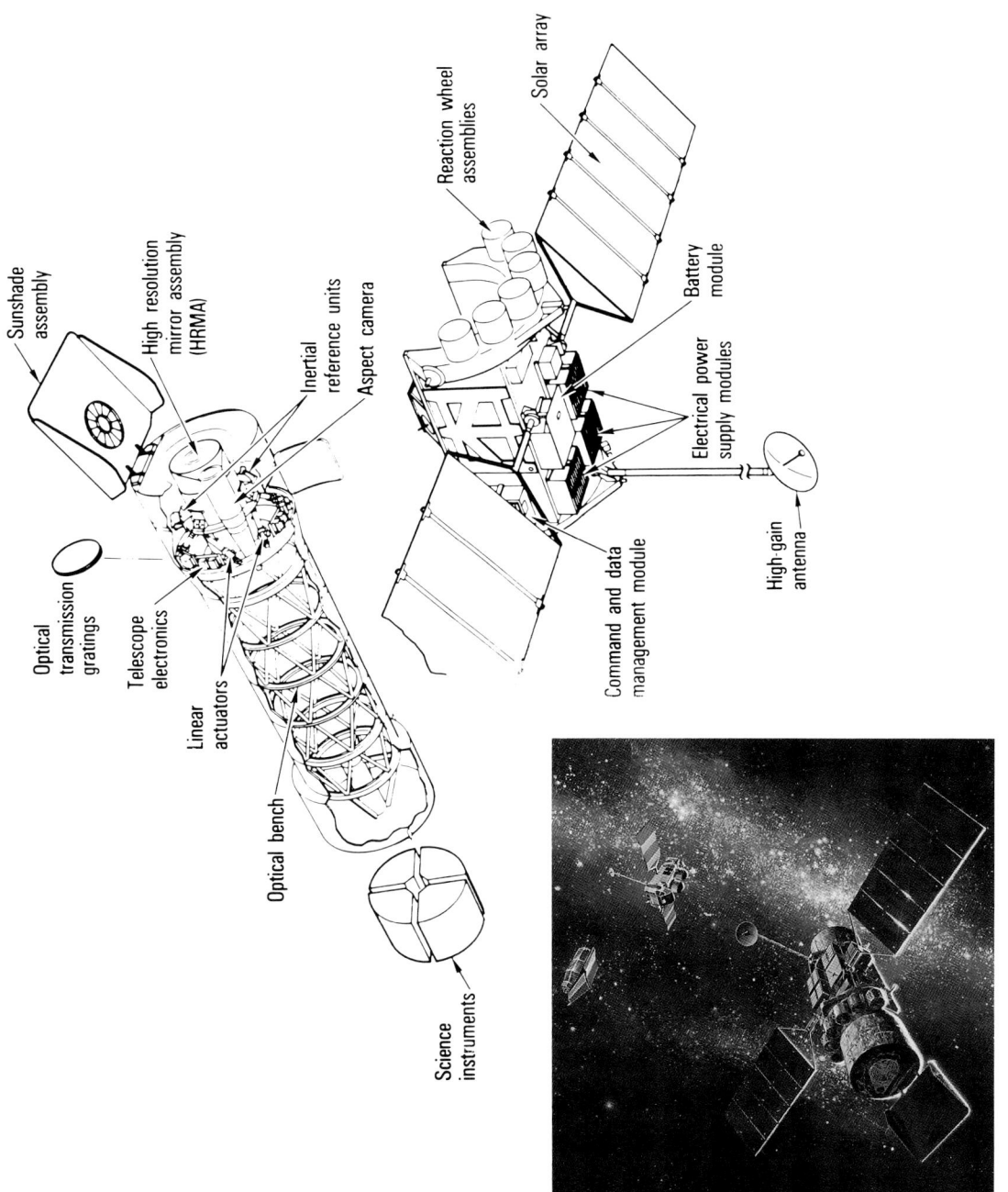

Figure 6.6 AXAF configuration. *Courtesy of TRW.*

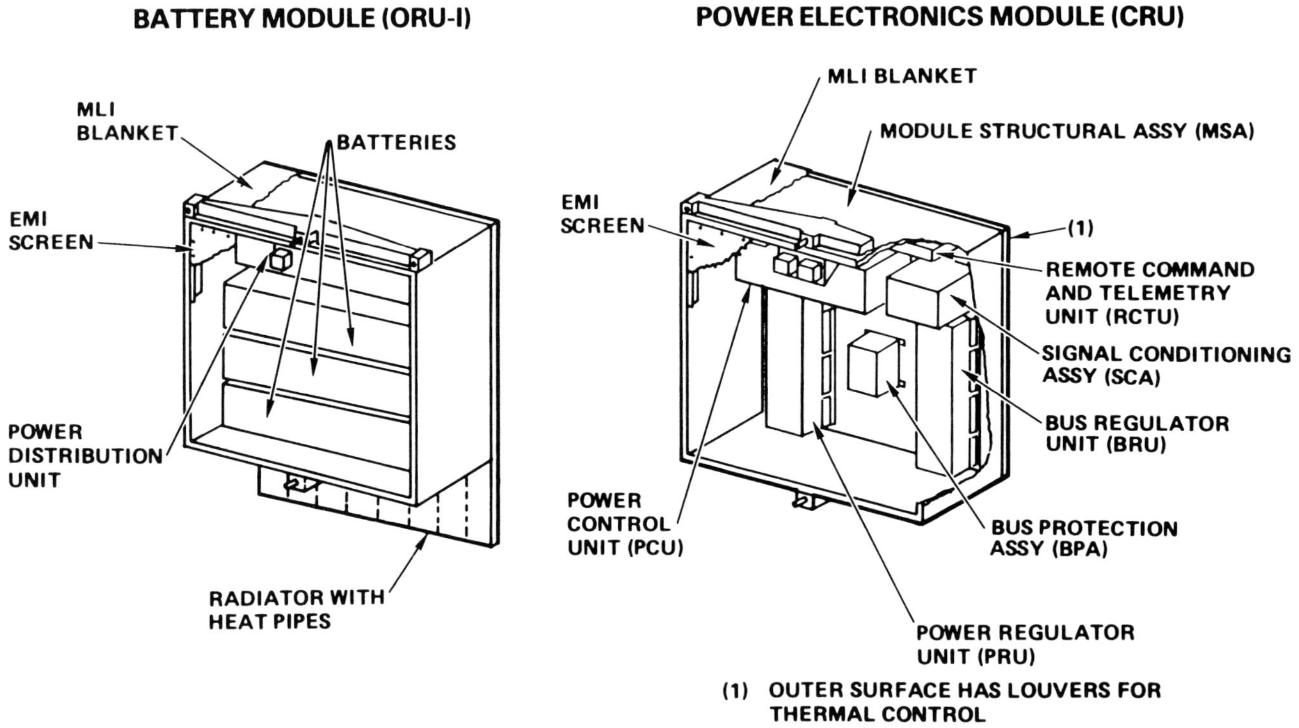

Figure 6.7 AXAF power system modules. *Courtesy of TRW.*

trolled with louvers. Fixed-conductance heat pipes are used in the battery module to maintain all NiH_2 cells at a uniform temperature and to conduct heat from the batteries to the outer surface of the radiator.

Blind mating, rectangular electrical connectors just below the upper retention mechanism provide the electrical interface between the modules and the spacecraft. An EMI screen between the module and the spacecraft provides protection from radiatively induced electromagnetic interference within the module. The module support tool (MST) can remove and replace these modules during EVA servicing.

The EVA serviceability of this design was well demonstrated during the SMM servicing mission.

AXAF Program Changes

Despite meeting all the objectives by Congress, AXAF is now undergoing detailed design and mission analysis to reduce total program costs. NASA AXAF managers lost funds to the Space Station Freedom program in a 1992 competition for development money. The original AXAF mission, described in previous paragraphs, has been shortened from 15 to 5 years. On-orbit mirror instrumentation changeout for product update will not occur. The main mirror size has been cut by one third. Further, two of the six original mirror instruments have been eliminated.

Plans to service the AXAF with the Space Shuttle or the Space Station Freedom have been dropped.

To counter somewhat these reductions, NASA MSFC and TRW have defined a separate, smaller mission to be launched in about 2001 which will include one of the eliminated instruments with a less expensive mirror.

The main AXAF spacecraft's orbit will now be established at an altitude between 10,000 and 100,000 km (5,396 and 53,960 nmi). In the low Earth orbit previously planned, AXAF would spend only 60 percent of its time viewing the sky, due to blocking by the Earth. In this higher orbit, 90 percent of the time is available. High Earth orbit has additional cost-cutting features. The large batteries needed for Earth eclipse can be eliminated, the solar arrays can be made smaller, and the heavy magnetic torques can be replaced with small gas jets. Communication can be simplified through use of the Deep Space Network.

But there will be a big loss as well. In high Earth orbit, AXAF can not be serviced or maintained by the Shuttle or Space Station Freedom. This program characteristic was vital to AXAF's 15-year lifetime. There is no planned replacement of cryogenic stores. New instruments will not be placed on AXAF once it is in orbit. In addition, if something goes wrong with the satellite in this high orbit, the consequences will be far more serious now that it may not be able to be fixed, unless NASA greatly accelerates its plans to develop automated (telerobotic) servicing with an OMV type vehicle/front end kit combination (Chapter 8).

This is undeniably a big shift in the mission—AXAF is

no longer a long-term observatory, but a short-term flight. However, AXAF scientists justified the idea in two ways. First, if the craft's basic capabilities remain intact, most of the key scientific projects can be accomplished in the first five years. Second, there are strong hopes that the AXAF, if built properly, might last far longer than its design life, as have many other successful NASA space systems.

NASA calculates that AXAF restructuring will save $600 million, reducing the total program cost to about $1.5 billion. For around $300 million of the saved $600 million, NASA AXAF managers intend to conduct a second mission to carry the calorimeter, dropped from the main mission, in concert with a small, simpler, lower cost mirror. Launching this second AXAF 1 to 2 years after the first AXAF will recover some of the lost astronomy X-Ray science.

In August 1992, the NASA Administrator, Daniel Goldin, approved the restructured AXAF program.

Eric J. Lerner gives an excellent summary of how AXAF was replanned to survive the Congressional budget ax in the October 1992 issue of the AIAA magazine *Aerospace America*. Most of the above information was derived from this article.

In his analysis of the AXAF restructuring, Lerner asks, "Does the new plan basically preserve AXAF scientific potential?" NASA AXAF managers think so. They are confident, Lerner says, that the spacecraft will survive the first year of the mission—the infant mortality phase—and operate longer than five years.

Without doubt, the loss of servicing due to lack of funds and difficulty of remote servicing in the high Earth orbit, places greater importance on spacecraft reliability and fault correction via telemetry (electronic work-around scheme).

As Lerner points out " . . . it is ironic that the Space Station Freedom and the Space Shuttle programs in NASA were both justified as a means of on-orbit servicing of science missions such as AXAF. Now, these manned programs' costs preclude this servicing from being funded, and could endanger the very existence of such missions in the future."

Gamma Ray Observatory

The $550 million Compton Gamma Ray Observatory (GRO) spacecraft is a science project at the NASA Goddard Space Flight Center. It was developed by the Federal Systems Division of TRW's Space and Technology Group, Redondo Beach, California. Launched in early April 1991 by the Space Shuttle, the GRO's instruments continue to explore a relatively unexplored portion of the electromagnetic spectrum, which could reveal new details about celestial forces that have shaped galaxies and the universe itself.

The Compton GRO is named after the American physicist Arthur Compton whose work on the interaction of high-energy radiation and matter played a key role in the development of modern physics.

The GRO was not initially designed for on-orbit servicing. However, GRO's use of MMS modules as ORUs renders the spacecraft serviceable on orbit. NASA now plans to replace MMS modules and replenish the GRO propellants as required to achieve an extended mission life. The nature of GRO's massive, complex, and intricately designed science instruments makes them impossible to replace on orbit. But, the GRO instruments have inherently long design lifetimes and extensive scientific potential, and hence are compatible with the NASA Great Observatories concept.

The GRO spacecraft, Figure 6.8, with a launch weight of 15,876 kg (35,000 lb), is a critical link in space science studies because of the large size of the gamma ray segment of the spectrum compared to the visible, infrared, ultraviolet, radio and x-ray portions. Observations in the gamma ray wavelengths are expected to provide information on such unusual celestial objects as quasars, pulsars and black holes.

There have been a half-dozen gamma ray experiments launched on smaller spacecraft or placed in high-altitude balloons since the early 1960s, but those missions have not approached the scope of the GRO mission or the sophistication and sensitivity of the spacecraft's four gamma ray instruments, some of which weigh 2 tons [5]. GRO is designed to conduct the first full-sky survey, covering the entire known gamma ray spectrum with instruments that are up to 20 times more sensitive than previous experiments.

Gamma rays convey critical information on the forces of creation and destruction that are continually shaping the universe. The high-energy radiation is unusual in that since it is stopped or deflected by little as it crosses the far reaches of the universe, it provides information not available in other wavelengths [5].

GRO instruments, and the large spacecraft required to support and align the instruments, need a launch system, like the Space Shuttle, with the payload volume and weight capabilities of present day transportation systems.

The GRO was the major payload of the Orbiter *Atlantis* on NSTS flight 37 which took off from Pad 39B at the NASA Kennedy Space Center on April 5, 1991. The 43 hour countdown for this launch never varied more than 10 minutes and was one of the best countdowns in the history of the Space Shuttle program, according to NASA's Robert Sieck, launch director for this mission. *Atlantis*'s climb on a direct insertion profile took it to 450 km (243 nmi), the altitude controllers wanted for GRO deployment.

On flight day three, GRO was deployed from the *Atlantis* payload bay using the RMS to lift the spacecraft out of the Shuttle's payload bay. With GRO on the end of the RMS, the spacecraft's two sets of folded solar array panels were

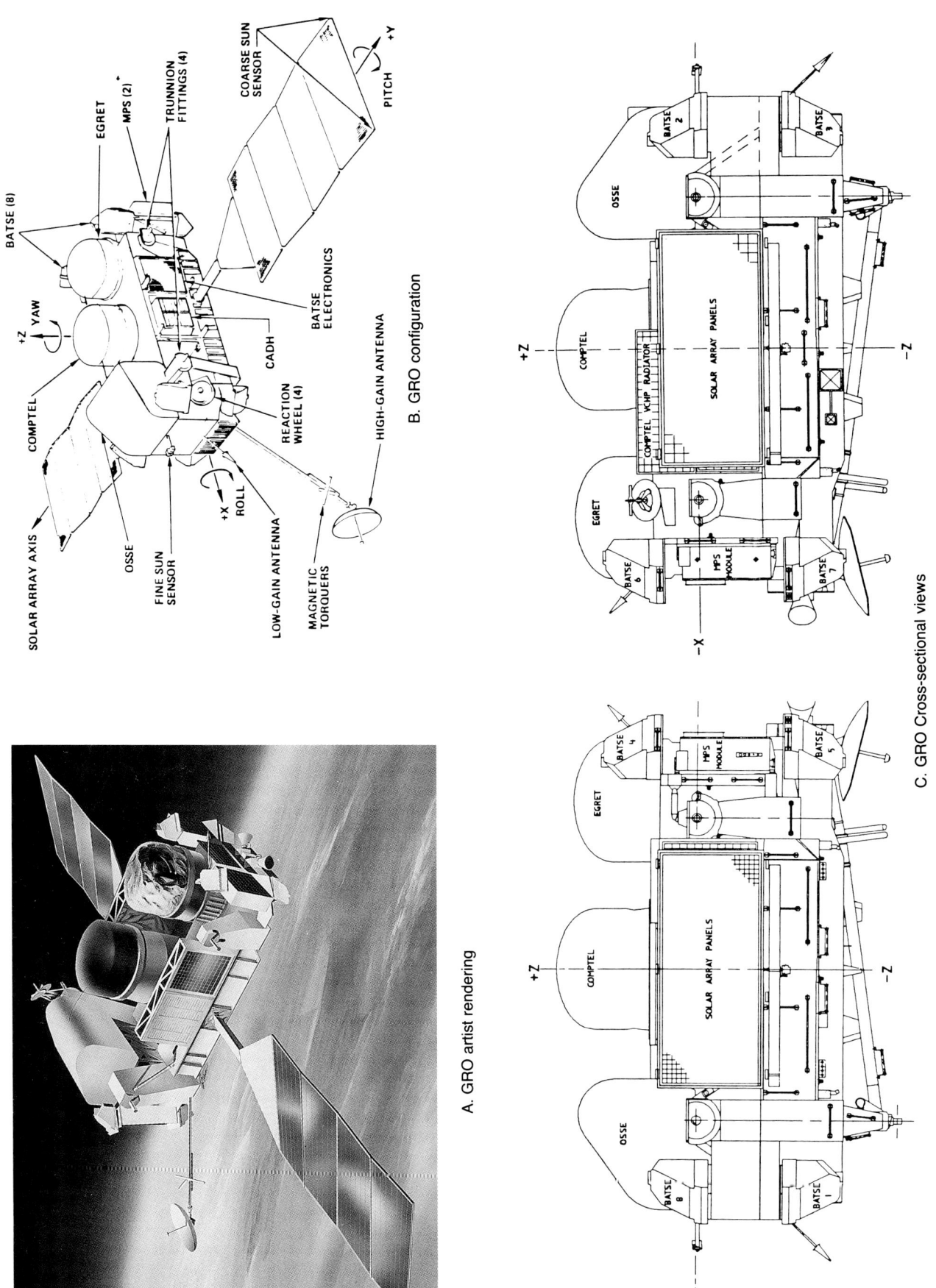

Figure 6.8 Gamma Ray Observatory. Instruments include: OSSE 3 Oriented Scintillation Spectrometer Experiment (Built by Ball Aerospace); COMPTEL 3 Compton Telescope (Built by MBB); EGRET 3 Energetic Gamma Ray Experiment (Built at Goddard SFC); BATSE 3 Burst and Transient Source Experiment (Built at Marshall SFC). Courtesy of TRW.

deployed via command from Goddard Space Flight Center. Then a problem developed. Despite repeated commands from Goddard, the GRO high-gain antenna would not swing out into its deployed position, though its activator was getting electric power and indications showed its latch to be open. Figure 6.8 shows the high gain antenna as it would be in its deployed position. Mission Control at Goddard decided to use the Shuttle Orbiter itself to try to shake the antenna free, first with a roll then with a pitch. But the satellite and its 5.0 meter (16.5 foot) boom antenna moved as a unit, no matter how *Atlantis* shook them. The decision was made to send astronaut mission specialists, Air Force Lt. Col. Jerry L. Ross and Jerome (Jay) Apt outside the Shuttle to perform a 1.5 hour GRO deployment procedure they had simulated underwater four times. With the crew's second pilot, Marine Lt. Col. Kenneth D. Cameron, orchestrating the EVA from the Orbiter flight deck, the space-suited pair began depressurizing *Atlantis*'s airlock.

When they entered the payload bay, Ross went immediately to the base of the antenna. Crawling over the satellite, he verified that the latch actually was open. After only 17 minutes into the EVA, as *Atlantis* was flying over Indonesia, the astronauts had loosened the balky boom on GRO and were able to assure Goddard Control they could manually move the high-gain antenna into place.

To complete the task, Ross used a ratchet to back out one bolt to disengage the actuator. Next he turned a pin to unlock the linkage to the boom. Then Ross eased the antenna out with one hand to the deployed position. And finally he re-locked the second pin. The astronauts completed the deployment work within 45 minutes of beginning the EVA.

Officials at Goddard and TRW, Inc., say the cause of the antenna snag may never be known. The best theory seems to be that insulation or some other material came loose during launch and was jammed in a way that prevented the boom from opening.

This EVA was the first U.S. extravehicular activity in more than 5 years. Besides saving the mission to deliver the second of the four "Great Observatories" NASA plans to orbit, the early EVA on GRO allowed NASA to cite the versatility of the Shuttle Space Transportation System, and the value of designing the GRO to be a serviceable spacecraft.

GRO is one of the first science satellites with on-orbit refueling capability. It carries about 1,814 kg (4,000 lb) of hydrazine propellant, with thrust provided by four 45 kg (100 lb) orbital adjustment thrusters installed in pairs. Electrical power is from two solar cell arrays designed to provide 3,980 watts at the end of the mission and six nickel cadmium batteries. The Orbiters *Atlantis* and *Discovery* could take the spacecraft to 510–556 km (275–300 nmi) if the Gamma Ray Observatory were the only major payload carried on the mission. The spacecraft, with a design life of 2.5 years, should have enough propellant to maintain operations on orbit for more than 8 years. Mission planners are scheduling observation periods of 14 days and plan to cover the universe during the first 2 years.

The presently planned GRO mission features these milestones:

- Launch and activation. With release of the GRO by the NSTS at 450 km (243 nmi), science mission altitude, this phase will take approximately 17 days. After this time GRO will maneuver to its first science target.
- Orbit maintenance. This event will occur at a minimum of every 65 days, assuming a mission life altitude band of 6 nautical miles. A 450 km (243 nmi) orbit has been chosen because it provides a long lifetime with the efficient use of the GRO fuel. The 65 day minimum between orbit adjusts assumes maximum expected solar flux. Orbit maintenance operations require approximately one day's loss of science data.
- Science retargeting. GRO will maintain its attitude for 14 days; therefore a maneuver to a new science target will be required after each science target period is over. Retargeting to a new attitude and attitude convergence requires approximately 5 hours.
- Refuel and repair. This event, should it be needed, requires the GRO to descend to 371 km (200 nmi) and rendezvous with the NSTS. The NSTS will dock with GRO, grapple it with the RMS, and berth it on the Multimission Modular Spacecraft flight support system (FSS) A-prime cradle, Figure 6.9. Either refueling or repair will be accomplished via EVA. A nominal refueling event takes approximately 10 1/2 hours. Repair mission times are a function of repairs needed. The GRO on-orbit replaceable units are the communications and data handling module and the two power system modules. Figure 6.10 is an operations flow of events for the on-orbit replacement of a GRO modular power system ORU.
- Rendezvous or controlled reentry. The GRO mission will end either by NSTS retrieval or by controlled reentry. For planning purposes the retrieval and controlled reentry phase are allocated a 3 month period. If retrieval is selected, the Shuttle Orbiter will rendezvous with the GRO, and following safing, grapple it. The appendages will then be stowed by EVA and the GRO placed in the Shuttle's cargo bay for the return to Earth.

Space Infrared Telescope Facility

The Space Infrared Telescope Facility (SIRTF) is a cryogenically cooled, long-life, one meter class space telescope which will be operated by NASA as a free-flying observatory for infrared astronomy [6]. To achieve its 5-year lifetime requirement (10 year goal), SIRTF must be replenished periodically with cryogenic helium and have its life-limited modular subsystems replaced; capability for

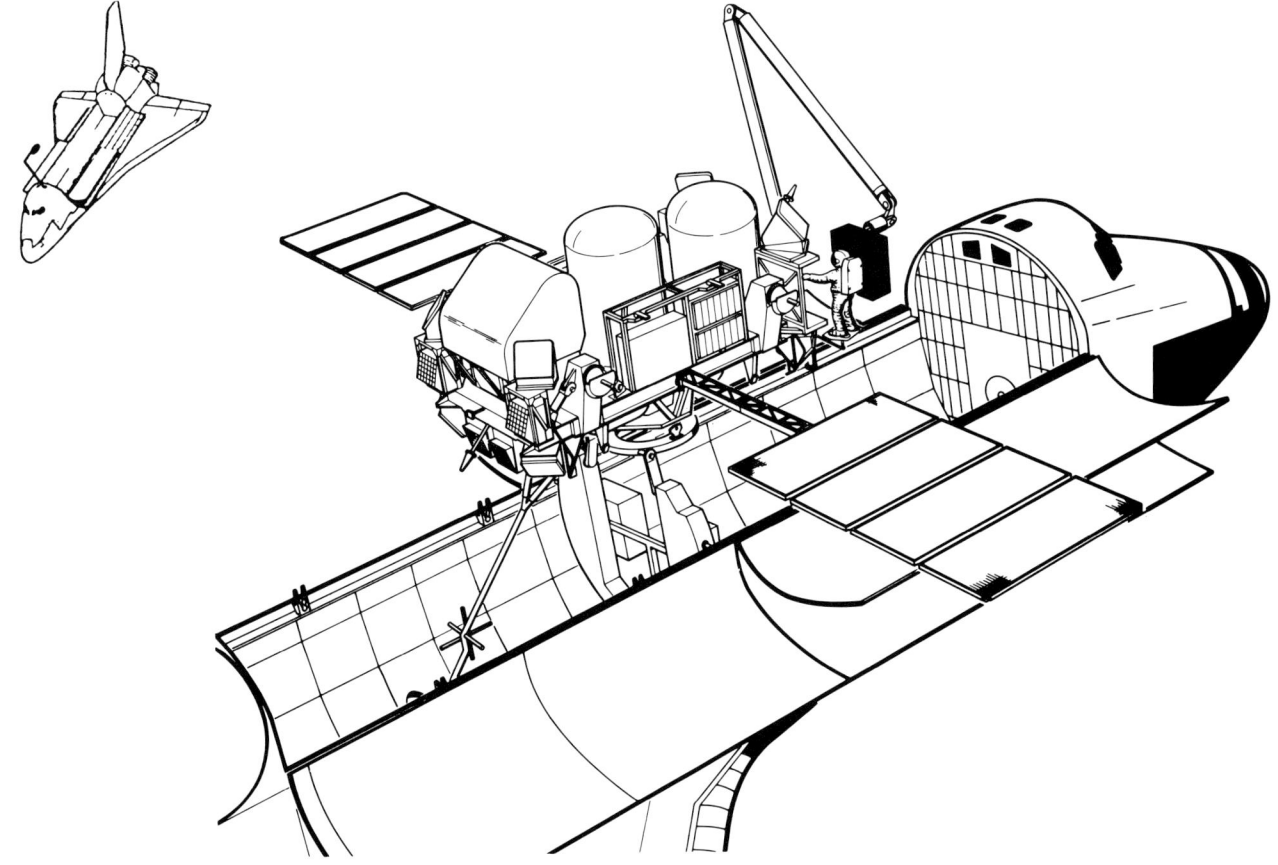

Figure 6.9 GRO showing ORU changeout while docked to the Space Shuttle. *Courtesy of TRW.*

contingency repair of warm components will also be provided in the observatory design. SIRTF is currently under advanced study at several NASA centers. NASA would like to see it operational in space by the year 2000.

The main features of the SIRTF, Figure 6.11, include the optical subsystem, the cryogen subsystem, and the structural subsystem.

Infrared light or infrared radiation, sometimes called heat radiation, was first identified two centuries ago by Sir William Herschel, a British astronomer best known for his discovery of the planet Uranus [7]. Herschel sent beams of sunlight through a prism, which dispersed or separated the light into the rainbow of visible colors: violet, indigo, blue, green, yellow, orange, and red. He used a thermometer to measure the energy carried by each color of light. But the temperature rose even if the thermometer was placed in the part of the spectrum beyond the red. Something was reaching the thermometer—but what?

Herschel realized that something like visible light, but invisible to human eyes, must have formed part of the beam of sunlight. He called this radiation beyond the red "infrared." Today we recognize that infrared radiation has longer wavelengths and lower frequencies (fewer vibrations per second) than red visible light. Human eyes are blind to the infrared, but we have made detectors sensitive to infrared radiation, often composed of crystals of germanium or silicon which are "doped" with small quantities of other elements. Infrared radiation shining on these crystals produces electrical signals which allow us to determine how much infrared is reaching them. An array of thousands of such detectors provides a sensitive camera for imaging the universe in the infrared [7].

Infrared astronomy from the Earth's surface is severely hampered by the life-giving water vapor that pervades our atmosphere. To avoid the absorption produced by water vapor, astronomers have gone to great lengths—and great heights. They have established infrared observatories on mountains so high that altitude sickness can be a problem. They have taken aircraft 18.5 km (10 nmi) above the Earth's surface and they have launched balloons to still higher altitudes. These efforts have yielded improved knowledge of the cosmos, but the full range of infrared observations can be realized only from space.

Serviceable Spacecraft

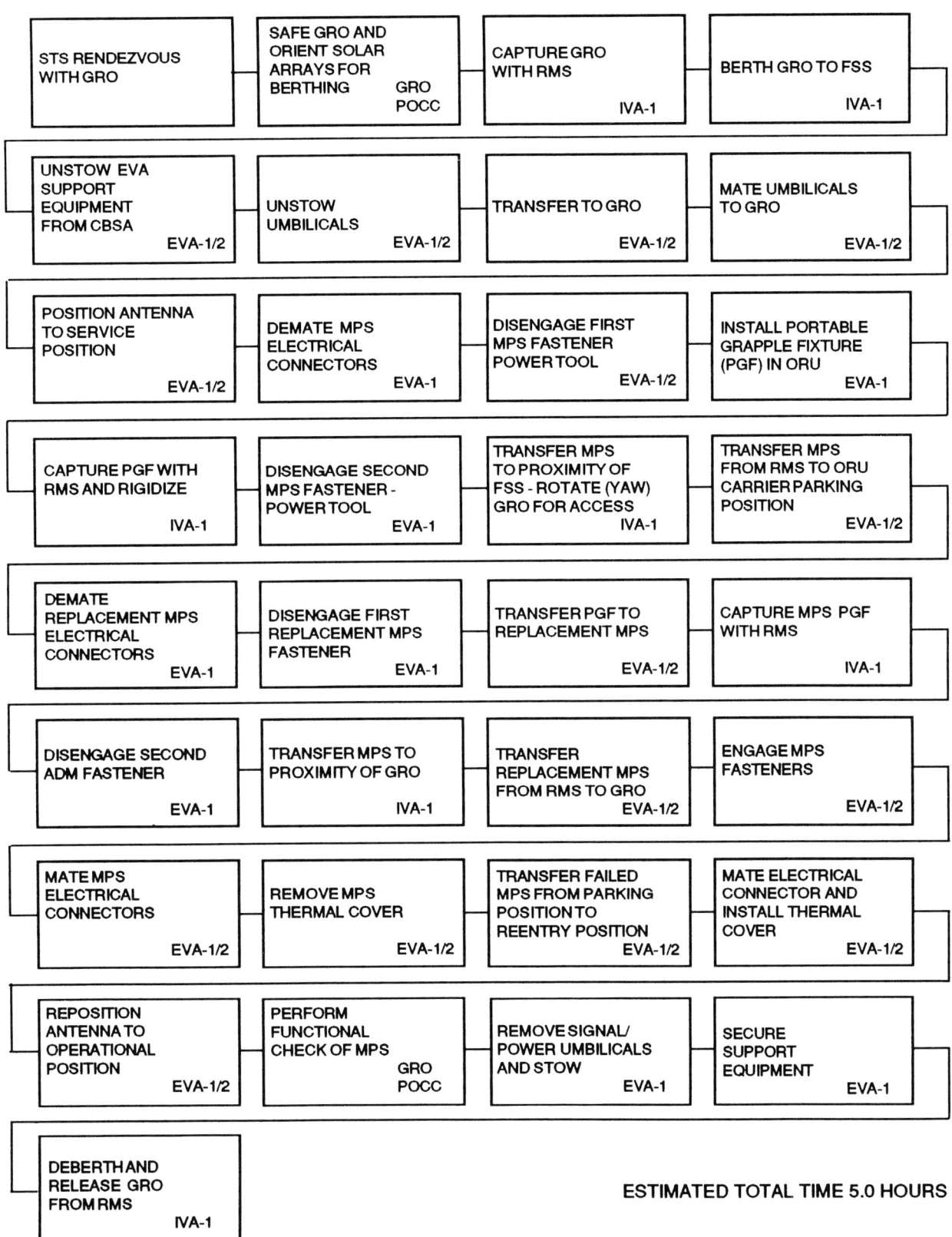

Figure 6.10 GRO operations flow chart of the on-orbit replacement of the modular power subsystem ORU. The IVA and EVA estimated times in hours are shown in each box of the 33-step sequence. *Courtesy of TRW.*

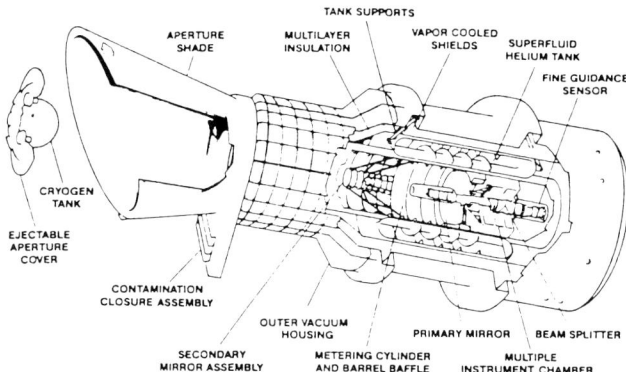

Figure 6.11 SIRTF telescope configuration. *Courtesy of NASA.*

The unavoidable infrared brightness of the atmosphere and a warm telescope drown weak cosmic infrared sources in a sea of local radiation. As a result, infrared observations from within the atmosphere are as inefficient as optical observations in broad daylight. In space, free of the atmosphere, we can cool the telescope to just a few degrees above absolute zero, so that its own infrared radiation is very weak. The supercooled telescope in space thus becomes 1,000 to 10,000 times more sensitive to infrared radiation than the best telescopes on or near the Earth's surface.

In 1983, the great veil of the cosmos was briefly lifted when the Infrared Astronomical Satellite (IRAS) scanned the skies for 10 months [7]. IRAS, a joint effort of the United States, the United Kingdom, and the Netherlands, was sensitive to four different frequency regions within the infrared portion of the spectrum. Helium in its extremely cold superfluid form was used to cool the telescope and its detectors to 4 degrees above absolute zero (−454 degrees F). Freed from Earth's interfering atmosphere, IRAS surveyed almost the entire sky twice between its launch on January 25, 1983, and November 21, 1983. On that date, three-quarters of the way through a third sky survey, the supply of helium coolant was depleted, the IRAS telescope warmed up, and its infrared detectors ceased to function.

Galaxies are the building blocks of the universe. IRAS provided the first assessment of their total energy output by detecting infrared radiation from some 20,000 galaxies and showing that most galaxies like our own shine as brightly in the infrared as in visible light.

IRAS's spectacular achievement was a beginning, not an end. We can think of IRAS as a quick, blurry first look through a new window on the universe. IRAS was primarily intended to survey the infrared sky, not to make detailed studies of the objects if discovered. A more powerful infrared observatory in space is needed to answer the questions raised by IRAS's discoveries and to continue the exploration of the universe through the infrared window. The Space Infrared Telescope Facility to be launched by NASA in 1998 will do this.

The SIRTF spacecraft concept has evolved from an earlier mission-unique system to a "common mission" configuration, for which several options presently are under consideration. The best defined concept, which is the result of a study performed by ARC and GSFC in 1985 [6], makes use of the Multimission Modular Spacecraft (MMS). Various modules of this design are also planned for use on UARS, GRO, and other future NASA missions. For the SIRTF application, an integral MMS is end-mounted on the SIRTF telescope via a transition adapter, which is attached to an "elephant-stand" cylindrical truss structure surrounding the aft portion of the telescope and bolted to the aft girth ring. The three MMS modules (power, command and data, and attitude control) require only minimal modification for SIRTF use. The high gain antennas are slightly modified Landsat 4 assemblies, and the solar arrays are inherited from the UARS program. A noteworthy feature is the addition of three MMS modules containing "warm electronics" for the science instruments and telescope mechanisms (cryo control valves, secondary mirror, and contamination closure); the electronics provide power, commands, and data handling for these subsystems, and their boxes are mounted around the girth ring. In this configuration, the 5,600 kg (12,348 lb) observatory attaches directly to the STS sill and keel fittings for launch. It can interface with the (1) flight support system cradle for STS-based servicing and (2) an OMV at the aft end of the MMS, for in-orbit transportation.

Servicing Philosophy and Plans [7]

All of the SIRTF spacecraft options under consideration are amenable to on-orbit servicing, due to their use of orbital replacement unit modules, which are designed for easy removal and replacement by an EVA astronaut. SIRTF concepts also include consideration of servicing mission basing; although Space Station Freedom-based maintenance and refurbishment are baselined, servicing compatibility with the NSTS is also provided as a contingency, including interfaces with the FSS Cradle A', and the RMS. Indeed, it is envisioned that the Space Station Freedom concepts will be developed to ensure that Space Station (SS) servicing facility designs allow payload compatibility with both STS- and SS-based servicing; for example, compatible mounting structures, power and communication interfaces for servicing payloads are needed.

SIRTF assumes a 10 year observation lifetime. While its SIRTF cryogen supply is sized to permit at least 2 years of operation on-orbit, it is presently planned that the first liquid helium replenishment will be performed about 18 months after launch. Uncertainties exist in the operational and orbital environments, warranting this conservative approach. At the time of first refill, the helium replenishment quantity will be measured, and this will guide subsequent planning for replenishment intervals.

Serviceable Spacecraft

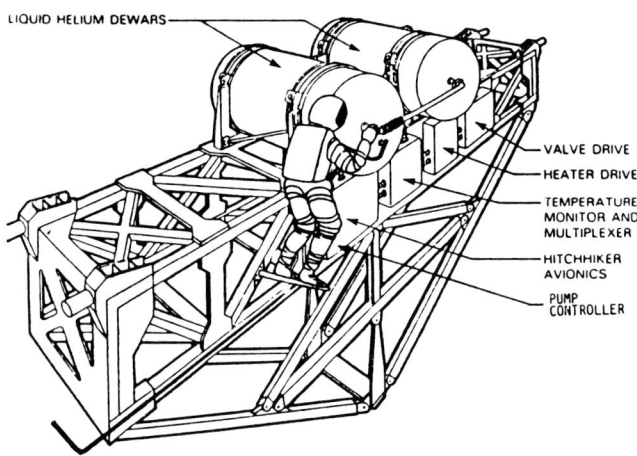

Figure 6.12 Liquid helium transfer project flight demonstration. *Courtesy of NASA.*

This first and all subsequent replenishment must be timed to coincide with SIRTF/SSF co-planarity opportunities, since any OMV will have limited capabilities for plane change during SIRTF orbital transfer missions. The schedule of these co-planarity events is dictated by the relative circular orbit altitudes of SIRTF and SSF. For the nominal 1041 km (562 nmi) SIRTF orbit and an assumed SSF constant altitude of 463 km (250 nmi), the co-planarity interval is approximately 9 months. Depending on SIRTF cryogenic consumption as measured during the first replenishment, the subsequent servicing interval may be 18 or 27 months. An overall SIRTF servicing baseline schedule assuming 27 month intervals for all but the first servicing is depicted in Figure 6.12.

Servicing Mission Scenario [6]

The SIRTF nominal servicing mission starts with delivery to Space Station Freedom of the servicing aerospace support equipment via STS flight. The aerospace support equipment includes as a minimum the cryogen replenishment tanker and contamination covers for the aperture shade and Sun/Earth sensors, plus any replacement ORU modules or contingency module components needed for a particular servicing mission. If an OMV is not based at the Station, one will have to be transported there by the NSTS. As the servicing date (co-planarity opportunity) may be up to 90 days subsequent to the NSTS delivery mission (assuming quarterly standard NSTS visits to SSF), the tanker must be sized to allow for any cryogen boiloff which may occur during that period. In addition, the contingency situation of SIRTF running "dry" and warming up must be considered, so the tanker must be adequately sized to enable cooldown and replenishment of a warm SIRTF. The nominal mission would replenish a still cold SIRTF, and the cryogenic helium quantity in the tanker would be sized accordingly (approximately 6,600 liters in an 11,750 liter tanker). Once delivered to the SSF, the tanker and replacement modules/components are stored in a suitable location protected from solar exposure and contamination. If stored in a bay, the tanker should only be exposed to low emissivity surfaces whose temperatures are 300 K or lower. The replacement modules and components should also be stored in environmentally protected areas. Power/communication interfaces are desirable to enable health monitoring and occasional checkout of these devices.

At the first subsequent co-planarity opportunity, an OMV is dispatched from SSF to maneuver to the SIRTF orbit and rendezvous with it. The detailed operational sequence of OMV/SIRTF rendezvous and docking are not fully established; however, a preliminary SIRTF/OMV Interface Requirements Document, including a general operations description, has been developed jointly by NASA/ARC and NASA/MSFC. The OMV is controlled by the ground or SSF until it has departed the SSF control zone, then it performs an automatic maneuvering sequence to transfer it to SIRTF's vicinity. The final rendezvous and docking maneuvers with SIRTF are again ground-controlled, using radar and television on the OMV in a "telepresence" mode. SIRTF's attitude control system is inhibited just prior to final docking, having put SIRTF into an inertially stable attitude. The docking interface is presently envisioned as the three "towel rack" configuration used on the GSFC Flight Support System Cradle A'; this interface has also been implemented on the HST, and requires a suitable docking adapter on an OMV. This interface is located on the aft end of SIRTF's spacecraft, as previously implemented on the Solar Maximum Mission's MMS. It includes an electrical power/communications adapter, to which the OMV umbilical adapter is connected via a motorized plug-in subsequent to docking. It is presently envisioned that OMV will supply power (SIRTF standby mode) and communications (SIRTF engineering telemetry) during all attached phases; it may be necessary to retract the SIRTF appendages prior to OMV powered flight due to excessive g-loads.

The rendezvous sequence must also consider SIRTF's high susceptibility to contamination. Although the SIRTF contamination (aperture) closure will be commanded shut prior to rendezvous, SIRTF's external high-emissivity finishes are subject to degradation due to bipropellant effluent impingement, with consequent impacts on SIRTF thermal performance and lifetime. Furthermore, any contaminants adhering to SIRTF's exterior might subsequently migrate to the cold telescope optics (following final redeployment), degrading optical performance. Therefore, all final rendezvous and docking OMV maneuvers are to be accomplished using the cold gas (GN_2) thrusters incorporated on OMV; also, all OMV attitude control functions during the attached transit phases must use inert gas. The main delta-V maneuvering thrusters on OMV are pointed away from the observatory, so their effluents will pose minimal contamination problems, provided there is no bipropellant leakage. Following docking, electrical connection, and appendage retraction, the OMV will transfer SIRTF to the SSF vicinity using its automatic maneuvering mode. The entire OMV retrieval mission is estimated to take about 16 hours.

Helium Transfer Technology [6]

Ames Research Center's Infrared Astronomy Projects Office, which manages SIRTF for NASA, has initiated a flight demonstration program to develop and prove the technologies required for the efficient transfer and replenishment of liquid helium on orbit. The Liquid Helium Transfer Project (LHTP) is a combined NASA ARC/GSFC effort to demonstrate cryogen transfer on two flights in the NSTS bay in the early 1990s. In summary, they consist of: the pump mechanism, for which two candidates are under development—a thermomechanical ("fountain effect") pump, and a submerged centrifugal mechanical pump; the liquid acquisition device, which maintains liquid (versus gas) at the pump inlet, for which candidates include fins, sponges, and galleries; thermometry and pressure gauging; leak tight motorized cryogenic valves; and level sensing (quantity gauges) and flow measurement devices. All of these are in various stages of development ranging from concept only, to previously flight-tested similar components. In addition, an EVA-compatible transfer line coupler, which meets the stringent STS requirements for fluid couplers, including redundant seals, must be designed and tested. JSC is presently developing this technology.

The LHTP experiment apparatus, Figure 6.12, is envisioned as being mounted on a mission peculiar equipment support structure (MPESS) developed by NASA/MSFC as part of the Hitchhiker program. It consists of: two identical liquid helium dewars connected by a vacuum-jacketed transfer line; mission-peculiar control electronics; and a Hitchhiker-M avionics module for interface with the Orbiter. The dewars measure 1.38 meters long (4.53 feet) by 79.4 centimeters (31.2 inches) in diameter, with a 210 liter capacity each. They are located side by side, with the transfer line mounted at the end of each dewar. The dewars consist of an aluminum outer shell, two vapor cooled shields, and the inner helium tank supported by fiberglass straps from the outer shell. Each dewar has four penetrations, one for a warm gas vent, one for a warm gas "high" vent, one for a burst disc safety pressure release, and one for the transfer line to the other dewar. Within the dewars are numerous motorized valves, copper slits, bayonet couplers, pressure gauges, check valves, burst discs, bellows, heat exchangers, and either a porous plug (thermomechanical) pump or a centrifugal pump. The dewars and mechanisms are in preliminary design.

The current LHTP is planned as a two-flight program, with the first flight a component qualification/technology validation demonstrator. For this flight, the transfer line is pre-mated on the ground and remains so throughout the mission. The transfer operations will be conducted manually from a ground POCC, with an astronaut monitor and override capability from the STS aft flight deck. This mission is intended to demonstrate component capabilities, mechanical and thermal performance, etc., to validate analytical models and qualify hardware. Both tanks are launched full of normal liquid helium, which is converted to the superfluid state by venting after orbit is achieved. The remaining fluid (50–60 percent) is to be transferred back and forth several times, with each dewar alternatively mimicking a supply and receiver tank. The second flight is a system demonstration, incorporating any changes or upgrades required based on the first flight results. The dewars are again launched with the transfer line connected, with an initial demonstration of an automated "expert system" transfer controller incorporated in the hardware. Then an astronaut EVA is planned, to demonstrate manual disconnect and reconnect of the bayonet transfer line/tank coupler. Subsequent transfers include a "warm" demonstration, where one dewar is initially depleted of liquid and actively warmed to 150 K; it is then cooled by helium from the "supply" dewar and finally replenished with liquid. This second and final flight is intended to demonstrate all the technologies, systems, and operations to be incorporated in the future resupply tanker needed by SIRTF and other cryogenic spacecraft.

Logistics/Spares Philosophy [6]

The SIRTF program is in an early stage of definition. This status, combined with the current uncertainty over the spacecraft design and heritage previously mentioned, does not allow an in-depth maintenance/logistics strategy to be developed. However, certain considerations and preferences can be addressed at this time. First, it is almost certain that some type of modular, maintainable spacecraft will be employed, and that some of its modular subsystems (ORUs) will require planned servicing. Other subsystems or components may also fail during SIRTF's lifetime, requiring contingency maintenance. At present there are several ongoing studies (both government-funded and contractor IR&D) to assess component reliability, failure rates, and lifetimes in the space environment, and as the results become available, the data will be used to develop a specific servicing plan for SIRTF. However, based on the results of the SIRTF/MMS study performed by GSFC and ARC, it is envisioned that as a minimum the batteries would require replacement at approximately midlife, 4–5 years. As these are located within the MPS module, the preferred approach is a simple changeout of the entire module during a planned servicing mission. The replaced module would be subsequently returned and refurbished on the ground, for later reuse on SIRTF or another compatible spacecraft. For such "planned replacement" ORUs, the best approach is most likely to have the module entirely assembled, integrated and stored on the

Serviceable Spacecraft

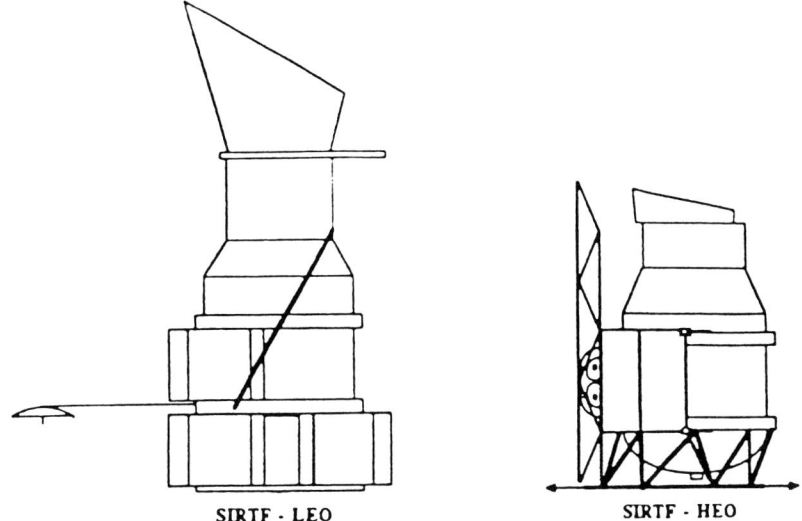

Figure 6.13 SIRTF observatory concepts for LEO and HEO missions. *Courtesy of NASA.*

ground, or at SSF, prior to its need date, as part of the SIRTF hardware complement.

SIRTF in High Earth Orbit [8,9]

In the spring of 1989 NASA looked at a new approach to the SIRTF mission and studied the feasibility of using an expendable launch vehicle to place SIRTF into a 100,000 km (53,960 nmi) high Earth orbit (HEO). This HEO puts the SIRTF above the Van Allen radiation belt whose trapped energetic particles could degrade the performance of the science instruments. SIRTF operation in this HEO mission is much simpler and more straightforward than in the LEO mission. The new mission uses a Titan IV/Centaur to launch a 4.6 meter (15 feet) long, approximately 4,536 kg (10,000 lb) SIRTF into a 100,000 km (53,960 nmi) altitude, 28.5 degree inclination orbit with a period of 100 hours. Seen from this altitude, the Earth subtends an angle of less than 7 degrees (versus 122 degrees in LEO), so it is possible to maintain a flexible viewing strategy while constraining the telescope to point no closer than 80 degrees to the Earth or Sun line. This orbit allows the use of a fixed solar panel, which shades the telescope and lowers the predicted temperature of the telescope outer shell to 110 K. Eclipses are infrequent in this orbit; the time and date of launch can be selected to orient the orbit so that the spacecraft passes through the Earth's shadow only a few days each year. Reaction wheels provide slewing and fine pointing capability. Environmental disturbance torques are small, but momentum control requires the use of cold gas because the Earth's magnetic field is too weak for momentum dumping at 100,000 km (53,960 nmi). The helium boiloff from the cryogen system and a small tank of compressed helium gas will be used for this purpose.

Two major system level performance parameters, cryogen lifetime and launch mass, Figure 6.13, were evaluated. A 5 year cryogen lifetime goal was established for the HFO mission because cryogen replenishment and servicing are not, according to NASA, feasible in HEO for the SIRTF.

The consensus of the SIRTF Study Office at NASA/ARC, and other NASA centers which participated in the Mission Options Study, and the SIRTF Science Working Group is that the high orbit offers significant scientific and engineering advantage for SIRTF [8]. However, the high orbit precludes NSTS or SSF servicing—a big consideration should a large on-orbit spacecraft problem arise.

Conclusions [9]

The SIRTF all-superfluid helium dewar, based on heritage from IRAS and COBE, can achieve a 5-year (with 20 percent margin) lifetime for the 100,000 km (53,960 nmi) HEO mission. For the previous 900 km (486 nmi) LEO mission, the dewar has a cryogen lifetime of 2 years (with 25 percent margin) which would require at least two on-orbit cryogen replenishment missions in order to meet the 5-year mission lifetime requirement. Further design optimization of the cryogenic system and observatory configuration is needed in order to enhance dewar lifetime for the HEO mission. Methods to reduce the dewar outer shell temperature should be considered. Use of advanced materials such as alumina/epoxy or alumina/poly-ether-ether-ketone (PEEK) in cryogen tank supports may improve the cryogen lifetime by about 10 percent. Reduction of launch loads to the tank supports will decrease the required tank support size, thus improving cryogen lifetime. The use of auxiliary cooler, hybrid cryogen system, or additional vapor-cooled-shield may improve lifetime, but possibly at the expense of

Platform	Altitude (km)	Equator Crossing	Payload Instruments	Mass (kg)	Power (kw)
Eos-A (NASA)	705	1:30 pm	19	13,000	6
Eos-B (NASA)	705	1:30 pm	10	14,000	8.7
EPOP-1 (ESA)	824	10/10:30 am	9	–	–
EPOP-2 (ESA)	705	10/10:30 am	5	–	–
JPOP (NASDA)	800	–	6	–	–

Figure 6.14 Polar orbiting platform missions. *Courtesy of NASA.*

adding complexity, mass, and power consumption to the existing design. For the current design, lifetime increases by 2 days for every kilogram of helium plus tank mass added to the system.

Earth Observing System

The polar orbiting platforms which will support EOS, the Earth Observing System, present a particularly challenging environment for servicing via expendable launch vehicles. Requirements for telerobotic or robotic servicing support, existence of a high ambient radiation environment, and the use of modular, serviceable engineering components, many of which come from the Space Station Freedom program, represent some of the mission characteristics which impact satellite design for servicing.

The Earth Observing System is named for Eos, Greek goddess of the dawn. It is part of an ambitious and comprehensive study of the planet Earth, its natural processes, and man's influence on these [10]. Data will be assembled for multidisciplinary measurements of atmospheric, oceanic, geologic, and biologic activity over an extended period of time in greater detail than has been previously attempted. More significantly, these measurements have been proposed as interrelated parts which will allow the development of correlations among measurement sets from different instruments on the same and on different platforms. Assembly of a comprehensive data set over a period of 10 to 15 years is also a key element of EOS. This, coupled with the usage of very large, expensive platforms and payloads, leads to consideration of the potential role of servicing in maintaining the science instruments and engineering subsystems of the platforms. The EOS program will operate with a combination of large polar orbiting platforms (POPs), individual sortie or free-flyer missions, and payloads in low inclination orbit attached to the Space Station Freedom manned base. Program participation will include NASA, NOAA, and international partners. EOS began in 1991. The first polar orbiting platform is scheduled for launch in December of 1997.

EOS Polar Orbiting Platforms

Payloads in several separate orbits fulfill the basic polar orbiting platform (POP) mission. International cooperation allows two platforms to be provided by NASA, and other platforms to be supplied by the European Space Agency (ESA) and the Japanese Space Development Agency (NASDA). Two ESA platforms with different payloads are under consideration. Figure 6.14 gives details of the platforms.

A basic Sun-synchronous orbit is required for all platforms. Repeat coverage requirements and phasing between EOS-A and EOS-B orbits constrain the NASA platforms. NASDA and EPOP platforms will operate at different orbital altitudes. The net effect of these constraints is that four to five operational platforms will be operating in distinctly different polar orbits, requiring significant maneuvering by a servicer vehicle traveling from one platform's orbit to another's.

Of the NASA platforms, EOS-A is under the management of Goddard Space Flight Center. The focus for this discussion is on EOS-B which is managed by the Jet Propulsion Laboratory, with the proviso that maximum commonality will be attempted with EOS-A.

Platform Configurations

Figure 6.15 illustrates a potential configuration for the EOS-B platform. Both NASA platforms are marked by an integral propulsion module at the aft end, payload instruments primarily arranged on the nadir side of the platform, and engineering orbit replaceable units (ORUs) containing support subsystem components on the zenith side. The configuration is largely driven by optical field-of-view requirements for the payloads and thermal requirements for engineering ORUs.

The deployed platform is approximately 22 meters (72 feet) long and 4 meters (13 feet) across, with a deployed solar array length of 29 meters (95 feet). The remote fields and particles platform booms are 25 to 17 meters (82 to 55 feet) from the platform. Figure 6.16 lists proposed EOS-B platform instrument payloads. Of special interest is the wide range of payload sizes and masses. The payloads range from items with tens of kilograms and modest sizes of 1 cubic meter or less to items occupying several cubic meters and massing up to 771 kg (1,700 lb).

Servicing Ground Rules and Environment

Early plans for POP servicing are based on Shuttle-based extravehicular activity, with the POP's maneuvering down

Serviceable Spacecraft

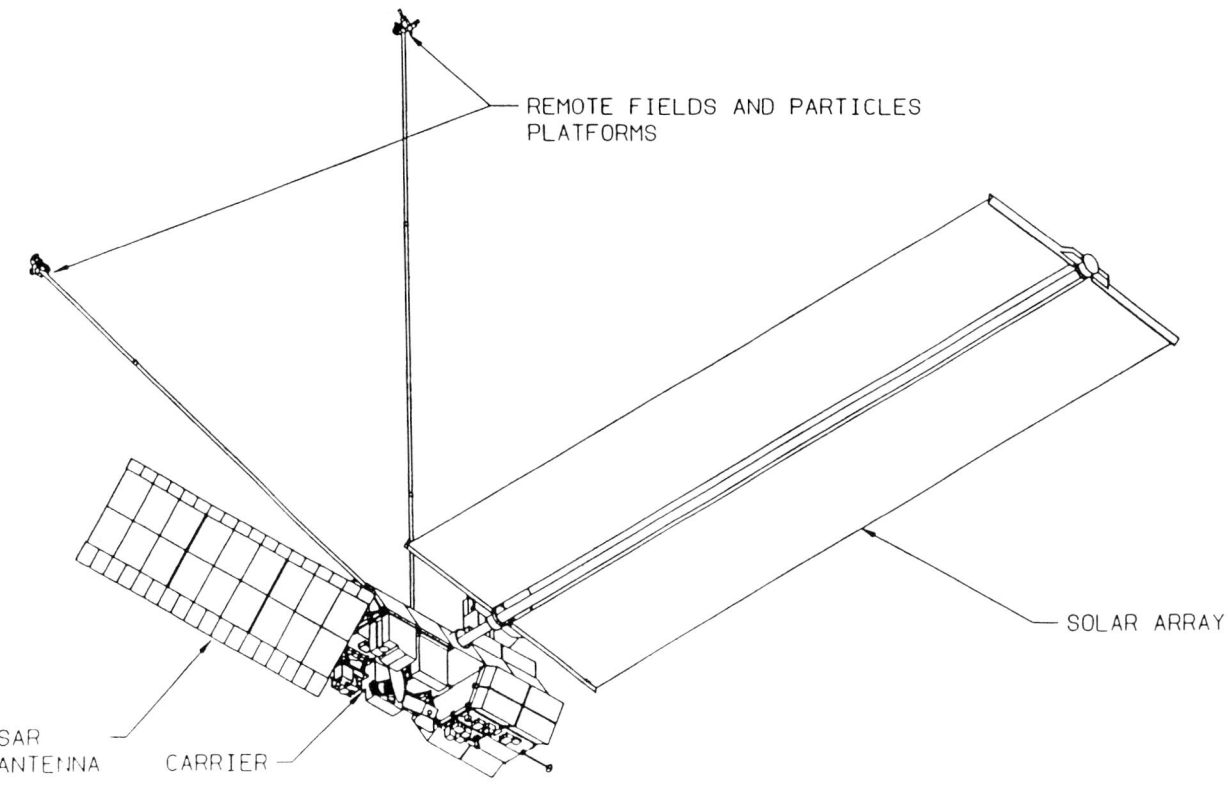

Figure 6.15 EOS-B on-orbit configuration. *Courtesy of NASA.*

to an orbit accessible to the Shuttle from a Western Test Range (WTR) launch. As there are no current plans for WTR Shuttle operations, this concept has been abandoned for the prospective use of expendable launch vehicle (ELV) servicing. A Joint Servicing Study was carried out under the auspices of the Space Station Freedom office for Program Engineering for transfer vehicles, launch vehicles, and robotics. Some of the ground rules for ELV servicing developed in this study were:

1. EOS platforms will not be maneuvered from their operational orbits to rendezvous with service vehicles.
2. All replaced instruments or engineering ORUs will be disposed of through controlled reentry. No provisions are available for refurbishment or reuse of replaced components.
3. Each individual platform will be serviced by dedicated missions.
4. Servicing in lieu of platform replication will require full ORU/payload replacement over projected component life span.

Servicing Mission Concepts

The two basic modes of servicing of instruments and engineering ORUs are use of an add-on module and robotic replacement of individual items. While neither concept has yet been chosen as a sole servicing design feature, platform design for serviceability can accommodate either option.

Add-on servicing provides a rapid addition of new or replacement components with good provisions for pre-mission integration and test. It does, however, limit expansion of the platform due to attitude control dynamics limits and has severe conflicts with space already occupied by instruments and the propulsion module at the ends of the platform.

Robotic servicing allows servicing flexibility at the cost of scars to the platform and risks associated with an unverified technology with no manned EVA backup.

Platform Design Features for Servicing

The overall platform accommodates servicing by use of a modular design concept illustrated in Figure 6.17. The basic features include use of an inter-module connector to accommodate service vehicle docking and add-on service modules, use of standard Space Station Freedom ORU designs for housing engineering subsystems, and use of standardized payload plates for mounting instruments to the structure via standard interface connectors.

The inter-module connector (IMC) serves as a docking adaptor capable of accepting an add-on service module or an exchange carrier module brought up by a service vehicle. The IMC must provide for resource connections for power, data, and thermal services with the platform. It latches the

Instrument	Mass (kg)	Peak Power (kw)	Average Power (kw)	Peak Data (Mbps)	Average Data (Mbps)
SAR	1670	9.	2.2	270.	20.
TES	491	.654	.446	30.	5.
SAFIRE	304	.350	.304	9.	9.
MLS	468	.887	.645	1.	.9
SWIRLS	90	.201	.196	.001	.001
IPEI	10	.006	.006	.001	.001
GOS	35	.110	.087	.008	.008
XIE	85	.035	.035	.024	.012
GGI	35	.090	.090	.050	.020
LIS	8	.015	.015	.003	.001
COMM	51	.027	.018	–	–
	3247				

Figure 6.16 EOS-B payload mass and power summary. *Courtesy of NASA.*

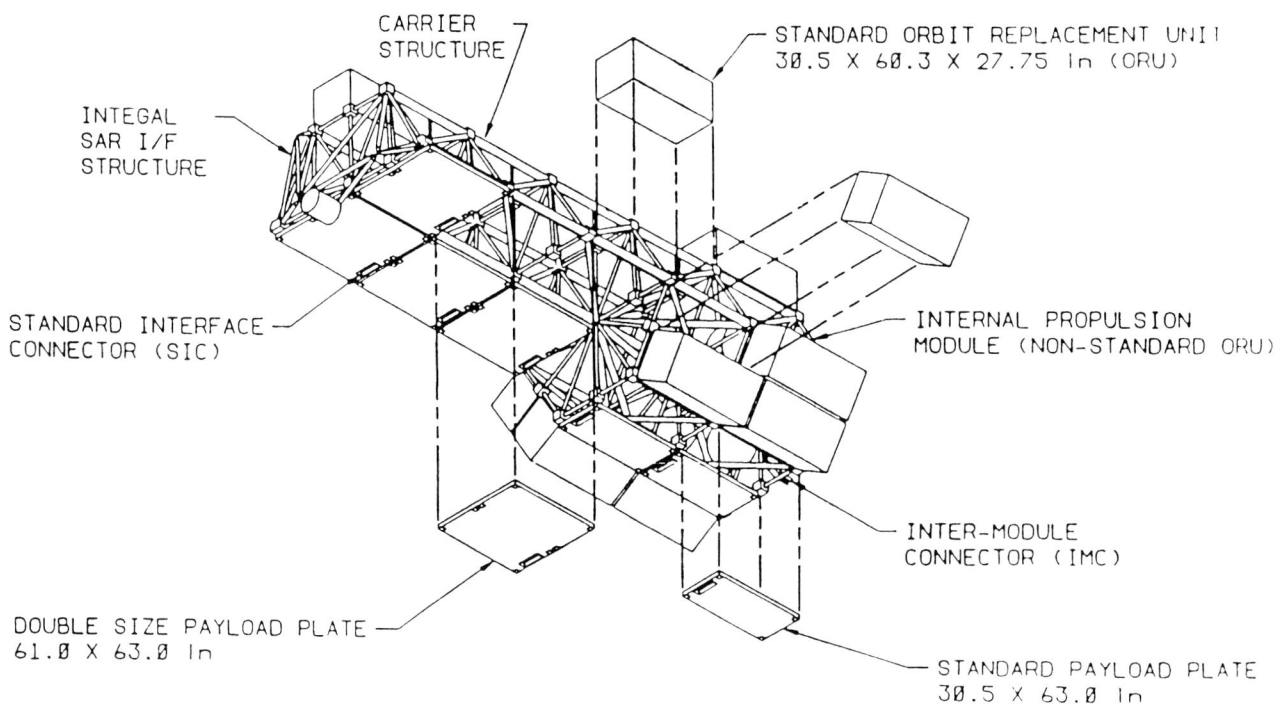

Figure 6.17 Modular carrier concept. *Courtesy of JPL.*

added module with sufficient rigidity and positioning accuracy to allow proper dynamic stiffness and reference axis alignment for instrument payloads. For an exchange service carrier, it must provide sufficient rigidity and positioning accuracy to maintain the robotic servicing arm/s in reference to platform ORU/payload references.

The ORU designs allow engineering subsystems to be mounted and changed out with a minimum requirement for dexterity and specialized tools. The current design of EOS-B has six standard power ORUs, two core ORUs, two recorder/HRM ORUs, two reaction wheel assembly ORUs, one GN&C sensor ORU, and two command & telemetry/data management system ORUs. These are common designs with EOS-A ORUs. The TDRS antenna boom and solar array size and deployment mechanism are different on the two platforms.

Payload instruments are mounted on standardized plates which attach to the platform via standard interface connectors. These plates represent the level at which exchange servicing would take place. The plates are available in two sizes, single and double, to accommodate the range of instrument sizes on the platforms. Several small instruments or components may be mounted on an individual plate to take maximum advantage of available surface area, though this may require changeout of a good instrument to replace a failed one on the same plate. The SIC interface serves to

Serviceable Spacecraft

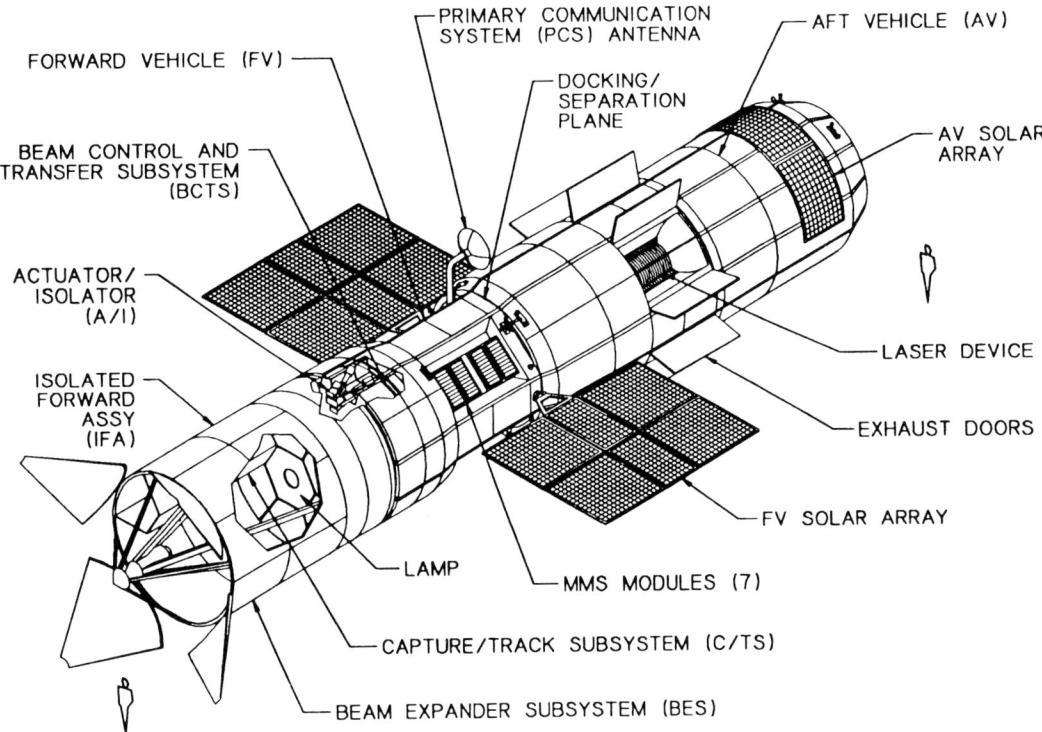

Figure 6.18 Zenith Star space vehicle. *Courtesy of Martin Marietta, Lockheed, and TRW.*

allow rigid, precise alignment of the plate and instruments to the platform while allowing for a simple tool interface to a robot servicer.

Some key components such as structure and system harnesses are not intended for servicing. Other major components such as solar arrays, TDRS/RF equipment, and payload items such as large SAR antennas would require non-standard servicing techniques and are not being considered for servicing at this time.

NASA has not yet approved final design of the EOS spacecraft. Rather than large platforms, considerable attention is currently directed at a series of small spacecraft to conduct the various Earth resources, meteorology, and downward looking Earth's atmospheric measurements in the EOS program. If it is decided to perform the EOS program with a variety of small platforms carrying a few instruments on each flight, the individual spacecraft may be so small, and therefore inexpensive, so as to invalidate on-orbit servicing as an economically attractive life cycle cost program option. If the price for mounting a servicing mission exceeds the spacecraft replacement cost, then obviously it is more cost-effective to replace the asset rather than repair it.

Zenith Star Program

The Air Force Zenith Star program objective is to resolve key technical issues required for an operational directed energy weapon systems by demonstrating technology in space related to capture tracking and pointing, beam generation, plume phenomenology, and space operations (docking, refueling, servicing) [11]. It is scheduled for initial launch in the mid 1990s.

The Zenith Star system, Figure 6.18, is large. It consists of forward and aft vehicles, each 12 meters (40 feet) long and each weighing about 22,680 kg (50,000 lb). The aft vehicle contains a modified Alpha laser, while the forward vehicle has the beam control and lamp mirror beam expander. The aft vehicle lifetime is planned for 90 days, but the forward vehicle lifetime is set at 1 year with life extension through resupply. It will be launched by the Titan IV with upgraded solids.

The Zenith Star servicing and resupply design attributes are shown on Figure 6.19. The design features:

1. Avionics packaged in serviceable MMS modules
2. Interfaces based on HST side of OMV/HST modules
3. Remote mateable—demateable connectors
4. A rendezvous and docking subsystem
5. RMS attach points which are populated when servicing reach envelopes are defined
6. Fluid servicing by modular replacement
7. Fluid servicing in three subsystems
   - Reactants for high energy laser
   - Pressurants for mirror coolant
   - Propellants and pressurants for propulsion

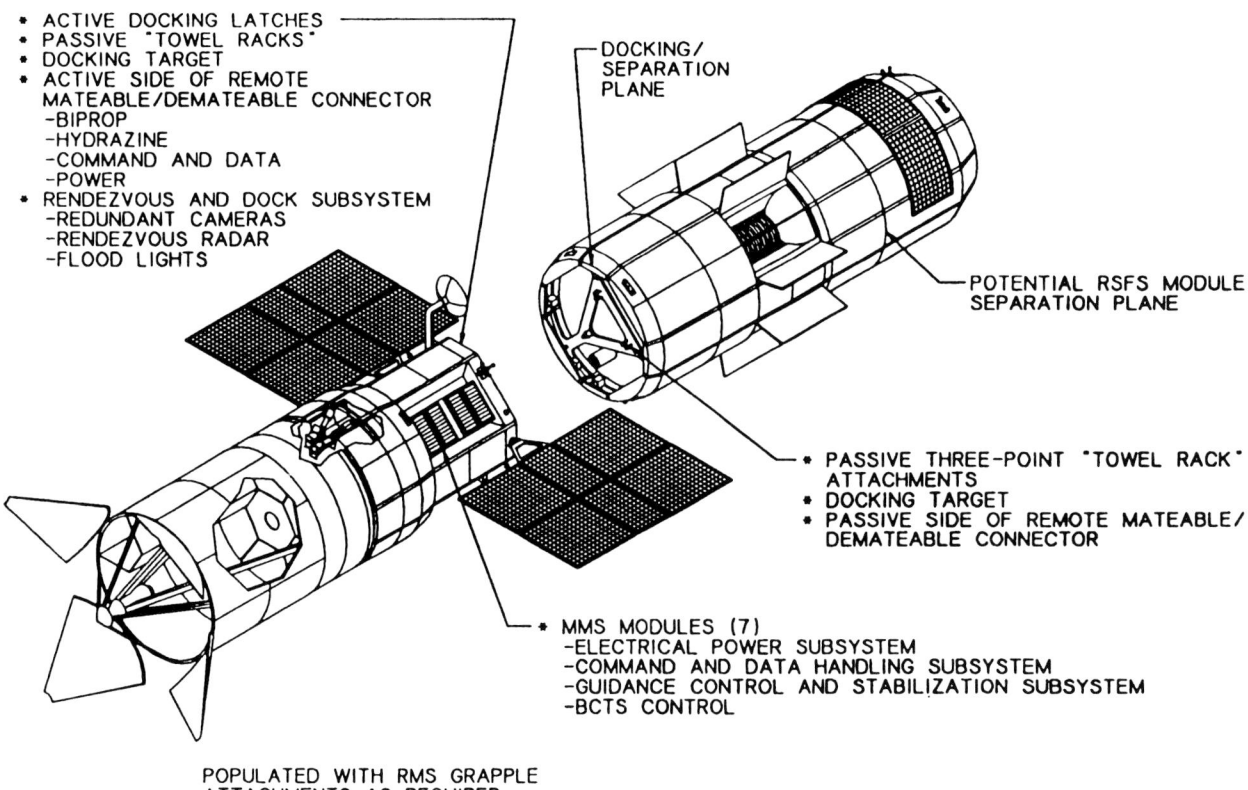

Figure 6.19 Zenith Star Servicing and resupply design attributes. *Courtesy of Martin Marietta, Lockheed, and TRW.*

Orbital Maneuvering Vehicle (OMV)

As discussed in Chapter 3, the OMV was under advanced design and development for NASA Marshall Space Flight Center by TRW's Federal Systems Division of their Space and Technology Group, Redondo Beach, California, until it was canceled for NASA budget constraint reasons in mid 1990. However, since the OMV was designed to be on-orbit serviceable, as well as designed to serve as an integral part of the U.S. satellite servicing infrastructure over the next two decades, it is appropriate to the overall intent of this book to present a review of its serviceable characteristics. Besides the NASA decision to cancel the OMV may have an evanescence quality as evolving NASA and Air Force special requirements for on-orbit servicing in the next few years may revive an OMV concept into some form of cargo transfer vehicle (CTV) for in-orbit transportation. Ideas for future servicing programs and events are presented in Chapter 10.

OMV Background

The very first sentence of this book stated . . . "On-orbit servicing is work in space" . . . Working in space means moving in space. The needs of the Space Station and other civil and military users guarantee that an OMV will play a key role in long-term space plans.

Space construction and satellite servicing require a reliable, reusable "space tug" to move structures and systems. That tug must adapt to payloads and missions not yet devised, and make minimal demands on a limited Space Transportation System. TRW designed a modular OMV, controlled from the ground or the Space Station and readily serviced on orbit.

Schemes for an OMV began under the shadow of Skylab in 1974. The Teleoperator Retrieval System (TRS) was a single-mission vehicle intended to ride the Space Shuttle to orbit and reboost the falling Skylab space station. Martin Marietta based the TRS on Viking hardware, but the Shuttle had yet to fly when Skylab reentered the atmosphere in July 1979, and the rescue vehicle was abandoned.

The idea of a ground-controlled spacecraft to deliver, retrieve, reboost and deorbit space systems remained attractive. TRW, LTV, and Martin Marietta completed Phase B design studies of an Orbital Maneuvering Vehicle for NASA in August 1985. The request for OMV proposals appeared that November and TRW was selected to develop the OMV in June 1986.

Serviceable Spacecraft

The contract awarded by the Marshall Space Flight Center in November 1986 covered a single OMV and an option for a second. The NASA contract also contained a separate U.S. Department of Defense task to identify possible changes required for military OMV missions, possibly aboard a Titan IV expendable launch vehicle. The OMV had no firm role in the Strategic Defense Initiative at this time, but it was considered to have great potential for military satellite servicing and inspection.

Plans originally called for an OMV demonstration mission to retrieve the Solar Max spacecraft in July or August 1991, in preparation for a reboost of the Hubble Space Telescope that November. However, telescope mission planners decided to shape their schedule around solar cycles and remain independent of the OMV, and the mission then of the space tug was tied to the International Space Station Freedom.

NASA defined 12 OMV reference missions:

OMV Reference Mission	Payload Weight	
	Kilograms	Pounds
1. Large Observtory Service	11,340	25,000
2. Satellite Placement	1,588	3,500
3. Satellite Retrieval	4,900	11,000
4. Satellite Reboost	11,340	25,000
5. Satellite Deboost	34,020	75,000
6. Satellite Viewing	91	200
7. OMV Acts Like a Subsatellite	2,268	5,000
8. Servicing Satellite	2,268	5,000
9. Satellite Place and Retrieve	4,536	10,000
10. Logistics Module Transfer To SSF	22,680	50,000
11. SSF Support	454	1,000
12. OTV/Payload Transfer	34,020	75,000

Shaping the OMV

Besides Space Station construction, there are satellite servicing tasks that cannot be done without an OMV, and there are free-flying experiment missions that don't require an Orbiter or dedicated satellites. The initial OMV was to be limited to a maximum orbital plane change of 6 degrees with the minimum 227 kg (500 lb) payload (Shuttle altitudes) and it could not reach polar orbits on a Cape Canaveral Shuttle launch. Geosynchronous orbit is more the realm of the larger, cryogenically fuelled Space or Orbital Transfer Vehicle. But the OMV was planned to grow and change to suit many ambitious missions.

A large observatory servicing mission calls for an OMV to retrieve a 11,340 kg (25,000 lb) satellite from an orbit 240 km (130 nmi) above the Shuttle or Space Station, and reboost it to its original orbit. The OMV can deliver a 1,588 kg (3,500 lb) payload to an orbit 630 km (340 nm) above the Shuttle and bring it back to the Orbiter bay if the satellite fails to work.

A payload retrieval mission calls for recovering a 4,990 kg (11,000 lb) satellite from an orbit 408 km (220 nmi) above the Shuttle or Space Station. A typical payload reboost includes a rendezvous 185 km (100 nmi) above Shuttle or Station to push an 11,340 kg (25,000 lb) payload back to an orbit 408 km (220 nmi) above base altitude. Useful as it is for saving space systems, the OMV can also de-orbit up to 34,020 kg (75,000 lb) of "space junk" for a safe reentry.

Some OMV missions do not require docking with a distant target: a payload viewing excursion circumnavigates a satellite 11,557 km (840 nmi) above the Shuttle or Station and returns video or high resolution photographic imagery. Other missions require some delicate work under ground control. For example, the OMV could take a 2,268 kg (5,000 lb) servicing kit to a free-flying platform 741 km (400 nmi) above the Space Station, dock, exchange modules, and return to base. The OMV could also routinely inspect the outside of Space Station Freedom and occasionally, for a week, deploy experiments from the Station.

The velocity increments required for orbital plane and altitude changes, and the duration of typical missions, established the size of the OMV fuel and electrical power systems. The confines of the Orbiter bay and requirements for autonomy, redundancy and on-orbit servicing helped determine the shape of the package and the systems aboard. The result was a modular spacecraft 4.5 meters (15 ft) in diameter but just 142 cm (56 inches) deep with a fully loaded weight of 9,027 kg (19,900 lb), Figure 6.20.

The OMV System [12]

The elements of the MSFC/TRW OMV System are identified on Figure 6.21. They include the OMV flight vehicle, payload accommodation equipment (PAE), the ground control console (GCC), ground support and test equipment, and ASE to interface with the NSTS. The OMV interfaces are shown, including the launch system, NSTS, the orbiting payloads it supports, such as the Hubble Space Telescope (HST), and the TDRSS for communications to the ground. The OMV could support the Space Station for logistics and other missions.

The RMS grapple docking mechanism (RGDM) and the three point docking mechanism (TPDM) provided two of the three methods of mating the OMV and the payloads that it supports. The third method was to hard mount, i.e., attach with bolts, payloads that were attached to the OMV during launch.

The OMV communications link through TDRSS to the

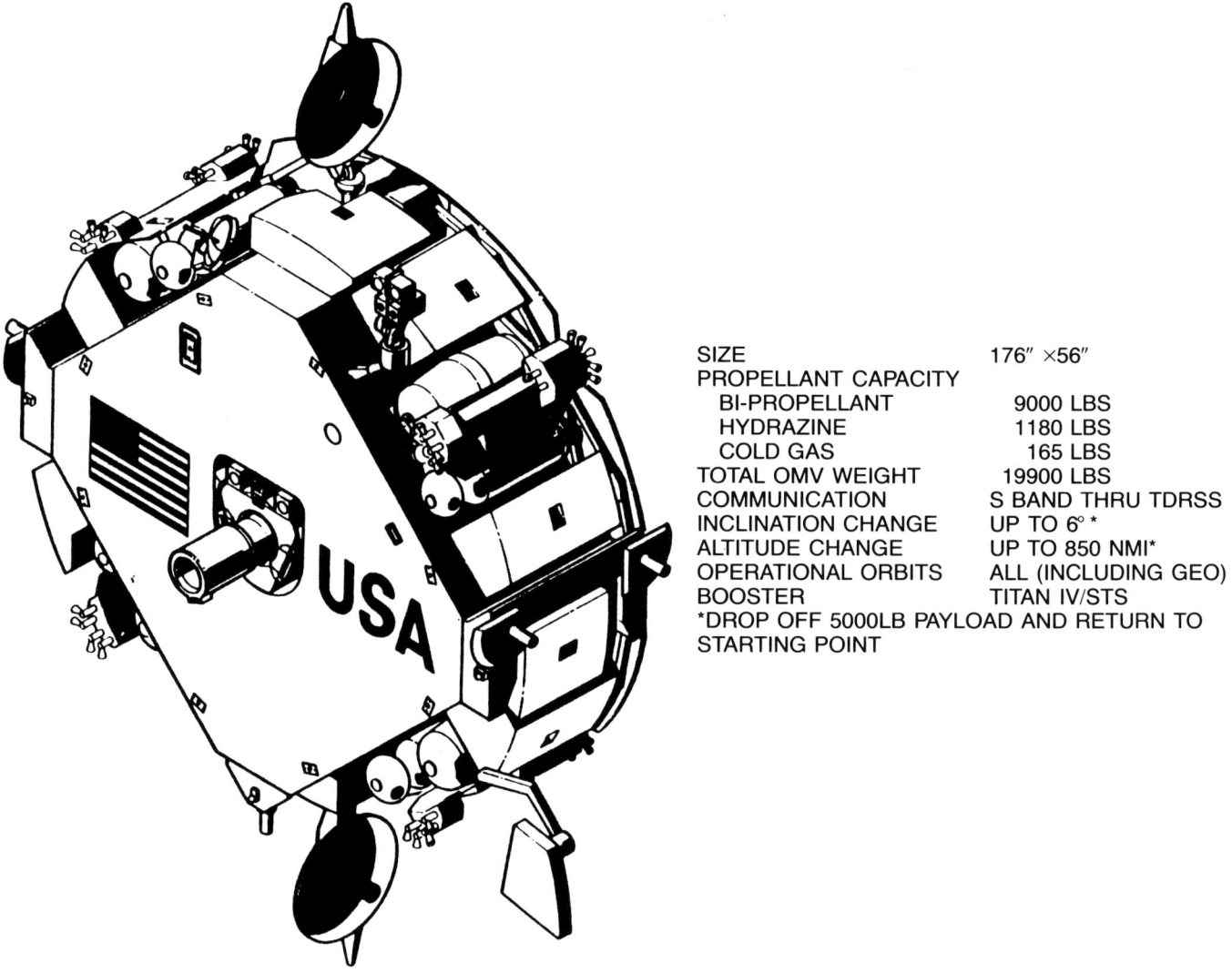

Figure 6.20 Orbital maneuvering vehicle characteristics. *Courtesy of TRW.*

GCC provided monitoring and control of the OMV missions. The OMV was to be capable of automatic rendezvous with a target satellite, to within 305 meters (1,000 ft). Ground controlled docking was accomplished with pilot assistance. Recent autonomous docking studies show the docking operation could be automated with appropriate changes to the OMV GNC software and the incorporation of a docking sensor to supplant the docking information supplied by the pilot from the ground.

OMV Configuration

The MSFC/TRW designed OMV, Figure 6.22, consists of the short range vehicle (SRV) and the propulsion module (PM) as illustrated on this figure, which also shows the orbital replacement unit (ORU) locations on the SRV. The ORUs house the vehicle equipment and are designed for replacement by EVA or robotics. The SRV is a fully operable vehicle with the PM removable because the hydrazine thruster, normally used for reaction control, can also be used for orbital maneuvers.

The PM consists of four variable thrust engines (VTE) fed from four NTO/MMH fuel tanks with total fuel capacity of 4,082 kg (9,000 lb). The reaction control system consists of a hydrazine system with 28 12-lbf thrusters with 540 kg (1,190 lb) of useable N_2H_4 plus a cold gas (G_2) system with 24 5-lbf thrusters with 75 kg (165 lb) of useable GN_2. The cold gas system is to be used for maneuvers in close proximity to sensitive payloads to avoid contamination. The four RCS ORUs are equipped with automatic fluid couplings at the ORU/SRV interface that manifolds the hydrazine tanks and connects to an EVA refueling coupling on the front surface. The GN_2 pressurant and cold gas RCS tanks can be refueled on orbit through refueling couplings located at the forward end of each of the RCS ORUs. The PM can be replaced on orbit by release of the latch mechanisms located in the SRV. There are no fluid connections

Serviceable Spacecraft

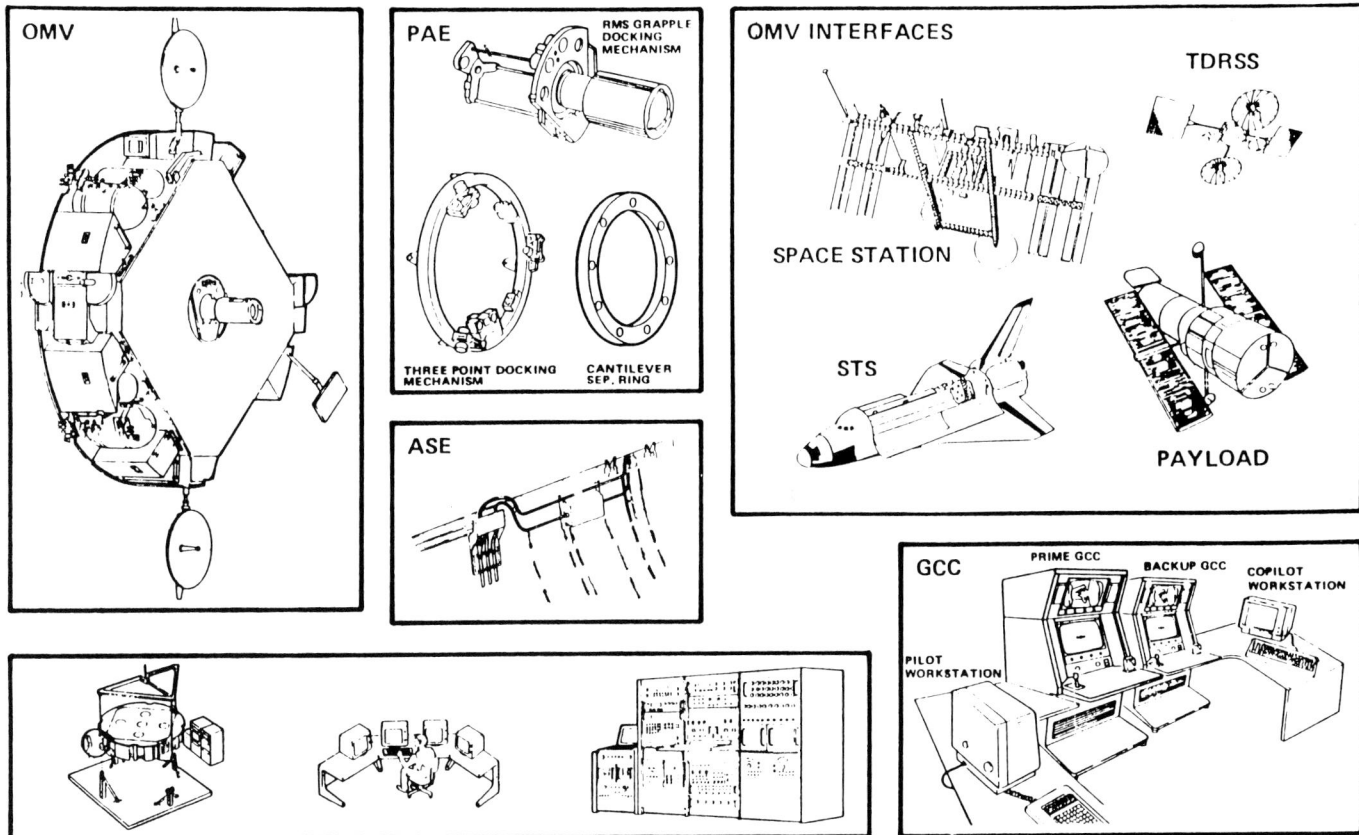

Figure 6.21 The orbital maneuvering vehicle system. *Courtesy of TRW.*

between the SRV and PM, although the PM was scarred to accommodate refueling of the tanks in the future.

Viewing of payloads and docking information to assist the pilot was provided by RV cameras, two located just above the RGDM and two pan-tilt-zoom cameras mounted on a deployable boom. The high gain antennas and the radar antenna were also located on deployable booms.

The payload electrical connector, noted on the front side, provided the electrical interface between the OMV and payloads that were attached with the TPDM or bolted to the eight attach points located on the perimeter of the front side (135 inch diameter bolt circle). Electrical connectors are also provided on the RGDM for payloads mated to the OMV by that method.

The OMV was to be a uniquely adaptable vehicle that provides propulsion, power, communications, attitude control and rendezvous, and docking systems to support attached servicer kits or mission specific payloads. Using these features for remote satellite servicing is discussed in Chapter 8.

OMV Servicing

The OMV was designed to provide servicing flexibility at the launch site(s) and on-orbit. The vehicle is modular; see Figure 6.22. The main delta velocity propulsion module (PM) is removable allowing the bipropellant system to be serviced and refueled in parallel with the short range vehicle (SRV) during pre-launch or post-launch processing. The avionics ORUs have mechanical and electrical connectors to the OMV that allow removal and replacement by either robotic or manual methods. Additionally, the manifolded reaction control system (RCS) ORUs are scarred for fluid disconnects in the hydrazine system. The ORU designs drive toward easily removable internal black boxes. This allows replacement of failed units during prelaunch processing and leads to servicing at an orbiting facility.

The PM design, which permits replacement of the total bipropellant delta velocity system, allows the OMV to be space based without requiring on-orbit biprop fuel transfer. The PM has only mechanical and electrical interfaces with the SRV. There are no propellant lines across the interface. In this way, resupply of a fully fueled PM to a space-based OMV gives mission flexibility to the program prior to the development of a space qualified biprop umbilical connector. Full scale mockups were used to demonstrate the replacement capability. The Grumman Aerospace Corporation large amplitude space simulator (LASS) was used for removal and replacement tests and the capability of astronauts to remove and replace the PM with EVA was demon-

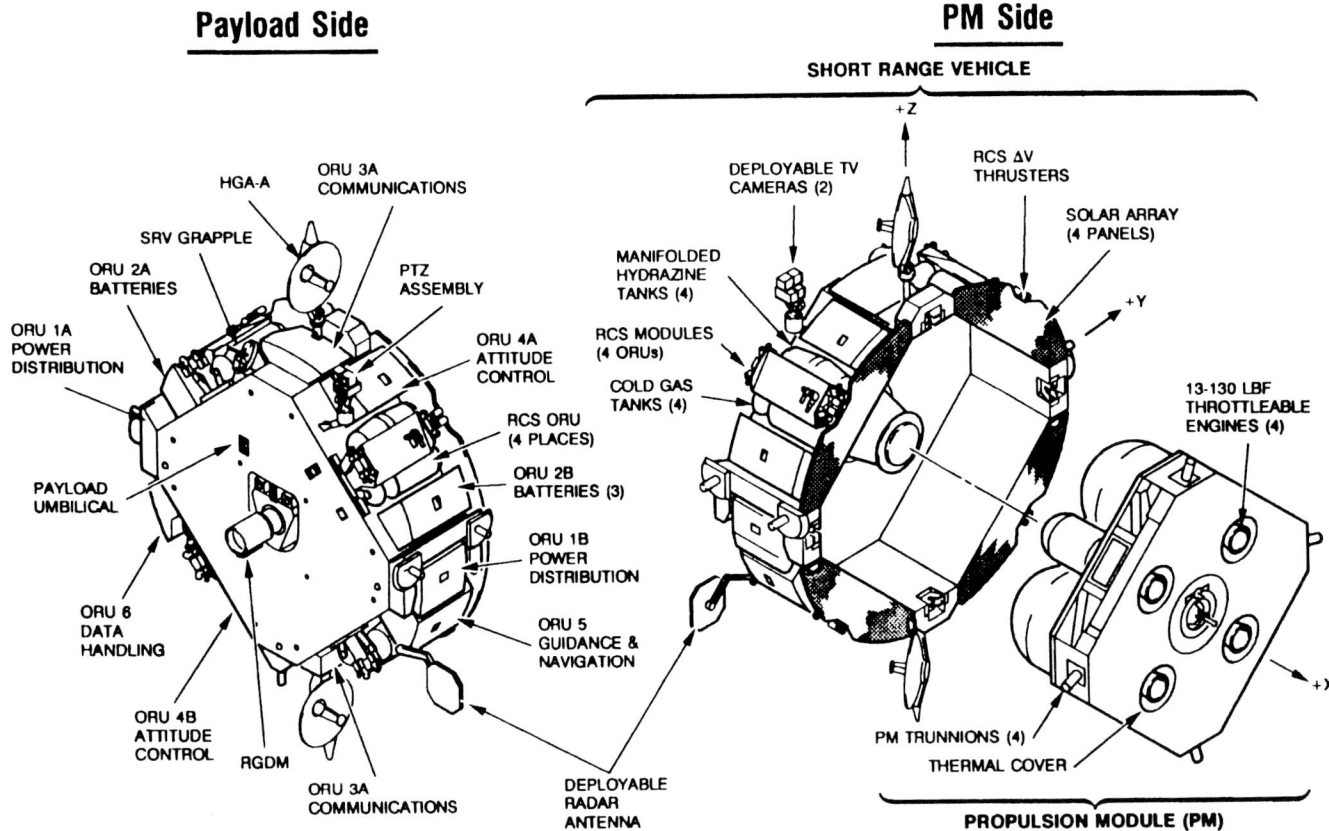

Figure 6.22 OMV, showing its fully modular design. *Courtesy of TRW.*

strated in MSFC's neutral buoyancy facility, as shown in Figure 6.23.

The ORUs are designed to have simple electrical and mechanical interfaces with the OMV. Each ORU is thermally independent. On-orbit failures in one ORU will not thermally degrade an adjacent ORU. The simple interfaces between ORUs and the OMV platform allow easy changeout of ORUs both on the ground and on orbit. During OMV development, ORU changeout was demonstrated in the MSFC neutral buoyancy facility. Suited astronauts removed and replaced an ORU on a full scale mockup of the OMV.

The design of the avionics ORUs permits removal of black boxes internal to the ORU with minimum effort. The ORU structure consists of a central spine for mounting equipment and supporting the external thermal control surface. The spine includes electrical connectors at the bottom and the attach mechanism for replacement/removal. For ground servicing, the black boxes are easily accessible. For on-orbit servicing in the Space Station era, the ORU is removed from the OMV and transported to a shirt-sleeve environment servicing area where similar servicing can be performed. The RCS ORUs have field joints at the fluid interface in the initial version of the OMV; however, the design is scarred to permit on-orbit changeout when a space qualified rematable fluid connector is available. Thrusters are to be welded to the propellant feed lines in the baseline design.

The design of ORUs allows changeout to be accomplished on-orbit with a remote manipulator system (RMS) arm and a universal servicer tool (UST) attachment. The attach mechanism utilizes a one bolt operated mechanism for attachment and a multimission spacecraft module compatible mechanism for the UST interface.

Launch Site Processing

The OMV was designed to accommodate either vertical or horizontal processing at Kennedy Space Center (KSC) and at the Western Test Range (WTR). Because of its modular design the OMV could support parallel pre-launch and post-launch activities. Servicing—propellant loading, pressurization, maintenance, and refurbishment—could be performed on the PM and SRV at different facilities, if required. Also the OMV could have been integrated into the STS launch flow either at the Orbiter Processing Facility (OPF) for horizontal processing or at the Rotating Service Structure (RSS) for vertical processing. For safety considerations the vertical processing was recommended.

On-Orbit Servicing

The OMV could have been serviced post-mission at the launch site(s) and normal maintenance/refurbishment

Serviceable Spacecraft

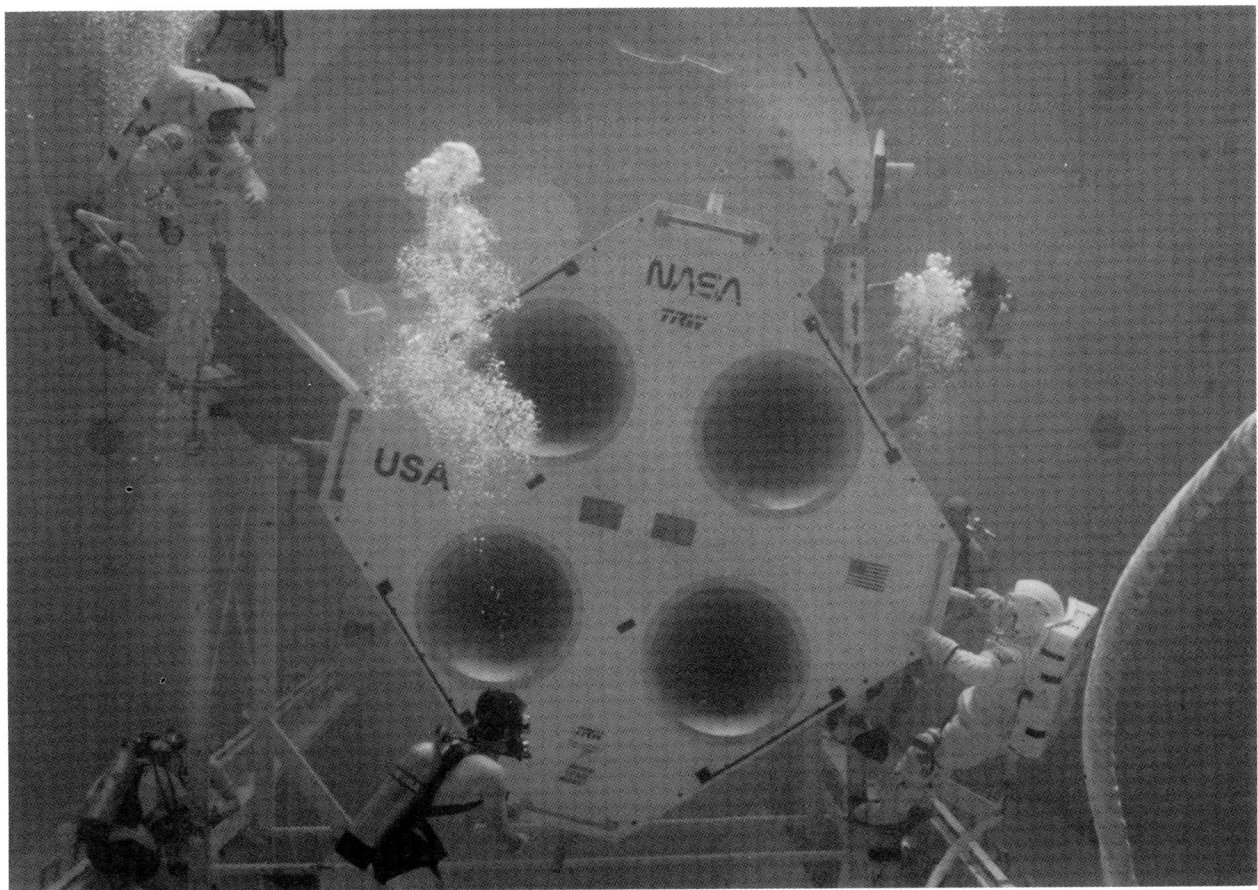

Figure 6.23 Leo Stytle of TRW, upper left, with OMV in the MSFC neutral buoyancy simulator performing changeout of the propulsion module. *Courtesy of TRW.*

tasks, in addition to replacement of equipment for an on-orbit anomalous performance, could have been accomplished with ORU/black box replacement. The addition of a cradle to transport replacement ORUs to orbit and to hold the OMV above the STS payload bay would permit on-orbit servicing of the OMV by changeout of ORUs and the PM. A space-based OMV could be serviced in the same manner as an STS-based OMV. That is, it could be returned to the launch site for servicing or it could be serviced by replacements utilizing an ASE cradle. In the Space Station era, servicing could expand from ORU/PM replacement to equipment replacement in a SSF servicing area. The PM could be changed out, or with a space qualified refueling coupler biprop, fluid transfer could be accomplished. The hydrazine system could be either refueled or RCS modules could be replaced.

Space Station Freedom Customer Servicing/Spacecraft Interactions

The International Space Station Freedom is an integral element in the U.S. infrastructure for satellite servicing. Accommodating a variety of customers, it will provide a true "servicing facility in the sky." This new national capability to our nations space effort will be initially available in the late 1990s. An OMV type vehicle should be operationally available in the same time frame, to move payloads or spacecraft to and from the Space Station and/or to act as the orbit-to-orbit vehicle to service spacecraft in situ, using the Station as its base.

This combined capability (Space Station and OMV) enables many spacecraft servicing strategy options. Now the key to implementing this robust capability is to match the Station's capability to the needs of the spacecraft user or customer. It isn't one way. The spacecraft owner/designer must be aware and responsive to the capability and constraints of the Station—especially the early Station. Figure 6.24 names the principal interactions. Requirements of both the Station and spacecraft must be understood and negotiated. With both the Station design evolution and the serviceable spacecraft eras just starting, we have the rare opportunity to effectively manage this interface to the technical and economic agreement of both.

The key to this optimum agreement is the employment of system engineering integration techniques successfully used on past space programs.

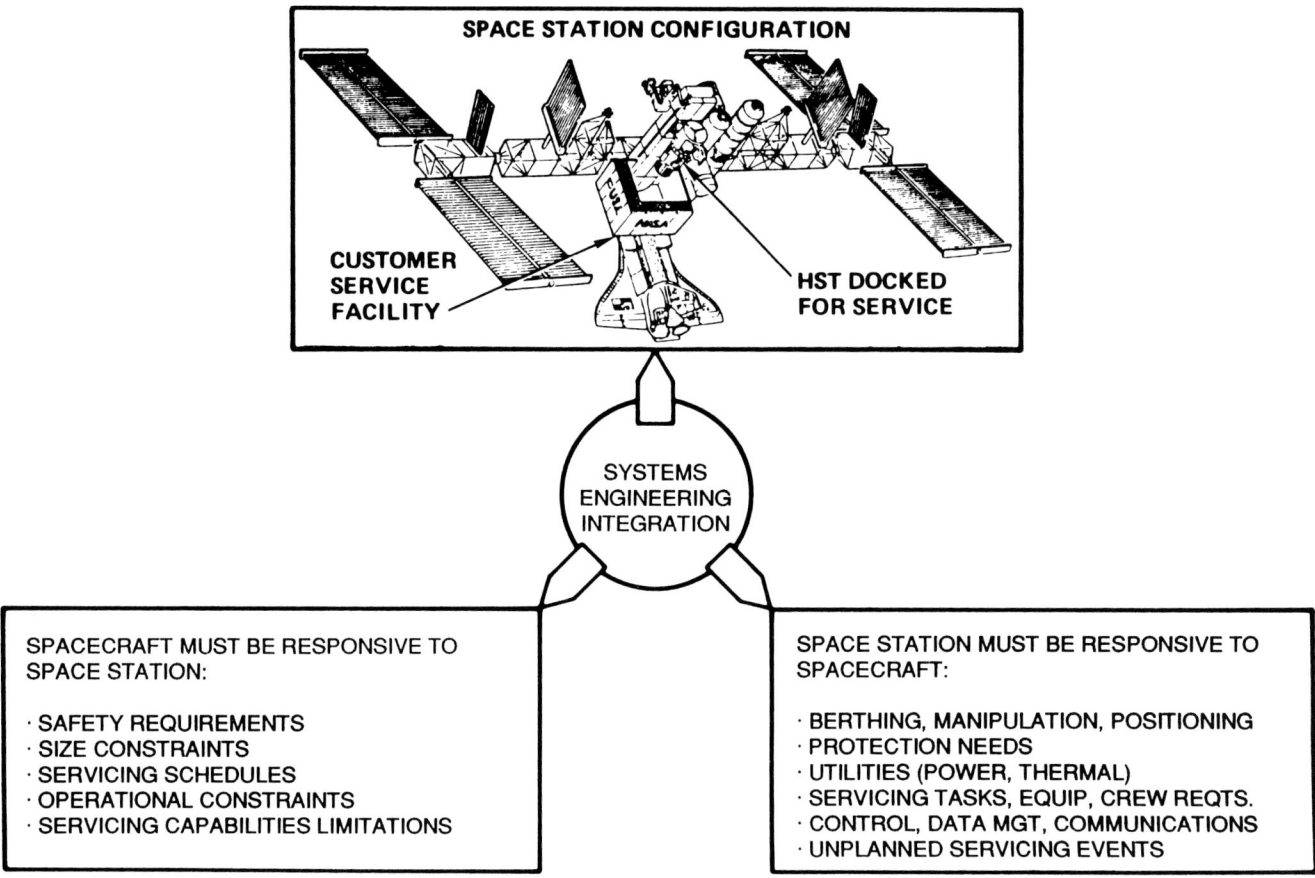

Figure 6.24 Space station customer servicing/spacecraft interactions.

Servicing the Non-Serviceable Spacecraft—The Rescue of Intelsat 6/F3

When the Space Shuttle *Endeavour* landed at Edwards Air Force Base, California, on May 16, 1992, with its seven-member crew, it marked the end of a historic, 9-day satellite servicing mission. The record-breaking satellite rescue and Space Station development mission demonstrated an unprecedented rendezvous and EVA capability that will be important to advancing the space operations of all nations. It was the 47th flight of the Space Shuttle in 11 years of operations and the 22nd flight following the *Challenger* disaster. *Endeavour*, launched May 7, 1992, from NASA/KSC, Florida, was built at a cost of $2 billion by Rockwell International to replace *Challenger*, lost January 28, 1986, in a takeoff explosion that killed seven crew members and stalled the U.S. manned space program for 2 1/2 years. With *Endeavour*, NASA once again has a fleet of four Space Shuttles.

The astronaut's dramatic capture of the marooned Intelsat 6/F3 communication satellite again embellished NASA's CAN DO reputation.

The crew conducted three separate rendezvous involving dozens of Orbiter maneuvers and extensive mission control support. It included several hours of formation flight with the Intelsat, much of it with an astronaut atop the Shuttle's manipulator arm, making multiple contacts with the satellite. It was the first U.S. orbital mission with four EVAs and the first U.S. or Soviet mission to involve three EVA crewmembers working simultaneously outside their spacecraft.

The mission had two goals: (1) send the errant satellite, owned by the International Telecommunications Satellite (Intelsat) Organization, into its intended geosynchronous Earth orbit, and (2) practice assembly techniques needed to build NASA's Space Station Freedom starting in 1996. Both goals were achieved, but not quite as NASA had planned.

It took a record three EVA's to rescue the $157 million Intelsat 6/F3 instead of the one that had been scheduled. The successful rescue effort occurred on May 13, 1992, with an unprecedented 8 hour and 29 minute EVA by three astronauts. This was the longest EVA in space history, breaking a 20 year record set on the Moon by Apollo astronauts Eugene A. Cernan and Harrison H. Schmitt. The three crewmembers grabbed the 4,218 kilogram (9,300 pound) Intelsat 6/F3 with their gloved hands and attached a capture bar specifically designed for this mission. This capture bar then became the handle for the Shuttle's remote manipulator system (RMS) arm to maneuver the 3.7 by 5.2 meter (12 by

17 foot) Intelsat satellite into the Shuttle's cargo bay. The capture was accomplished just as *Endeavour* passed to the southwest of Hawaii, 417 kilometers (225 nmi) high and traveling at 7,823 meters per second (25,666 feet per second or 17,500 mph).

During the 4 1/2 hours after this satellite retrieval, the astronauts attached a United Technologies Chemical Systems Division Orbis 21S solid rocket motor stage to the Intelsat 6/F3. The Intelsat, with its newly attached propulsion stage, was then launched by the Shuttle. After the Shuttle had separated from the Intelsat to a safe distance, a ground command fired the Orbis 21S motor on May 14, sending Intelsat to its intended geosynchronous orbit, 41,322 kilometers (22,300 nmi) over the Atlantic Ocean.

The manual capture of the satellite, accomplished after two frustrating days of failure, was conducted by Navy Commander Pierre J. Thuot, Navy Commander Richard J. Hieb, and Air Force Lt. Col. Thomas D. Akers. Coast Guard Commander Bruce E. Melnick operated the Shuttle's RMS. Navy Captain Daniel C. Brandenstein, mission commander, and Air Force Lt. Col. Kevin P. Chilton, the pilot, skillfully maneuvered the 78,473 kilogram (173,000 pound) Shuttle within 1 or 2 meters of the slowly pitching Intelsat satellite so that it could be hand caught by the three astronauts. For the catch Hieb was strapped to a foothold on the side of the Shuttle's open cargo bay, Akers was balanced on a pole straddling the bay, and Thuot was on a work platform held by the RMS. All three men were anchored to their posts by foot restraints. This manual capture technique placed great emphasis on the integrity of the astronauts' space suits and gloves.

After Brandenstein had piloted the Shuttle to a position where the Intelstat 6/F3 was literally inside the bay and between the three EVA crewmembers, Hieb locked the 4.6 meter (15 foot) capture bar onto one side of the satellite, and then Thuot clamped on the other end of the bar. After waiting for almost a half hour for the satellite to come into the right position, Hieb said "Let's do it!" Almost as one, the three put their hands on the satellite. They surrounded it like three legs of a tripod. The operation required extraordinary delicacy; any jarring motion could have caused the fuel inside the satellite to start it rocking.

The capture bar had failed to work for Thuot during the two earlier rescue attempts on May 10 and 11. On these attempts, Thuot was repeatedly frustrated as he worked alone trying to snag the satellite with the capture bar. The Intelsat proved much more sensitive to force than NASA had expected and spun away every time Thuot attempted to snare it with the bar. That prompted the decision to send three astronauts outside the Shuttle on May 13 to grab the satellite.

The Johnson Space Center simulator developed to train Thuot in capture techniques is a spinning, 3.7 meter (12 foot) wide wheel suspended from posts on a platform that rests on a special "air-bearing floor." Air that gushes from jets in the floor suspends the platform, and the wheel, in an attempt to duplicate weightlessness. However, engineers concluded that the actual satellite, orbiting in space, is at least 10 times more sensitive to forces applied to it than the simulated satellite on which Thuot practiced. A fundamental problem, too, was that even with the previous Shuttle salvage operations—with the Solar Maximum, Palapa/Westar and Syncom spacecraft—ground simulators could not duplicate accurately all aspects of the Intelsat flight. The *Endeavour* experience will lead to better simulators.

Neither the United States nor Russia has ever had three people outside a ship in space at one time. More than two complicates communications and tethering. But the *Endeavour* did not have enough fuel for a fourth rendezvous. If the May 13 rescue mission had not succeeded, the communications satellite would have been left to languish in low Earth orbit. The satellite was supposed to have been placed in a geosynchronous orbit in March 1990, but a mis-wired Titan rocket left it in a useless 556 kilometer (300 nmi) orbit.

Intelsat, a consortium of 122 nations, hopes the satellite will generate about $1 billion in income over the next 11 years. They spent $131 million to buy it and $115 million to launch it on the unmanned Titan vehicle. Replacing the stranded unit with a new satellite and rocket would have cost Intelsat an additional $250 million. But NASA agreed to rescue it for just $93 million—or less than one-fifth the cost of a Shuttle flight. In accepting the NASA rescue mission, Intelsat had to spend an additional $46 million for the new booster motor.

In April 1992 the *Orlando Sentinel* reported that NASA offered Intelsat a bargain flight aboard the new *Endeavour* because agency officials were eager to perform such a challenging, high-profile mission—and because the agency still was using an outdated formula for charging commercial customers who use the Shuttle. NASA said the formula will be reviewed by a panel led by Massachusetts Institute of Technology professor Eugene Covert and former astronaut Tom Stafford.

Intelsat intends to have its rescued communications satellite operating by July 1992 in time to relay broadcasts of the Summer Olympics in Barcelona, Spain. It will be able to simultaneously transmit 120,000 telephone calls and three television feeds. It is expected to have an on-orbit operational life of about 12 years.

The second goal of this mission—to practice Space Station assembly techniques—was accomplished on May 14 by physicist Kathryn C. Thornton and Thomas D. Akers. Thornton became only the third woman in history to conduct an EVA, following Russian cosmonaut Svetlana Savitskaya and NASA astronaut Kathryn Sullivan.

During their nearly 8-hour space cargo EVA, Thornton and Akers stowed a stubborn Shuttle antenna and then evaluated SSF component assembly and handling under the NASA/McDonnell Douglas assembly of station by EVA methods exercise.

The exercise was to see how long it would take to assembly pieces to be used to build Space Station Freedom. It took longer than NASA thought, but this EVA was directly applicable to planning for the attachment of the Space Station modular node preintegrated Station truss on Shuttle missions in 1996.

Akers also tested a wandlike propulsive unit being considered as a self-rescue device for Station crewmembers. The crew rescue techniques are important for the Station era, since the Shuttle will not always be available to speed after an astronaut who drifts away from the outpost.

More broadly, this *Endeavour* mission's four EVAs with a total of about 60 crew-hours of EVA are considered an important trial run for the many EVA tasks envisioned by NASA for space operations and lunar/Mars explorations in the late 1990s and the next century. The most significant lesson learned, the astronauts said, is that ground simulators cannot exactly duplicate conditions in space, especially in situations involving large masses such as the Intelsat 6/F3.

Kathryn C. Thornton said she had no regrets about staying inside the Shuttle during the Intelsat rescue. "Very truthfully, I was hoping those big guys were all going to fit in the air lock," said Thornton, who stands 5 feet, 4 inches tall. "It seemed to everybody to be a better idea to have three guys out there roughly the same size if they wanted to level out the satellite. I'm about a foot shorter than Rick."

The Intelsat 6/F3 spacecraft is an example of a space system that was designed and built without on-orbit servicing as an operational objective. There were no hand-holds, capture bar grippers, or plug-in attachment points incorporated into its configuration. Yet NASA and the contractors involved in the retrieval/relaunch of Intelsat 6/F3 were able to devise equipment and techniques for its orbital servicing. It would have required very few funds and added very little weight to have included servicing features into its initial design.

Summary

Four summary comments to serviceable spacecraft design are offered:

1. The spacecraft design process must employ the system engineering disciplines to match the serviceable spacecraft design requirements to the interfaces and capabilities of the servicer system, and to negotiate as much customer friendly options as possible into the servicer hardware.
2. Modularity and standard interfaces are key to serviceable spacecraft design.
3. Serviceable spacecraft may offer an affordable option to satellite replacement.
4. Technology readiness, user needs, and economics are the "big three" influences on the decision to incorporate servicing into spacecraft design.

With the proper technology thrust, user motivation, and adequate funding, on-orbit servicing support capability should be operationally feasible in:

- Low Earth orbit by 1995 to 1999
- Geosynchronous orbit by 2000 to 2005

The reason for the band of years indicated in each orbit category is that operational feasibility will evolve from early demonstrations to a recognized operational status.

The satellite servicing overall support concept development will be an iterative process. The servicing strategies for NASA and DoD will differ, but both will evolve as the servicing architecture evolves and is validated.

References

1. Cepollina, Frank, and R. Bridgers. *Multimission Modular Spacecraft Systems Specification*. NASA Goddard Space Flight Center, S-700-190 Revision A, April 1986.
2. Asker, James R. *Hubble to Require Most Extensive EVA Yet on a Single Shuttle Mission*. Aviation Week & Space Technology, November 19, 1990.
3. NASA. *Designing An Observatory For Maintenance In Orbit*. Published by the Space Telescope Project Office, NASA/Marshall Space Flight Office, Huntsville, AL (MSFC) 1186, 1986.
4. Lillie, Charles F. *AXAF Servicing Design Concepts*. TRW, NASA JSC Satellite Services Workshop IV, June 21-23, 1989.
5. *Gamma Ray Observatory to Study Celestial Forces That Shaped Universe*. Aviation Week & Space Technology, March 5, 1990.
6. Wiltsee, Christopher B., and Larry A. Manning. *Servicing Operations For The SIRTF Observatory at the Space Station*. Goddard Space Flight Center Satellite Servicing Workshop III, June 1987.
7. Irvine, Nelson J., Cynthia E. Irvine, Donald Goldsmith, and Michael W. Werner. *SIRTF, A Window on Cosmic Birth*. NASA Ames Research Center, October 1989.
8. Werner, Michael W. Walter F. Brooks, Larry A. Manning, and Peter Eisenhardt. *SIRTF in High Earth Orbit*. NASA Ames Research Center, 3rd Infrared Detector Technology Workshop, 1989.
9. Lee, J. H., Y. S. Ng, and S. S. Maa. *Comparison of SIRTF Dewar Performance in the 900 KM and 100,000 KM Orbits*. NASA Ames Research Center, Cyrogenics Engineering Conference, UCLA, July, 1989.
10. Turner, Philip R., and Wayne F. Zimmerman. *Servicing Polar Orbiting Platforms*. NASA JSC Satellite Services Workshop IV, June 21-23, 1989.
11. Miller, Bill, Doug Beekman, and Ernie Littler. *Implementation of Resupply and Servicing Requirements on the Zenith Star Program*. Martin Marietta, Lockheed, and TRW, NASA JSC Satellite Services Workshop IV, June 21-23, 1989.
12. Bell, Gordon, James Walker, and Arthur Stephenson. *OMV Utilization in Satellite Servicing*. TRW, NASA JSC Satellite Services Workshop IV, June 21-23, 1989.

Chapter 7

Manned Servicing

This chapter and the next discuss the hardware, equipment, and tools that support on-orbit space systems assembly, maintenance, and servicing. The support equipment used on manned servicing missions is presented in this chapter; automated servicing equipment is described in Chapter 8.

Information common to both servicing modes is offered at the beginning of this chapter to place into perspective the application of manned and automated servicing to servicing missions.

Serviceable spacecraft design is keyed to the functional capabilities and safety assurance of the servicing hardware and supporting equipment. In turn, servicing strategy and servicing hardware, tools, and support equipment designs are influenced by the answers to these issues which must be addressed for *each* spacecraft program:

1. Crew and spacecraft safety
2. Servicing mission cost constraints or guidelines
3. Spacecraft servicing planned interval and time phasing
4. Number of spacecraft serviced on one servicing mission
5. Orbital location where the servicing activity will be performed
6. Availability of space-based servicing infrastructure elements (Shuttle, Space Station, OMV, OTV)
7. Security classification, if the servicing mission involves military space systems
8. Servicing tasks to be performed. Echelon level of tasks. Constraints on time spacecraft can be "off-operations" while servicing is taking place.
9. Spacecraft configuration features:
   - Overall dimensions and shape factors
   - Number, size, type, and location of the ORUs
   - Location of refueling ports
   - Location of servicer access and docking ports
   - Location of crew handholds and foot restraint devices
   - Location of appendages. Are they retractable?
   - Structural, electrical, fluid, mechanical, and thermal interface standards used in design of the spacecraft
   - On-board self-check and operational verification systems
   - Spacecraft electrical charging characteristics
   - Spacecraft shutdown sequence
   - Other as needed
10. Natural and enemy induced (if military servicing mission) environmental issues important to orbital operations
11. Potential availability of servicing equipment used successfully on prior servicing missions for similar type spacecraft
12. Other as specific to the servicing mission of the program

Generally, a trade between the servicer (man or machine) and spacecraft is in order to establish simplicity and reliability of servicing events. Since a variety of spacecraft servicing functions will make use of a particular servicing tool or support device, it is reasonable and economically advantageous to allocate greater simplicity to the serviceability design features of the spacecraft than to the servicer and its hardware. However, there may be some exceptions to this rule as dictated by specific manipulation requirements of particular servicing tasks that would require the extra complexity and cost of incorporating additional servicing design features on the spacecraft.

The parallel development of spacecraft and its related servicing hardware/tools equipment must occur so that spacecraft on-orbit servicing can be planned for implementation when the associated servicing, assembly, and maintenance hardware/tools are available. In some cases the servicing support equipment might be used in the initial assembly, integration, and testing of the spacecraft in the contractor's facility.

The NASA servicing hardware/tools and technologies

should be evaluated for DoD requirements for the sake of the cost-effectiveness of shared servicing equipment. NASA and DoD serviceable spacecraft have these areas of commonality and differences:

Commonality

- Philosophy of safety
- Need to make servicing an important parameter in the program life cycle cost goal
- Shared elements of SAMS infrastructure
- Shared astronaut crew training techniques and facilities for servicing operations
- Engineering approach to modularity, replenishment, repair
- Standard interfaces, fit, form, function
- Inside out location of serviceable subsystems
- Launch and transportation vehicles' interactions
- Common servicing hardware/tools/equipment, but recognizing some equipment may have to be mission unique
- Planning and execution of certain segments of the servicing mission scenarios
- Logistics systems operations (ground and orbital)

Differences

- Survivability—potential enemy-induced threat environment resulting in shielding, hardening, ASAT, and quick action maneuverability for DoD space programs
- Security—program classification with encryption unique to military operations
- Mission availability—operational continuity, mean time between critical failure, and mean time to repair requirements tend to be more demanding for military space systems

Hardware/Tools Selection Criteria

The criteria below should be applied to each specific satellite program as factors for servicing equipment selection. These criteria could be weighted and subjected to a quantitative discriminating process. Obviously safety and cost will rank high in any weighted structure.

Safety: How does the servicing hardware tool or support equipment affect personnel and spacecraft safety?

Cost: Cost is a combination of development, production, and operational costs. The cost rating between servicing hardware options needs to consider all three components.

Productivity: Productivity is the quantity of work performed per unit time.

Flexibility: Flexibility is how well the hardware or tool adjusts to change, or can be modified or adapted to unexpected developments in its usage. For concepts involving many tools, the option which is more able to adapt to unplanned situations is more flexible.

Human Factors and Ergonomics: Human factors focuses on the interaction between the servicing tool and the human operator. Specific issues for evaluation and analysis are training requirements, physical and mental workloads associated with tool/concept complexity of operating procedures, supporting aids, and equipment requirements and the susceptibility to human error.

Commonality: Commonality is akin to flexibility in most minds but is a separate and clear judgment criteria. Commonality is rated by how many times a tool or tool option can be used in different servicing roles. We may not be able to completely rate commonality until most of the work is done.

Reliability: Reliability is judged by the dependability of the servicing hardware or tool to perform as expected. Complexity of components, sensitivity to the environment, fatigue, and performance history are aspects of reliability.

Technical Risk: Technical risk is judged by asking what is the probability that the projected productivity, reliability, etc., will provide a tool/concept that accomplishes the objective? This includes risk associated with damaging or contaminating the spacecraft or hardware/tool.

Spacecraft Design: The impact that servicing hardware design has on spacecraft design can be important. A high rating goes to servicing hardware/tools that can handle a wide range of spacecraft configurations and operations. A low rating is assigned to servicing equipment which will require significant weight and cost increases to a satellite program.

Manned versus Automated Servicing

The selection of the manned or the automated servicing mode is decided on the basis of personnel safety, economics (particularly transportation costs), and the capabilities of the selected mode [1]. The costs associated with transporting personnel and their life support systems to locations other than those visited regularly by the Space Shuttle are currently projected to be too high to make manned maintenance at those locations a practical near term option. One alternative approach is the remote retrieval of satellites to be serviced by an orbital maneuvering vehicle (OMV) or an orbital transfer vehicle (OTV). The retriever vehicle sorties out from the Space Station Freedom or Space Shuttle, then performs rendezvous and docking with the subject satellite, and returns it to the Space Shuttle or Space Station where the servicing is performed. After man-performed servicing tasks are complete, the OMV or OTV redelivers the satellite to its desired orbit. The capability of the OMV limits these excursions to a few degrees of plane change, and about 1,852 to 3,704 km (1,000 to 2,000 nmi) of altitude change. Future versions of an OMV or a manned OMV with greater propulsion capacity may extend this capability.

A primary consideration that limits the use of EVA in future space activities is personnel safety [1]. The offshore oil industry's use of divers is a remarkably similar application of manned versus remote work activity, and it is the current philosophy of that industry to eliminate the planned use of any divers whenever possible in any new procedures. Divers are used only as backup, to work around unexpected occurrences. This approach was adopted because of the cost of diving operations and the safety issues involved. Certainly the space environment presents the same issues for civil flight. Military missions may have other considerations and priorities that dictate manned operations.

The third factor that impacts the selection of EVA or remote maintenance techniques is the capability of the servicing mode [2]. At present, a human in EVA is far more dexterous and capable than any remote system currently available or in design. Spacecraft missions requiring servicing are therefore planned for orbits that can be reached by the Space Shuttle or Shuttle/OMV, even when there is some compromise to mission performance. In the future, spacecraft will be designed to allow exchange of ORUs using relatively simple exchange mechanisms. The preferred mode of maintenance for straightforward replacement of ORUs will be the use of simple and robust remote servicers. However, when unexpected events preclude successful maintenance by the simple servicer, man in EVA will be employed as a backup solution as long as the orbit is within reach. Human capabilities to observe and reason, be adaptive, and handle delicate parts and materials will allow astronauts to overcome unanticipated difficulties and to make repairs not possible by remote servicing. EVA activity of this type will continue to be the accepted approach for the near term. When something like a flight telerobotic servicer (FTS) system becomes available, it will be used as a more capable tool than a simple remote servicer, thus extending the capability of that system, and extending the capability of the FTS in terms of reach and arm strength.

Some of the comparative characteristics of the manned and automated, or remote, systems are summarized. They are [1]:

Versatility: Since robot developers have not as yet developed units with the capability of humans, even the limited capability of humans in space suits, satellites designed to be serviced by EVA can have more variety in geometry, access, ORU attachment schemes, and insulation than can a satellite designed for remote servicing. Astronauts, either in EVA or operating a telepresence servicer, will be more capable of handling unplanned or unexpected events and anomalies. This may be a good reason for including teleoperation capability even on a normally autonomous servicer system.

Precision: Mechanized servicers can be made to locate and move in a much more precise manner than can be accomplished by humans. EVA operations therefore require more tolerant alignment aids than do mechanized devices.

Reach, Strength, and Size: Mechanized servicers can be designed to have more or less capability than EVA operators in these areas; however, one can assume that the mechanized device will be adequately designed for the task at hand. Of particular interest is reach, since manipulator arms, especially in a weightless environment, can have extensive reach. EVA operations require multiple foot restraint positions and translation rails to provide equivalent capability.

Task Allocation and Decision Rules [3]

The initial Space Station Freedom may not be significantly more autonomous from the ground than present manned United States and C.I.S. systems, but over time there will be a gradual shift in locale of control. By the year 2005, there could be substantial Space Station autonomy as confidence in automated systems grows with increased use. NASA will try to automate as much as possible. Some suggested decision rules for reallocating servicing tasks locale from ground to space could come from these issues: (1) Can the service task be performed only in space with the required reliability? (2) Is the immediate judgment of the space crew necessary for the task? (3) Is it less expensive to do the servicing task in space with the required reliability?

The series of ten questions/answers to follow help put into perspective the problem of allocation of work in space to humans, to machines, or to a blend of both.

Question 1. What is the nature of a servicing task that determines whether it is appropriate for automation? What type of tasks should be allocated to humans? What combinations of humans and machines will be most effective?

There is no best systematic approach to the allocation of servicing functions to machine and human operators. Tables of satellite servicing tasks best performed by humans or machines have been compiled, but they are incomplete. Some monitoring and control systems can be automated with current technology. Tasks requiring complex levels of decision-making repair probably will not be automated until about the year 2000; functions requiring judgment and interpretation of unexpected events will be automated only in the long term. Tasks demanding human-like dexterity will be difficult to automate with current technology unless they are repetitive and very limited in their requirements for fine manipulation.

In general, machines tend to be quite reliable but lack flexibility while humans tend to be less reliable than machines but far more flexible. If the service subtasks remaining after automation (such as watching monitors) are more boring than the original task, it is better not to automate and to let astronauts perform the task in its entirety. Humans have the ability to supervise and control and should not have to perform menial subtasks which subordinate people to machines. An effective human/machine combination is teleoperation or telepresence systems. In these systems the human remains in a safe environment and performs tasks which may otherwise: (1) be unsafe, (2) require strength beyond human capability, or (3) require prohibitively expensive EVA or vehicle life support systems or development of an autonomous machine beyond the reach of current technology. Present-day end-effectors are barely adequate but aggressive development in this area seems more practical in the near-term than pursuing a purely artificial intelligence-based approach.

Question 2. What are the decision rules for allocating servicing functions between humans and automated systems, whether in space or on the ground?

One approach to devising decision rules is to create an expert system. An expert system is an artificial intelligence (AI) approach to decision making, which builds up evidence for choices by asking users questions based on an established set of rules.

Strong evidence for the decision to automate servicing may exist (1) if the task requires perceptual abilities outside the range of human limits; (2) if the task involves safety or health risks outside tolerable limits for humans; (3) if the task requires computing ability; (4) if the task entails detection of infrequent or rate events; and (5) if the task requires continuous monitoring of systems.

Weaker evidence for favoring automation arises (1) if it is technically feasible to automate the task; (2) if it is economically feasible to automate the task; (3) if the task involves storing and recalling large amounts of precise data for short periods of time; (4) if the task involves routine repetitive precise tasks; (5) if the task regularly requires an attention span of more than 20 minutes; and (6) if humans don't like to do the task.

Strong evidence favoring humans for a task may exist (1) if the task requires deductive reasoning ability; (2) if humans like to do the task; (3) if the task requires the ability to arrive at new and completely different solutions to problems; (4) if the task requires the ability to detect signals in high noise environments; (5) if the task requires ability to use judgment; and (6) if the task entails many unexpected or unpredictable events.

Weaker evidence for using people may arise (1) if the task requires EVA; (2) if the task requires the ability to profit from experience; and (3) if the task cannot easily be reduced to a series of preset procedures.

Question 3. What are the decision rules for determining whether a servicing function can be performed better in space or on the ground?

Allocation of activities between ground and space may also be discussed in terms of possible decision rules for an expert system.

Strong evidence for allocating a task to be done in space exists (1) if the task requires the space environment; (2) if time delays cannot be tolerated; (3) if the task requires the physical response of the crew; or (4) if the task involves crew leadership, requiring a common sense of sharing a stake in the situation.

Weaker evidence favoring space arises if the task can be done less expensively in space.

Strong evidence favoring ground allocation may exist (1) if the task does not require the immediate action of the crew or (2) if large space or heavy machinery is needed.

Weaker evidence for performing the task on the ground may arise if the task costs less when performed on the ground.

Question 4. What is the astronauts' role with respect to on-board autonomous subsystems? In what operational modes do human serve best?

The astronaut will function as supervisor or manager and must understand basic system behavior, diagnose faults, and repair or replace faulty components. However, many subsystems will be self-contained and will operate independently. With automated Space Station monitoring, subsystem abnormalities will cause a higher level system (machine or human) to be alerted. Using fault-tolerant computing and redundant systems, many faults can be handled without huuman intervention. If the troubleshooting procedure for the detected fault is well specified, then the computer should complete as many of the steps as possible before alerting the crew. This avoids the inefficient current practice of human review and execution of an entire troubleshooting procedure which is largely routine. Of course, if a critical system must be shut down or a redundant system started up, humans should be consulted or informed so that there is an opportunity to approve or disapprove the action.

Of course there are many faults which are unanticipated or for which no simple step-by-step procedure can be written. In these cases, helps, hints, and operational information should be provided by the Station data management and information retrieval systems but the human must make the decisions, perform the troubleshooting, and make the repair. Ideally, the crew could still repair faults in critical systems, such as communications, autonomously.

Question 5. What are the management principles for operation of autonomous servicing equipment, particularly as a function of machine intelligence?

They are largely unknown. Intelligent systems are currently most adept at dealing with symbols rather than material objects, and can work with sets of rules as in expert systems. If the operation of the equipment, which may include fault detection and resolution, can be reduced to a specific set of conditions and remedial actions, then the system can be managed by machine intelligence. If the system requires changes in operation based on expected or unpredictable results, then state-of-the-art AI techniques are inadequate.

Current expert systems produce very impressive results, but these packages generally are used by people whose expertise is comparable to that embodied in the software. Expert operators are required, both to ensure the "common sense" of results and to modify the system's rules as new expert knowledge accumulates (although learning and automated theory formation are reasonable goals for the future). For the initial Space Station design, prudence suggests limiting deployment of expert systems to domains in which they are known to work, such as monitoring and fault diagnosis of power systems or interactive real-time crew scheduling. As other working systems are demonstrated and evaluated they should be added to the evolving International Space Station Freedom. Caution is advised, but it should be possible to identify potential domains where an expert system might be suitable for future Station implementation.

Question 6. How does one determine when human intervention is required? What are the principles which determine how to provide status information to the human? How can unsafe human interventions be prevented?

Humans should be involved in the control of an action or decision which is irrevocable or which significantly affects another system. The level of action to be taken and the seriousness of the event requiring action determine how status information will be presented. A major failure should attract attention immediately, probably through both audio and visual alarms. Additional information describing the cause and nature of the failure should be displayed on a CRT. But printed warning messages are less effective than using both audio (e.g., voice or sound) and visual signals (e.g., a flashing light). Minor events should activate a small visual indicator or log a message for later review.

The two main concerns with unsafe human intervention are that (1) an unauthorized person might interact with the system, and (2) an authorized person could make a mistake adversely affecting the system or other systems. Fail-safe interlocks and passwords can prevent unauthorized action. Good training and a basic understanding of the systems provide significant assurance against mistakes. Other steps can also be taken. For example, if an action could cause major damage, the consent of more than one person might be required—perhaps that of a crewmember and another person on the ground. Computers could perform a contingency analysis for the crew or request that crucial commands be repeated, prior to taking action.

Question 7. What new skills do people need in dealing with autonomous subsystems? What skills (organizational, personal, and physical) need further development?

The needed skills are similar to those presently required for the overall U.S. astronaut program. People who deal with autonomous subsystems must be comfortable working with automation technology and must thoroughly understand the displays and information presented by Station systems. This requires intensive training and an ability to maintain high levels of familiarity with the technology. Strong decision-making skills are essential, such as when serious component failures or other stressful situations necessitate rapid assessment of the accuracy of autonomous subsystem feedback—especially if this information conflicts with intuition or common sense.

Organizational and personal skills needing development are (1) the ability to live (and thrive) in a cramped, fragile, artificial habitat located in a hostile environment from which immediate escape is impossible and (2) the ability to design and operate decentralized social systems (i.e., greater autonomy for organizational subunits), multimode computer-augmented interpersonal communications networks, and evolutionary human/machine systems.

Question 8. What are the decision rules which apply to extravehicular operations? What advancements in technology are required to shift the task allocation?

There is strong evidence favoring manned EVA (1) if the task can be done with safety or (2) if the task requires working with nonstandard fasteners and tools; and weaker evidence (1) if the task cannot be reduced to a series of preset procedures or (2) if the task requires sensitivity to a wide variety of stimuli. There is strong evidence that a human/machine system should perform the EVA (1) if the task is dangerous or (2) if the task is repetitive and requires limited dexterity; and weaker evidence (1) if the task must be done immediately or (2) if the task requires continuous work of 4 hours or more.

Technologically the primary components of an early telepresence system are available but the integration of these components is necessary in order to provide an operational system in the near future. Ground-based telepresence has limited application because of the delay problem. A larger variety of end effectors with greater effectiveness and dexterity must be developed, and tactile sensors must be improved. However, standardization of connectors, fasteners, attachment methods, module configuration, and tools could accelerate the use of telepresence as an operational system even without the aforementioned advances.

Robotics will take advantage of gains in telepresence systems, but major significant improvements must be made in artificial intelligence systems before robots will become an effective part of the Space Station or satellite servicing system. Limited use of supervisory control should be possible in the 1990s.

Question 9. How can the man/machine mix be optimized for extra Station activity? What evaluation criteria apply?

Manned EVA is useful in many situations because intelligence and flexibility are important human characteristics. However, the space environment places severe restrictions on human activities (e.g., reduced dexterity, short operational time, bulky life support systems). With the limited abilities of available intelligent machines, the use of teleoperated systems may provide an effective and, with foreseeable technology, near optimal human/machine mix. With the astronaut as operator, telepresence employs human judgment and manipulative skills, takes advantage of machine durability and mechanical performance, and can incorporate autonomous robotic technology as it becomes available.

As human/machine capabilities are developed it may be useful to use a weighting function in the decision process which includes the importance of the task, the effectiveness of the human/machine system, and the cost to support the system.

Question 10. What is a feasible evolution of human/machine systems in space over the next 20 to 30 years? How will the human/machine interaction change over time? What is the role of people in human/machine systems as these systems evolve with technological advances?

When Space Station Freedom is operational in the late 1990s, people will still play the dominant role in almost all human/machine servicing related systems. Manned EVA will be used in its construction and for satellite servicing. Mechanical manipulators with limited dexterity and sensory feedback also will be employed. These will be teleoperators or telepresence devices with human controllers and decision makers. Monitoring will be done by computers of limited "intelligence" (e.g., fault-tolerant systems), but under human supervision. Much of the decision-making control will shift from ground to Space Station and the crew will receive intelligent assistance from onboard computers. The major computers for monitoring and mission operations will remain on the ground together with a limited number of operators and experts.

This mode of operation will change dramatically during the 20 years after SSF IOC. Information will become much more available and cheaper, just as most other resources will become more expensive. The human/machine interface will become more permeable, allowing easier transfer of information. This process is already underway in terminal design, relational database organization, attempts at natural language front ends, expert systems, and head-up displays.

It is unknown how intelligent machines can become. The conservative assumption is that problems in developing basic AI theory will prove as intractable as those of turbulent flow, but, to extend the analogy, that some very useful systems will be flown nevertheless. In all likelihood, advances in AI will allow truly intelligent machines to exist. Highly developed sensory capabilities will extend the uses of autonomous robots. Intelligent assistants and monitoring systems will be created and installed on the Space Station. Nearly all space system activities ultimately may be controlled from an expanded Space Station.

The use of autonomous, intelligent machines will not reduce the amount of work that humans do but rather will permit the effective performance of an ever-increasing number of more complex and productive servicing tasks.

The following are general conclusions:

1. Machines and/or equipment will not replace humans in space. Rather, they will free us for more productive endeavors. People and machines in space will demonstrate new types of interactions and will thrive, not just survive. The people and machines must be viewed as an integral system from the first stages of conceptualization and design.

2. Artificial intelligence systems will not have a major impact on the initial Space Station design. There are expert systems that can be employed in specific areas but it will take at least another 5 to 10 years before highly autonomous intelligent machines become available. An evolutionary Station with servicing facilities should be designed with this future possibility in mind.

3. Two areas of human/machine interaction appear most promising: (a) using computers for monitoring with humans serving in a supervisory capacity and (b) direct interaction in the form of teleoperation and telepresence. No major technological breakthroughs are necessary to develop effective teleoperation systems. These systems eliminate the near-term need for extensive intelligent AI systems, and the development of superior end effectors will provide exceptional physical capability to perform many servicing functions. Furthermore, as artificial intelligence systems emerge, the advances which have been made in teleoperator systems can be used to create more efficient and effective robots for servicing jobs.

4. Sophisticated monitoring systems can be developed to sharply reduce ground personnel requirements. However, use of these systems will not increase ground/space station autonomy because monitor computers will be located on the ground so they can be improved and developed as technology advances. Later, though, most of the human control will shift to the Space Station.

The following are recommendations:

1. Major effort and funding should go into the development of manned EVA, teleoperator/telepresence, and robotic servicing systems. NASA should develop servicing related EVA suits, tools, and capabilities for near-term and mid-term use; and invest in robotics for mid-and long-term use.

2. Using the latest technology, high-level monitoring systems for servicing tasks should be established on the ground, with onboard microcomputers maintaining the normal operation of many Station systems and taking over many routine decisions formerly made by humans. Astronauts must retain ultimate authority, making the highest level decisions of which machines are incapable. Databases should be developed with eventual AI uses in mind, and accessible by all users. One or more computer networks should be employed on board the Space Station, enabling critical servicing functions to be separated from scientific and other uses. Every effort should be made to take advantage of the capabilities of commercial systems, particularly in the areas of computer hardware and software development, natural language and expert systems.

3. To counteract the psychological and social negatives of living and working in a highly automated, relatively isolated artificial environment, the Space Station should be designed from the outset with extraterrestrial setting factors, communications factors, and organizational factors in mind. Interdisciplinary teams should address problems of work and setting design for human/machine interaction. The Station should be previewed as a facility instead of a flight. NASA should encourage an up-to-date examination of issues and findings in social sciences research of possible relevance both to Space Station organizational and physical design and to a long-term human presence in space.

Manned Servicing Equipment and Tools

The learning process associated with qualifying EVA as an operational technique on the Shuttle forced development in areas of space suits and independent life support systems. Equally important were the development of man-machine interfaces, body restraints, physiological tolerance of crewmembers to the atmospheric environments of space vehicle cabins and space suits, workload planning, simulation, and training.

Shuttle EVA provisions and some basic carry-on equipment required for EVA are baselined for each Shuttle mission, to provide EVA capability on every flight, for contingencies, crew rescue, and on-orbit servicing. In addition to these mandatory Shuttle EVA provisions and crew equipment, consumables and expendables must be provided for future long duration operations on the Space Station Freedom and other missions such as manned lunar and planetary explorations and the Strategic Defense Initiative (SDI) space-based servicing missions.

Satellite servicing EVA requirements flow directly from a functional analysis of planned and contingency mission operations. The definition of EVA operations and the associated procedures is vital so that human physiological needs, support equipment, and crew interfaces can be understood and specified. Issues that relate to EVA mission operations include identification of EVA duties and duty cycles, Space Station support operations, logistics, rescue and response procedures, crew communications and data management, and EVA servicing task skill requirements. The requirements for on-orbit routine maintenance and servicing of EVA and support equipment are issues that also must be addressed.

Hardware Requirements

Satellite servicing EVA hardware includes: space suits, the crew enclosures, portable life support systems, powered maneuvering capability, specific safety-related equipment, satellite support devices, and crew support equipment such as tools, work stations, and restraints.

EVA hardware directly influences the level of human productivity during mission operations. To optimize potential beneficial effects on productivity, the specific hardware components must be designed to be versatile and easy to operate. Features such as simplicity, standardization, and durability that foster human productivity also lead to high standards for reliability and maintainability.

Other EVA servicing issues to be addressed as part of a servicing hardware requirements analysis include: radiation exposure, meteoroid/space debris, spacecraft charging, physical hazards from the sharp corners of satellites and Space Station equipment, contamination of the external environment and its impact on fragile sensors, EVA system power requirements, propulsion methods, thermal management, guidance and attitude control requirements, and communication links.

Present EVA Equipment

Various types of equipment have been developed for the NSTS to provide a wide range of EVA capabilities. Figure 7.1 shows the flight manifested servicing tools for Space Shuttle flights over the past few years. Of the 56 tools listed, 42 were made by ILC Space Systems Division, Houston, Texas. The two right side columns on Figure 7.1 indicate ILC's judgment of the tools' EVA compatibility and adaptability for robotic operations.

In addition to the standard EVA tools of Figure 7.1, many special tools have been developed for the Space Shuttle satellite retrieval/repair missions. They are [4]:

- Stinger—Developed for the Palapa and Westar communications satellite retrieval mission, Figure 7.2.
- Leasat bars—For the Leasat salvage mission, four special bars were developed to capture, hold, grapple, and later spin the Syncom IV satellite.
- Toolboards—These devices provided a convenient means of storing and presenting tools for EVA use.
- EVA power tool—A self-contained, battery powered power tool was developed for repetitive rotations such

TITLE OF TOOL	FLIGHT MANIFEST										EVA COMPAT-IBILITY	ADAPTABLE FOR ROBOTIC COMPATIBILITY
	STS 5	STS 6	STS 41-B	STS 41-C	STS 41-G	STS 51-A	STS 51-I	STS 61-B	STS 31	HST M&R		
POWER TOOL, EVA (HST)									X	X	HR	
POWER TOOL, MINI											HR	
PROBE	X	X	X	X	X	X	X	X	X	X	R	
RETRACTABLE TETHER	X	X	X	X	X	X	X	X	X	X	HR	
SAFETY TETHER	X	X	X	X	X	X	X	X	X	X	HR	
SCISSORS, MODIFIED				X		X	X				R	
SHROUDED FLEX SCREWDRIVER										X	R	X
SHROUDED RIGID SCREWDRIVER										X	R	X
TAPE CADDY, KAPTON				X	X	X	X	X	X	X	NR	
TENSION BUCKLE				X		X	X				R	
THERMAL MITTENS	X	X	X	X	X	X	X	X	X	X	NR	
TOOL BOARD			X	X		X	X	X		X	HR	X
TOOL CADDY	X	X	X	X	X	X	X	X	X	X	HR	X
TOOL STOWAGE BAG					X						R	
TRASH BAG, LARGE			X	X		X	X			X	R	X
TRASH BAG, SMALL	X	X	X	X	X	X	X	X	X	X	R	X
TUBE CUTTER	X	X	X	X	X	X	X	X	X	X	R	X
VELCRO CADDY			X	X	X	X	X	X	X	X	R	
VISE-GRIP PLIERS, ONE-HANDED											R	
WAIST TETHER	X	X	X	X	X	X	X	X	X	X	HR	
WRIST MIRROR	X	X	X	X	X	X	X	X	X	X	R	
WRIST TETHER	X	X	X	X	X	X	X	X	X	X	HR	
WRIST TETHER, ADJUSTABLE	X	X	X	X	X	X	X	X	X	X	HR	
1/2" OPEN END WRENCH											R	X
1/2" RATCHETING BOX END WRENCH			X	X	X	X	X	X	X	X	HR	
1/4" ALLEN WRENCH EXT.-3/8 DRIVE	X	X	X	X	X	X	X	X	X	X	R	X
15/16" WRENCH WITH CHEATER BAR											NR	
3/8" DRIVE McTETHER RATCHET									X	X	HR	X
5/16" HEX SOCKET FOR 3/8" DRIVE						X					R	X
5/16" HEX SOCKET-3/8" DRIVE (HST)										X	R	X
7/16" HEX SOCKET EXT.-3/8" DRIVE			X	X		X	X				HR	X
7/16" HEX SCKT.EXT.-3/8" DRIVE(HST)										X	HR	X
ADJUSTABLE WRENCH	X	X	X	X	X	X	X	X	X	X	NR	X
BOLT PULLER	X	X	X	X	X	X	X	X	X	X	R	X
CABLE CUTTER, LARGE									X	X	R	
CIRCULAR CONNECTOR TOOL									X	X	R	X
D-CONNECTOR DEMATE TOOL										X	R	X
D-CONNECTOR MATE TOOL										X	R	X
DIAGONAL CUTTERS	X	X	X	X	X	X	X	X	X	X	NR	
ELEC. CONNECTOR PIN STRAIGHTENER										X	R	
FLASHLIGHT				X		X	X	X			R	
FORCEPS	X	X	X	X	X	X	X	X	X	X	R	
HAMMER	X	X	X	X	X	X	X	X	X	X	R	
HARPOON SPACE STATION TRUSS NODE											R	X
HOOK, TRUSS STRUT											R	X
LEVER WRENCH	X	X									NR	
MINI WORK STATION	X	X	X	X	X	X	X	X	X	X	HR	
NEEDLE NOSE PLIERS	X	X	X	X	X	X	X	X	X	X	R	
PAYLOAD RETENTION DEVICE	X	X	X	X	X	X	X	X	X	X	R	
PFR WORK STATION STANCHION							X				HR	X
PORTABLE FLOOD LIGHT									X	X	R	X
PORTABLE ORU HANDLE, LARGE										X	R	X
PORTABLE ORU HANDLE, SMALL										X	R	X
POWER TOOL BATTERY BAG, EVA			X	X		X	X		X	X	R	
POWER TOOL BATTERY, EVA			X	X		X	X		X	X	R	
POWER TOOL BATTERY, MINI											R	

HR = HIGHLY RECOMMENDED

Follows good human factors design for EVA operation with a pressurized suit or gloved hand. Easily operated with convenient adjustments to accommodate different crewmembers. True one-handed operation possible.

R = RECOMMENDED

Provides acceptable accommodations for EVA operation. Nominally meets all design requirements.

NR = NOT RATED

Used on a mission requirements basis, not usually manifested for flight. Considered as contingency item.

Figure 7.1 Servicing tools manifested on recent Space Shuttle flights. *Courtesy of ILC Space Systems.*

Manned Servicing

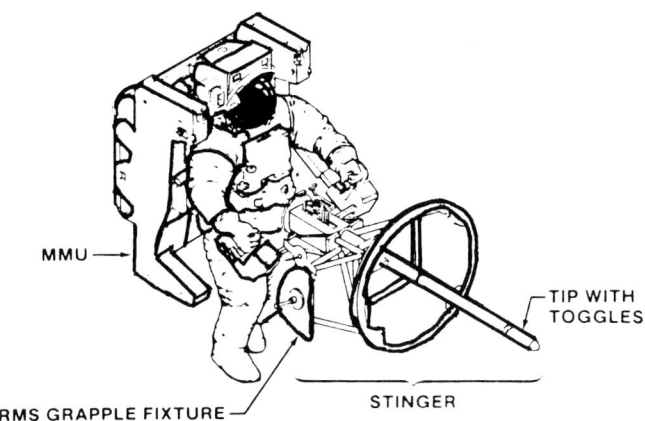

Figure 7.2 Stinger attached to the manned maneuvering unit. *Courtesy of NASA.*

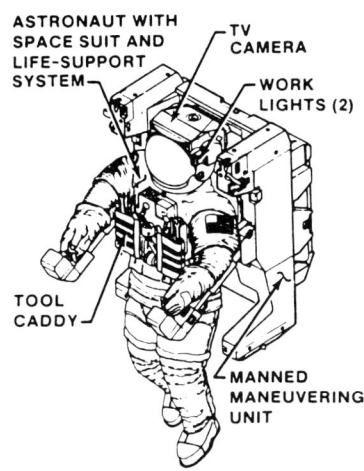

Figure 7.3 EVA serviceman. *Courtesy of NASA.*

as bolt and nut installation and removal and for backup mechanical drives.

- EVA power package—Provided an electrical impulse to open a protective sunshield or to close a tiedown latch.

The *EVA Catalog of Tools and Equipment* [5] contains a complete list and description of 230 items of EVA equipment already developed for the Space Shuttle. A wide range of groups should find this document useful in the pursuit of EVA hardware information. Designers can note the construction of recommended items based on existing hardware. Servicing users and mission planners can take advantage of tools presently in the NASA inventory and save the expense of "reinventing the wheel." The catalog contains a mixture of tools and equipment used throughout the Space Shuttle based EVA program. Promising items which have reached the prototype stage of development are also included. Each item is described with a photo, a written discussion, technical specifications, dimensional drawings, and points of contact for additional information. Overall guidelines for EVA equipment design and operations can be found in Reference 6.

Tools such as shown in Figures 7.1 and 7.2 and discussed above, when attached to a space suit equipped with a television camera, work lights, and the EMU for transportation, give the astronaut the capability to make on-site repairs of malfunctioning satellites [4], Figure 7.3.

Although most items of equipment were developed to satisfy specific needs, the full range of applications has yet to be determined. The present NSTS extravehicular mobility unit (EMU) is an independent anthropomorphic system that provides environmental protection, mobility, life support, and communications for the Shuttle crewmember to perform EVA in Earth orbit. Two EMUs are included in each baseline Orbiter mission, and consumables are provided for three two-man, 6-hour EVAs. Two EVAs are available for payload use, and the third is reserved for Shuttle Orbiter unscheduled safety-critical EVA.

The EMU consists of a space suit assembly that includes the basic pressure garment components, a primary life support system (PLSS), a backup life support system for emergency use, an ultrahigh-frequency (UHF) radio communication system, and the displays and controls required to operate them.

Space Suit Technology [7]

In the near-perfect vacuum of earth orbit, the pressure suit must maintain a minimum pressure of 3.1 psia to protect the EVA crewmember from hypoxia. The Shuttle EMU operates at 4.3 psia and requires a prebreathe time of 210 minutes when the Shuttle Orbiter cabin is maintained at 14.7 psia. A prebreathe time of 40 minutes is required when the Orbiter cabin is maintained at approximately 2 psi. Currently under development is a higher pressure (8 psia) suit which will completely eliminate the need of prebreathing.

The NASA-developed space suit configuration for Project Mercury and the Gemini Program originated from high-altitude-aircraft full-pressure-suit technology. These early suits lacked sophisticated mobility systems, since the suit served primarily as a backup system against the loss of cabin pressure and required limited pressurized intravehicular mobility functions for a return capability. Beginning with the Gemini program, enhanced mobility systems were developed to enable crewmembers to perform useful tasks outside the spacecraft. The ILC Dover, Inc. zero-prebreathe Mark III (ZPS Mk III) model of a higher operating pressure, 8.3 psi space suit assembly represents a significant phase in the evolutionary development of a candidate operational space suit system for the Space Station program [7].

For the SSF program, the extravehicular activity mode of operation will be expanded to provide numerous unique on-orbit service capabilities not fully achieved in previous space program operations. It is planned to conduct Space Station EVAs on a routine basis over the course of the Station lifetime. Future space exploration operations beyond the current SSF in Earth orbit and on the lunar and planetary surfaces may call for up to 2,000 man-hours per year of EVA. On-orbit EVA operations will include, but not be limited to, such activities as

- Satellite assembly, maintenance, and servicing on planned and contingency missions
- Installation, removal and transfer of Space Station, payload, and satellite orbital replaceable units
- Inspection and remedial repair and replacement of structural elements, solar panels, and thermal/meteoroid protective panels
- Propellant transfer and refueling operations
- Large structure erection and assembly

The capability to perform EVAs routinely and economically will be a key element in achieving the full operational potential of the present Space Station and future space systems that are even more sophisticated. An EVA involves all activities in space in which a crewmember dons a space suit with integral life support system (together called an extravehicular mobility unit (EMU)) and performs operations in unpressurized environments. The significant numbers of EVAs projected will establish EMU design and performance capabilities relative to the Space Station mission objectives. Since EVA will be a routine activity, EVA hardware must minimize overhead time, be easy to use, unencumbering, serviceable on orbit, and maintainable on orbit. EVA systems must limit consumable usage and EVA restraint systems must support a variety of tasks.

For future planned manned orbital operations in support of the Space Shuttle and Space Station programs, it is recognized that EVA preparations or overhead activities need to be reduced to fully utilize EVA as a viable capability. One of the most significant EVA overhead issues is the elimination of prebreathing operations. If the current 4.3 psia Space Shuttle EMU assembly is used, Space Station EVA crewmembers adapted to a 14.7 psia cabin environment (21% oxygen, 79% nitrogen) will be required to breathe 100% oxygen during a "washout" period (referred to as denitrogenation or prebreathing) before performing EVA operations. This procedure will be necessary to eliminate nitrogen gas from their bodies in order to preclude potential physiological effects of a decompression from the sea-level-equivalent cabin pressure to the 4.3 psia operational suit pressure of a Space-Shuttle-type EMU.

Operationally, Space-Shuttle-type prebreathing procedures are mission-time consuming and place a number of additional constraints on spacecraft systems, onboard experiments, and crewmember operations. Since Space-Shuttle-related EVA operations are of a relatively short duration (approximately 21 hours of planned EVA per flight), prebreathing timeline activities are accommodated in mission planning.

To maximize experience gained from the previous phases of development activity and to incorporate pertinent Space Station requirements, the current ILC Dover, Inc. ZPS Mk III 8.3 psi suit program was initiated along the lines of an integrated systems engineered approach. The JSC in-house and contractor (ILC Dover, Inc.) development efforts were directed toward concept configuration definition of alternate body seal closure configurations for improved don/doff characteristics. High-fidelity shell torso mockups were fabricated for both rear entry and dual-plane body seal closure configurations.

Although both closure concepts were evaluated and deemed to be feasible, the rear entry configuration was selected for the ZPS Mk III suit on the basis of its being less complex in design and less costly to develop and fabricate than the dual-plane configuration. Additionally, the dual-plane closure concept (because of the size of associated hardware mechanisms) appeared to reduce vertical torso sizing adjustment capabilities, to limit or restrict the optimization of shoulder bearing size and angular orientation, and possibly to affect the waist joint range of travel. It was also determined through subjective evaluations that ground-based don/doff activities with a dual-plane closure-torso arrangement would be more cumbersome.

In support of the "zero prebreathe" ZPS Mk III suit design, Figure 7.4, concept and configuration development, JSC designed and fabricated various test stands that were utilized both by JSC and ILC Dover, Inc. to define the following critical don/doff and operational performance features:

- Rear entry closure geometry and critical dimensions
- Scye angles and helmet angle. The scye angle, which defines the opening around the shoulder, is a term from anthropometry—the science dealing with measurements of the human body to determine differences in individuals.
- Scye bearing diameters and helmet size
- Feasibility of incorporation of a waist joint feature without loss of torso length sizing capability
- Concept for a combined waist capture/retention ring and EVA tether brackets including a universal don/doff support stand configuration
- Support of KC-135 aircraft simulated zero-g don/doff validation of rear entry closure concept and self-engagement/capture of waist ring retention system

In addition, design activities were implemented during the ZPS Mk III effort to further investigate the feasibility of developing and incorporating higher operating pressure 8.3 psi all-fabric technology-based arm and leg mobility systems. A critical aspect of this development activity was to identify any potential technology limits associated with the all-fabric

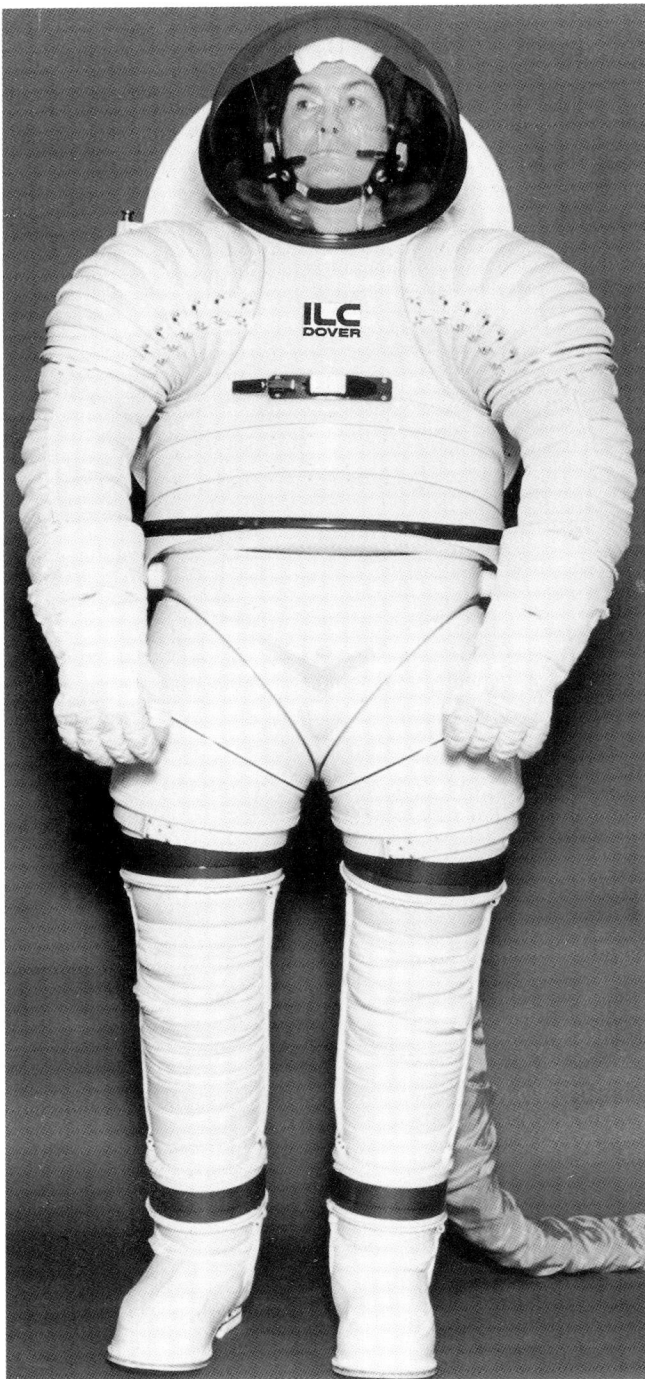

Figure 7.4 The ZPS Mark III 8.3 psi spacesuit. *Courtesy of ILC Dover, Inc.*

approach for elbow, knee, and ankle mobility joints operating at 8.3 psi. As a result of earlier evaluations of various Phase I and Phase II mobility systems incorporating hard elements, it was determined that a number of potential advantages would be possible through the utilization of soft (fabric) joint elements over hard, mechanical joint systems. The potential advantages include:

- Less complexity (simple construction features)
- Less bulk and weight
- Less cost to fabricate
- Easier stowage (improved logistics)
- High degree of mobility and comfort (user acceptability)
- Functionally unlimited mobility range
- Better "wearability" (high life cycle potential; can withstand high day-to-day use/handling activities)
- High resistance to impact damage

To complete the integrated systems engineered approach for the ZPS Mk III suit development activities and in anticipation of Space Station EVA assembly operations conducted over several years, design requirements were implemented regarding protection against potential EVA hazards. Along with the basic protective aspects of the typical Space-Shuttle-type integrated thermal/micrometeroid garment (TMG) such as thermal/abrasion and micrometeoroid protection, specialized provisions for chemical (propellant) protection, radiation shielding, electrostatic charge control, increased impact protection against orbital debris, considerations for possible atomic oxygen degradation and improved long-term service life characteristics were investigated. Advanced TMG material layups were developed to provide the necessary EVA-hazards protective capabilities [8].

Subsequent to the initial torso closure configuration definition phase of the ZPS Mk III program, extensive hardware design considerations were formulated. The design considerations were based on these factors:

- Maintainability
- Sizing
- Material selection
- Hard upper torso assembly
- Closure mechanism and actuator
- Lower arm assembly
- Leg and boot assemblies
- Suit mobility [9]
- Comfort
- Don/doff
- Stress analysis
- Helmet assembly
- Shoulder assembly
- Waist assembly
- Hip/thigh assembly
- Advanced thermal/micrometeroid garment

The ZPS Mk III suit test activities are planned to provide both qualitative and quantitative information regarding suit functional operation and performance capabilities. The established baseline to which the ZPS Mk III suit will be compared during the suit evaluation testing activities is the current Space Shuttle suit.

New and Future EVA Tools in Development

Servicing support tools in active development at NASA/JSC are:

- EVA tools and equipment to repair the Hubble Space Telescope

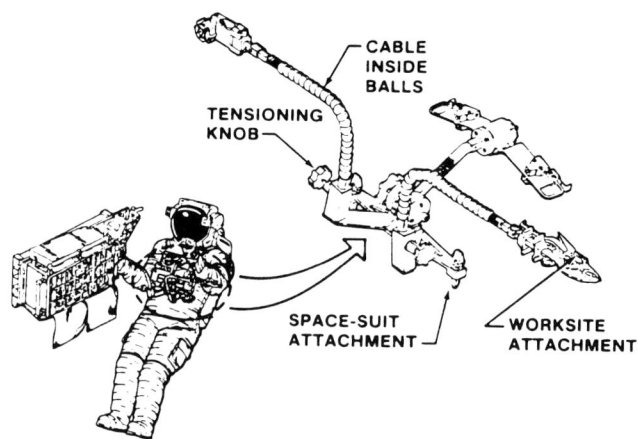

Figure 7.5 Space suit waist tether which can provide rigid support. *Courtesy of ILC Dover, Inc.*

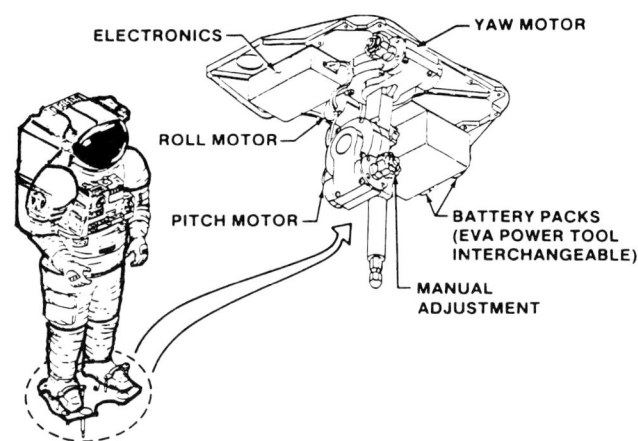

Figure 7.6 Powered portable foot restraint. *Courtesy of NASA.*

- EVA tools and equipment to maintain Space Station Freedom
- Portable foot restraints
- Smart power tool
- Powered ratchet tool
- Torque multiplier
- Rigidizable waist tether, Figure 7.5
- Powered portable foot restraint, Figure 7.6
- Smart cuff checklist
- Smart work restraint
- Smart tool boxes
- Heads-up displays

Space Station Freedom Maintenance

The Space Station is following the traditional path in maturation, as seen in Apollo and other projects. Weight and power consumption, for example, tend to balloon—as they have with the Station—and then fall back into control as attention is focused on the problems. The same curve is holding true for exterior maintenance requirements.

Central to the operations of the Space Station Freedom is the capacity to maintain its systems in good working order. Maintenance of the Station will consist of preventive and corrective maintenance. Preventive maintenance includes all scheduled maintenance actions performed to retain a system or end item in a specified condition. These actions include periodic inspection, condition monitoring, critical item replacements, and calibration. In addition, lubrication and fueling are considered a part of scheduled maintenance [10].

External equipment maintenance is accomplished by EVA or use of telerobotics. EVA maintenance time is a limited resource on the Station because planned EVA is Shuttle based and performed while the Orbiter is present and only contingency maintenance is performed from the Station between Orbiter visits [10].

NASA has addressed the problem of excessive astronaut EVA maintenance operations on SSF. In July 1990, NASA released the results of a 7 month study [11] managed by Dr. William J. Fisher, an astronaut, and Charles R. Price, chief of the Robotics Systems Development Branch at NASA/JSC. The study by the twelve member External Maintenance Task Team (EMTT) determined that as currently planned, SSF would require an average of 3,276 EVA man-hours per year over its baselined 35 years lifetime. This equates to 273 two-man EVAs annually over these 35 years or about five two-man EVAs each week. The numbers are far higher than the first informal NASA goal of 500 man-hours per year.

A still more serious problem appears to be EVA required for repairs during assembly. The group said 6,267 hours of EVA would be required for external maintenance before the Station could house a crew permanently. Performing the equivalent of four EVAs per week with astronauts visiting only via the Space Shuttle is clearly impossible.

The Fisher-Price panel strongly urged NASA to include in the Station program a new space suit designed to eliminate the necessity for "prebreathing." And a "solutions team" in Houston made some 100 recommendations aimed at lowering the expected EVA requirement.

Fisher-Price [11] identified 8,158 SSF orbital replacement units located external to the Station's pressurized modules. The breakdown of this ORU total is as follows:

Electronic	327
Electrical	1312
Electro-mechanical	868
Mechanical	1046
Structural-mechanical	3925
Structural	680
TOTAL	8158

Manned Servicing

Description	No. of External SSF Failures/Yr Requiring EVA	No. of EVA Maintenance Actions/Yr	Total EVA Man-Hrs Per Year
EMTT Baseline Assessment	249	507	3276 ± 400
All EVAS Changes & All Task Times ≤ 1 Hour	249	507	1014
Robotic Compatible ORUs	125	253	507

Figure 7.7 The external maintenance task team's estimate of post-assembly on-orbit servicing required by Space Station Freedom. *Courtesy of NASA.*

AVERAGE POST ASSEMBLY COMPLETE

Description	EMTT Assessment of total EVA man-hours per year	EMST Assessment of total EVA man-hours per year
Initial Baseline Demand	3276 ± 400	3519
Projected Demand Based on Incorporation of Recommendations	507	485

Figure 7.8 External maintenance summary status. *Courtesy of NASA.*

The report [11] stated that by adopting 20 recommendations to reduce EVA and mandating that most exterior items be designed for changeout in 1 hour or less, the total EVA time would drop to 1,014 man-hours per year. The greatest reduction would come in "overhead," the time needed to move to a worksite, set up for a task, and then finish off the EVA once the task is complete.

The recommendations also stressed use of robots to perform maintenance. If 50 percent of the ORUs could be changed out by the Station's fleet of robots, the total EVA time would further drop to 507 man-hours per year, Figure 7.7.

A second panel, independent from Fisher-Price's EMTT but working from the same data, ran in parallel, further refining the time estimates and proposing additional solutions. The External Maintenance Solutions Team, headed by Dr. William E. Simon, deputy manager of the Project Integration Office at JSC, further reduced the total EVA time per year to 485 hours, Figure 7.8. In addition, the solutions panel felt that a new overall approach to Station design would further reduce the time, to reach the new NASA goal of 150 EVA man-hours per year.

The "facility maintenance approach" proposed by the solutions team would involve a radical shift in design philosophy. Previously, NASA's overriding motto could be stated, "Always carry a backup system." The new approach would eliminate backup systems for those parts deemed unlikely to fail or not critical if they failed. Backups would be maintained or increased for systems analyzed as likely to fail [12].

The approach would reduce the amount of hardware on the Station, and fewer parts translate to fewer items which could fail. "We feel with the full application of this approach to the design and operation of the Station, we can realize an additional 30 to 40 percent reduction of total external maintenance," Simon said, " . . . We do believe it has some tremendous potential."

Price said, "We think that the post-assembly, the external maintenance is a manageable requirement through the application of EVA and robotics that are in the program, given the proviso that the recommendations that we bring . . . are incorporated into the design."

"But you still have a problem prior to the construction of Space Station," Fisher commented. The Fisher-Price study found that as currently planned, 941 maintenance tasks would occur in the first 30 months of Station construction, a period when the Station would not be permanently occupied, requiring 522 two-man EVAs, the equivalent of four spacewalks per week. "The significance of this is, of course, that there is no one up there to do all those EVAs. If you kept the Station as it is now, they would have to be done from the Space Shuttle, and that's a pretty hefty load," Fisher noted.

Simon's solutions panel found that a backlog of 500 exterior maintenance tasks would accumulate before the Station is permanently manned—equating to a full year of repair/replacement work. "The assembly problem is still with us," he said.

The Fisher-Price study, using an outside contractor to develop failure rates, found that 0.02 failure per hour would occur, equating to 175 failures per year.

The number of 175 failures per year is not a complete picture. Other factors, called the "K-factor," must be taken into account, such as when one failure causes another part to fail. The K-factor is expressed as a ratio of the number of parts that fail to the actual number of maintenance tasks.

A key component in the K-factor is the human factor. "K-factor just accounts for the fact that nobody's perfect," Fisher said [12].

The solutions team headed by Simon took the recommendations of the Fisher-Price study one step further. Numbers used in the estimates were refined and additional recommendations were made.

The solutions team based their figures on the Station design as updated since the design of January 1990 that Fisher-Price used as a basis for their study. Refinements in the design since the start of the year reduced the number of ORUs, thereby decreasing the number of potential failures.

The solutions team also refined the K-factor, which also helped reduce the number of failures. Robot capability also was reassessed.

In total, the refinements of the solutions team resulted in a decrease of the Fisher-Price estimates for maintenance times that would result from adoption of their recommendations. The total dropped from 507 man-hours per year to 485 man-hours per year, Figure 7.8.

"This got us the majority of the way there, but there was a ways to go," Simon said. Additional reduction of the exterior maintenance time depended on a facility maintenance approach. Simon's report said:

"The Facility Maintenance Approach to maintaining the SSF is one which is entirely new to the NASA way of thinking. It reduces hardware redundancy and complexity by reducing the actual amount of hardware on the Station. This approach determines optimum redundancy by replacing the two-failure tolerant concept followed for all critical systems with a concept where redundancy is determined by failure probability. ORU modularization would also be changed where components having higher failure probabilities would have increased redundancy inside the ORU, while those components having very high reliability would require less multiple-ORU redundancy than under the present requirements.

In addressing "bottom-up" redundancy, it must be remembered that criticality alone is insufficient as a driver of redundancy requirements. New concepts not in the current program, such as accepting a certain amount of degradation, and temporary unavailability of certain functions, must be investigated on a case-by-case basis."

Implementing the new approach requires identifying the critical items in a system that may fail and building redundancy (backup systems) based on critical failures. "We think that instead of making everything redundant, we could end up with many less things to fail by just designing it a little smarter," Simon said. " . . . There are many things that may not need to be redundant."

New approaches to redundancy thinking may create more debate over the Space Station, as will the problem of maintenance while the Station is being constructed. Debate also will be sparked because many of the proposed changes involve cost increases.

NASA is addressing Space Station Freedom operational maintenance activities by using the principles and lessons learned from the Skylab flights in the 1970s and Space Shuttle missions during the 1980s and 1990s [13]. Skylab experience demonstrated the feasibility of performing in-flight repair and maintenance. The value of that capability was effectively portrayed during the first manned Skylab flight when successful repairs in effect saved the mission.

Spare components, tools, procedures, and training were provided for performing 160 different unscheduled tasks on Skylab flights. Provisions were made for three categories of in-flight maintenance: scheduled activites for normal cleaning and replacement tasks; unscheduled activities for anticipated repair and servicing of designated equipment; and a general capability for unexpected or contingency repairs.

There have been 141 IVA and several EVA maintenance and or repair procedures performed on Space Shuttle flights [13]. The IVA tasks ranged from the replacement of digital autopilot switches to the removal and replacement of general purpose computers. Seven out of a total of 15 EVAs have been conducted to perform maintenance tasks. Nearly all of the procedures performed were not anticipated prior to flight, in spite of intense efforts to predict likely candidates for on-orbit failure and replacement. Reference 13 provides an excellent account of both Skylab and Space Shuttle in-flight maintenance events. It also describes how NASA and contractors are applying the lessons learned from Skylab and Space Shuttle missions to the Space Station on-orbit maintenance program. In the case of the Space Shuttle, this experience log is still growing.

The director, Space Station Program and Operations, appointed a manager for on-orbit maintenance and directed that each NASA development project office and international partner assign an on-orbit maintenance manager at their locations. He authorized the formation of an In-Flight Maintenance (IFM) Working Group to solve and to funnel on-orbit maintenance issues to the program level, when solutions were not evident. The Robotics Working Group was formed for the same reason, and to ensure that the best robotic technologies available are used in the assembly and maintenance of the Space Station. Each of these actions does not guarantee that problems with on-orbit maintenance will not occur, but it illuminates that there is a consciousness of maintenance as a primary influence in the design of the Station and as a primary contender for available on-orbit crew time [13].

Conclusions

Satellite servicing tools and equipment have been under development and use since the mid 1970s. As a result, the hardware to perform manned servicing is relatively mature. As noted above, 56 servicing tools are available for manifesting on current Space Shuttle flights. The NASA *EVA Catalog of Tools and Equipment* contains a listing and description of 230 tools. NASA/JSC, with a contractor team consisting of McDonnell Douglas, Lockheed, and ILC Space Systems, is developing about 70 tools for Space Station Freedom assembly and maintenance. Some of these will be adaptions or derivatives of the current 230 tools in the EVA equipment catalog.

But more needs to be done to ensure equipment and tool availability for mission operations when potential satellite servicing users start to design their serviceable spacecrafts. A set of generic servicing equipment should be baselined with their capabilities made known to the potential customers. These customers will not design their vehicles for a particular type of servicing if there is a risk that the necessary equipment will not be available at the appropriate time.

Five hardware items are of particular interest to this community. These items include a payload interface panel, a second Shuttle Orbiter Bay RMS arm, a monopropellant tanker with at least 2,268 kg (5,000 lb) capacity, a remote controlled fluids coupler, and a heavy berthing/docking fixture. This equipment should be available for all Shuttle Orbiters to avoid manifesting limitations. As servicing expands beyond the Shuttle Orbiter, equipment features should be the same wherever possible for servicing done by an OMV, OTV or the Space Station. Customers will then not be forced to select the servicing vehicle when their spacecraft is designed.

With a set of baselined servicing equipment in place, the potential customers must be given a set of cost guidelines for the use of this equipment and for any other non-hardware related items which must be paid for (i.e., transportation to orbit and training). The user community places equal or greater value on this information than on hardware details since cost trades must be performed early in the design cycle.

There is the perception by the user community that NASA will lead both hardware development and on-orbit operations. The DoD is also assuming that a hardware development program is underway within NASA and that this hardware will be available when the DoD decides the type and volume of servicing it wishes to conduct. These two perceptions indicate that a dialogue between NASA and the non-NASA community must be opened to avoid misconceptions and to allow for appropriate planning by all involved. The needs and roles of all organizations who will eventually be involved with on-orbit servicing are currently in a fluid state. A regularly scheduled forum where needs, desires, and constraints are aired will help each organization to define its own options and allow for cooperative endeavors to be arranged.

Finally, interfaces for servicing equipment must be standardized. The satellite servicing user customer needs this kind of information to properly design a spacecraft for compatibility. Standardization allows the customer to use the same equipment regardless of orbit location (LEO, polar, GEO) or servicing vehicle (Shuttle Orbiter, OMV, OTV, or Space Station). The details of these interfacer standards are one of the tasks which could be worked out by the government/industry forums.

References

1. Chappelear, D. N., and T. R. Danielson. *Space Assembly, Maintenance and Servicing Design Concepts Handbook*. Advanced Systems, Astronautics Division, Lockheed Missiles & Space Company, Inc., July 1, 1988.
2. Dellacamera, R. J. *EVA-Planned or Contingency*. MDC G9065, McDonnell Douglas Space Systems Company, Huntington Beach, CA, presented at 14th Intersociety Conference on Environmental Systems, San Diego, CA, July 16, 1984.
3. *Space Station Automation Study*. Satellite Servicing Final Report, TRW contract NAS 8-35081, Z410.1-84-175, December 20, 1984.
4. Whitsett, C. E. *New Tools for EVA Operations*. SAE Technical Paper 871499, 17th Intersociety Conference on Environmental Systems, Seattle, WA, July 13–15, 1987.
5. *EVA Catalog of Tools and Equipment*. NASA/JSC 20446 Rev. A, April 1989.
6. *Space Transportation System EVA Description and Design Criteria*. JSC-10615, Rev. A, NASA JSC, 1983.
7. Kosmo, Joseph J., William E. Spenny, Rob Gray, and Phil Spampinato. *Development of the NASAZPS Mark III (8.3 psi) Space Suit*. SAE Technical Paper Series 881101, 18th Intersociety Conference on Environmental Systems, San Francisco, CA, July 11–13, 1988.
8. Chodack, Jeff, and Phil Spampinato, ILC Dover, Inc. *Spacesuit Glove Thermal Micrometeoroid Garmet Protection Versus Human Factors Design Parameters*. Space Station and Advanced EVA (SP-872), 911383, SAE Technical Paper, 21st International Conference on Environmental Systems, San Francisco, CA, July 15–18, 1991.
9. Welch, Joseph V., ILC Dover, Inc. *Analysis of Space Suit Mobility Bearings Using the Finite Element Method*. Space Station and Advanced EVA (SP-872), 911385, SAE Technical Paper, 21st International Conference on Environmental Systems, San Francisco, CA, July 15–18, 1991.
10. Morata, L. P., and F. D. Riel. *Space Station Freedom Assembly and Operations*. McDonnell Douglas Space Systems Company, Space Systems Division, MDC H7024, 28th Space Congress, Cocoa Beach, FL, April 1991.
11. Fisher, William F., and Charles R. Price. *Space Station Freedom External Maintenance Task Team*. Final Report, July 1990.
12. *NASA presents its plan for Freedom maintenance*. Countdown Publication, September 1990, p. 14.
13. Accola, Anne L., Gerald E. Johnson, and Richard Robbins. *Designing For On-Orbit Maintenance*. NASA Headquarters Space Station Freedom Program Office, Reston, VA, IAF Paper No. 91-091, 42nd Congress of the International Astronautical Federation, October 5–11, 1991/Montreal, Canada.

Chapter 8

Automated Servicing

This chapter discusses the equipment and tools to support automated satellite servicing. Automated systems include: teleoperators, with man in the control loop; robots performing programmed or self-adaptive task sequences, with man retaining supervisory control and override capability; and machine intelligence assisting man in planning, decision making, monitoring, and troubleshooting.

Figure 8.1 illustrates servicing evolution from hands-on through teleoperated to robotic operations. Teleoperation, which uses the human operator's sensing, cognitive, and decision-making abilities, is preferred for servicing functions that involve unforeseen tasks and require impromptu responses. Fully automated operations by robots, including the use of machine intelligence, are preferred for servicing missions where remote control by teleoperation would involve long signal transmission delays (e.g., servicing a satellite in situ). The common element in teleoperation and robotic servicing is the dexterous manipulator controlled either by the human operator or by computer signals. Robotic operation may include some degree of machine intelligence to adapt the control action to changing conditions recognized by visual or tactile sensors usually located on strategic positions on the robotic arms.

Remote Mode-Of-Operation Considerations [1]

The three modes of space robotics operations are defined on Figure 8.2 in terms of their order of evolution. They range from a simple system completely dependent on human instruction to an interim system which works closely with the human controller to relieve him of routine and tedious tasks. In the far term, the robotic system will evolve to execute tasks without human supervision. Control is ultimately from the ground regardless of the operational mode.

Figure 8.3 outlines the advantages and disadvantages of each level of technology. The teleoperated mode is most representative of the current state of the art and is appropriate to accomplish simple tasks. The chief advantage to the teleoperated approach is that the operator is in control of the operations at all times. Because the system is human-controlled, it is easier to design and requires less controller software than in the other two modes. In the teleoperated mode the user remotely controls all the mobility, manipulative, and sensing capabilities of the robotic system. The robot arm and manipulative capabilities are merely an extension of the operator's capabilities.

Supervised autonomy is possible with current state-of-the-art technology but the availability of this technology is limited by its specific industrial application. The supervised autonomy mode improves on the teleoperated mode by freeing the user from specific, routine, time consuming, and tedious tasks. The human operator directs the task to a point where a preprogrammed subroutine is appropriate. At this point the human operator transfers the control to the robotic system and the robotic system executes the predefined task. When the robot has completed its portion of the task, the human operator assesses the success of the task and either takes over the more intricate and complicated tasks or commands the robotic system to continue with the next phase of the work. This continues until the total job is completed. The benefit of this approach is that the human operator can be doing several tasks at once thus increasing space operations productivity.

The autonomous mode is the most challenging of the three modes in terms of reliability, programmability, and predictability. There are many advantages in being able to operate the robot in the autonomous mode. Decisions made by the robot are executed in real time, thus mitigating the delay time disadvantage of the other two modes. The robot, given a more advanced sensor fusion capability, as well as a higher level of "reasoning capability," accomplishes its tasks with little human intervention. A robot such as this will accomplish the routine maintenance and servicing tasks as well as the hazardous tasks while allowing greater human involvement with more complex and cognitive tasks.

In the autonomous mode, the key technology challenges lie primarily in the controller. The controller needs to be enhanced with capabilities such as real time processing, large memory storage capability, clever algorithms and some

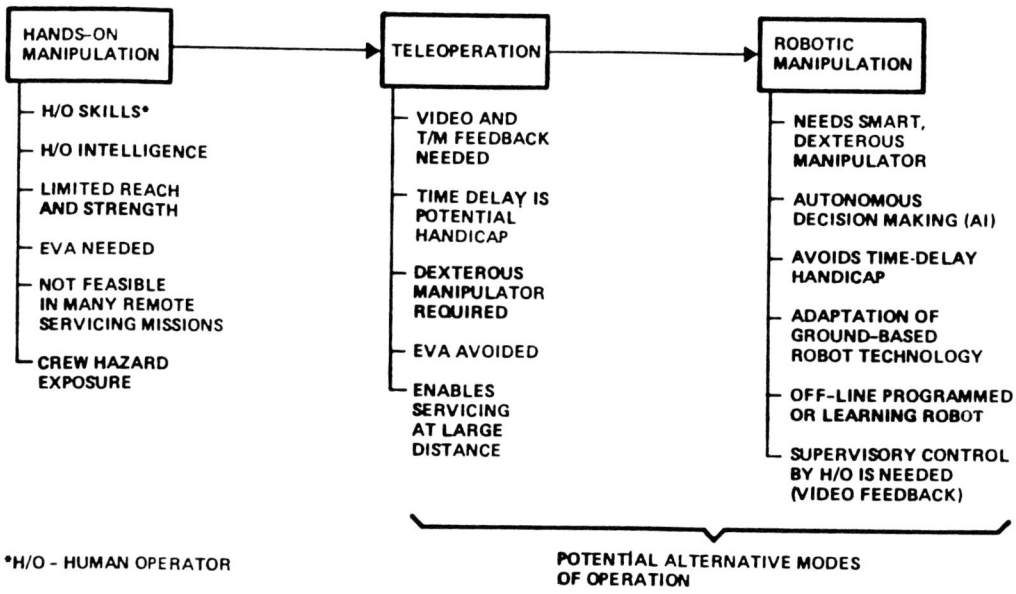

Figure 8.1 Evolution of manipulation modes in satellite servicing. *Courtesy of TRW.*

form of decision making such as artificial intelligence combined with a neural network system.

A comparison of the three modes of robotic operation applied to an ORU changeout scenario is detailed in Figure 8.4.

Current and Future Remote Servicing Tools [2]

At present, a flight qualified on-orbit remote servicer does not exist. Therefore, there are no existing remote servicing tools. The Shuttle RMS has and will be used during servicing operations, but it does not use any hardware which could be classified as automated servicing tools. Currently, the nuclear and marine industries are using remote systems to perform servicing and salvaging operations in harsh environments. These hardware items are not specifically designed for on-orbit use and would require some modification to become flight qualified.

Since there is a lack of available flight qualified remote servicing tools, a need exists to determine requirements for future tool developments. Figure 8.5 identifies 11 remote tools necessary to perform the same tasks performed by EVA methods. The needs for these 11 tools are discussed below [2]:

1. Three grapple end effectors are required. One is required to interface with the SAMS Study [2] design module IF grapple fixture. The large grapple is needed to grasp EVA handholds, tools, and structural members. The small end effector should be designed to acquire and hold bolts, small cables and connectors, or any other small/fragile items.

2. The rotary socket is the actual interface between a rotational drive source and the bolt or drive unit to be acted upon. The design may include various extensions and various socket sizes, assuming that all such interfaces on the spacecraft are not identical.

3. A rotary knife is needed to cut away thermal insulation to gain access to an ORU. Such a device is the powered equivalent of the EVA scissors.

4. The probe is a contingency tool used to pry away insulation, move small items, or gain access to tight areas for inspection.

5. Electrical connector tools are used to remove and replace electrical connectors on an ORU. There are three basic types of connectors currently utilized on spacecraft: round "Cannon" connectors, coax connectors, and rectangular connectors. Tools must be designed to handle all types and to take into account any access problems encountered on the spacecraft. The connectors to be removed may be in a position that requires access from a specific angle and may therefore require a different tool.

6. A line retention device acts as a tether system in most scenarios, but could be used to maintain the retraction rate of a restowing solar array during its replacement. Such a device should have a spring loaded, lockable reel and a grapple at the end of the line which could be operated by the remote system.

7. Cutting shears are needed to cut away appendages, cables, or whatever else is necessary from a spacecraft. They should be capable of severing any spacecraft material and also produce as little contamination as possible.

8. The fuel coupling tools refer to whatever tools are

Automated Servicing

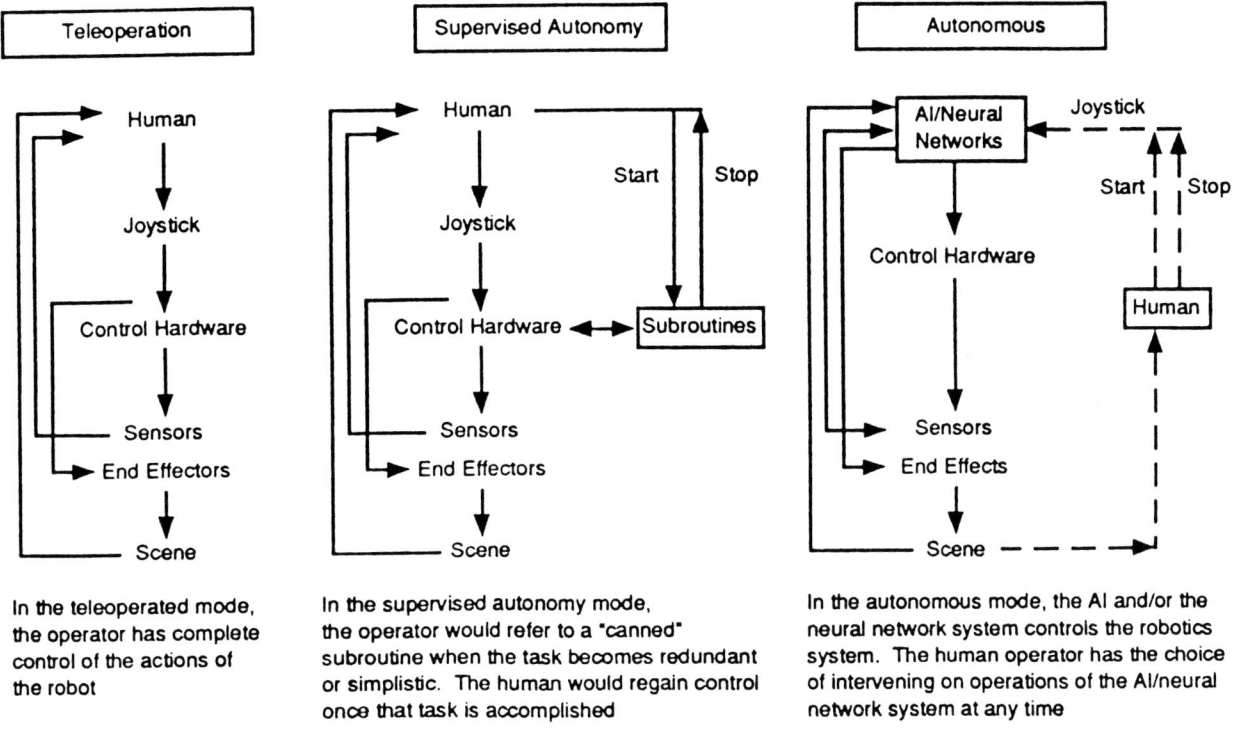

Figure 8.2 Remote modes of operation. *Courtesy of McDonnell Douglas Space Systems Co.*

Operational Model	Advantages	Disadvantages
Teleoperation	Good for simple tasks Minimal software development Minimal complexity	Visual info to human may not be sufficient Execution and feedback delay
Supervised Autonomy	Operator has discriminate control of robot Frees operator during repetitious tasks	Requires sense and depth perception feedback to operator Requires sophisticated force control and force feedback
Autonomy	Increased human productivity Decreased human risk Decreased operational costs Decreased mission time	Technology not yet mature Initially limited to simple tasks

Figure 8.3 Comparison levels of robotics technologies. *Courtesy of McDonnell Douglas Space Systems Co.*

required for the remote system to interface with the on-orbit refueling system. This tool design naturally depends on the design of the fueling system itself.

9. A fuel detector is required to detect leaks during a refueling operation or assembly of a fuel system. The current detector for EVA use depends on the presence of atmosphere to operate. Since the remote system is not likely to encounter adequate atmosphere, a different type of sensor will need to be developed.

10. Debris collection is necessary in many servicing scenarios, so some type of debris retention container is needed that can be operated by the remote system.

11. Alignment aids are necessary during the changeout of a non-modular ORU or during assembly procedures. Alignment aids could be physical or optical, depending on the application.

This list of required tools was developed for space applications, but the nuclear and marine industries have similar requirements for servicing, maintaining, and assembling systems in their respective harsh environments. These industries have developed systems to accomplish inspection, cutting, cleaning, prying, drilling, tapping, and welding. Some of the specific tools developed include a spreader system, a cable cutter, an impact wrench, an abrasive saw, sock-

Servicing procedure	Teleoperation	Supervised autonomy	Autonomous
Receive failure signal	Operator receives signal	Operator receives signal	Failure code fed to robot
Determine repair approach	Operator decision	Preprogrammed repair algorithm to robot	Robot uses smart repair algorithms to determine repair approach
Gather the necessary tools	Operator attaches appropriate robot end-effector	Operator attaches appropriate robot end-effector	Robot selects predetermined tool set, appropriate tools, and end-effectors
Identify target approach	Long range proximity sensors (V, IR)* Operator-guided	Long range vision sensor Automated approach process	Feedback from proximity sensors performs feedback recognition
Approach target/dock	Proximity sensor Middle/short range vision sensor Operator guided	Proximity sensor Middle/short range vision sensor Sensor info feedback to approach algorithm until "STOP"	Middle/short range sensors (V, IR)* Sensor info feedback to smart algorithm and directs approach speed, distance, proximity and pattern recognition
Grasp failed ORU	Human identifies failed unit Human assists gripper to failed unit Forces sensor (roll, pitch, yaw)	Human assists approach of failed unit Identification of failed unit written to software Forces sensor feeds back to robot (roll, pitch, yaw)	Smart algorithm assists approach to failed unit Identification of failed unit written to software Forces sensor feeds back to robot (roll, pitch, yaw)
Remove failed ORU	Use appropriate gripper to remove failed ORU Torque sensor Translational motion	Use appropriate gripper to remove failed ORU Force sensor Translational motion	Smart algorithm directs procedure for removing failed ORU with the aid of force sensors Translational motion
Transport failed ORU to spares carrier	Translational motion Proximity sensor Human assisted motion Middle/long range vision sensor	Translational motion Proximity sensor Preprogrammed motion Middle/long range vision sensor	Translational motion Proximity sensor Smart algorithm defines motion
Place failed ORU in stowage location	Proximity sensor Lighting Short range sensor (V, IR)* Yaw, pitch, roll	Proximity sensor Lighting Short range sensor (V,IR)* Yaw, pitch, roll	Proximity sensor Lighting Short range sensor (V, IR)* Smart algorithm directs motions (Yaw, pitch, roll)
Secure interfaces	Use appropriate gripper Operator controls motion (yaw, pitch, roll)	Use appropriate gripper Preprogrammed algorithm (yaw, pitch, roll)	Use appropriate gripper Smart algorithm (yaw, pitch, roll)
Undo interfaces	Use appropriate gripper Operator controls motion (yaw, pitch, roll)	Use appropriate gripper Preprogrammed motion (yaw, pitch, roll)	Smart algorithm selects correct gripper and directs motion (yaw, pitch, roll)
Grasp replacement ORU	Tension sensor Operator controls motion (yaw, pitch, roll)	Tension sensor Operator directs motion (yaw, pitch, roll)	Tension sensor Smart algorithm directs motion (yaw, pitch, roll)
Remove from storage carrier	Translational motion Operator controlled	Translational motion Operator supervises preprogrammed steps	Translational motion Smart algorithm directs motion
Install ORU	Translational motion Operator controlled (yaw, pitch, roll)	Translational motion Operator supervises preprogrammed steps	Translational motion Smart algorithm directs motion
Verify correct installation	Operator tests connectivity/functional tests	Embedded connectivity/functional tests	Embedded connectivity/functional tests
Return to standby mode	Proximity sensor Translational motion Operator returns robot to standby mode	Proximity sensor Translational motion Operator supervises preprogrammed steps	Proximity sensor Translational motion Robot returns to standby mode at job completion

*V = Visible Camera Sensors
IR = Infrared

Figure 8.4 Comparison of robotic modes in ORU changeout operations. *Courtesy of McDonnell Douglas Space Systems Co.*

TOOL	SAMSS TASKS						
	MODULAR ORU CHANGEOUT	BOLTED ORU CHANGEOUT	APPENDAGE ORU CHANGEOUT	APPENDAGE REMOVAL	REFUELING	INSTRUMENT CHANGEOUT	ASSEMBLY PROCEDURE
Grapple							
SAMSS IF	X	-	-	-	-	-	-
Large	-	X	X	X	-	X	X
Small	-	X	-	-	-	X	X
Rotary Socket	X	X	X	X	-	X	X
Rotary knife	-	X	-	-	-	-	-
Probe	-	X	-	-	-	-	-
Elec. Conn. Tools	-	X	X	-	-	X	X
Line Retention Device	-	-	-	X	-	-	X
Cutting Shears	-	-	-	X	-	-	X
Fuel Coupling Tool	-	-	-	-	X	-	-
Fuel Detector	-	-	-	-	X	-	-
Debris Collection	-	X	-	X	-	-	X
Alignment Aids	-	X	-	-	-	-	-

Figure 8.5 Remote servicing tools needed to perform the same tasks performed by EVA. *Courtesy of Lockheed Missiles & Space Co.*

ets, hole saws, a chipping hammer, a rope knife, and a tool storage rack that allows manipulator access to any tool. Some of these might be modified for on-orbit operation.

Robotic Servicer Concepts

Robotic servicer systems are best described as automated work stations attached to an OMV, to other advanced orbit-to-orbit transportation systems or stages, to the Shuttle's remote manipulator system, or to mechanical arms on orbiting manned tended or automated servicing platforms. Their role is to conduct supervised telerobotic assembly, maintenance, and servicing functions on unmanned spacecraft.

A satellite servicer system (SSS) provides capability for:
- Retrieval (retrieve/deploy or reboost, reposition)
- Resupply (replenish consumables)
- Repair (return failed or degraded units to operation)
- Revision (experiment or subsystems ORU changeout, satellite upgrade)

Study efforts as far back as 1974 have indicated that a servicer system can be designed to accommodate a wide range of spacecraft configurations. The technology is currently available to build such a system to operate in remote orbits on an OMV or OTV or other advanced stages. Several satellite servicer concepts are presented below.

OMV Utilized as a Tanker [3]

The design concept for utilizing the OMV as a tanker (resupply vehicle), Figure 8.6A, adds a fluid transfer kit to the front face of an OMV. The kit consists of control valves, interconnecting fluid lines, flow meters, pumps, and command/telemetry avionics associated with the control and monitoring of the fluid transfer operation.

In addition to the fluid transfer control kit, the fluid couplings necessary to mate with the user spacecraft are added circumferentially on the baseline OMV remote grapple docking mechanism. An engagement mechanism is provided to make the final mating and open the fluid couplings.

OMV fluids are manifolded, as necessary, and routed to the front face and attached to the fluid transfer kit. This concept minimizes the OMV modifications and provides a convenient fluids interface for supplying fluids from an OMV to a user spacecraft and also provides refueling capability for all OMV fluids.

OMV plus Tanker Kit [3]

The large volume tanker design adapts the OMV propulsion module (PM) to serve as an OMV front end kit, Figure 8.6B. This concept was chosen by TRW during their OMV contract work, after a review and assessment of the adaptability of the large volume tanker concepts developed in the NASA Phase B Orbital Spacecraft Consumables Resupply System (OSCRS) study.

The configuration utilizes the PM structure with added beams and trunnion/keel pins to support the tanker in the STS cargo bay. For ELV launch, these supports are not installed and the aft end of the PM structure is attached to the OMV at the four corner point. The PM corner points are on a 356 centimeter (140 inch) diameter. The OMV structure includes eight payload attach points on a 343 centimeters (135 inch) diameter. Local fittings can be used to make the transition between these diameters.

The tanker can be used for monopropellant or bipropellant missions with usable loads of 3,402 or 4,082 kg (7,500 lb or 9,000 lb) respectively. The baseline OMV PM propellant management device (PMD), as designed by TRW, was retained to facilitate propellant transfer.

The tanker/user spacecraft fluid interface utilizes the same fluid transfer kit described in the above OMV tanker

A
OMV UTILIZED
AS TANKER

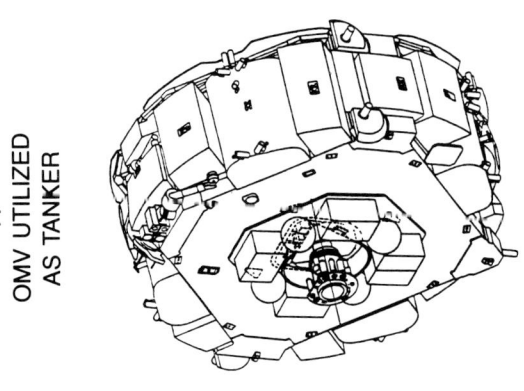

FEATURES

- SIMPLE/DIRECT FLUID UMBILICAL
- ALL OMV FLUIDS AVAILABLE FOR RESUPPLY
- BIDIRECTIONAL FLUID FLOW
- STRAIGHTFORWARD FLUID TRANSFER KIT DESIGN
- UTILIZES PROPULSION MODULE REFUEL SCARS
- MINIMIZES OMV MODIFICATIONS
- REDUNDANT FLUID COUPLING SEALS

B
OMV + TANKER KIT

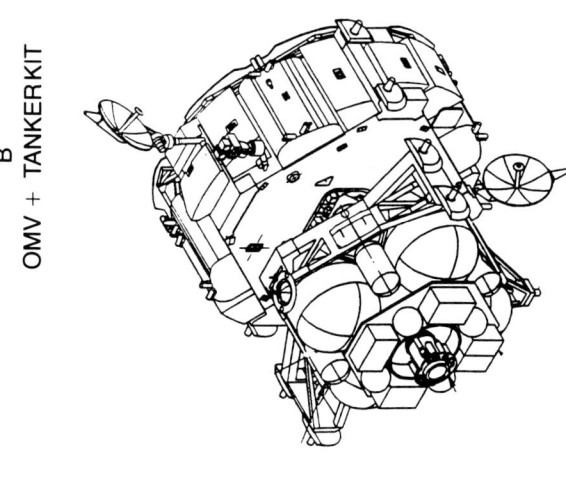

FEATURES

- UTILIZES OMV PM (TANKS AND PRIMARY STRUCTURE
- CAPACITY:
  LARGE VOL MONO   7,500 LB
  LARGE VOL BI PROP 9,000 LB
- EMPLOYS THE SAME FLUID INTERFACE FORWARD (USER) AND AFT (OMV)
- TRUNNION/KEEL PIN SUPPORT STRUCTURE REMOVED FOR ELV LAUNCH

C
OMV + RESUPPLY
SERVICER KIT

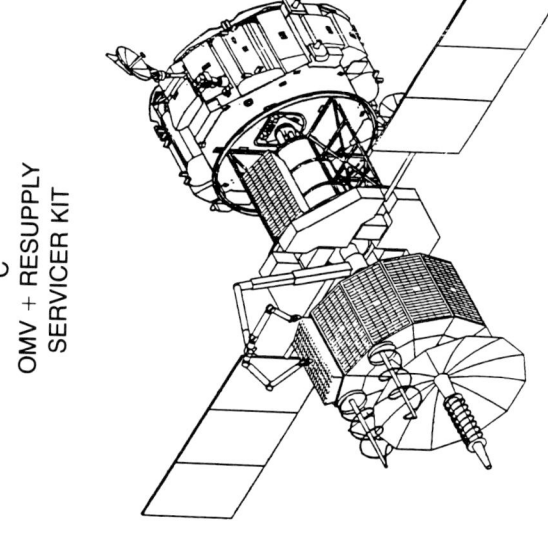

FEATURES

- IOSS TYPE POST PROVIDES:
  - USER SPACECRAFT ATTACHMENT RGDM/FLUID COUPLINGS)
  - SUPPORT AND INDEXING OF MANIPULATOR SYSTEM
  - SINGLE (IOSS) OR DUA (FB) ARM MANIPULATOR COMPATIBILITY
- POWER SYSTEM TO AUGMENT OMV POWER SUBSYSTEM
- STORAGE MODULE IS SEPARABLE AND CAN BE CONFIGURED TO MEET MISSION REQUIREMENTS

Figure 8.6 TRW OMV tanker concepts for satellite servicing. *Courtesy of TRW.*

Automated Servicing

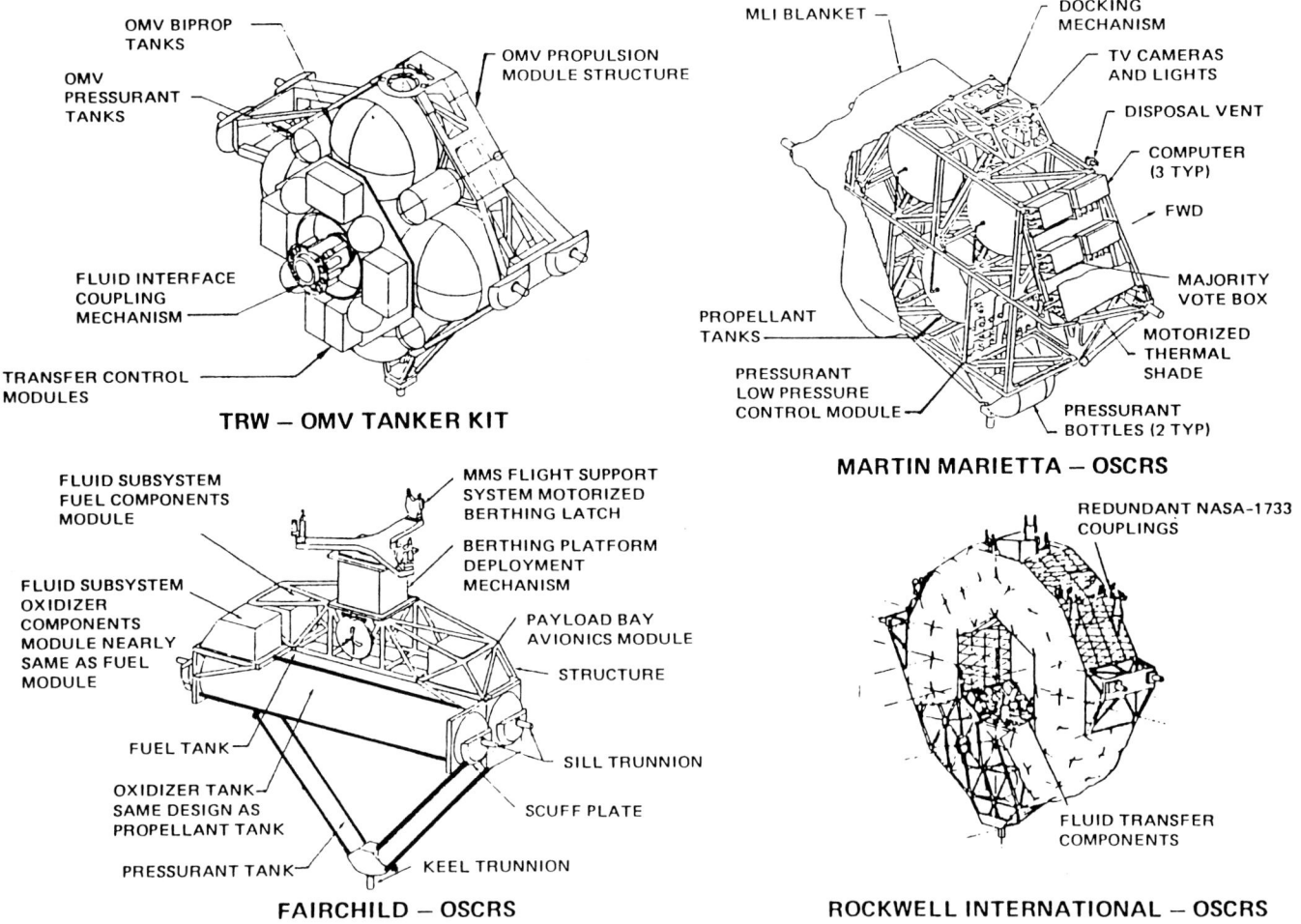

Figure 8.7 Large volume tanker options. *Courtesy of TRW.*

design except that it is located on the forward face of the tanker kit instead of being located on an OMV. To provide the transfer of fluids between an OMV and the tanker kit, the OMV fluid lines are attached to the aft end of the tanker. The attachment of the two can be "hard line" with flex provisions to accommodate relative movement, or automatic couplings can be used to provide full modularity.

The results of the NASA Orbital Spacecraft Consumables Resupply System (OSCRS) Phase B studies conducted by Fairchild Space Company, Martin Marietta Corporation, and Rockwell International, were included in the hardware design for the OMV front end kit.

The OSCRS design concepts from each of the three Phase B contractors are shown with the selected OMV tanker kit concept in Figure 8.7. A comparison of the options is presented in Figure 8.8.

An alternate approach to the large volume propellant resupply mission was developed that combines a medium volume mono propellant front end kit attached directly to an OMV with the option of supplying bipropellant from the OMV, Figure 8.9. This kit launch weight can be held within the 2,268 kg (5,000 lb) limit for STS launch on the baseline OMV design. There are no separate attachments required for STS mounting. The concept includes a propellant transfer system design, four TDRS hydrazine tanks plus pressurant tanks from the OMV propulsion module. The TDRS tanks can carry 340 kg (750 lb) each of the 1,360 kg (3,000 lb) total. The kit launch weight is estimated at 2,215 kg (4,883 lb).

OMV plus Resupply/Servicer Kit [3]

The objective of this concept is to combine on-orbit consumables resupply with orbital replacement unit (ORU) exchange. In this system, the mission is performed by integrating an OMV front end kit, designed for mission peculiar requirements, with an OMV in a manner that satisfies the overall resupply/servicing requirements.

The selected servicer kit concept features a NASA/MSFC Integrated Orbital Servicing System (IOSS type) central docking post that performs several functions, Figure 8.6C.

DESIGN BASIS	MAX CAPACITY (lb)		TANK HERITAGE		MASS FRACTION		MASS BALANCE ATTACHED TO OMV
	MONO	BIPROP	MONO	BIPROP			
TRW - OMV PROPULSION MODULE	7,500	9,000	OMV	OMV	0.66	0.70	GOOD
FAIRCHILD - TRIGON OSCRS	6,000	7,400	NEW	NEW	0.74	0.75	POOR
MARTIN MARIETTA - OSCRS	4,850	11,400	TDRS	L-SAT	0.66	0.76	FAIR
ROCKWELL INTERNATIONAL - OSCRS	7,440	8,545	GRO	GRO	0.70	0.72	GOOD

Figure 8.8 Large volume tanker comparison. *Courtesy of TRW.*

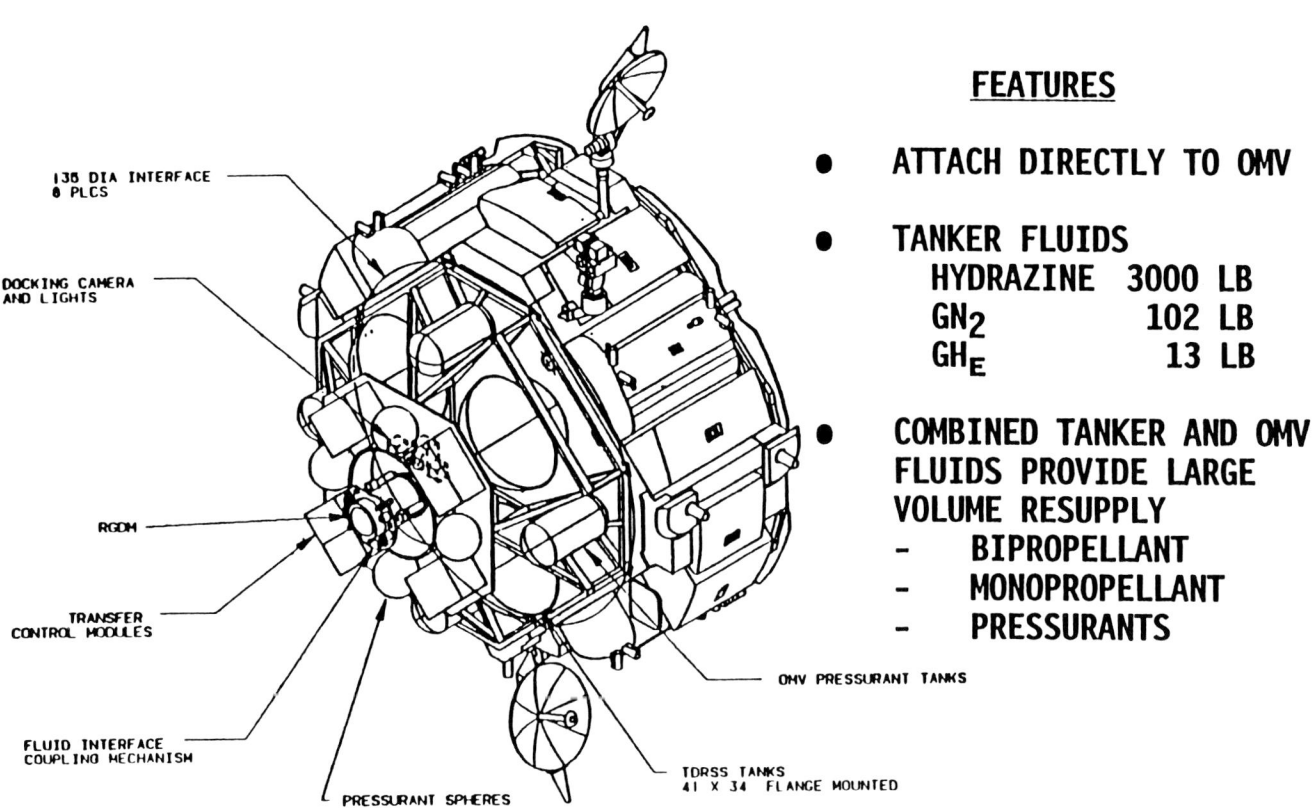

Figure 8.9 Medium volume monopropellant tanker kit. *Courtesy of TRW.*

Automated Servicing

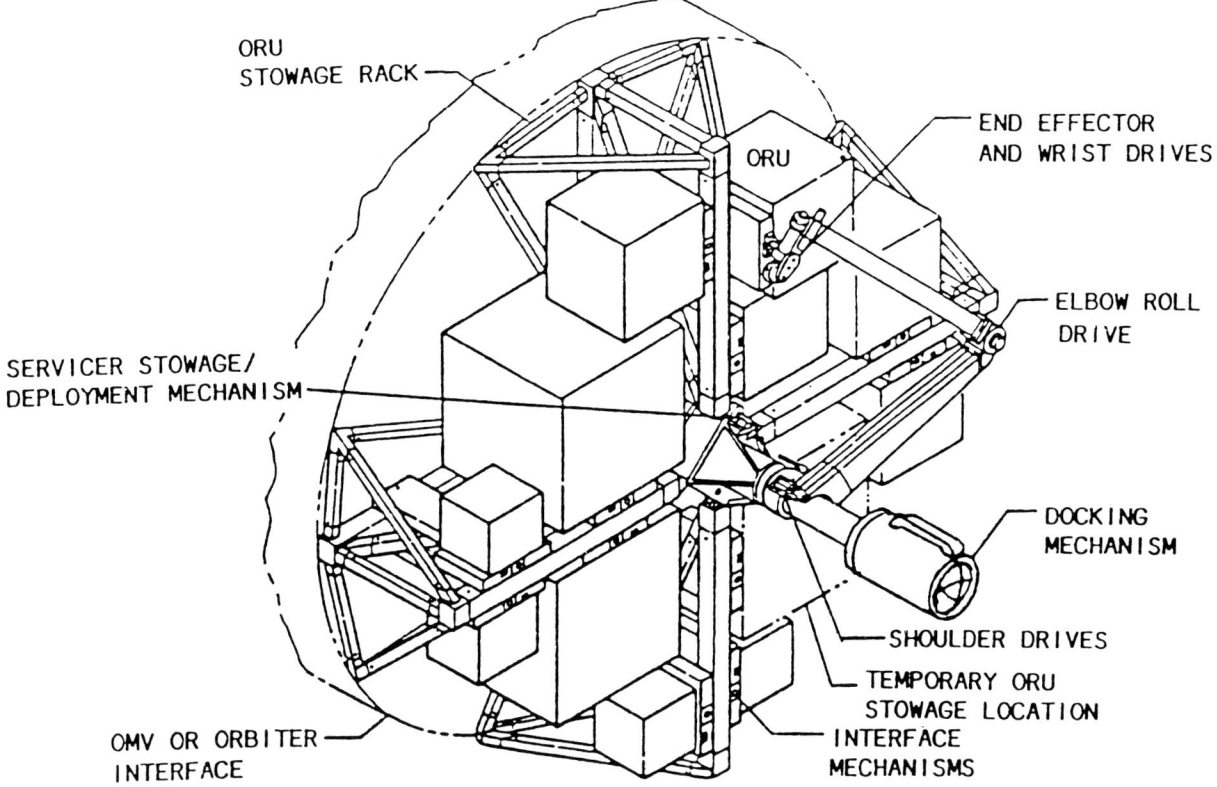

Figure 8.10 IOSS on-orbit servicer configuration. *Courtesy of Martin Marietta Corporation.*

- The post establishes a rigid attachment with the user spacecraft by utilizing the remote grapple docking mechanism (RGDM), the same as that on the baseline OMV. Fluid and electrical connections between the servicer and user spacecraft are made as a part of the docking operation, thus eliminating the use of the manipulator system to perform this function.

- The aft (lower) end of the post is equipped with an indexing mechanism that provides rotational capability for the manipulator.

- The indexing/post concept is compatible with a single arm (IOSS type) or dual arm manipulator system. The manipulator, or robot, is attached to the outboard end of the indexing arm.

The docking post is attached to an equipment platform that is used to mount the manipulator control electronics, the OMV power augmentation components (batteries and solar array), and the fluid transfer system control modules. The platform with these items installed is called the servicer bus and is separable from the servicer storage rack.

The OMV power augmentation equipment necessary varies, depending on mission duration, from a battery pack consisting of two OMV AgZn batteries (one plus one redundant), plus a small solar array 4.6 square meters (50 square feet) to a large solar array 18.6 square meters (200 square feet).

The Integrated Orbital Servicing System [4]

The Integrated Orbital Servicing System (IOSS) was designed by the Martin Marietta Corporation for NASA/MSFC. It is a candidate front end kit on an orbital maneuvering vehicle. The IOSS automated and teleoperated module exchange capabilities can be extended by the addition of appropriate sensors, logic control, and control and display devices to become a telepresence system. The control logic can be further extended to include artificial intelligence functions.

The IOSS configuration shown on Figure 8.10 evolved through a series of iterations during which a very wide range of alternatives was considered. The design is compatible with maintenance of most spacecraft of the NSTS era. Adapters are used to accommodate support structure differences across the applications. The design has only two major components: a servicer mechanism and a stowage rack for module transport. A docking mechanism is also shown for reference and so that the mechanical interface aspects may be more readily visualized. Stowage racks can be configured and loaded for particular flights prior to attachment to the carrier vehicle. It may be desirable to have several stowage racks available for this purpose. The stowage rack shown mounts directly to an OMV.

The primary characteristics of the IOSS on-orbit servicer are listed on Figure 8.10 along with some accommodations required from the OMV. The servicer mechanism has the capability to replace modules in both the axial (parallel to the docking post) and radial (perpendicular to the docking post) directions as well as off-axis directions and combinations of directions. The module size limits shown are representative for cubes. Smaller sizes can be handled, but they are not efficient. Larger sizes can be handled at lower rates and with appropriate stowage rack configurations.

The design philosophy of the IOSS rests upon the fundamentals listed below:

- Servicing activities are defined before the mission is started.
- Module exchange is a major servicing activity with extensive applications.
- Module exchange is useful for repair and maintenance, fluid resupply, upgrading equipment, and product return.
- Module exchange task is basically simple: Remove, flip, relocate, and insert.
- Automated module exchange is technically feasible.
- Modules designed for mechanized exchange can also be exchanged while on EVA.
- A variety of interface mechanisms are acceptable.
- The interface with the servicer mechanism and the stowage rack should be standardized.

The high cost of a Shuttle launch indicates that any on-orbit maintenance activity should be well understood before the Shuttle is launched. This means that the specific failure, the problem component or module, associated repair activities, and all other aspects of the maintenance mission are known. This implies that the mission can be fully planned and that the only unknowns are those due to subsequent failures in the system or operations. The IOSS design takes full advantage of this prior knowledge and ability to preplan. One example is that the coordinates of the failed spacecraft module are loaded into the servicer's memory before launch; also the module exchange task is basically simple and thus is very amenable to being standardized and automated. Note that it is not necessary to standardize the interface mechanisms; these should be left to the spacecraft designers choice. However, the interface between the servicer and interface mechanisms should be standardized.

The servicing system design was based on these items:

- Applicable to many spacecraft
- Incorporate existing technology
- Places few requirements on spacecraft
- Remotely operable
  — Autonomous for most activities
  — Teleoperation as alternative
- ORU interfaces operable by
  — Servicer system
  — Astronauts on EVA
  — Ground handling equipment
- Supported by carrier vehicle
  — Rendezvous and docking
  — Attitude control
  — Electrical power
  — Two-way communications

The selected servicer system can be applied to most spacecraft that will be launched in the Shuttle because its size, degrees of freedom, and joint ordering were carefully selected to match this class of spacecraft. In some cases, more than one docking may be necessary and good judgment should be used in locating the modules and their attachment interfaces.

The estimate of weight of the IOSS is 285 kg (629 lb). Its main components are:

- Servicer mechanism
- Docking probe structure
- Docking probe end effector
- Stowage rack
- ORU interface mechanism guides (12)
- OMV adapter (3 point)
- IOSS electronics
- Microprocessor and signal conditioner
- System wiring

The primary mission of the IOSS is to perform on-orbit servicing of spacecraft by using the servicer to remotely exchange failed spacecraft modules with operating modular orbital replacement units. This is a significant, cost-effective capability that, with the other servicing functions of fluid resupply, remote inspection, and product retrieval, makes the IOSS a valuable tool for servicing a range of spacecraft. The IOSS with the appropriate carrier vehicle can service spacecraft in orbits ranging from low Earth orbits through geosynchronous orbits.

Flight Telerobotic Servicer [5]

NASA is leading the way in space robotic technology. Through Goddard Space Flight Center, the agency is building the flight telerobotic servicer (FTS), a robotic device that can be teleoperated under the constant command of a human operator or run by itself under human supervision. Plans call for the FTS to assist the astronauts in the assembly, maintenance, servicing, and inspection of Space Station Freedom. The FTS project began in 1986 when Congress asked NASA to develop a sophisticated telerobotic system as part of the Space Station Freedom automation and robotics program. The need for space robotics was identified in the Freedom definition studies and strategic plans for servicing Earth-orbiting spacecraft beyond the reach of current manned systems.

As this book was in preparation, a series of discussions were taking place within NASA to determine if NASA had sufficient funds to carry the FTS into hardware development. The text to follow addresses the FTS, however, as if it will be a going program. The FTS project forms the basis for combining teleoperational and robotics technologies and for applying the evolving technologies to government and commercial ventures in space or on Earth. It is driven by five major objectives: to reduce space station dependence on crew EVA, improve crew safety, enhance crew utilization, provide remote servicing capabilities for platforms, and accelerate technology transfer from NASA sponsored research to U.S. industry.

The project will satisfy the first two objectives by providing a reliable, spaceflight quality telerobot with its end effectors and tools; operator workstations for operating the FTS from the Station and the Shuttle; and operating software and procedures. The third and fourth objectives are being met by designing the servicer to accommodate advanced technologies, such as those that increase its autonomy, as they become available. As for the fifth objective, technology transfer among U.S. industry, universities, and other government agencies is integral to the project.

The FTS will have the basic capabilities to support any task it might be assigned, although the design is derived from six specific design reference tasks:

1. Install and remove truss members.
2. Install a structural interface adaptor on the truss.
3. Change and replace orbital replacement units.
4. Mate thermal utility connectors.
5. Perform inspection tasks.
6. Assemble and maintain the electrical power system radiator assembly.

Even though the FTS system is the most visible component of NASA's robotics initiatives, it is only one part of a significantly broader program. The Office of Aeronautics and Space Technology sponsors, under the leadership of NASA's Jet Propulsion Laboratory, a program to develop advanced robotics technology. In addition, Johnson Space Center is leading a study on adapting a significant portion of the FTS concept for satellite servicing missions. Marshall Space Flight Center will use portions of the FTS system as the "smart front end" for the orbital maneuvering vehicle.

Another significant part of the FTS project is a ground system that will support operations and system evolution. Problems occurring during orbital operations will be resolved under the engineering test system, which will be housed in a ground facility called the Robotic Assembly and Servicing Simulation Facility. NASA-Johnson, supported by Goddard and the FTS contractor, Martin Marietta, will train the astronauts. In addition, many other NASA centers are examining the application of robotic technologies to their specific responsibilities. Goddard's Development, Integration, and Test Facility will be used to develop and test new technologies for the FTS, ensure that new technologies developed elsewhere can be integrated into the servicer during evolution, work with users to make sure their systems are compatible with the FTS, and perform operational assembly and servicing scenarios.

NASA is planning for the future with the FTS project. The servicer is essential for developing more sophisticated robotic systems—systems that can accompany and more adequately support humans during their exploration of the Moon and Mars. Robots on the Moon and Mars will have to be more autonomous, if only because of the long time delays with Earth and Earth-based human operators. Development of the initial FTS is leading the way to significantly more advanced technologies, which can be used in servicing satellites in orbits not accessible to the Shuttle, or establishing a manned base on the Moon. Not only will the FTS provide a needed operational capability during the assembly and operation of Space Station Freedom, it also will provide an expanding foundation for proving more advanced robotic and telepresence concepts in space.

FTS Design

In June 1989, the Martin Marietta Space Systems Company, Denver, Colorado, became the prime contractor to build the flight telerobotic servicer (FTS) system, Figure 8.11. Martin's design for the system includes the hardware, software, subsystems, training and simulation equipment, and the evolutionary "scars and hooks" that will make the telerobot autonomous.

The FTS has two manipulators, each with 7 degrees of freedom (DOF). It also has one 5-DOF attachment stabilization and positioning system mounted on a compact body, which serves as a leg support. The body contains internal electronics consisting of black boxes that provide the power, data management, processing, and communications functions. The internal components, manipulators, and leg are modular orbital replacement units and can be accessed easily through hinged panels. Also mounted on the body are two Ku-band antennas for communication, a camera positioning assembly with two head cameras, and holsters for storing tools and end effectors, which act as the robot's hands.

The other two parts of the FTS system include the Shuttle and Space Station workstations, which will provide the operator with similar interfaces, including three color video displays with split screen and test and graphic overlay capability and two hand controllers for operating the manipulator with two arms, Figure 8.11. The video display can be

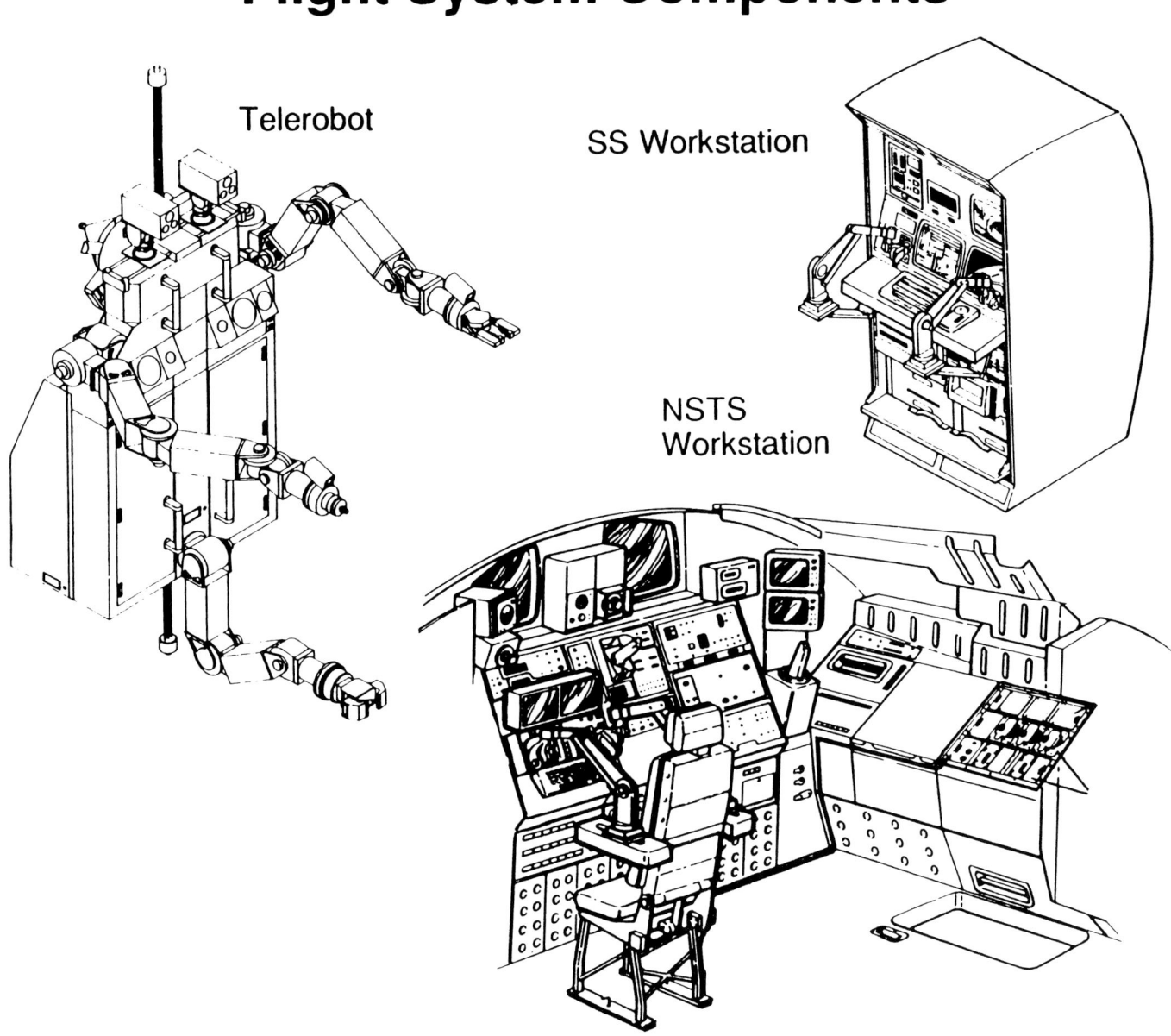

Figure 8.11 The Martin Marietta Corporation's flight telerobotic servicer system elements. *Courtesy of Martin Marietta Corporation.*

converted to a command data panel (with a graphics display and command menus) from which an operator can control the telerobot in all its modes. The operator can also verbally control the cameras and lights, which allows simultaneous hand control of the telerobot.

The camera views allow the operator to perform tasks without viewing the worksite directly. The operator selects telerobot control modes and issues discrete commands through a set of variable function control switches. As commands are issued, the graphic display and graphic video overlay change and provide status information, command menus, and subsystem schematics. Anomalous events result in automatic caution and prioritized warnings. During servicing operations, the sequence of events is displayed, which the operator will use as a checklist. Hard-wired switches allow the operator to control the manipulators, tooling, cameras, and other safety-critical components during computer failure.

The FTS system is designed to evolve, Figure 8.12. Hooks and scars are design features that permit the future addition of software (hooks) and hardware (scars). The hooks and scars required to accommodate evolution consist of identified memory (75 percent) average and through-put (131 percent) margins, four spare slots within the computer chas-

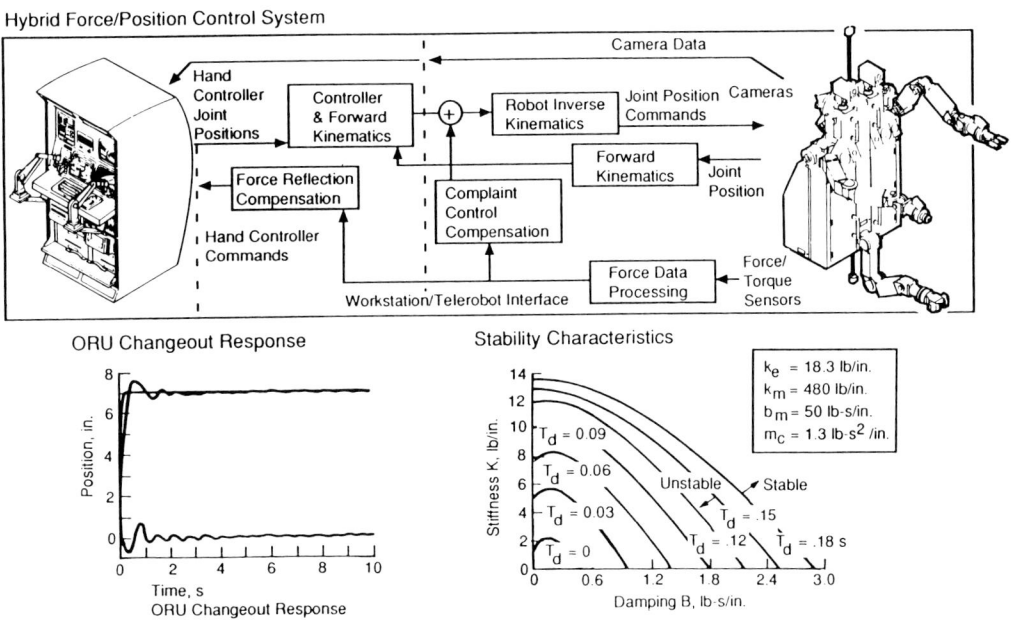

Figure 8.11 *Continued*

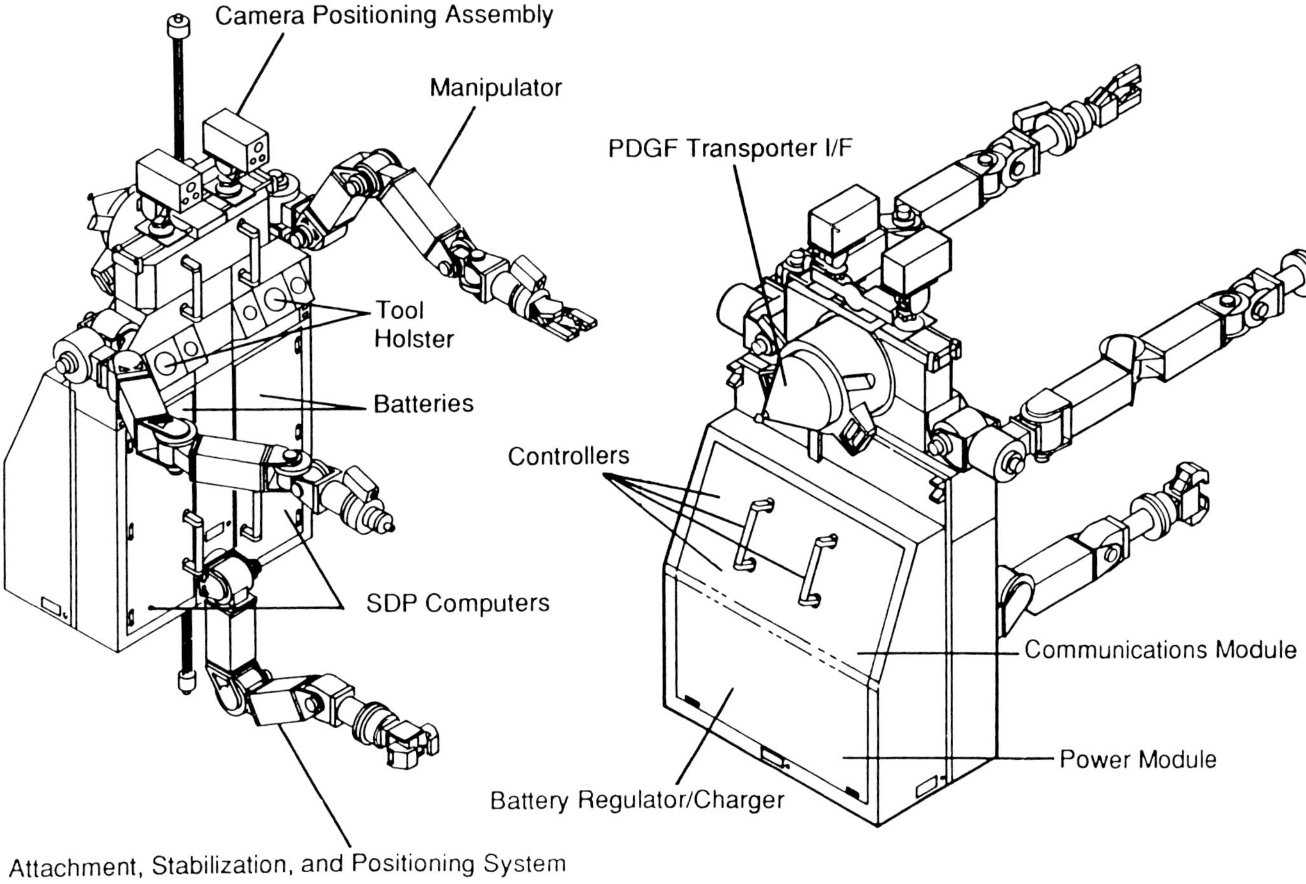

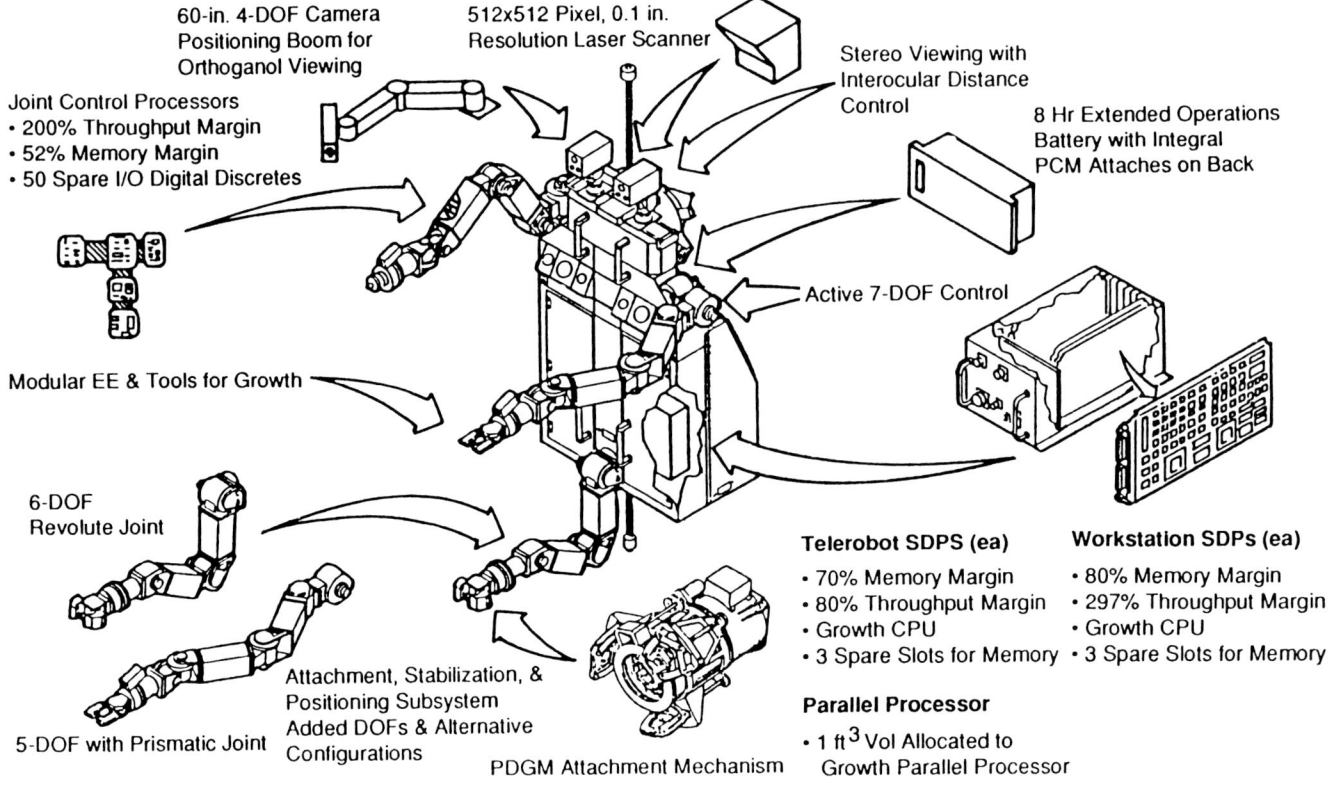

Figure 8.12 Flight telerobotic servicer growth accommodations. *Courtesy of Martin Marietta Corporation.*

sis to accommodate additional memory, cabling, structural interfaces for sensors and batteries, and a 30.5 centimeter (1 foot) space allocation for an image processing computer. The end effector interfaces have been standardized to permit the addition of new tools.

Simulators will provide a real-time graphic display of simulated telerobot operations. They will be used to familiarize the crew initially and to develop mission timelines and operation sequences. The telerobot promises to be a useful, reliable, and safe tool to assist the astronauts in performing assembly, maintenance, servicing, and inspection tasks on Space Station Freedom and the Space Shuttle.

FTS Operations

The earliest planned use of the flight telerobotic servicer (FTS) in the Space Station Freedom program is for the initial assembly of the station. Before Freedom is permanently manned, the FTS will operate out of the Shuttle bay. The Shuttle remote manipulator system will pick up the telerobot by its grapple fixture and transport it to worksites in the Shuttle bay or on Freedom's truss. The needed resources will come from the Shuttle through the grapple fixture. Astronauts on the aft flight deck of the Shuttle will provide the control.

The FTS will be stored on orbit between assembly flights at a storage accommodation site that will be attached to one of the truss bays on the nadir-racing side of the truss. Here it will be shielded from cold space and warmed by the reflected rays from the Earth. It will be within easy reach of the Shuttle RMS when the Shuttle is docked. Storing the FTS on orbit means that it will not have to be transported back and forth.

After initial assembly is complete, the 18.3 meter (60 foot) long arm of the Canadian mobile servicing center (MSC) will transport the FTS to the worksites. The space station RMS, which is part of the servicing center, will retrieve the telerobot from its storage bay and place it on the MSC or at a worksite. While attached to the RMS, the telerobot can be operated while connected to power, data, and video through the RMS.

The FTS has three operating modes: dependent, transporter-attached, and independent. When dependent, the telerobot is attached to one of eight "improved worksites" or through an umbilical to one of 12 MSC utility ports. At the "improved worksites," the telerobot will receive utilities through a worksite attachment fixture.

Planning is under way to identify and design the interfaces for the FTS on the Shuttle and Space Station. Accommodations for mechanical power, data, and video will be part of Freedom's design right from the start. Hardware will be designed with interfaces for the FTS and the astronauts. Future tasks are being analyzed and verified by hardware performance in the lab. Such intensive up-front planning ensures that the FTS will be a safe, reliable, and valuable tool for the astronaut crews of Freedom.

FTS Advanced Technology

Evolution toward more autonomy is part of the flight telerobotic servicer (FTS) design. NASA-Goddard has established a Development Integration and Test Facility where new technologies can be developed and applied to operational scenarios. Technologies in control strategies for an arm with 7 DOF, including a safety system containing skin sensors for obstacle avoidance, are being developed. Supporting this effort is the development of a dynamic model for more precise simulations. Innovative and effectors, built for grasping in zero g, are being developed specifically for autonomous servicing.

An important aspect of the development effort is to make orbital replacement units on scientific satellites "robot friendly." Robot compatibility must be taken into consideration well in advance of final design. Under way now is the design of an instrument called ASTROMAG, a large, superconducting cosmic ray facility that will be assembled by the telerobot as an attached payload on Freedom. A mockup will be built to examine the limits of the instrument design, verify the robot vision sensing system, determine compatibility of the gripper and latching design, and evaluate instrument servicing scenarios.

The vision sensing system consists of cameras on the robot and distinct high intensity reflecting targets on the ORUs. Using a low level processing algorithm, it determines orientation and range and relays the data to the robot controller. Langley Research Center has a broad program in telerobotics research, including control of robotic manipulators, short range (proximity) sensing, autonomous space structure assembly, and advanced systems architectures and mechanisms. Langley engineers in the Intelligent Systems Research Laboratory have developed a framework for extending single-arm control to dual-arm coordination. Using resolved motion rate control for simple tasks and tactile force/torque and vision feedback methods for automating parts of the tasks, it allows the operator to control two arms to perform a task.

Research over the last 5 years has led to the development of laser scanning and ranging systems, which use coherent semiconductor laser diodes for short-range sensing. One system provides a 256×256 pixel range and image map of the visual scene. Another couples the laser to a fiber optic switch with the range computed from a number of individual fibers. Langley engineers have proposed a coherent laser system with switched optical fibers as a possible sensor for the telerobot.

Robotics and structural engineers at Langley's Automated Structural Assembly Laboratory are investigating the possibility of using a robot to autonomously assemble space structures. Already developed are small graphite epoxy truss elements, nodes, carrier racks, and a unique end effector to build a large, lightweight, tetrahedral truss that embodies high stiffness. Force/torque sensing and control methods have been used to enable the end effector to precisely insert one part into another, allowing the telerobot to perform automated assembly tasks on the Space Station Freedom.

NASA Langley Research Center engineers have developed the laboratory telerobotic manipulator, a dual-arm telerobot with traction (friction) drive rollers at each joint rather than gears, a joint processor embedded in each link, and fiber optic signal lines for high-speed data. Each arm has 7 DOF, allowing the researcher to optimize the robot's movements. The telerobot can be used as a dual-arm, full-size replica, master-slave teleoperated system, or the full-size master arms can be replaced with miniature master controllers or by joysticks. A comprehensive evaluation of telerobot operations and control is planned by NASA for the 1992–1995 time frame. This evaluation will increase our knowledge of how manipulators behave in the teleoperated mode.

The Jet Propulsion Laboratory's program in telerobotic technology development varies from basic research to complex system implementation and flight experiments. The development of manipulator control algorithms addresses the challenge of controlling the FTS's two manipulators, each of which is approximately 168 cm (66 in) long and weighs 57 kg (125 lb). These arms are difficult to control because for each end-effector position, an infinite number of configurations are possible.

A control framework is being developed at JPL that allows direct global control of the manipulator. This configuration control meets the traditional goals of manipulator redundancy but also lets engineers increase the manipulator's mechanical advantage, optimize its inertial properties, or minimize reaction forces during a task. The research is being extended to multiple manipulator control where closed kinematic chains are formed during contact tasks.

Another thrust of the program at JPL is the design and development of systems for controlling telerobots remotely from Earth or anywhere communications delays exist be-

Figure 8.13 TRW's satellite servicer system conceptual design. *Courtesy of TRW.*

tween the operator and telerobot. Researchers are developing systems that permit an operator to quickly reconfigure the telerobot to do new tasks safely despite communication delays of up to 6 sec. A development system was used in recently to execute a simple inspection task at the Kennedy Space Center, Florida, when the operator was at JPL, Pasadena, California.

Satellite Servicer System

Remote autonomous satellite servicing capabilities will have a significant role in NASA's Space Station Freedom (SSF) program and its Space Science and Applications programs. The satellite servicer system (SSS), Figure 8.13, is capable of natural evolution into a system that is compatible with the future requirements of these programs [6].

Automation has historically been a significant part of NASA's space missions. Pioneer, Viking, Voyager, Magellan, and Surveyor are examples of unmanned automated spacecraft. The Space Shuttle, in travelling from Earth to orbit, is an automated system with the capability for human intervention. Without a high degree of automation, the operation of complex manned vehicles such as the Space Shuttle and the planetary mission vehicles would be technically impossible or prohibitively expensive. However, present automation is traditionally "preprogrammed," i.e., rigid and inflexible. Future automation will have the capability to: (1) adapt to a changing and uncertain environment; (2) decompose high level commands into those that a machine can execute; (3) develop plans to accomplish tasks, monitor the execution of those plans, and dynamically replan as necessary; and (4) know what and when to report back to its human supervisor. In short, the next generation of automation will be far more autonomous and flexible or robust than the current generation. These advances will enable the system to assess performance, predict possible failures, and diagnose and troubleshoot problems. This added power and flexibility will free scarce human resources from a myriad of tasks that are dangerous or repetitive. These developments in autonomous operating capabilities are vital to the widespread use and cost effective application of space-based servicing [6].

The overall objective of the joint NASA/SDI Satellite Servicer System Flight Demonstration (SSSFD) program is to initiate and develop an evolving and autonomous on-orbit satellite servicing capability to protect the national investment in future civilian and DoD satellite programs. The

NASA/SDI partnership was formed because of the similar objectives and anticipated requirements for the development of an on-orbit satellite servicing capability that could be used in remote locations.

The specific objectives are to verify, by several on-orbit demonstration missions in the mid to late 1990s, the capability to perform: autonomous rendezvous and docking, ORU exchange, fluid transfer, and Space Station proximity operations, utilizing supervised autonomous control [6]. The major elements required for the SSSFD program include: (1) the satellite servicer system (SSS), (2) the target vehicle, (3) the Space Shuttle, (4) the on-orbit control work station, and (5) the ground control work station.

In June 1990, NASA/JSC as a result of an industry competition, selected TRW's Space and Technology Group and the Martin Marietta Corporation to conduct Phase B studies on the SSS. The studies were aimed at flight demonstrations to satisfy the specific objectives mentioned above, but due to NASA's overall funding problems in 1990, the start of these studies has been put on hold.

SSS Design [7]

The TRW Satellite Servicer System referenced for the mid-1990s flight demonstration program is shown on Figure 8.14. This SSS features the use of an OMV, selected parts of the flight telerobotic servicer (FTS), a support module (SM), and a 5 degree-of-freedom, 6.0 meter (19.8 foot) reach servicer position arm (SPA). The design makes use of existing and developing NASA hardware: Shuttle remote manipulator system (SRMS) joints are used for the SPA; electrical power subsystems are from the OMV support module; command and data handling units are from the OMV project; and fluid storage is provided by existing tanks.

The reference support module design consists of two primary assemblies: the servicer equipment assembly (SEA) and the carrier assembly (CA). The servicer equipment assembly contains all the support subsystems for remote servicing and provides all the required mechanical, electrical, and data interfaces with the OMV, Shuttle, and target vehicles. The carrier assembly houses ORUs and hydrazine tanks.

The support module (SM) design general arrangement, Figure 8.15, supports SSSFD ORU accommodation, robotic manipulation, autonomous rendezvous and docking, and fluid transfer while providing mechanical and electrical/data interfaces to the OMV, Space Shuttle, and the demonstration target vehicle. Support module subsystems are housed in the SEA, Figure 8.15, and the CA supports the SEA and contains the SM/NSTS interface attachments. The SM is 4 meters (13 feet) long and weighs about 2,722 kg (6,000 lb), including replacement ORUs.

The modularity of the design with clean interfaces between the servicer equipment assembly/carrier assemblies enables a cost-effective building block approach to satisfy varying mission requirements. For example, all flight demonstration objectives can be satisfied by the optional configuration of the reference design that uses only the servicer equipment assembly to demonstrate fluid transfer and exchange ORUs. Demonstration of fluid transfer is accomplished by taking hydrazine from the OMV storage tanks. ORU exchange is demonstrated by manipulating target vehicle ORUs. This flexible approach allows NASA to defer ORU/tank carrier assembly development costs from the SSSFD program to the operational program without adversely impacting the latter. Preliminary estimates indicate a savings of over $25 million in development/launch costs.

The reference design can easily be reconfigured with a family of carrier assemblies to accommodate many flight demonstrations, leading to full operational capability. The building block approach is more appropriate than a fixed ORU carrier concept because ORU/tank size and quantity requirements are expected to vary mission-to-mission.

Conceptually designed on-orbit Space Shuttle and ground work stations enable OMV and FTS supervised control. Accounting for tight space on the Orbiter flight deck, the on-orbit work station is electrically reconfigurable to allow control of the OMV or SPA/FTS from the same work station. The ground work station adapts SPA/FTS control to NASA's OMV ground control console.

ORU Servicing Sequences [7]

NASA intends, as part of the SSS flight demonstration program, to validate the SSS concept for the remote changeout of orbital replacement units on a serviceable spacecraft. The SSSFD program will employ a target vehicle to simulate an operational NASA or DoD satellite. Figure 8.16 depicts a sequence of SSSFD target vehicle servicing showing autonomous docking and ORU exchange. Neutral buoyancy tank tests indicate that roughly 40 minutes will be required for supervised autonomous removal and replacement of an MMS type ORU.

TRW performed a mission analysis functional definition study to determine the ORU exchange positioning system reach, range, and degree-of-freedom required for the SSS to verify ORU changeout on demonstration target vehicles. This kinematic analysis, which concluded that a reach of about 6.1 meters (20 feet), excluding the FTS, was needed to service the target vehicles, was then extrapolated to examine the ORU reach needed on 14 candidate operational satellites. The results appear on Figure 8.17. Note the required reach goes from 4.5 meters (15 feet) for the NASA MMS to 16 meters (53 feet) for the DoD Zenith Star. Obviously multiple docking points on the larger satellites will enable shorter reach arms to be used.

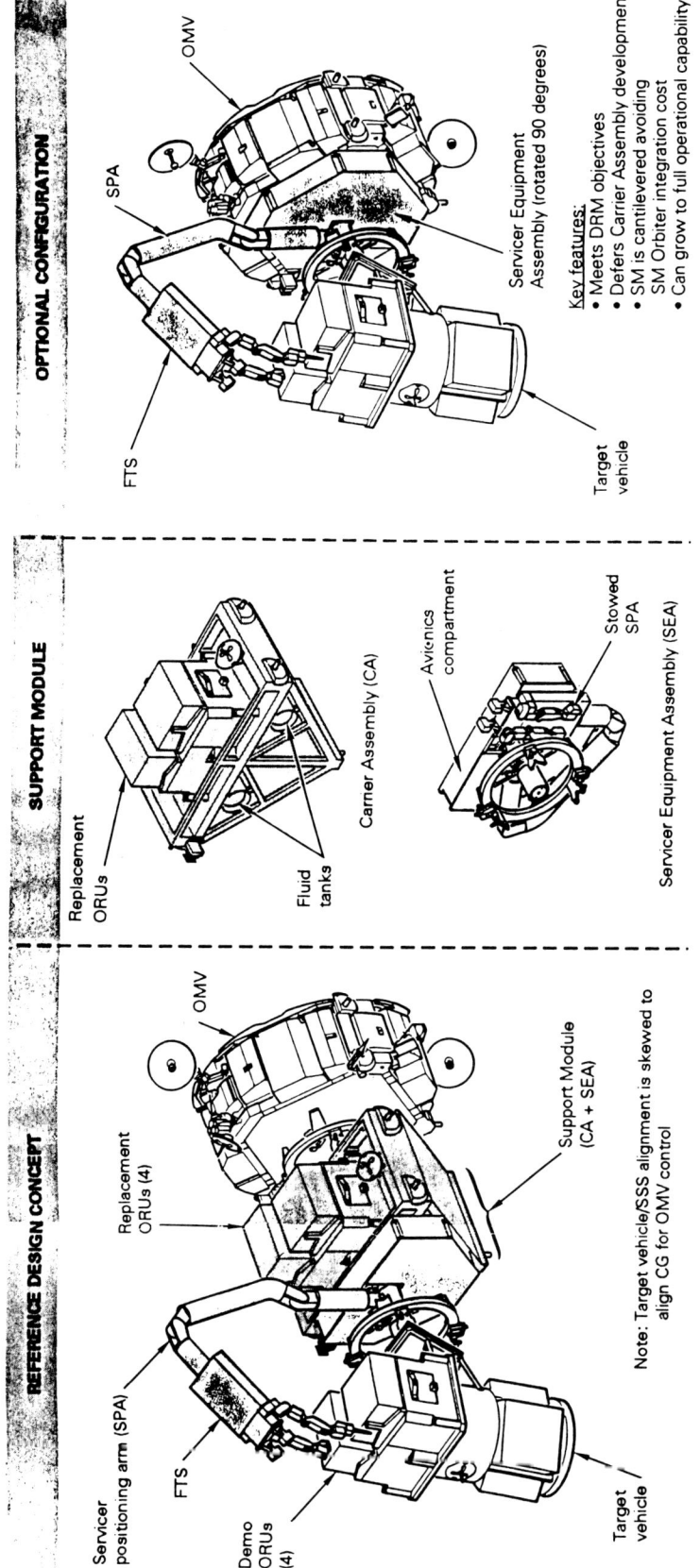

Figure 8.14 Satellite servicer system (SSS) and target vehicle. *Courtesy of TRW.*

Automated Servicing

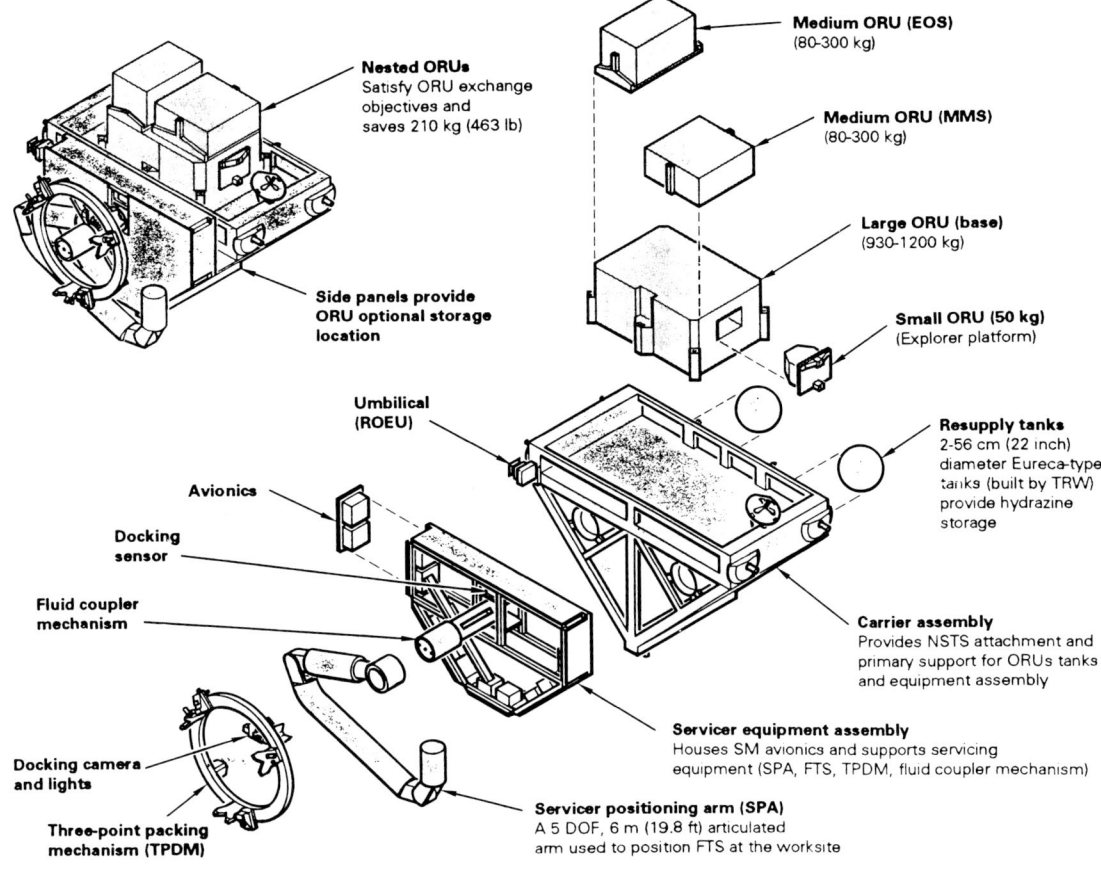

REQUIREMENTS

- Mechanical/structural
  - Support subsystems, ORUs, resupply fluid, FTS
  - Docking interface
  - Design loads per JSC 07700, Vol XIV
  - FTS translation
- Thermal control
  - Maintain temperature limits
- Fluid transfer
  - Safety per NSTS 1700.7B
  - Fluid (N2H4) quantity TBD
- Electrical power
  - Provide support module subsystems and FTS power
- Command and data handling
  - Augment OMV C&DH
- Control system
  - Docking sensor
  - Robotic system control

CAPABILITIES

- Modular design provides separable structural sections
- Provided by TPDM
- Structure sized for Vol XIV loads and maximum ORU mass
- Provided by SPA

- MLI and VCHP control temperatures with lowest heater power
- Comply
- 181 kg (400 lb)

- Battery powered system (3 to 350 AHr with space for 4)

- SDP computer and remote units provide augmentation

- LDS provides inputs to OMV GN&C for auto docking
- Control mode coordinator sequences FTS/SPA operations

WEIGHT SUMMARY

Support Module (SM)	(kg)	(lb)
Mechanical/structure	1049	2308
SPA	300	660
Servicer equipment assembly	219	482
Carrier assembly	396	871
TPDM assembly	75	165
Retention/release mechanisms	59	130
ORUs (4 nested)	930	2046
Thermal control	35	77
Fluid transfer	74	163
Avionics	320	704
Contingency	336	737
Support module—total weight (dry)	2743	6036

Figure 8.15 TRW's support module reference design for the satellite servicer system flight demonstration program. *Courtesy of TRW.*

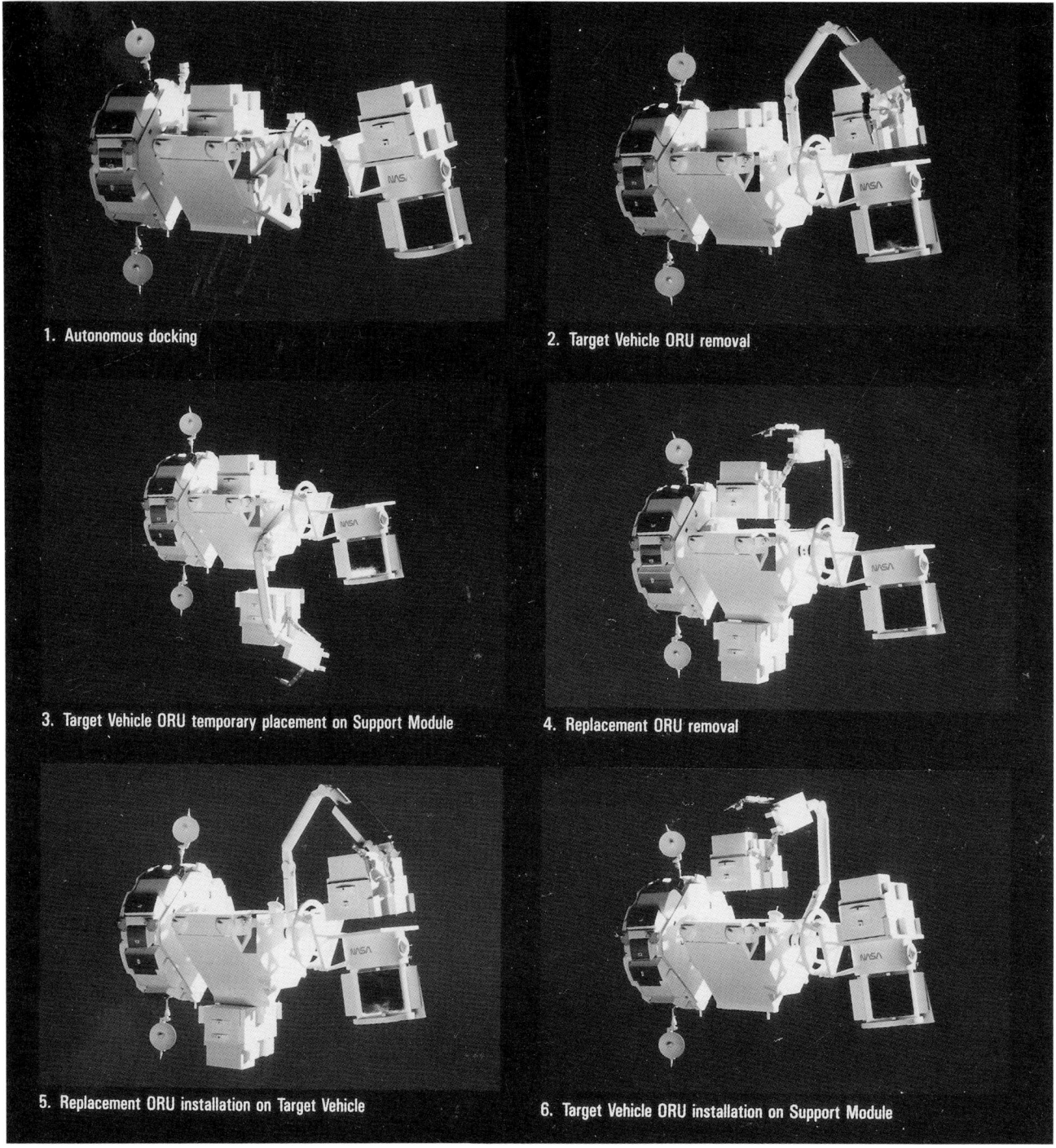

Figure 8.16 Target vehicle servicing sequence for the satellite servicer system flight demonstration. *Courtesy of TRW.*

SSS Flight Demonstration Program [7]

When (and if) NASA orders the SSSFD program to start, the contractors (TRW and Martin Marietta) will find that the support module (SM) has the same general design considerations that stem from mission and operational requirements similar to an OMV program. Specifically, the similarity goes to the requirements for multiple launches and landings and short duration missions. The NASA SSSFD program will capitalize on the MSFC/TRW OMV experience in meeting these requirements by implementing an SM structure designed with bolted, instead of welded, joints. This design can effectively withstand the multiple loading cycles, satisfy fracture mechanics criteria, and simplify the post-flight inspection process. Key technical issue requirements in the program are:

Automated Servicing

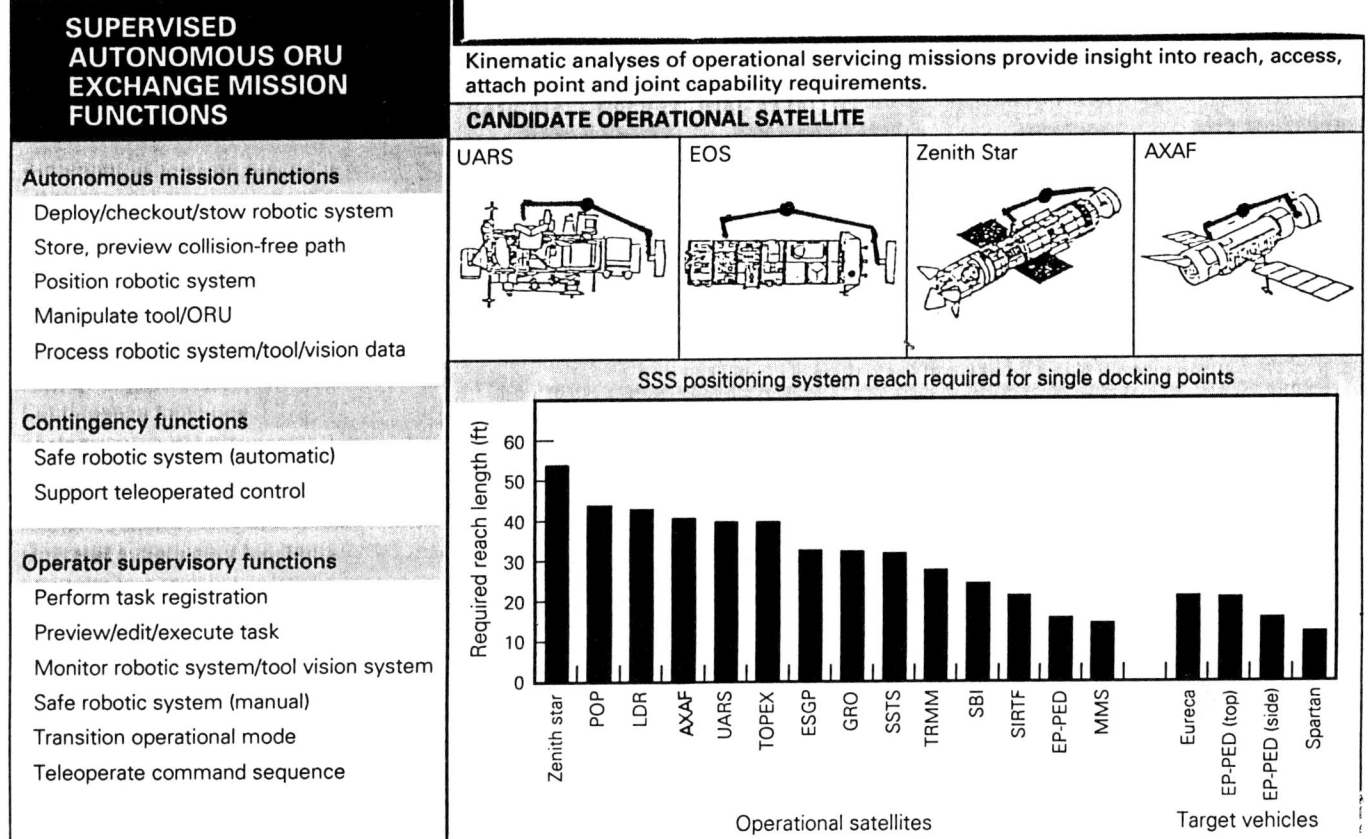

Figure 8.17 Satellite servicer system arm kinematic analysis for operational ORU exchange. ORU exchange range and degree of freedom can be determined by this analysis. *Courtesy of TRW.*

1. Structural support on the SM of the ORUs, subsystems, and FTS elements
2. FTS transition to the worksite on the target (or operational) vehicle and launch/landing stowage
3. OMV interfaces (mechanical, electrical, thermal, and fluid)
4. Target vehicle and operational spacecraft docking mechanism interface with the SM
5. Fluid coupler interface with the target vehicle and operational spacecraft to ensure automatic engagement and release

Support module configuration mission flexibility is central to development of a design that satisfies SSSFD program objectives and limits cost risk by minimizing dead-end hardware.

Support Equipment

Figure 8.18 is one way of showing the organization of satellite servicing tools, hardware, and support equipment. Most of the support equipment is concentrated in the boxes titled Special Fixtures and Service Systems.

Table 8.1 lists examples of servicing support equipment. Most of the items are described in Reference 8 which was generated under the direction of NASA/JSC by the Lockheed Engineering and Management Services Company, Inc., ILC Space Systems, and the OMNIPLAN Corporation. The document was funded by the NASA Headquarters Office of Space Flight, Satellite Servicing Branch.

Some equipment is adaptable to use in both the EVA and robotic servicing mode. Modifications may be required to render the hardware completely suitable to the selected mode and this should be a design consideration when serviceable space systems and the servicing equipment are in their early concept design stages. Here is a list of the major items of servicing hardware and equipment that could serve both operational modes:

- Orbital spacecraft consumables resupply system (OSCRS)
- Space Shuttle based remote manipulator systems with appropriate work stations or end effectors
- Space Station Freedom based remote manipulator systems with appropriate work stations or end effectors
- Rendezvous and docking aids including laser docking systems
- Portable berthing devices that can be placed at several places on the spacecraft being serviced so as to facili-

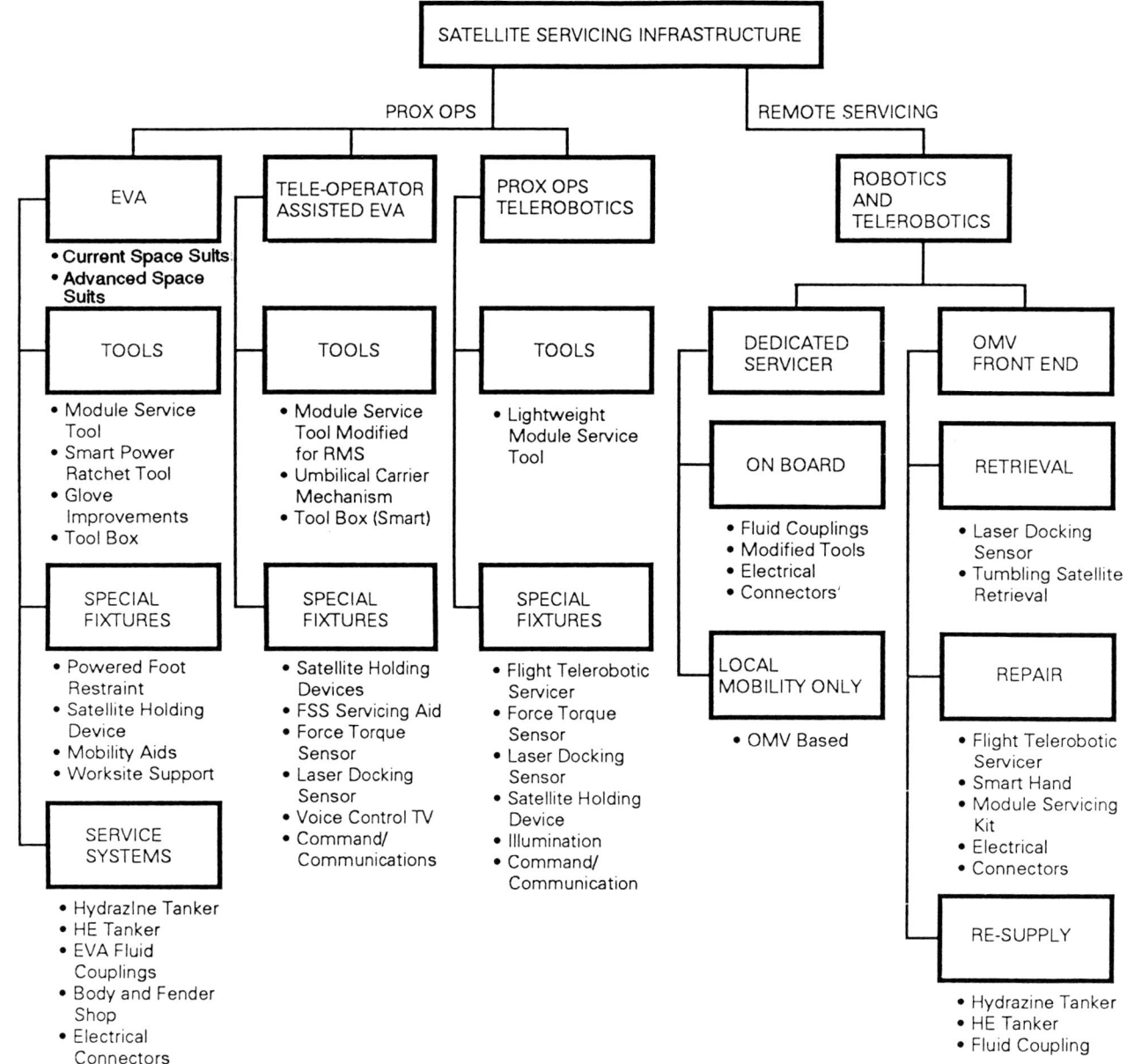

Figure 8.18 Servicing hardware, tools, and support equipment organization for operations close to the Space Shuttle or Space Station and at locations remote from the Space Shuttle or Space Station. *Courtesy of NASA.*

tate reach of the servicing work site and to maximize safety of operations

- Payload heat exchangers to alleviate thermal problems while a space system is being serviced
- Portable sun shades that can be erected and placed at various servicing locations on the spacecraft or servicer to protect men and equipment against the direct sunlight
- De-orbit propulsion packs that can be attached on-orbit to a space system
- Carriers, pallets, and platforms to transport or store new or used ORUs to and from orbit and also transport ORUs from orbit to orbit
- Umbilical connectors
- Display consoles in space (STS or SSF) or at ground stations to aid in the command and control of the servicing activities
- Inspection, gauging, and measuring devices to determine the status of a satellite before and after the servicing operations

On-Orbit Refueling Equipment

Satellite servicing studies [7, 10, 11] have shown that fluid resupply integrates very well with ORU exchange to enable cost-effective satellite servicing operations.

Automated Servicing

Table 8.1 Servicing Support Equipment

EVA Mode:

- Work stations
- Control stations on STS and Space Station Freedom
- Tool caddy
- Safety tethers and restraint units
- Light sources
- Module service tools
- Flight support system [9]
- Manipulator foot restraint
- Cameras
- Repair tape
- Extensions/sockets
- Sensor systems
- Extravehicular mobility unit
- Manned maneuvering unit
- Heads-up displays
- Portable grapple fittings
- Tool storage boxes
- Power rachet tools
- Payload berthing system
- Payload retention system
- Holding and positioning aids
- Fluid couplers
- Trash bags

Robotic Mode:

- Dexterous manipulation arms and support module containing servicing utilities
- Fluid management and transfer systems
- Satellite holding devices
- Vision systems
- Computer systems
- Supporting software
- Electronics and control units
- Tool selection and handling devices

A large number of potential fluids resupply missions exist through the 1990s. Below is a partial list of spacecraft where on-orbit refueling is a consideration. See Appendix A for spacecraft program names.

1. SPACE STATION
2. GRO
3. OMV and OTV
4. HST
5. AXAF
6. SIRTF
7. EOS
8. UARS
9. SSTS
10. MPS 19.
11. COLUMBUS
12. RSSS
13. MMS
14. LEASECRAFT
15. RADARSAT
16. LDR
17. Zenith Star
18. ESGP
19. TOPEX
20. TRMM
21. DMSP
22. LANDSAT
23. GEO PLATFORM
24. SPARTAN
25. EURECA
26. NLS/SSTO

This list is neither complete nor final. The development of robotic refueling operations in the mid-1990s will cause many of the GEO comsats and DoD surveillance sats to be added.

Top level requirements for satellite refueling can be summarized as follows:

1. Resupply consumables consist of fluids, gases, and pressurants.
2. Consensus of several studies is that between 150 and 200 fluid resupply missions for NASA, DoD, international, and commercial users will be needed from 1989 to 2000, not including the space-based SSF, OMV, and OTV. Most will probably be Space Shuttle tended due to concurrent servicing requirements.
3. Long-term space experience will define the deterministic demand for consumables resupply and thus the relative emphasis on space-basing resupply modules, fuel depots, warehouses, and tankers.
4. The four general mission characteristics used for program screening for potential on-orbit consumables supply are [11]:
   — Continuous, or very long, mission lifetime objectives
   — High unit cost satellites
   — On-orbit useful life of satellites in the program severely limited by depletion of expendables
   — Relatively low-cost access for on-orbit resupply missions

The first fluid transfer requirement will be the resupply of Earth storable propellants, primarily because these propellants will be required sooner and in greater quantities than other fluids. This is partially because of the need to transfer spacecraft from operational orbits to Orbiter accessible ones or to the Space Station for servicing and the orbital maneuvers required for such rendezvous. Mission objectives that require satellite operation at low altitudes or require repeated maneuvers also generate large impulse, and therefore propellant, requirements. The reusable upper stages, OMV and OTV, will inherently require large quantities of propellant because of their intended missions [12].

When the Space Station Freedom reaches operational ca-

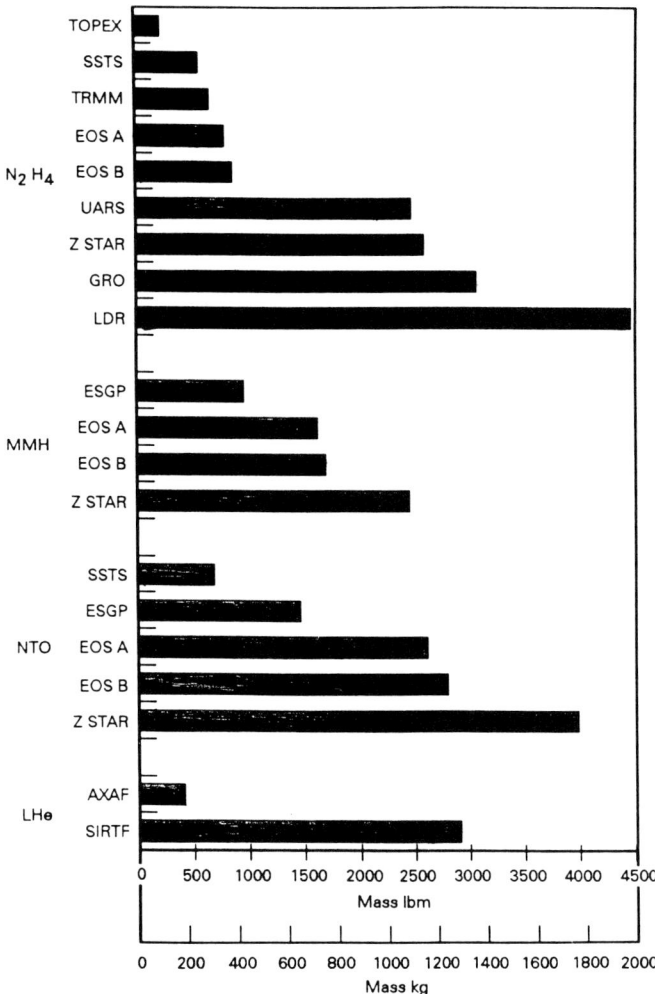

Figure 8.19 Fluids requirements analysis. This TRW fluid transfer study determined the mass and types of fluids likely to be needed on operational missions by the spacecraft listed on the left. *Courtesy of TRW.*

pability, many additional requirements for resupply will result. These will include the resupply of water for the crew and the subsequent return to Earth of the crew-produced liquid wastes. Cryogenic resupply and the transfer of biological processing—raw materials and harvested products—represent long-term requirements.

Studies by TRW's Federal Systems Division, Space and Technology Group [13] and RI's Space Flight Center [11] produced the fluids requirements as depicted on Figure 8.19 and the fluid transfer parametric requirements as shown on Figure 8.20. A total of 20 fluid types are identified as necessary for various resupply events. They are listed below [14]

— Aerozine-50 (MMH/N_2O_4)
— Ammonia (NH_4)
— Argon (gas)
— Carbon Dioxide (gas)
— Helium (gas)
— Helium (liquid)
— Monomethylhydrazine (MMH)
— Nitrogen (gas)
— Nitrogen Tetroxide (N_2O_4)
— Oxygen (gas)
— Oxygen (liquid)
— Helium (superfluid)
— Hydrazine (N_2H4)
— Hydrogen (liquid)
— Water (H_2O)
— Xenon

Military Space Systems Refueling

The U.S. Air Force is interested in developing a capability to perform onorbit spacecraft refueling because it gives an element of operational flexibility in planning the deployment and use of space-based assets. Refueling capability allows a number of unique mission opportunities and enhancements:

1. It overcomes launch weight constraints currently affecting programs
2. It allows launch of spacecraft with more useful payload, since only 5 years of fuel is needed rather than 10
3. Replacing the spacecraft's fuel load enables it to restore its military capability and extend its use lifetime through rapid repositioning, elusive maneuvering, evasive maneuvering, storage in secure orbits, and launch anomaly recovery
4. It permits more robust survivability measures, for example, launching with oversized, but partially filled fuel tanks.

The Air Force is pursuing a telerobotic approach to refueling that features:

1. Performance of refueling missions in situ.
2. Modification of satellites to accommodate refueling.
3. Autonomously docking the tanker with the receiving spacecraft; the fuel, oxidizer, and pressurant are then transferred as needed from the tanker to the spacecraft.
4. Tanker separation from the spacecraft; the tanker then stored in space for reuse or injected into a disposal orbit. As an alternative, the tanker might remain attached to the spacecraft.

These considerations are important in modifying satellites to allow on-orbit refueling:

1. Attachment of interface hardware to the spacecraft for docking and fuel transfe
2. Re-routing of fuel lines
3. Software modifications for control of refueler/override
4. Thermal management and control as the total refueling event will impact the spacecraft/payload thermal balance
5. Ensuring the safe dissipation of the electro-static charge
6. Addition of fault tolerances "Failsafe" provisions to the spacecrft's data management and reporting system
7. Decision as to whether or not the satellite will be functional during refueling event

USER AND CONSUMABLE TYPE	1989	1990	1991	1992	1993	1994	1995	1996	1997	1998	1999	2000	2001
NASA													
HYDRAZINE	—	400	3,830	1,100	5,430	1,100	6,694	5,100	1,100	1,100	5,100	1,100	1,100
LIQUID HELIUM	—	—	—	—	160	—	1,389	—	992	1,320	992	—	2,312
WATER	—	5,000	15,000	25,000	30,000	30,000	25,000	15,000	5,000	—	—	—	—
Xe-METHANE MIXTURE	—	—	—	—	30	85	—	—	85	—	—	85	—
COMMERCIAL (POTENTIAL)													
BI-PROP (N_2O_4/MMH)	—	—	—	—	—	—	818	2,482	2,307	3,318	4,694	2,414	4,694
HYDRAZINE	—	—	—	—	—	—	—	500	1,000	500	500	500	500
DOD*													
BI-PROP (N_2O_4/A-50)	—	14,000	14,000	7,000	21,000	14,000	28,000	—	35,000	7,000	7,000	28,000	14,000
HYDRAZINE	—	—	70	70	70	17,470	17,470	70	8,770	70	17,470	17,470	70
LIQUID HELIUM	—	—	—	—	—	2,400	2,400	—	1,200	—	2,400	2,400	—
DOD* (POTENTIAL)													
BI-PROP (N_2O_4/A-50)	—	—	—	—	—	—	8,000	15,000	2,000	1,000	15,000	2,000	14,000
HYDRAZINE	—	—	—	—	—	—	—	950	250	600	950	600	950
CONTINGENCY**	—	—	1,500	—	—	1,500	—	—	1,500	—	—	1,500	—

*SMALL QUANTITIES OF PRESSURANTS ALSO REQUIRED FOR SOME SATELLITES — IMPORTANT FOR DESIGN REQT
**ESTIMATE OF PROPELLANT RESUPPLY REQUIRED IN CONTINGENCY SITUATIONS

- HYDRAZINE USED FOR THE MOST NUMBER OF UMBILICAL ENGAGEMENTS
- GREATEST MASS OF FLUID FOR UMBILICAL TRANSFER IS BI-PROP
- WATER & SPECIAL GASES ARE TRANSFERRED VIA MODULE CHANGEOUT

Figure 8.20 Fluid transfer parametric requirements resupply requirements summary. *Courtesy of Rockwell International, Inc.*

The Air Force is considering an autonomous mini-refueler tanker concept that: will carry fuel, oxidizer, and pressurant consumables for transfer; has avionics, a docking device, and a small propulsion system; is launched directly at the receiving spacecraft; and can stay attached or be ejected and separated from the receiving spacecraft after the refueling operations are complete.

Reference 7 pointed out these refueling technologies as being very important to refueling missions:

1. Leak detection and repair
2. Refueling hardware: EVA, robotic, and blended
3. Fluid management
4. Cryo material handling
5. Fluids logistics
6. Fluids transfer
7. Variable set-point regulators and relief valves
8. Monopropellant catalytic vent life with long burn times and high concentrations of noncondensible gases and pulsed operations
9. Pressurant solubility effects during fill
10. Contamination control during venting
11. Adiabatic compression heating in surface tension tankage
12. Automatic fluid couplings
13. Resupply mechanism to make and break the fluid coupling
14. Tank quantity gaging system
15. Oxidizer burner and fuel burner that can accept high concentrations of noncondensible gases, and pulsed operation, especially a burner that can handle both simultaneously or separately
16. Separation of gas/vapor from liquid during venting (required for ullage exchange and vent/fill repressurize transfer methods to be effective)
17. No-vent fill

NASA'S Orbital Fluid Resupply Development Activities

The Propulsion and Power Division (PPD) at the NASA Johnson Space Center has been active in the development of orbital fluid resupply capabilities since 1982 when the ORS (orbital refueling system) flight experiment was initiated. Subsequent to the flight of the ORS in 1984, PPD conducted a number of study contracts as well as prototype and flight hardware development contracts. Conceptual and preliminary design studies for the orbital spacecraft consumables resupply system (OSCRS), an Earth storable fluid tanker, have been completed. OSCRS is discussed in the next section. Conceptual design studies of the superfluid helium tanker (SFHT) have also been completed. An EVA hydrazine resupply coupling was successfully developed and

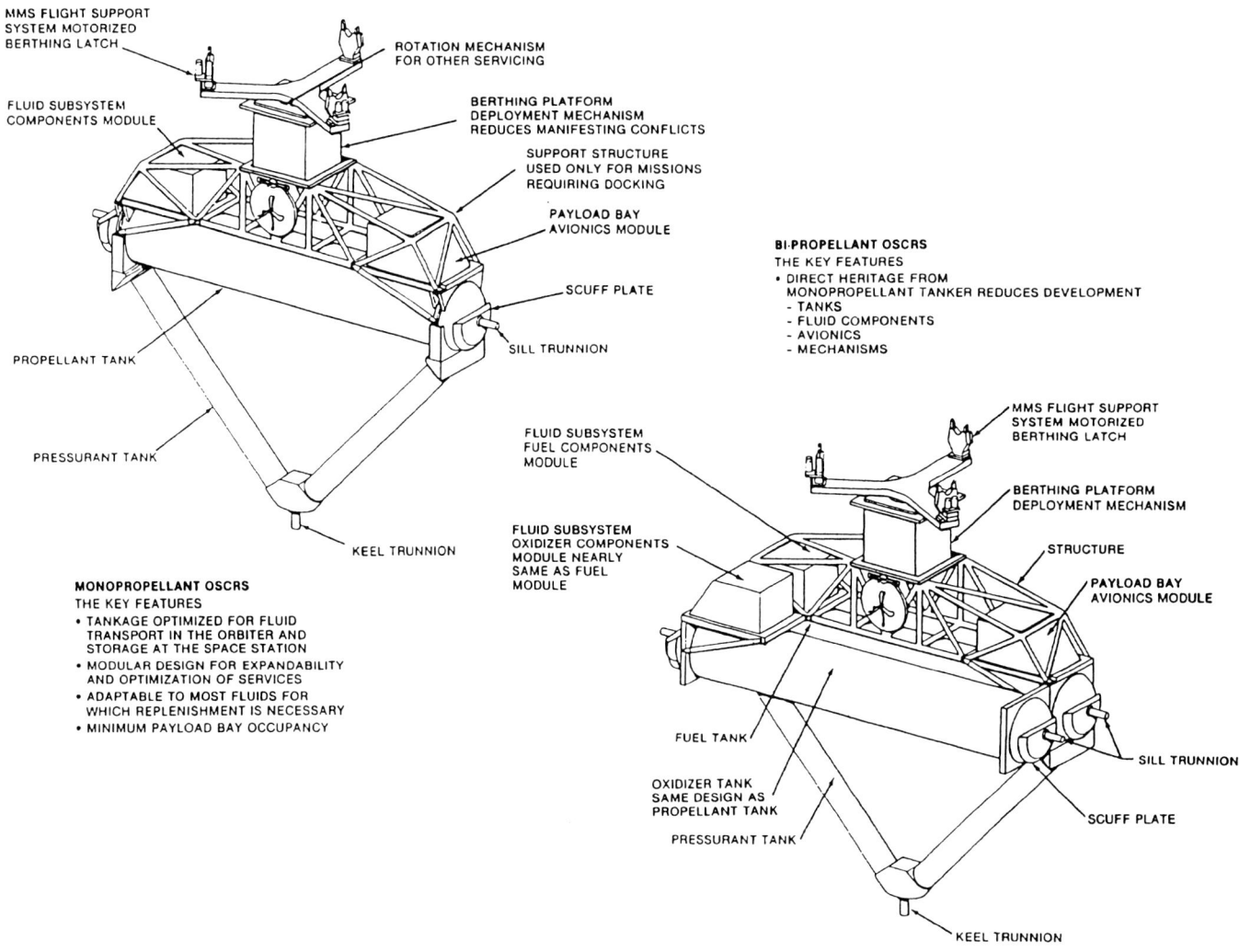

Figure 8.21 Fairchild's Trigon OSCRS [12]. *Courtesy of Fairchild Space Company.*

flight certified, and is now installed on the Gamma Ray Observatory. An EVA superfluid helium resupply coupling is currently under development for use on the superfluid helium on-orbit transfer (SHOOT) flight experiment, scheduled for launch in 1992. The prototype development of a unique electronically variable pressure regulator is underway, as is the design of high-pressure gas compressors compatible with long-term orbital operations associated with tankers and Space Station Freedom. Plans are being formulated for the development and flight certification of an automatically operated hydrazine resupply coupling applicable to the satellite servicer system and an OMV.

The NASA OSCRS Resupply System

In 1986 the Fairchild Space Company, Martin Marietta Corporation, and Rockwell International Corporation each performed a refueler design Phase B study for the NASA Johnson Space Center. The report was titled *Orbital Spacecraft Consumables Resupply System Study*. This section draws heavily on the contents of the Fairchild Space Company's OSCRS study final report [12].

OSCRS is a spacecraft consumables resupply system that can be used with the Space Shuttle, Space Station Freedom, or an OMV to refuel, on-orbit, satellites with both monopropellants and bipropellants. The three OSCRS contractors employed a design approach featuring: low weight structure, modularity, commonality, simplicity, and standardization to drive their OSCRS configurations. The Fairchild trigon design incorporates features that will make the OSCRS cost effective. Their design, shown in Figure 8.21, represents a low cost way to acquire an initial refueling capability for the GRO spacecraft, and others, that can be expanded incrementally as need arises and budget allows.

By making resupply an economically attractive alternative, OSCRS has the potential to extend the useful life of a wide variety of spacecraft. The demand for OSCRS already exists; in fact, the user community is assuming that a resupply capability will be available in the mid-1990s. The

Automated Servicing

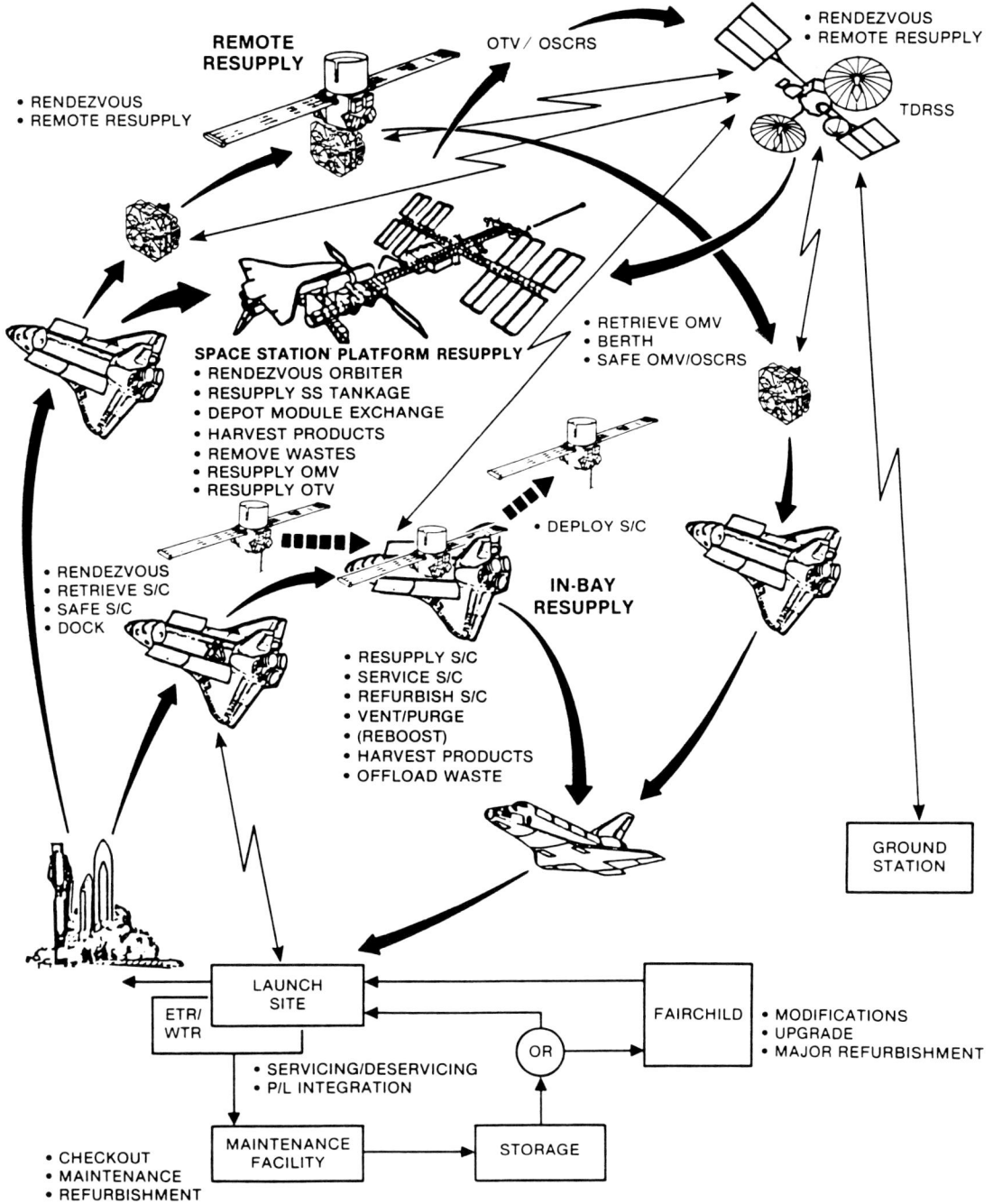

Figure 8.22 Orbital spacecraft consumables resupply system architecture [12]. *Courtesy of Fairchild Space Company.*

OSCRS will evolve into an integral part of the space infrastructure in the Space Station era. This system, illustrated in Fairchild's suggested architecture, Figure 8.22, provides fluid transportation to and from low Earth orbit. As illustrated, this will include operation in the STS Orbiter payload bay, as an adjunct to both the OMV and the OTV, and at the Space Station. Because OSCRS is not simply an isolated piece of hardware or a relatively independent satellite, there are important economic considerations that can only be evaluated on the basis of the OSCRS role as part of a system.

The design of system interfaces will proceed from an understanding of how the OSCRS will function within the system, and for this reason it requires a systems approach.

The existence of a viable resupply capability will have a significant impact on spacecraft design philosophy over the next 25 years. To accommodate its role, the design of OSCRS must anticipate the future. To minimize the total systemic costs, OSCRS must also be designed for low impact on the spacecraft for refueling. Compatibility require-

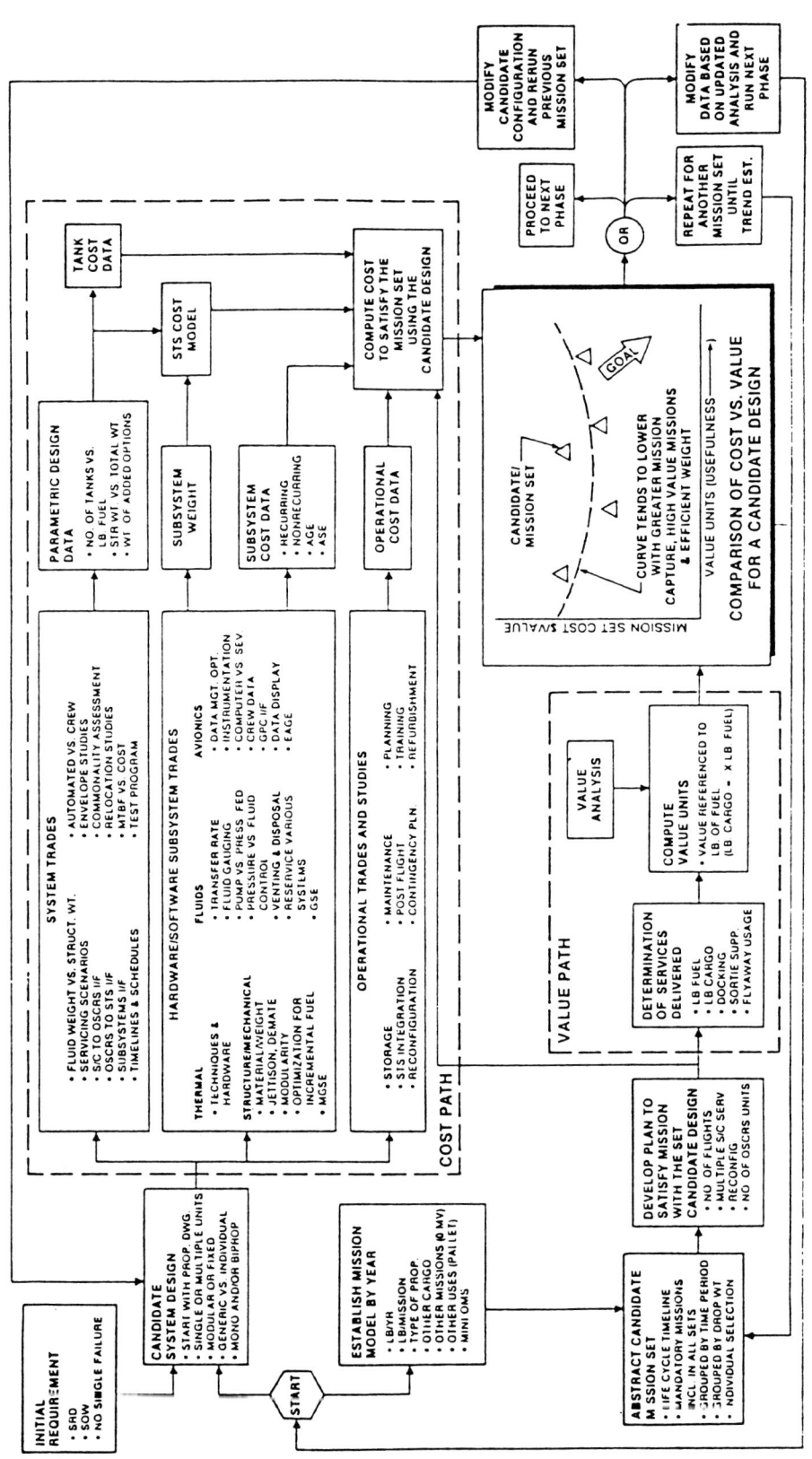

Figure 8.23 OSCRS design selection based on program life cycle cost analysis [12]. *Courtesy of Fairchild Space Company.*

ments are of special concern, to optimize a design that both accommodates known, existing interfaces and incorporates generic provisions for adaptability.

Early standardization of spacecraft and vehicle interfaces and refueling/reservicing operations will simplify the compatibility requirements, reducing the number of mission-specific interfaces/operations and the associated development efforts.

Figure 8.23 shows a comprehensive technical and economic examination of all facets of the OSCRS program from development through operations, maintenance, and transportation. It is based on a life cycle cost (LCC) analysis technique. The four monopropellant candidate OSCRS systems examined by Fairchild, Figure 8.24, provided a range for analysis in total weight and degree of modularity.

Five spacecraft missions for a monopropellant OSCRS design and five more for a bipropellant OSCRS concept were correlated to provide a set of specific performance requirements [12], Figures 8.25 and 8.26.

OMV/IOSS/OSCRS Combinations

The Martin Marietta Astronautics Group, Space Systems, in a 1987 study [10] put together an interesting refueling concept by combining the orbital maneuvering vehicle, the integrated orbital servicing system, and the orbital spacecraft consumables resupply system.

A trade study was performed to examine the candidates for tanks and tankers. Based on this study, tracking and data relay satellite system (TDRSS) monopropellant tanks, and OSCRS monopropellant and bipropellant tankers were recommended. Additionally, the combination of these elements with the IOSS and an OMV was introduced to provide a fluid resupply capability.

A sketch of a candidate system combining fluid resupply and module exchange is shown in Figure 8.27. The MMC recommended approach is to develop a series of building blocks that can be assembled in different configurations depending on the mission requirements. In all cases, an OMV is a part of the configuration as it is needed to transport the IOSS and the fluid resupply elements to the spacecraft to be serviced. The IOSS is also part of each mission as it is required for orbital replacement unit (ORU) transfer and for positioning the fluid resupply umbilicals. For missions that require a small amount of fluid to be transferred, the fluid is stored in one or two tanks in the IOSS stowage rack. Larger fluid quantities are stored in the OSCRS tanker. The IOSS stowage rack can be configured to hold up to three monopropellant tanks. Two OSCRS configurations are recommended: one for monopropellants and one for bipropellants. For missions requiring even larger amounts of propellant, two OSCRS-type tankers could be used. Another alternative is to configure tanks as ORUs that can be exchanged by the IOSS servicer mechanism, using the same procedures involved in the exchange of any other ORU.

Potential OMV front end servicer kits that provide fluid resupply are broken down into four configuration types (A through D) [10]. Each configuration type is bordered by the IOSS and the OMV. Configuration Type A includes only the IOSS and the OMV. Type B consists of the IOSS, an OSCRS monopropellant tanker, and the OMV. Type C has the IOSS, an OSCRS bipropellant tanker, and the OMV. Type D includes all four elements—the IOSS, both types of OSCRS tankers, and the OMV. The four configurations are shown on Figure 8.28. Fluid capacities for each configuration are summarized on Figure 8.29.

Automation Technology Transfer to Ground-Based Applications

The development of space-based automation can benefit the industrial automation field in two ways [15]:

1. It will provide a strong research stimulus and economic subsidy to advancing the ground-based state-of-the-art for automation technology directed to industrial applications.
2. Robotic capabilities peculiar to space-based servicing needs will be developed, tested, and applied operationally on the International Space Station. They include the adaptability and flexibility to deal economically with "one-of-a-kind" servicing functions. Such flexibility will be much in demand in the factory of the future and a direct technology spinoff potential is evident.

Of particular interest in the industrial/manufacturing field are robots designed to perform in a highly flexible and adaptable fashion under greatly diversified situations. Such robots are envisioned to operate in sophisticated manufacturing facilities in the next decade. Admittedly the subject of totally automated factories is a fervent and polemical topic among industrial engineers and managers. However, the following future robotic applications are being mentioned that would benefit from the space-based automation, and particularly servicing automation, technology:

1. Adaptable machines with flexible, as opposed to fixed, automation
2. Reprogrammable machines (by keystroke) for diversified tasks
3. Responsiveness to new situations, eliminating obsolescence
4. Economic production of "quantities of one" (or at least, small quantities) and mixed batches
5. Low inventory/zero inventory trends
6. Proliferation of models
7. Software linkage between diversified computers

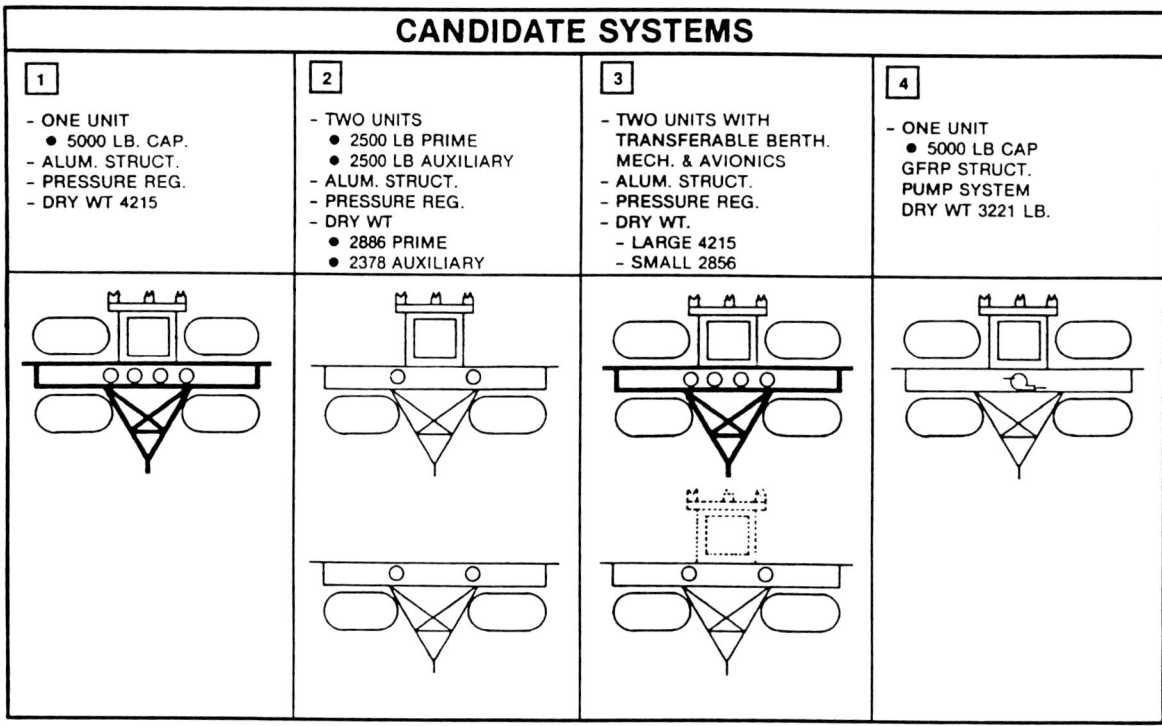

Figure 8.24 Monopropellant candidate system comparison [12]. *Courtesy of Fairchild Space Company.*

Automated Servicing

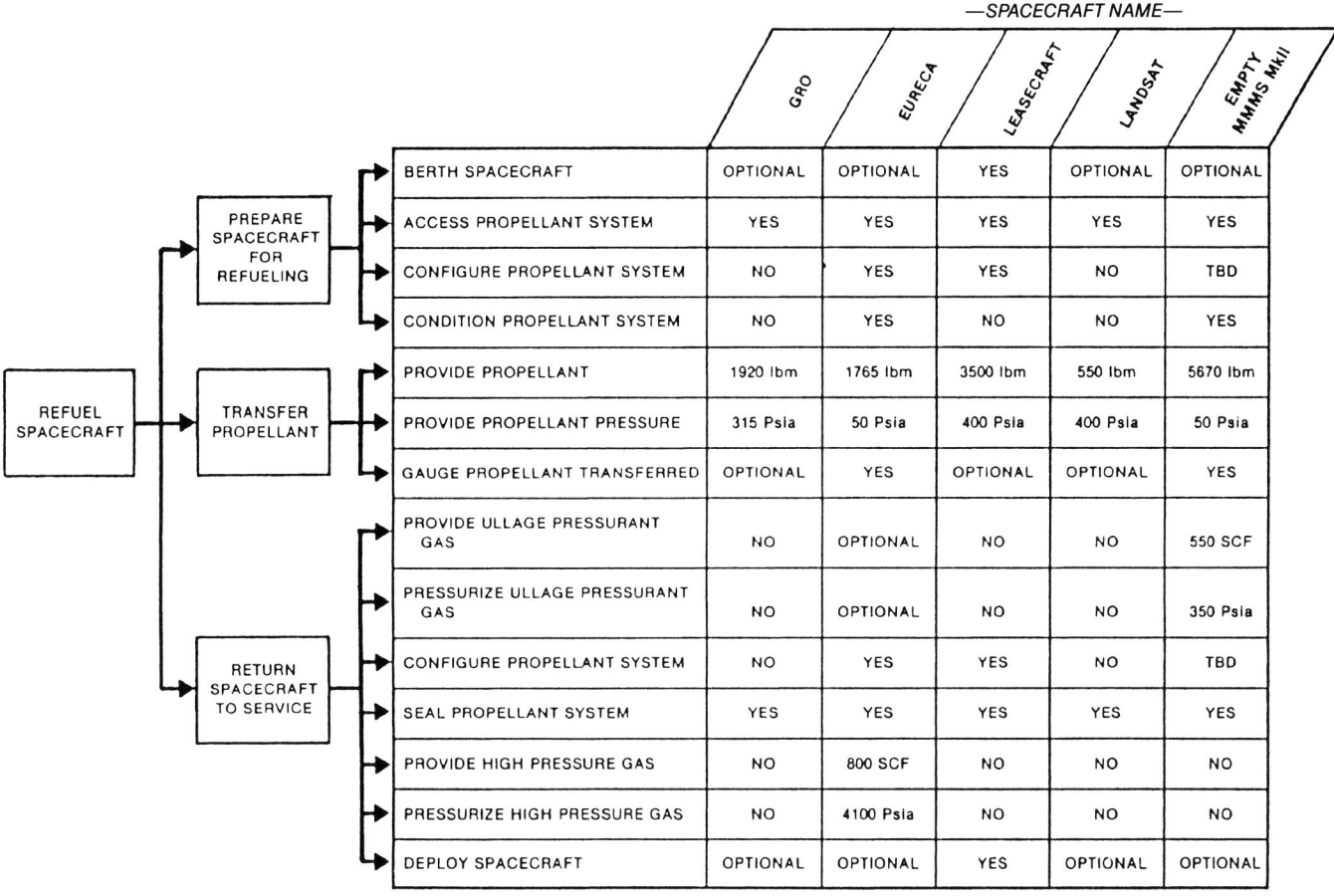

Figure 8.25 Monopropellant OSCRS requirements [12]. *Courtesy of Fairchild Space Company.*

Other potential transfer of automation technology developed for space-based servicing may include ground-based applications in hostile or unsafe environments such as deep mining, underwater operations, nuclear power plant emergency activities, and near proximity to explosives. Examples include: robots designed for window cleaning on skyscrapers, fire fighting (currently under development in Japan), defusing or neutralization of bombs placed by terrorists (a technology currently in use by security forces in Israel), smart robots that respond to contingency conditions that pose a major safety problem to the immediate area, and robots that transfer equipment and supplies as instructed.

Conclusions and Summary

All three principal automation disciplines—teleoperation, robotics, and artificial intelligence—will be needed in on-orbit servicing missions. Teleoperation will be utilized more widely tnan fully robotic systems, at least during the early Space Station years, owing to the diversity and the unpredictability of many servicing tasks which call for the human operator's skills, resourcefulness, and decision-making ability. In situ servicing in low and geostationary Earth orbits is the principal driver toward fully automated, robotic manipulation techniques.

As in all other Space Station Freedom automation functions, there will be heavy dependence on a sophisticated, flexible, readily accessible, high-speed and high-capacity data management system, which can provide artificial intelligence support as required in diagnostics, troubleshooting, configuration control, decision making, task scheduling, and mission planning. Thus, the Space Station data system will play a key role in providing comprehensive support functions in all phases of satellite servicing.

Twelve automation devices or technologies are key to space servicing [15]:

1. Dexterous manipulators
2. Servicing-compatible spacecraft
3. Space-qualified robots, robotic servicing
4. Data system servicing support
5. Advanced man-machine interfaces
6. Advanced fluid transfer systems
7. Robot vision

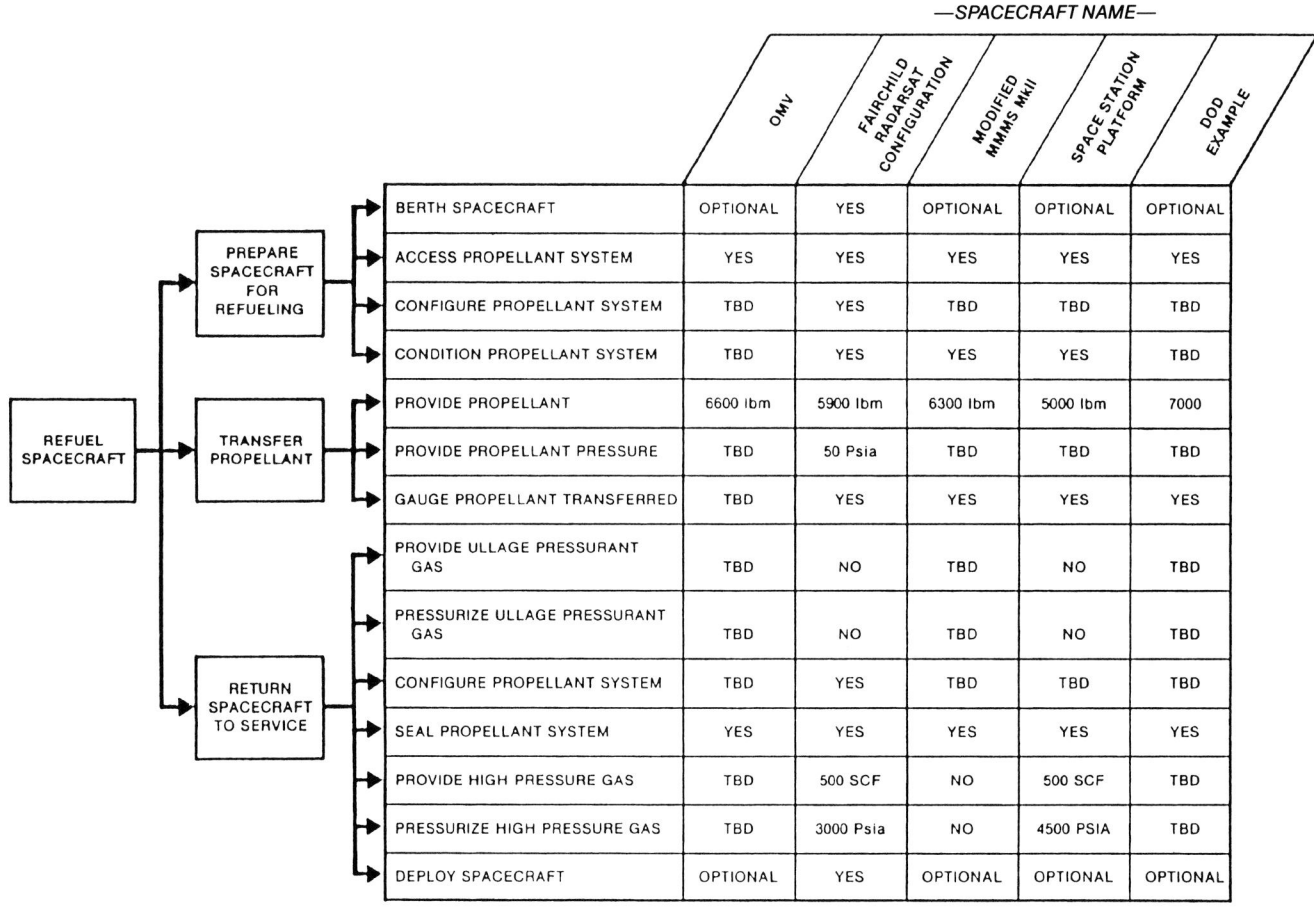

Figure 8.26 Bipropellant OSCRS requirements [12]. *Courtesy of Fairchild Space Company.*

8. Automated load handling/transfer
9. Automated rendezvous/berthing
10. OMV with smart front end
11. Knowledge-based system support
12. Reusable OTV

Space-based servicing will draw on current developments in automation technology such as advanced robotics, expert systems, robotic vision, speech recognition, natural language, data processing and display, fault detection/recovery, computing, and software. However, practical application of this technology to space servicing automation objectives requires a continuing major development effort.

Automated satellite servicing capabilities will be required on the Space Station Freedom to maximize crew productivity, to reduce the frequency and duration of extravehicular activity, and hence, to reduce crew exposure to hazardous conditions. Study results [15] show that 40 to 60 percent of the crew time can be saved by using automated support if it is developed and implemented. Automation also will speed up servicing schedules and thus help reduce any backlog that may develop due to growing demands for maintenance, repair, and refurbishment of satellites in low and high Earth orbit as well as servicing of the Space Station itself, its subsystems, and attached payloads.

A significant degree of commonality exists between the automation requirements of various servicing functions and a generally high utilization rate of automated design features, once they are implemented.

Principal conclusions from this chapter can be summarized as follows:

- Many satellite servicing functions benefit from, or rely on, automation support.
- Automation will expedite on-orbit satellite servicing and will increase productivity of crew operations.
- Orbital servicing of satellites and of the Space Station itself is a principal driver of automation technology development. Technology evolution, in turn, will greatly expand servicing capabilities.
- Satellite servicing requires more teleoperation and less robotics than other automated Space Station activities.
- Teleoperation or fully automated (robotic) use of the same manipulators offers flexibility and adaptability.

Automated Servicing

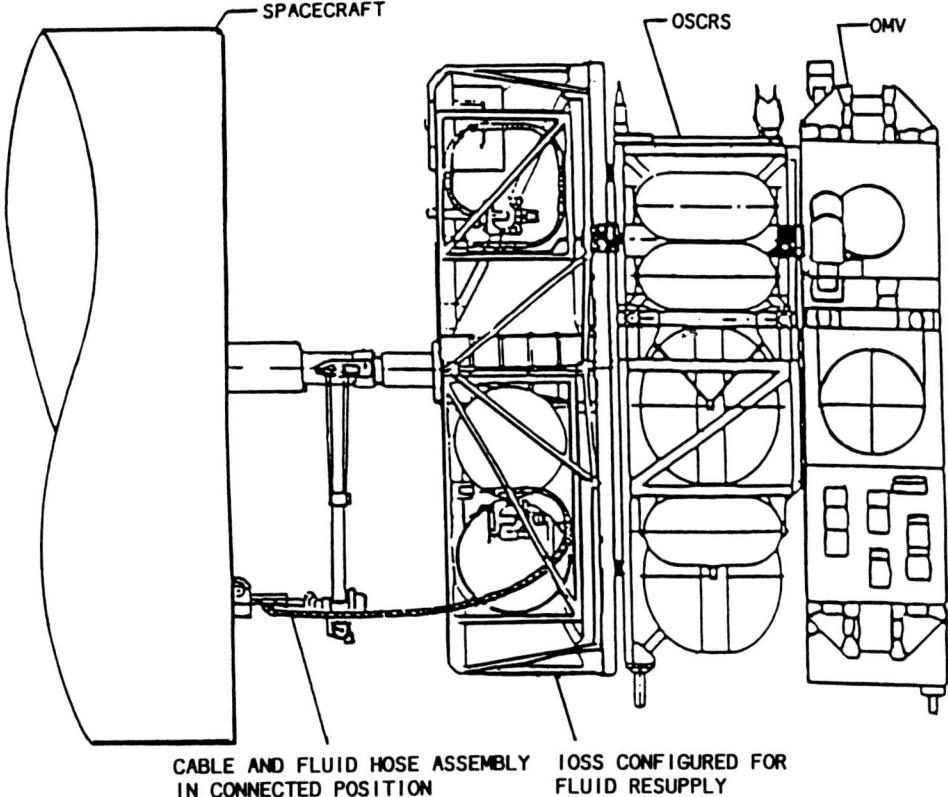

Figure 8.27 Candidate systems that combines fluid resupply and module exchange. *Courtesy of Martin Marietta Corporation.*

- Robotic servicing development is driven by in situ, particularly geostationary, satellite servicing objectives.
- In situ servicing by teleoperation will be feasible only if transmission delays are reasonably small depending on characteristics of the task.
- The transmission delay (feedback control delay) in remote servicing missions can be greatly reduced by communication during direct line-of-sight contact intervals rather than via relay satellite.
- Major data system support is essential for planning, scheduling, execution, monitoring, and other servicing functions.
- Servicing support by artificial intelligence will expand with Space Station evolution.
- Ground-based automation technology will be applicable to satellite servicing.
- Servicing automation, in turn, will benefit ground applications.

References

1. Chucker, S. M. *Space Asset Support System Implementation Plan.* McDonnell Douglas Space Systems Company, Huntington Beach, CA, Contract No. SDI 084-8C-0020, MDC H5541, October 1989.
2. *Space Assembly, Maintenance and Servicing Study (SAMS)* Final Report, Volume I, Executive Summary, TRW No. SAMSS-196,, Volume II, System Analysis, TRW No. SAMSS-195, Volume III, Design Concepts, TRW No. SAMSS-197, Volume IV, Concept Development Plan, TRW No. SAMSS-198, Volume V, Neutral Buoyancy Simulation, TRW No. SAMSS 199. TRW, Redondo Beach, CA, July 6, 1988.
   and
   Space Assembly, Maintenance, and Servicing Study (SAMS) Final Report, Volume I, Executive Summary, Volume II, System Analysis, Volume III, Design Concepts, Volume IV, Concept Development Plan, Volume V, Simulation Report. Lockheed Missiles and Space Company, Inc., Sunnyvale, CA, July 6, 1988.
3. *Remote Tanker and Servicer Analysis Study.* TRW contract NAS8-36800, TRW report 51000.89, TD005-003, January 23, 1988.
4. *Servicer System Users Guide.* Martin Marietta Denver Aerospace, Contract NAS8-35625, MCR-86-1339, July 1986.
5. McCain, Harry G. *NASA's First Dexterous Space Robot.* Aerospace America Magazine, February 1990, p. 13.
6. Moore, James S., George M. Levin, and Lt. Col. Neal Eli. *Satellite Servicer System Flight Demonstration Program.* Paper presented at the Space Logistics Symposium, Colorado Springs, CO, May 1990.
7. Technical conversations with TRW people—Art Stephenson, Gordon Bell, and Leo Stytle—on the Satellite Servicer System Flight Demonstration Program, September 1990 through March 1991.
8. *EVA Catalog of Tools and Equipment.* NASA/JSC 20446, Rev. A, April 1989.
9. *Multimission Modular Spacecraft Flight Support System Description.* Document prepared by Fairchild Space Company for NASA/GSFC, Satellite Servicing Workshop III, June 1987.
10. DeRocher, W. L. Jr., and N. G. Smith. *Servicer System Dem-*

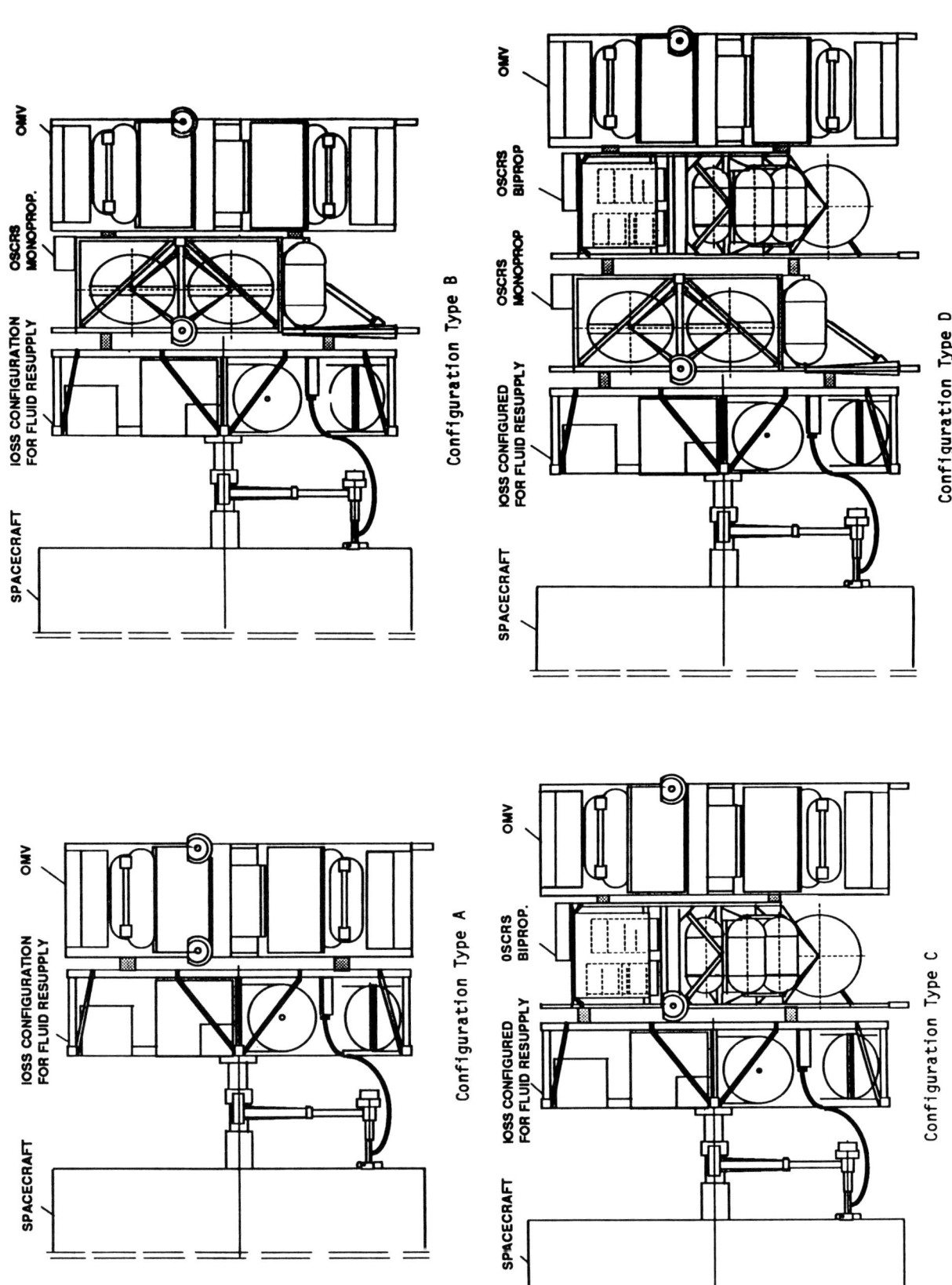

Figure 8.28 Potential OMV front-end servicing kits that provide fluid resupply. Each configuration is bordered by the IOSS and the OMV. *Courtesy of Martin Marietta Corporation.*

	Monopropellant	GN$_2$*	Bipropellants
Excluding OMV fluids			
Type A	2910	135	---
Type B	7760	135	---
Type C	2910	200	11400
Type D	7760	200	11400
Including OMV fluids			
Type A	4090	175	8775
Type B	8940	175	8775
Type C	4090	240	20175
Type D	8940	240	20175

*Assumes a four to one ratio of pressurant gas carried to pressurant gas resupplied, and full transfer of pressurant gas exchanged as an ORU tank set.

Figure 8.29 Fluid capacity summary. *Courtesy of Martin Marietta Corporation.*

onstration Plan and Capability Development. Martin Marietta Astronautics Group Space Systems, Contract NAS8-35626, MCR-87-1352, December 1987.

11. *Space Platform Expendables Resupply Concept Definition Study*. Rockwell International, Contract NAS8-8335618, July 19, 1984.

12. *Orbital Spacecraft Consumables Resupply System Study*. Final Report, Volume I, Executive Summary, Fairchild Space Company, Contract NAS9-17586, 339-FR-1000, November 7, 1986.

13. Mitschler, Larry. *ORU Exchange/Fluid Resupply Requirements*. TRW Document Z410.89.LM-013, December 8, 1989.

14. *An Investigation of Selected On-Orbit Satellite Servicing Issues*. Science Applications International Corporation, Contract No. NAS9-17207 from NASA/JSC, November 1986.

15. *Space Station Automation Study*. Satellite Servicing, Volume II, Technical Report, TRW, Contract NAS8-35081 from NASA/MSFC, December 20, 1984.

Chapter 9
Benefits and Economics

As NASA enters its fourth decade, more than 100 different satellites have been developed for over 250 scientific, applications, and technology missions. For these missions, the spacecraft were designed as unique configurations which, after launch and useful lifetime in orbit, became "space junk" to eventually experience orbit decay and burn up in the Earth's atmosphere. In many cases these programs represented an investment of hundreds of millions of dollars for a one-time mission with no reuse planned for either the launch vehicle or the satellite.

During the 1970s, when the Space Shuttle concept was evolving into a significant technique for producing an economical, reusable launch vehicle, NASA was in the process of developing a multifunction, economical spacecraft complete with ground and spaceborne support equipment, software, and necessary documentation to facilitate a large variety of space missions on a routine and cost-effective basis. To reduce the expense and time required for the development of customized spacecraft, NASA, with the help of the aerospace and electronic industries, conducted an extensive research program to identify new approaches to a reusable low-cost spacecraft design. These studies resulted in the development of the NASA/GSFC Multimission Modular Spacecraft (MMS), which is described in Chapter 6. At present, three MMS spacecraft—the Solar Maximum Mission (SMM) and the two Landsat-D Earth resources satellites, Landsat 4 and 5—are operating in orbit. All of these missions have utilized Delta expendable launch vehicles. SMM was launched on February 14, 1980, from the Kennedy Space Center in Florida. Landsat 4 and 5 were launched from the Vandenberg Air Force Base (Western Test Range) on July 16, 1982, and March 1, 1984, respectively.

Both SMM and Landsat 4 have achieved successful missions but each has experienced some spacecraft or payload instrument malfunction. Previously, such failure would have terminated or at least permanently curtailed these missions. With the advent and utilization of the repairable MMS concept, however, and the emerging capability of the Space Shuttle to rendezvous, capture, fix, and redeploy satellites, on-orbit repairs are now practicable. The first demonstration of this capability was the Solar Maximum Repair Mission (SMRM) launched April 6, 1984, the 11th Shuttle flight, now designated STS-41C (formerly called STS-13).

The Solar Maximum Repair Mission has stimulated a new era in space technology which will lead to routine on-orbit maintenance and repair. Not only has the repair mission been demonstrated but hardware with lengthy exposure to space environments is available for assessment of long-range degradation effects on components and materials. The mission has provided a springboard for the industrialization of space using permanent, mobile platforms in low Earth orbit and its successful completion gives added emphasis to NASA's announced goal of maintaining a permanent manned presence in space.

This chapter discusses the demonstrated and potential benefits of on-orbit satellite servicing and the cost considerations associated with the decision to incorporate servicing into satellite design and operations.

Assumptions

These assumptions strongly influence the benefits and economics of on-orbit servicing of space systems.

1. *National Policy*—A national level space servicing program, with sponsoring organizations, DoD, NASA, SDIO, creating a unified policy for servicing equipment development, spacecraft design standards, orbital operations, and approach to planning, will be established in the next 5 years.
2. *Space Shuttle*—The reusable National Space Transportation System (NSTS), the Space Shuttle, built by Rockwell International under contract with the NASA Johnson Space Center, will continue its successful performance of space missions. The next generation of Shuttle development will become a program reality resulting in block changes to the Shuttle with increased payload weight-to-orbit capability, lower

launch and operations costs, longer on-orbit stay times, and better crew accommodation facilities for conducting servicing tasks at or near the Shuttle.

3. *Space Station*—The International Space Station Freedom program, now under development by NASA and a large industry contractor team, will be operational in space by the year 2000. At some point in its growth capability, the Station will have facilities to accommodate satellite assembly, maintenance, or servicing.

4. *Orbital Maneuvering Vehicle (OMV)*—An OMV will be operational before 1997 to support space servicing missions into the first decade of the next century. It will strengthen the total capability of the NSTS and the Space Station. It is a vital link in the continuing evolution from specialized spacecraft to flexible flyers.

5. *Astronaut EVA Space Suits*—The next generation of astronaut space suits and related equipment will be operational and available in the late 1990s.

6. *Servicing Infrastructure*—The space-based and ground servicing infrastructure/tool inventory will evolve with customer requirements and mission needs. A logistics architecture will be planned and incorporated into this infrastructure.

7. *SDI Projects*—The U.S. implementation of Strategic Defense Initiative projects will drive certain space assembly, maintenance, and servicing requirements; influence the servicing infrastructure; and dominate the space logistics architecture. In fact, SDI will probably be the DoD forerunner for space-based servicing.

8. *Transportation Costs*—Since the largest portion of space servicing costs is the ground-to-orbit transportation cost, it is central to the economics of space operations that launch costs be reduced over the next 15 years. It is therefore assumed that some parts of the National Launch System will be selected for development with a main thrust directed at lowering the dollars per pound to orbit cost.

9. *Manned/Robotics Structure*—The near term space servicing activities (now to 1996) will feature labor intensive (manned EVA) on-orbit operations in low Earth orbits (LEO); but automation and robotic devices will gradually phase into mission operations, so that after the year 2000, space servicing will expand to GEO with heavy use of telerobotic hardware and automated work sites. 2000 is an arbitrary date. The phasing from manned to robotics dominated servicing missions will occur over about 5 years (1993 to 1998). However, even in the far term (after 2000) there will always be missions where human presence will be required to perform certain functions.

Potential Benefits of On-Orbit Servicing

The capability to perform on-orbit servicing of space systems will result in certain benefits to satellite programs. A list of potential benefits is given below. Not all of these benefits will accrue to every program. The degree to which benefits apply to a specific satellite program depends on the satellite's mission, the national need involved, and the servicing cost versus the value added to the satellite after it has received the servicing. Potential benefits are:

1. Extended satellite lifetime
2. Enhanced science from satellite payloads
3. More mission flexibility/availability
4. Improved performance and reliability of critical components
5. Enhanced military mission assurance
6. Reduced life cycle costs of large, long-term programs

An explanation follows of these six benefits.

Extended Satellite Lifetime

Spacecraft modular designs will allow rapid on-orbit payload, subsystems, instruments, and components changeout at planned or at failure points in the mission, thus extending the useful lifetime of the spacecraft. On-orbit resupply of expendables (fuel, coolants, film, batteries) will allow mission continuation. On-orbit final testing/checkout before release of the spacecraft by the Space Shuttle will reduce the effects of "infant mortality" problems as experienced by the Landsat, Leasecraft, and Seasat satellites. The explosive rate of advancement in the early 1980s of payload electronics is slowing down. This slower rate of obsolescence is generating a need to get a longer lifetime out of a satellite.

Enhanced Science from Satellite Payloads

Increased life of expensive LEO scientific satellites/platforms such as the HST, AXAF, and GRO requires on-orbit maintenance/servicing to get more science out of the satellite system. Typical experiment/instrument lead times will run 5 to 7 years in the future, compared to the current 2 to 4 year on-orbit life. Retrieval (harvesting) and restocking of materials processing products from free flyers will allow long durations for experiments to be conducted on microgravity materials processing in space science. Payload or subsystem modules changeout at planned points in a space science program to introduce technology updates into the satellite will expand the scientific scope of a project. This is called preplanned product improvement.

More Mission Flexibility/Availability

Modular spacecraft design coupled with orbital servicing will allow easy changeout/repair/upgrading of subsystems (power supply, communications, data handling, propulsion, etc); components (solar arrays, booms, antennas, sensors,

Benefits and Economics

etc); and payloads (optics, transponders, detectors, etc.). The changed modules could expand or alter mission objectives, scope of data acquired, or quality of information processed. Spare replacement modules could be carried in the Space Shuttle on an ORU carrier or stored at the Space Station or a space-based servicing depot.

Improved Performance and Reliability of Critical Components

Periodic servicing will keep satellite performance at peak levels. Such planned on-orbit testing and servicing will be used to: calibrate payloads; trim thermal control on-orbit prior to spacecraft release from Shuttle; clean optical, solar array, and sensor surfaces; and ensure deployment of booms, antennas, and solar arrays. This will improve the performance of the space system and enhance mission reliability of critical components.

Enhanced Military Mission Assurance

Orbital mechanics coupled with the demand for mission continuity may require on-orbit servicing with or from space-based facilities. Critical mission strategic operations, availability, and timing could depend on orbital servicing support. Consumable depletion time and the replacement interval impacts overall constellation operations. As with NASA satellites, preproduct improvement via modular replacements on military satellites at planned intervals could aid mission assurance.

Reduced Life Cycle Costs of Large, Long-Term Programs

Up to 30 percent of current spacecraft design is redundant to ensure reliability. We can probably cut this factor in half with onorbit servicing designed into the vehicle. Planned or contingency servicing costs may be lower than satellite replacment (fix the old rather than replace the new spacecraft). Reduced space system ground integration and test time and cost is possible through modularity and standardization which would be incorporated into a serviceable spacecraft. On-orbit pre-ejection testing could reduce total system test costs. The realistic environment testing in space could lower test cost for appendate deployment qualification, command/data management operational assurance, electric power flow, propulsion readiness, thermal balance, and payload performance. The number of space- and/or ground-based replacement spare satellites per constellation could be reduced and the number of launches during fleet operational lifetime could be reduced.

Economic Analysis

A primary, if not the most important, motive for developing a multiuser space assembly, maintenance, and servicing capability is economics. As a historical note, a basis for servicing and repair was established in 1986 when the underwriters of the disabled Westar and Palapa communication satellites opted to retrieve these spacecraft for repair and eventual relaunch based solely on an economic or cost-effectiveness analysis.

As a result of the motivation to reduce costs of complex spacecraft and systems, cost benefit analysis and related cost estimation were incorporated as an integral part of the TRW and Lockheed Contractor SAMS study [1]. This study is summarized in Appendix B. Beginning with (1) a careful construction of the study methodology and study approach, (2) the incorporation of a comprehensive and consistent WBS framework, and (3) a unifying set of ground rules, a series of SAMS study products and conclusions was generated which will be discussed in this chapter.

First, an exhaustive set of SAMS study related design reference missions (DRMs) was constructed and costed. The 11 TRW DRMs corresponding to 5 basic Air Force designated DRMs are discussed in Appendix B. They vary by orbit location, mission type, spacecraft size, timeframe, and technology impact. The first priority was to establish and assess the magnitude of the mission events costs and, where required, to develop comprehensive DRM life cycle costs. In the course of costing each of the SAMS DRMs, many issues surfaced that suggested opportunities for design, performance, or program level cost trades. As a result, additional effort was directed to the cost trade area, representing another major product of this study.

In addition to costing individual mission events and related trades, a comprehensive cost analysis of forecast missions was conducted within the context of a hypothesized evolutionary SAMS program. Forecasts were developed for proposed space missions. An aggregative level cost model was developed to (1) assess the evolutionary costs of such a program, (2) develop the savings resulting from a SAMS implementation, and (3) depict the cost benefit "cash" flow where the payback period could be assessed as well as the magnitude of the projected cost saving. This included both analysis of constant fiscal year as well as discounted (present value) cost/benefit streams. Finally, study conclusions were generated and reported [1].

SAMS Cost Assumptions, WBS, and Definitions

The cost and economic analysis of the SAMS study followed the basic work breakdown structure, definitions, and assumptions as specified by the USAF Space Division and its cost analysis support contractor, Tecolote Research, Inc. A summary of significant assumptions is shown on Figure 9.1.

The summary SAMS study work breakdown structure (WBS) is portrayed in Figure 9.2. The WBS defines the space assembly, maintenance and servicing (SAMS) total system as it refers to the complex of all hardware, software,

- The estimates for the SAMS Study are in 1986 constant dollars
- A 14% profit (Fee) shall be included in all SAMS contractual estimates during acquisition
- A Management Reserve (risk) of 15% shall be included during acquisition on all SAMS items
- While the accuracy of each individual life cycle cost estimate is important, the comparative accuracy of cost estimates are of primary concern to assure reliable determination of cost comparative ranking among the competing SAMS concepts.
- Acquisition of space transportation systems are considered sunk, national asset costs, and *NOT* a SAMS cost included in Life Cycle Cost estimates. The cost can be stated as an additive cost for a sensitivity analysis using the government furnished data.
- The operations and support costs of the space transportation system *ARE* considered SAMS cost, and are included in the Life Cycle Cost estimates.
- The explicit methodology, including CERs used to develop the cost estimates, shall be clearly stated
- The acquisition cost of existing C3 systems shall be considered sunk and not part of a SAMS System. The acquisition of any new C3 system required by the SAMS system shall be considered as a SAMS cost in the Life Cycle Cost Analysis.
- The operations and support cost for these C3 systems (existing and future) shall be included in the SAMS Life Cycle Cost Analysis. Existing systems estimates will be supplied as Government Furnished Data

Figure 9.1 SAMS study: Economics assumptions. *Courtesy of TRW Space and Technology Group.*

facilities, and associated personnel and support services required to develop and produce the capability for the assembly, maintenance, or servicing of satellite systems in space or on the ground. The SAMS elements include:

- The user segment (user or customer planning function and the user satellite)
- The SAMS segment (space servicer, DoD or NASA space platform or bay, mission control center and hardware/tools for space repair)
- Space transportation segment (launch vehicle, orbital transfer vehicles, orbital maneuvering vehicles)
- The integration services segment (all existing and new communications systems required for SAMS operations)

SAMS Study Life Cycle Costing Methodology [2]

The analytical approach to compare the cost effectiveness of various repair and replace strategies for failed satellites on orbit is illustrated in the methodology overview block diagram of Figure 9.3. The interaction of the various methodology elements, called modules, is described below.

Satellite Sizing Module [2,4]

The beginning step of the comparison is to identify the satellite total cost (first unit cost) and total dry weight of the baseline satellite. Thus the user or analyst describes this information before applying the comparison methodology. Given this data, the user applies the adjustments from the satellite sizing module to obtain the cost (first unit) and weight of the baseline, enhanced reliability, or modular satellite at a more detailed subsystem level. This sizing module also provides the cost (first unit) and weight of the ORU. The sizing relationships for the subsystem cost and weights of the satellite are obtained from the Air Force Space Division Unmanned Spacecraft Cost Model (USCM), Version 5 [3]. A diagram of the satellite sizing module is shown in Figure 9.4. Its input is the total dry weight and first unit cost of a particular satellite using today's nonrepairable or nonmodular technology.

Programmatic Module [2,4]

The programmatic module incorporates various quantity related inputs to determine the required output quantities, events, and timelines necessary for developing the satellite and ORU requirements for launch and orbital transfer vehicle manifesting. This module provides the satellite quantity information for the satellite systems identified in the satellite sizing module. Repair or replace events are identified based upon input satellite reliability and the economic lifetime of the analysis. The diagram of the programmatic module is shown in Figure 9.5. The purpose of this module is to calculate the various quantities required for the subsequent calculations in other modules (transportation and LCC modules).

First the size of the constellation of satellites in the analysis must be specified. A value of one is a baseline so that the life cycle cost comparison between a single satellite under either replacement or repair strategies can be addressed.

Next, the time period of the economic analysis in years is specified. A value of 10 or 20 years is usually selected for system trade studies. This value of economic life does not automatically update the infrastructure life cycle which must be calculated separately, but must be made consistent by the analyst.

Next a measure for satellite reliability is incorporated. The number of failure events is input during the 20 year lifetime. A value of zero means the original satellite constellation lasts the full economic lifetime. This value can be obtained from very rigorous off line reliability analysis of the satellite program under study.

Benefits and Economics

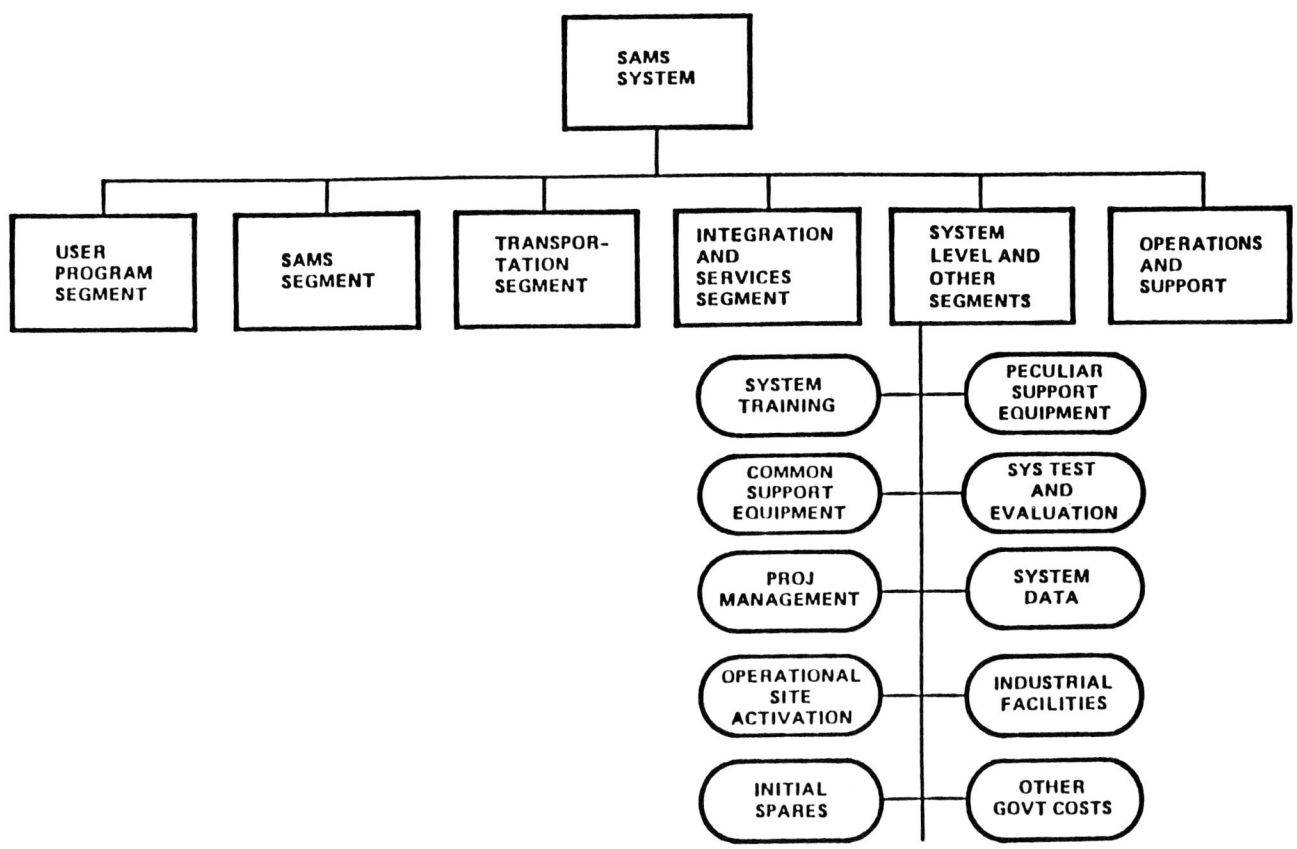

Figure 9.2 SAMS study: Work breakdown structure for the hardware development and operations phase. *Courtesy of TRW Space and Technology Group.*

The number of ORUs required in the economic lifetime (not counting spares) is treated as a single input variable. There is an ORU "set" in the basic repairable satellite and one for exchange at the first repairable event which occurs at the mean mission duration (MMD) time of the program life. The original set is brought back for refurbishment and the ORU set from this variable is used for the exchange. The use of a number greater than one implies that less refurbishment takes place.

The number of spare satellites (repairable and replaceable) which are needed in later LCC calculations is treated as a simple input in this variable. The initial spare ORU quantity is also a simple input number. The number of replenishment spare parts, expressed as a percentage, is used later in the production cost CERs in the LCC calculations. The programmatic module takes the inputs discussed above, calculates the necessary quantities and events, and outputs the items identified in block 8 of Figure 9.5. Launch timelines are developed using the mean mission duration and economic life. Using the information about satellite constellation size, we can determine the number of replacement (i.e., nonmodular) satellites which must be launched in the initial constellation. The number of failure/repair events times the number of satellites in the constellation determines the number of replacement satellite events or ORU repair events. The number of ORUs is equal to the number of satellites in the constellation times the number of ORU sets required.

Transportation Module [2]

The transportation module develops the unit transportation (Earth to LEO) cost for both repair and replace missions for the satellites and the necessary hardware (ROTV, OMV servicer, cargo bus, circularization stage, and support structure) required for each mission. A part of this calculation is the cost of the fuel for the ROTV (when needed) to transfer from LEO to the satellite's final orbit for each mission type. This module obtains the weight information of the infrastructure elements from the infrastructure discussed next. The diagram of the transportation module is shown in Figure 9.6. The purpose of this module is to calculate the ROTV fuel costs for a resuable orbital transfer vehicle for various missions and to calculate the LEO launch vehicle costs. To arrive at both of the above values, weight manifests for the replace mission and repair mission are required. The ROTV fuel reserve is a simple input of fuel reserve in kilo-

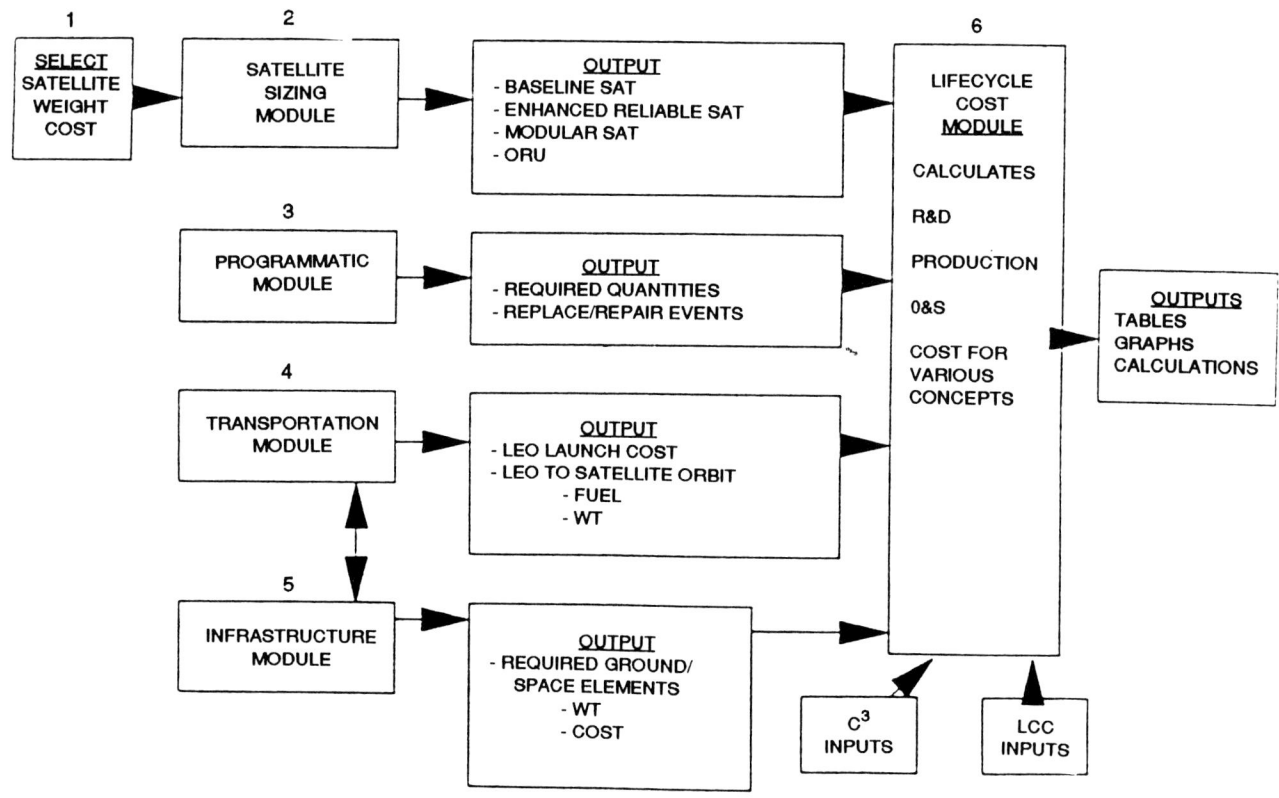

Figure 9.3 Satellite servicing life cycle cost estimating—methodology overview [4]. *Courtesy of Air Force Systems Command Space Division.*

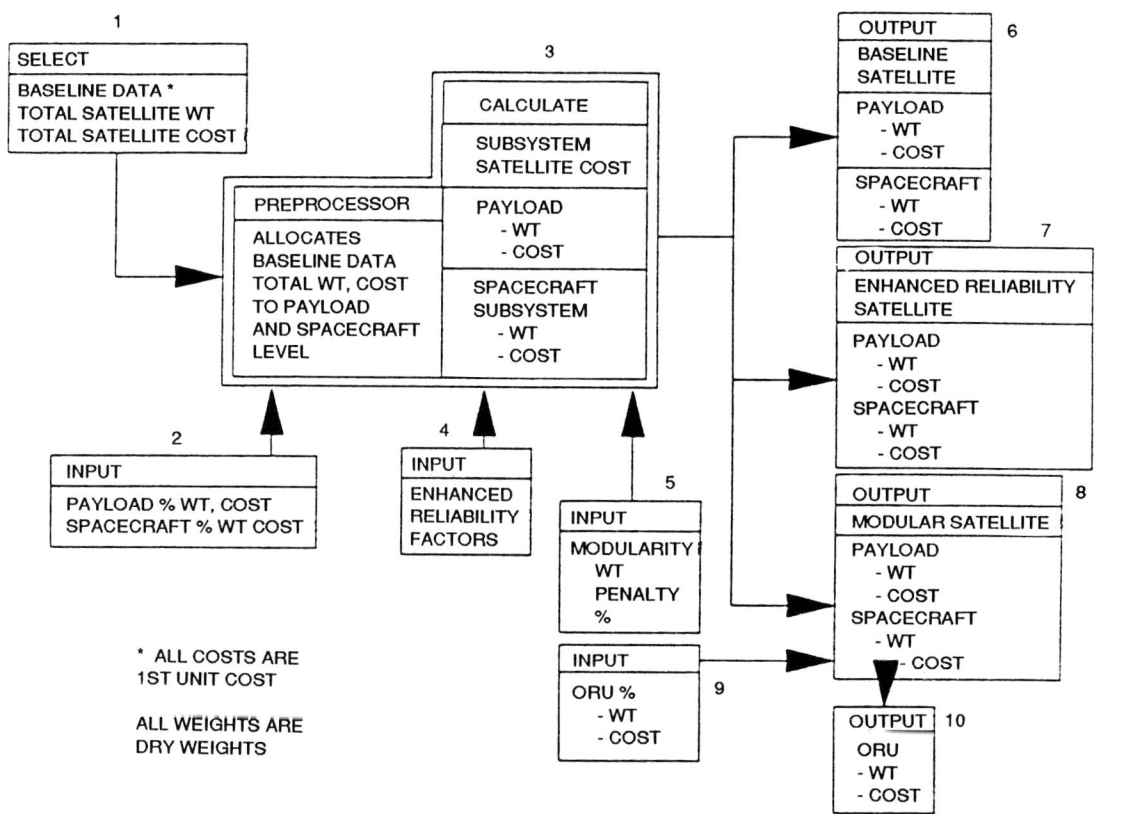

Figure 9.4 Satellite sizing module [4]. *Courtesy of Air Force Systems Command Space Division.*

Benefits and Economics

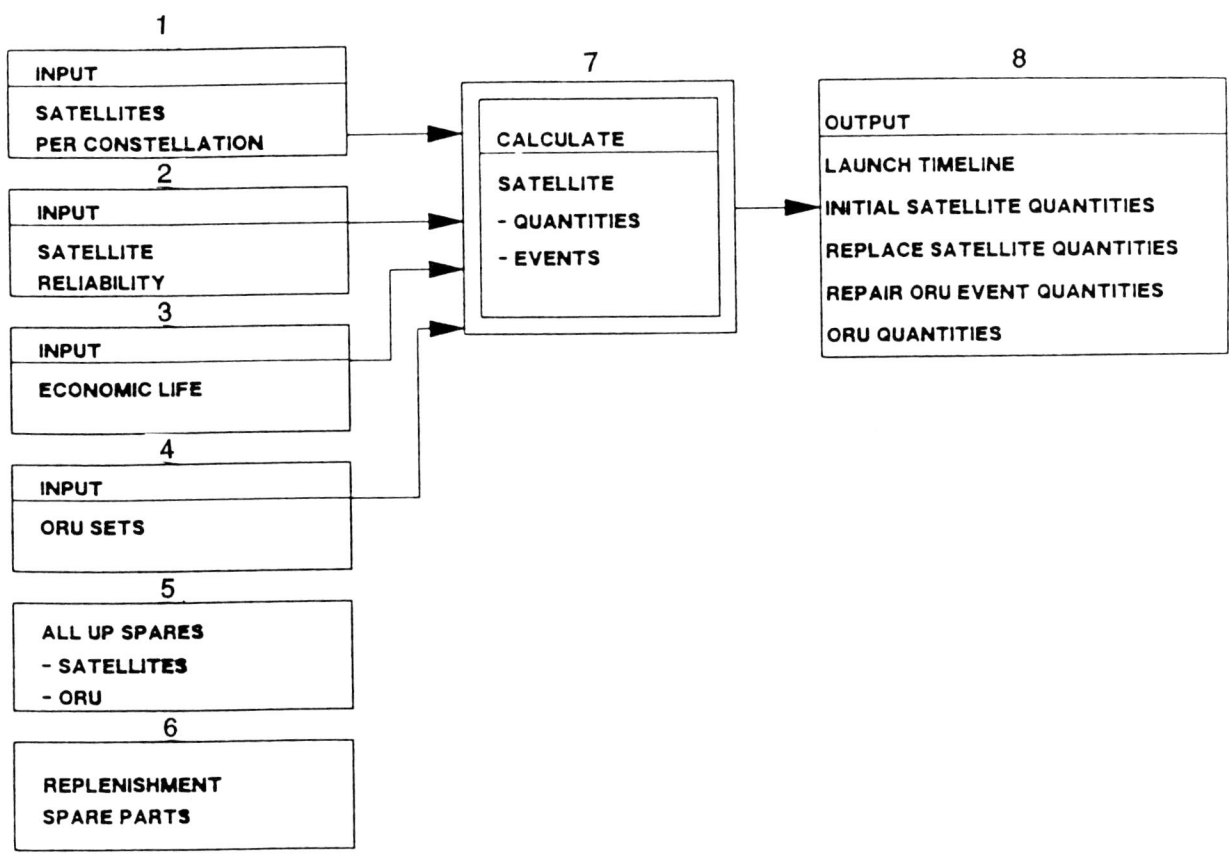

Figure 9.5 Programmatic module [4]. *Courtesy of Air Force Systems Command Space Division.*

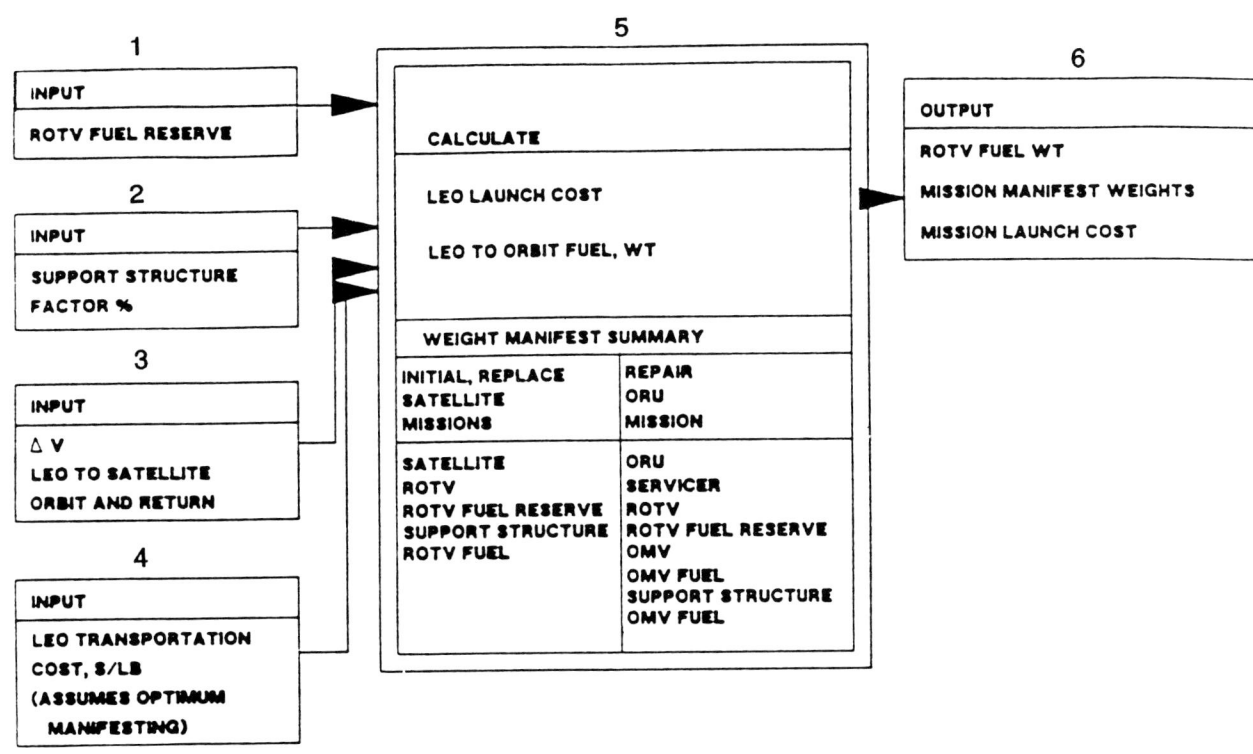

Figure 9.6 Transportation module [4]. *Courtesy of Air Force Systems Command Space Division.*

grams (pounds) of extra fuel over the mission requirements calculated below.

The support structure is an input percentage which is applied to all of the hardware (but not ROTV fuel) weight which is manifested to ground launch the replace and repair missions. For satellite replace missions, this factor is applied to the satellite, ROTV, and ROTV fuel reserve weights. For ORU missions, this factor is applied to the ORU, servicer, ROTV, ROTV fuel reserve, OMV, and OMV fuel. Note that this fuel does not have to be launched on the same mission as its hardware, but must be tracked for the specific mission. The fuel could be prepositioned in space or launched on a separate launch vehicle.

The LEO launch costs are treated as a dollars per unit of weight, kilograms or pounds, cost to low Earth orbit. Specific launch systems could be identified by using unique values of $/kg ($/lb). However, the use of $/kg ($/lb) for launch cost eliminates the need for identifying the specific launch vehicle. This value was treated parametrically in the analysis.

The specific value of launch cost to LEO for a single satellite or ORU mission is the LEO $/kg ($/lb) times the satellite or ORU manifested dry weight to orbit. The ROTV fuel cost is treated as the $/kg ($/lb) for the refueling mission times the weight of the fuel. The methodology of using launch cost as dollars per unit weight assumes optimal manifesting, i.e., only the cost of the actual weight required on orbit is estimated even if the payload bay is half empty. If the launch cost were estimated using discrete dollars per launch, the launch cost would be higher than the $/kg ($/lb) approach since the cost of the entire launch vehicle would be charged even if its payload bay were half empty.

The output of this section produces the launch cost for satellite and ORU missions. The ROTV fuel cost is presented separately from the LEO launch cost due to its magnitude. It is a significant weight and cost consideration.

Infrastructure Module [2]

The servicing infrastructure is expensive and in the broadest sense consists of these elements:

- The ROTV for high-energy delivery misions
- A robotic satellite servicer system for ORU replacement and fuel transfer
- Fuel tank farms and a tanker vehicle for liquid and gaseous consumable resupply
- Man-tended servicing platforms and warehouses for constellation servicing
- A cryo storage facility
- EVA servicing and assembly hardware, tools, and equipment to enable man servicing tasks
- OMV type vehicles for short-range maneuvers and proximity operations for cargo transfer, satellite servicer system transfer, and crew movement
- Space-based and ground mission control centers and work stations
- Ground facilities for launch operations, cargo integration, ORU repair, and crew training

The infrastructure module provides the definition of the above various elements for servicing missions in terms of weight and cost. Each elements will not be required for all missions, but each mission will require some of the elements. This module allocates the total life cycle cost (R&D, production, and O&S) to a per use cost for a given quantity of missions (amortization quantity) over the satellite's lifetime. Thus, even though single satellite programs are analyzed, an attempt is made to allocate the infrastructure cost to other user programs and charge only a portion to a single mission. This is an important concept to this methodology. The diagram of the infrastructure module is shown in Figure 9.7.

Life Cycle Cost Module [2]

The life cycle cost module uses inputs from the integration services or communications, command, and control (C^3) element and LCC factors. It then calculates the various R&D, production, and O&S costs of the satellites, repair or replace quantities, transportation costs, and infrastructure use charges to obtain LCC estimates for each concept. The output of the methodology is in the form of printed tables of cost, graphical comparison, and calculation results.

The life cycle cost module is shown in Figure 9.8. The LCC module draws upon all other modules for input, as illustrated in Figure 9.3. This module uses factors and cost estimating relationships (CERs) to calculate the R&D and production cost of satellites and ORUs. Quantities from the programmatics module are used along with the unit infrastructure and transportation costs to determine total infrastructure and transportation costs. These same quantities are used to obtain total C^3 costs based upon a unit C^3 cost which is input. Program level and management reserve costs are calculated as a percentage of the above totals. The costs are then row and column summed to give the life cycle cost by phase and by WBS element for both replacement and repair cases.

Air Force/NASA SAMS Study Cost Results

The SAMS study [1 through 5] generated cost estimates for:

- Conducting each of the 11 on-orbit servicing DRMs using general spacecraft categories as example space programs
- The servicing infrastructure stated in the infrastructure module above

Benefits and Economics

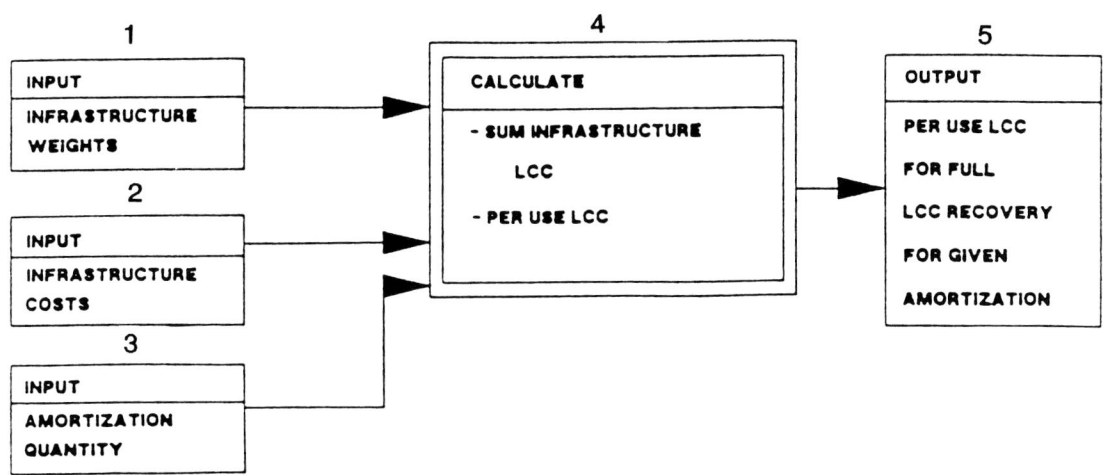

Figure 9.7 Infrastructure module [4]. *Courtesy of Air Force Systems Command Space Division.*

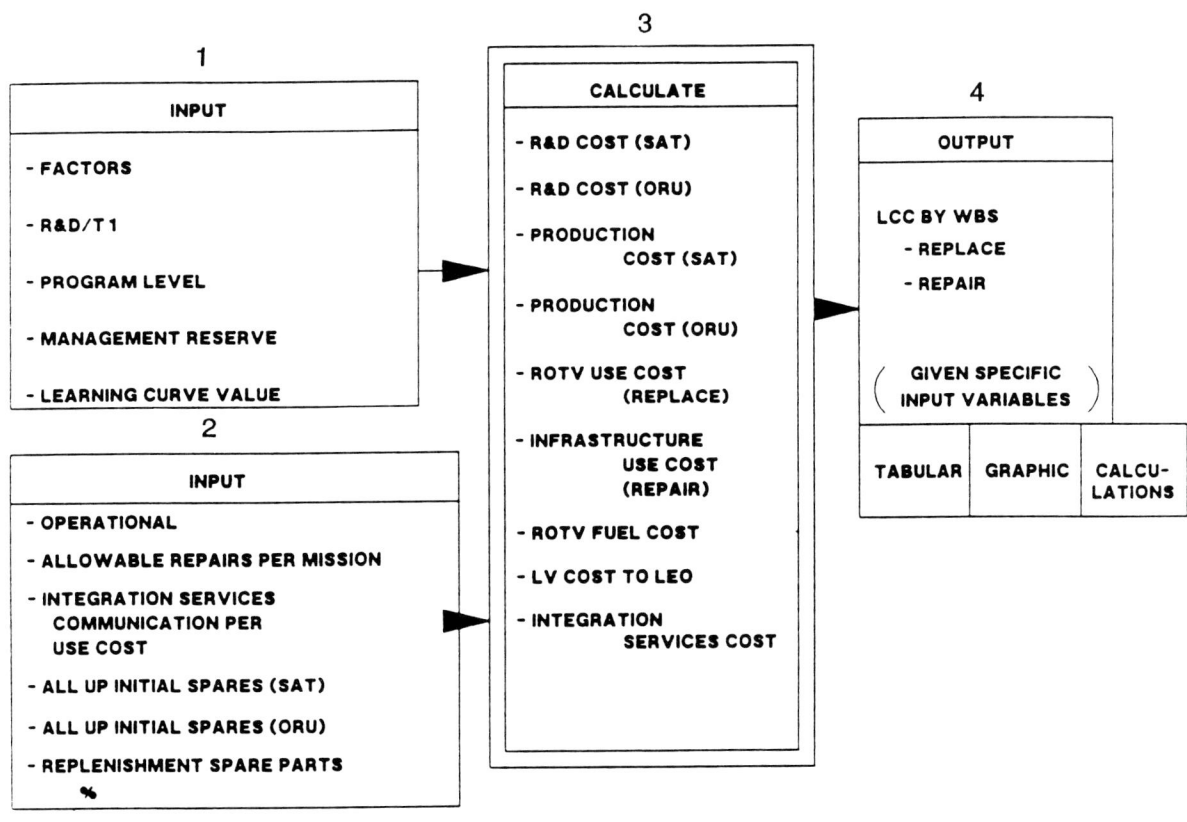

Figure 9.8 Life cycle cost module [4]. *Courtesy of Air Force Systems Command Space Division.*

- Incorporating servicing into the program life cycle cost of specific Air Force/SDI spacecraft such as:
  — Defense Satellite Communication System (DSCS)
  — Fleet Satellite Comunication System (FLTSATCOM)
  — Military, Strategic, Tactical and Relay (MILSTAR)
  — Defense Support Program (DSP)
  — Defense Meteorological Satellite Program (DMSP)
  — Global Positioning System (GPS)
  — Boost Surveillance and Tracking System (BSTS)
  — Space Surveillance and Tracking System (SSTS)
  — Space Based Interceptor (SBI)
  — Space Based Radar (SBR)

SCENARIO/DRM	1	2	3	4	5	6	7	8	9	10	11
SEGMENT											
USER	28	13	-	44	14	36	13	277	21	35	-
SAMS	5	2	40	5	23	9	107	151	9	12	418
TRANSPORTATION	16	55	488	36	217	74	157	220	74	75	745
INTEG/SERV	2	1	1	2	3	3	3	3	3	4	5
SYS LEVEL	4	5	35	6	17	8	17	41	7	8	70
O&S	1	-	-	1	3	1	3	3	1	1	3
TOTAL MISSION $M	56	76	564	94	276	131	300	694	115	135	1241

Figure 9.9 SAMS study: TRW's cost summary of 11 DRMs [1]. Segments are shown on Figure 9.2. DRMs defined on Figure 9.10. *Courtesy of TRW Space and Technology Group.*

The DMSP is a low Earth orbit satellite; the GPS, MILSTAR, and SSTS operate a mid-Earth orbit; the DSP, FLSATCOM, DSCS, and BSTS are geosynchronous birds; while the SBI and SBR are positioned in "specific case" orbits.

Design Reference Mission Costs

Figure 9.9 is a summary of the total cost for each of the 11 DRMs from the TRW SAMS study [1]. Costs are shown for each of WBS segments of Figure 9.2. The major characteristics of the 11 DRMs are given in Figure 9.10.

Infrastructure Costs

The costs of the individual elements of the servicing infrastructure were developed in the SAMS study [1 through 5]. The life cycle cost of the full infrastructure for a 20 year life cycle use is listed in Figure 9.11. The 20 year LCC of $10.08 billion in FY 91 dollars is considered realistic for the servicing the use models employed and the assumptions of Figure 9.1 applied. Additional point decision information and perspicacity for the infrastructure elements are required before a high confidence can be placed in any cost estimates for the elements.

It is estimated that a conservative number of infrastructure uses over the 20 year period, used for life cycle costing was 100. This assumes the Air Force and SDIO are satellite servicing users, at least to some degree. This then is an average of 5 servicing missions per year over the 20 year period. This quntity is referred to as the amortization quantity. Since these two items, infrastructure cost and amortization quantity, are relatively soft, they should be made variables within any analysis. Figure 9.12 shows how this is done. For the baseline infrastructure which costs $10B and is used on 100 programs, the per use cost is $100M per use. If only 50 uses of the infrastructure were attained, the per use cost for full recovery would be $200M or if 200 was attained the per use cost would be $50M. Thus, if we felt that the total LCC for the infrastructure of $10B was low, it could be easily changed within the model as could the amortization quantity. For example, should the infrastructure cost be $15B rather than the $10B, its per use cost at 100 missions is $150M.

Results Applicable to "Real Satellite Systems"

The repair/servicing concept can potentially be cost beneficial to NASA, Air Force, SDIO, and commercial satellite systems as shown in the SAMS studies by TRW, LMSC, Air Force Space Division, NASA, Tecolote Research, and Science Applications International Corporation [1 through 5]. In general, nominal savings of 20 to 30 percent and maximum savings of up to 50 percent can be realized.

Even though the TRW SAMS study and some parts of the follow-on Air Force SAMS study were purposely performed using generic satellite design parameters and orbits, the results can be employed to generate a first order screening for real NASA, Air Force, SDIO, and commercial satellite systems. Figure 9.13 identifies certain systems which could potentially realize a cost benefit by using a maintainable/servicing design. The referenced studies made no effort to quantify the potential cost savings over replacement of the entire spacecraft. This quantification calculation would be highly system specific and require proprietary cost data from the satellite contractors and their customer program offices. However, at a top level, Figure 9.13 does show that a large percentage of space systems could benefit from on-orbit servicing. In addition, the cost savings benefit could extend to the ground integration assembly and test activities for modular constructed satellites. The trend is favorable to the repair/servicing concept in general since infrastructure results showed that a large number of users were required to effectively amortize infrastructure costs.

DESIGN REFERENCE MISSION DESCRIPTION	EPOCH	ORBIT	NUMBER OF S/C SERVICED
1. Consumables resupply of a 30,000 lb. satellite in LEO. Example programs: GRO, HST, AXAF, SIRTF, SPARTAN, EOS, and SSF	Pre 1996	400 NM 98.5 degree	One
2. Satellite repair, ORU replacement, selected maintenance in LEO. Example programs: DMSP, Solar Max, Landsat, and GRO	Pre 1996	400 NM 98.5 degree	One
3. Assemble large space structure in LEO. Example programs: SSF, LDR, SSTS, and Space Test Program	Pre 1996	250 NM 28 degree	One
4. Emergency ORU replacement and consumables in resupply in LEO. Example programs: GRO, EOS, UARS, TOPEX, and Explorer Platform	Pre 1996	400 NM 28 degree	One
5. Constellation servicing and maintenance. Example programs: GPS, SSTS, DMSP, and EOS	Post 1996	38,643 NM 55 degree	20 to 100
6. Maintenance and consumables resupply on multiple platforms. Examples are: MILSTAR, DSP, DSCS, and Leasat	Post 2000	GEO	3 to 20
7. Satellite assembly at HEO. Examples are: SLCSAT, BSTS, and Servicing Platforms	Post 2000	1620 NM 65 degree	One
8. Multiple SAT servicing and maintenance at depots. Examples are: Zenith Star, SSTS, MPS, and GPS	Post 2000	1620 NM 65 degree	40 to 200
9. Emergency repair and replenishment in GEO. Examples are: Milstar, DSP, and GEO Platform	Post 1996	GEO 10 degree	One
10. Maintenance and servicing of satellites in GEO. Examples are: DSP, Milstar, and BSTS	Post 1996	GEO, 0 to 13 degree	One planned, and one emergency
11. Large spacecraft assembly in GEO. Examples are: GEO Platform, BSTS, and GEO Servicing	Post 2000	GEO	One to five

Figure 9.10 Design reference missions (DRM) characteristics [1]. *Courtesy of TRW Space and Technology Group.*

Cost Input to the Satellite Repair versus Replace Trade

The costing methodology described in this chapter from References 1 through 5 was developed to compare the mission and cost effectiveness of repairable and nonrepairable satellites in LEO, MEO, and GEO. The variables to a cost-effectiveness sensitivity analysis are given on Figure 9.14. Each variable can be assigned a range of values using a combination of real spacecraft data and best engineering es-

ELEMENT	DEVELOP	PROD	ANN O & S	20 YR O & S	TOTAL
ROTV	$900M	$400M	$50M	$1000M	$2300M
Servicer	250	45	40	800	1095
Tanker	300	80	20	400	780
Platform	500	200	30	600	1300
Cryo Facility	600	200	10	200	1000
EVA Hardware	300	190	5	100	590
OMV	300	80	30	600	980
Mission Control	150	55	20	400	605
Grd Facilities	400	150	40	800	1350
TOTAL $M	3700	1400	245	4900	$10000M

Quantity is one of each element item
Develop = development cost
Prod = production cost of one flight article
Ann = annual
O & S = operations and support cost

Figure 9.11 Estimated space assembly, maintenance, and servicing (SAMS) infrastructure cost in millions of 1991 dollars. The total LCC is $10 billion.

	BASELINE			HIGH		
Infrastructure LCC	$10B	$10B	$10B	15B	15B	15B
Amoritzation Use	50	100	200	50	100	200
Infrastructure Use Charge	$200M	$100M	$50M	$300M	$150M	$75M

Figure 9.12 Infrastructure amortization.

timates. Reference 4 is an excellent source of showing how methodology, input variables, and parametric analysis are used in determining, from a life cycle cost standpoint, whether satellites should be repaired or replaced. The repair or replace strategy must evolve by examination of all the input variables in a sensitivity analysis to show the degree of impact each input variable has on the LCC of replacement and repairable satellites. Here detailed cost estimating of all the elements is necesssary to accurately establish repair versus replace break-even points and cross-over trends in LCC curves.

Figure 9.14 makes the point that strategy for safe repair and or servicing versus satellite replacement for various classes of space systems is based on the four input parameter categories of satellite, launch vehicle, infrastructure, and operations.

In general, the overall trends which make satellite maintenance and servicing appealing are similar for all generic orbit regimes considered (GEO, MEO and LEO). There are, however, differences in the magnitude of potential savings in the different orbital regimes due largely to increases or decreases in infrastructure required to access the satellites and the absolute costs of the satellite in each orbital regime. Following is a brief description of the trends which make satellite maintenance and servicing appealing [4].

Satellite

The satellite input variable category describes the characteristics of the individual satellite. The first satellite characteristic that favors the repair concept is high complexity and therefore high cost satellites. The second satellite characteristic is heavy satellites which in most cases are also expensive satellites. The third satellite characteristic is low cost orbital replacement units. This result will impact sat-

> NASA SYSTEMS - Space Station Freedom, Hubble Space Telescope, Gamma Ray Observatory, Advanced X-Ray Astrophysics Facility, Space Infrared Telescope Facility, Large Deployable Reflector, Upper Atmosphere Research System, Earth Observing System, Orbiting Solar Laboratory, Orbital Maneuvering Vehicle, Spartan, Explorer Platforms, Ocean Topography System, Materials Processing, Orbital Transfer Vehicle, Solar Max Mission, GEO Platform
>
> DOD SYSTEMS - Defense Met Sat Program, Fleet Sat Comm Sat, Milstar, Defense Space Program, Global Positioning System, Space Based Radar, Defense Space Comm Sat, Submarine Laser Comm Sat, Space Test Program
>
> SDI SYSTEMS - Boost Surv. and Tracking Sys, Space Surv. and Tracking Sys, Spase Based Laser, Space Based Interceptor, Zenith Star, Battle Management Station
>
> COMMERCIAL SYSTEMS Leasecraft, Landsat, Westar, Spacenet, Telstar, Industrial Space Facility, Insat, Syncom
>
> INTERNATIONAL SYSTEMS Arabsat-Arab League; Aussat-Australia; Meteosat-European Space Agency; Eureca-European Space Agency; SPOT-France; Palapa-Indonesia; Columbus-European Space Agency; Skynet-Great Britain; China Broadcast Sat-China; Anik-Canada; Insat-India; Geostationary Metsat-Japan

Figure 9.13 Partial list of satellites that might benefit from on-orbit servicing.

ellite compartmentalization and ORU design for modular satellites. The fourth satellite characteristic is that the satellite have a relatively low mean mission duration. The more times a satellite needs to be replaced, the more the repair concept is favored. Thus, if high reliability can be assured at a reasonable cost, maintenance and servicing may not be worthwhile.

Launch Vehicles

The characteristic within the launch vehicle category that favors the repair concept is low transportation cost. This is largely because large payloads are required to be delivered to orbit for repair missions (just as for replacement missions) because of the large mass of hardware and fuel required for repair missions. Thus, transportation costs can be a big cost driver. At the same time, however, it is a big driver for replacement missions also. Thus, it is somewhat a relative issue. Even if transportation costs are high, repair missions may still be cost effective if other mission parameters override transportation costs.

Infrastructure

The infrastructure input variable category represents the elements required to support repair missions. Low infrastructure LCC and a high amortization level favors the repair concept. The analysis [1 through 5] shows that both GEO and MEO satellites could be repaired or serviced using the same infrastructure elements (including the same space node) which minimizes cost and operational complexity.

Additional cost savings could be realized by NASA and the DoD/SDIO sharing such facilities.

Concept of Operations

The characteristic within the operations category which favors service and repair missions most is the capability to visit more than one satellite per mission. A huge economy of scale can be exploited for such missions assuming a sufficient number of satellites are available at a given time for such missions. Because of fuel limitations such missions are limited to co-planar satellites. This is not a prohibitive restriction for GEO satellites, since they are close to being co-planar because of the unique feature of this orbit. Several MEO satellite systems also have multiple satellites per plane which can benefit from the multivisit technique. LEO satellites, however, probably cannot benefit from such a technique because there are a limited number of assets in LEO and they are usually not co-planar.

Space Insurance

Despite its short history, the space insurance industry is now viewed as mature and stable [6]. The industry's capacity has stabilized after the 1986 through 1987 hiatus in international launch activity and has now begun to grow. At the same time, premium rates have also stabilized. In today's technical environment, the risks of spaceflight are better understood, partly as a result of the intense scrutiny brought on by the spacecraft losses of the 1980s, see Figure 9.15. But today's risks continue to carry a relatively high degree of

IN-PUT VARIABLE SATELLITE	POSSIBLE RANGE
Spacecraft weight, Lbs.	2300 to 15000 (GEO & MEO) 2000 to 35000 (LEO)
Spacecraft cost	High and low values for LEO, MEO, LEO
Added to cost to make the S/C serviceable	2 to 8%
Modularity weight penalty	10 to 20%
Modularity cost	5 to 30% of total S/C cost
ORU cost	10 to 50% of total S/C cost
S/C development $/lb	25000 to 35000
Payload development $/lb	50000 to 60000
Constellation size	1 to 12 per orbit ring
Mean mission duration, years	2 to 15
Servicing interval, years	1/2 to 5
LAUNCH VEHICLE	
Launch vehicle type	NSTS, ELV, Orbit-to-orbit vehicle
Transportation cost to LEO, $/lb	600 (ALS goal) to 6000
Transportation cost to GEO, $/lb	7000 to 10000
INFRASTRUCTURE	
Infrastruction location	Ground, LEO, MEO, GEO; 28 degrees to polar inclinations
Infrastructure 20 year LLC, $/B	10 to 15 (fuel capacity)
Amortization, total uses	50 to 200
Infrastructure use charge	See Figure 6.12
SPACE OPERATIONS	
Orbit extremes	LEO to GEO
Number of sats repaired per mission	1 to 5
Basing concept	Most of the infrastructure in space based
Servicing missions per year	2 to 4 pre 1996 4 to 10 1996 to 2000 12 to 20 post 2000

Figure 9.14 Cost-effectiveness variables for use in sensitivity analysis and trade studies [4]. *Courtesy of Air Force Systems Command Space Division.*

Date	Spacecraft	Type of failure	Loss
9/77	OTS 1	Delta failure	$29M
2/79	AYAME 1	N-1 failure	$15M
12/79	Satcom 3	AKM failure	$77M
4/82	Insat 1A	Satellite failure	$70M
9/82	Marecs B	Ariane failure	$20M
2/84	Westar 6†	PAM failure	$105M
2/84	Palapa B-2†	PAM failure	$79M
6/84	Intelsat V F-9	Centaur failure	$102M
4/85	Syncom IV F-3†	Satellite failure	$84M
8/85	Syncom IV F-4	Satellite failure	$84M
9/85	ECS 3	Ariane failure	$80M
9/85	Spacenet 3	Ariane failure	$83M
5/86	Intelsat V F-14	Ariane failure	$82M
11/87	TV-Sat	Satellite failure	$53M
7/88	Insat 1C†	Satellite failure	$78M
9/88	G-Star III‡	Satellite failure	$72M

†Salvage through retrieval and resale or repair on-orbit moderated these losses.
‡Sums insured; loss and/or salvage potential pending.

Figure 9.15 Major space insurance losses over $10 million. *Courtesy of International Technology Underwriters, Inc.*

companies, an improvement in hardware and software reliability over that experienced in the past 10 years must be demonstrated. This is necessary for the insurance industry to break its pattern of losses and merit reduction in premium rates. If reliability is to improve in the decade of the 1990s, it must do so in an environment of continuing technological change. New, advanced spacecraft designs, with composite structures, microelectronic circuits, nuclear propulsion, and machine intelligence/robotics mechanisms, to name just a few upcoming technologies, will be introduced by spacecraft manufacturers. Traditional expendable launch vehicle developers (General Dynamics, McDonnell Douglas, and Martin Marietta) are designing upgraded versions of previously "known" launch vehicles [6]. New entrants are introducing unfamiliar launch systems for commercial use, including new small ELVs. Viewed collectively these thrusts represent a changing environment in which improving reliability will be a serious driver in mission and economic space operations strategy.

Risk Management

Insurance as a means of dealing with risk should be only one approach in a much broader scheme of risk management [6]. Where there is uncertainty, there is risk. To try to eliminate risk in space enterprise is futile. Risk is inherent in the commitment of present resources to future expectations. Risk is the probability and consequence of *not* achieving some defined program goal. Risk is the chance and consequence of being unable to obtain what we want, when we want it, for the amount of dollars we want to spend on it. But now comes risk management. Risk management is the process which encompasses the identification, assessment, tracking, control, and mitigation of program risks and

uncertainty. This was recently exemplified by the antenna deployment problems (solved by crewmember EVA actions) on the Gamma Ray Observatory launch from NSTS flight 37 (Atlantis) on April 7, 1991.

The next 10 years may see 50 to 80 insured *commercial* payloads launched into space. While this level of activity represents a more substantial base of business for insurance

Elements		
Cost Risk	Schedule Risk	Technical Performance Risk
Competitive optimism	Competitive optimism	Competitive optimism
Cost estimates • Accuracy • Uncertainty • Timing Affordability • Funding level • Funding profile • Contract type	Schedule slippage • Long-lead time materials or items • Critical components • Manpower availability • Manpower training requirements • The marching army problems • Facilities and equipment availability	Spacecraft complexity • Feasibility • Producibility Technology • Feasibility • Uncertainty • Obsolescence Engineering support • Capability • Availability • Fragmentation of responsibilities
Quantity • Number of spacecraft	Quantity • Number of spacecraft	Quantity • Number of spacecraft • Overlapping development of interdependent projects Material procurement • Availability • Long-lead time materials or items • Design changes
Customer uncertainty • Need and urgency • Funding level and profile • Contract provisions	Customer uncertainty • Need and urgency • Funding level and profile • Contract provisions	Customer uncertainty • Need and urgency • Mission and performance requirements • Contract provisions
Management control • Monthly accounting • Ongoing account tracking	Management control • Monthly scheduling • Ongoing schedule tracking	Management control • Design tracking • Performance requirements tracking • Material procurement tracking

Figure 9.16 Program risk elements and associated factors. *Courtesy of TRW Space and Technology Group.*

results in overt actions to accept known risks or to make program adjustments which avoid their potential consequences [7]. Program risk elements are shown on Figure 9.16, while Figure 9.17 lists, by program phase, risk analysis applications. All major aerospace government agencies and aerospace contractors have their own risk analysis tools and techniques for managing technical, cost, and schedule risks for large and small satellite programs. A risk management strategy must be established early and continually addressed throughout a program's life cycle.

The first step, identifying the risks, should be done at all levels of a project. Issues such as which subsystems should be *replaceable versus repairable on-orbit*, or how many layers of redundancy are to be built into certain systems, must be addressed. At a broader level, issues such as whether the design will permit alternative means of access to space, or how a crew might be rescued, are major concerns.

Once the risks are identified, they must be evaluated from two standpoints: how serious are the consequences, and how likely are they to occur? The most serious risks must be dealt with so as to make them as improbable as possible, and those with lesser degrees of severity must have their consequences accommodated in the least damaging manner.

The third step is to decide how to handle each identified risk. There are essentially four basic means for doing this: avoidance, reduction, transfer, or retention.

Avoidance is simply choosing to use another means of accomplishing an objective to avoid a risk, like using terrestrial communications technologies to avoid the risk of launching satellites. However, as with any avoidance strategy, this also means missing the benefits associated with the riskier strategy, which may suggest using other risk management techniques.

Reducing a risk can be done in one of three ways. The uncertainty that sometime will go wrong can be reduced by increasing the number of events under consideration, like designing a satellite with multiple transponders and backup amplifiers to accommodate the occasional failure. Alternatively, the actual hazard to be encountered can be reduced, as in using more reliable components in a satellite. Finally, the effects of a loss if one occurs can be reduced,, such as eliminating single point failures or *designing a satellite to be serviced* for certain types of failures or maintenance.

Risk can be transferred by a number of means, including hedging, using surety bonds, incorporating, or subcontracting. And, of course, by purchasing insurance.

Phase I	Phase II	Phase III	Phase IV
Concept Studies	Concept Validation	Full-Scale Engineering Development	Production

• Mission analysis
• Concept evaluation
• Trade studies
• Cost effectiveness analysis
• Risk identification
• Risk assessment
• Risk avoidance
• Risk control/mitigation plans
• Technology development planning for risk reduction
• Uncertainty in cost estimates
• Uncertainty in schedule estimates
	• Requirements review
	• Final design selection
	• Technology selection
	• Planning of activities for risk control/mitigation
	• Program contingency
		• Requirements review
		• Management reserve(s)
		  — Schedule
		  — Financial
		  — Technical performance margin
		• Network and schedule planning
		• Risk tracking and management
		• Lessons learned
			• Management reserve(s)
			  — Schedule
			  — Financial
			  — Technical performance margin
			• Network and schedule planning
			• Risk tracking and management
			• Change risk assessment
			• Lessons learned

Figure 9.17 Risk analysis applications by program phase. *Courtesy of TRW Space and Technology Group.*

Finally, those risks which are not avoided, reduced, or transferred are retained. Retained risks include those which have been reduced to an acceptable level, those with no identifiable means of risk management (hopefully few or none), and, unfortunately, those which may not have been identified.

As is true in any venture, there is an interaction among these approaches to dealing with risks in space ventures. Figure 9.18 illustrates varying tradeoffs in implementing these approaches for a range of space risks. While the table is generally designed to illustrate reduced levels of risk from left to right, only when considered in a context of a complete system can a particular approach be evaluated as "better" than another.

Fundamentally, therefore, in evaluating risks for purposes of insurance, insurers seek demonstration of a credible risk management plan. The premium rate will be a function of how well the company appears to have managed the risks with which it is faced, with the lowest possible rate going to those who appear to be managing the risks the best.

Given insurers' strong desire to minimize failure potential and properly fund for loss, and their inherent risk management skills, savvy companies tend to take advantage of their insurers' expertise as a way to improve their risk management. While it may appear counterintuitive, companies should be concerned if insurers aren't paying much attention to the details of the risk; if the insurers understood it better, they might be able to offer a helpful hint to avoid a problem or offer a lower premium rate for a well-managed project [6].

With the advent of more routine manned operations in space, the ability to carry out maintenance and salvage activities will increase significantly. Inasmuch as several satellite failures have been successfully serviced or salvaged, and others are believed to have been serviceable, this capability could well reduce the risks inherent in launching and operating satellites and thus reduce the cost of the associated insurance coverages.

Economic Conclusions

Point cost estimates, program life cycle cost analysis, and cost benefit determinators are so dependent on assumptions, so complicated due to the many driver inputs, so tied to technology enhancements, and so subjected to the vicissitudes of government fiscal year budgets that no chiseled-in-granite economic conclusions on satellite servicing can be stated.

However, some generalities will be attempted.

Cost Benefits

The SAMS study [1 through 5] found that the nominal life cycle cost savings of up to 33 percent could be obtained by

Example of risk	Approach A	Approach B	Approach C
Loss of subsystem unit(s)	Only existing unit	Ground spare exists	Second unit is on-orbit, spare on ground
Serviceable system failure	Non-serviceable design	Serviceable design	Serviceable design and servicing system available
Launch vehicle unavailable	Dependent on single launch vehicle for initial launch and resupply	Dependent on single launch vehicle for initial launch, has backup available for resupply	Backup launch vehicle for both launch and resupply
Damage to system during resupply	Automated docking system used	Docking with tether then pulled in	System parked and picked up by teleoperated vehicle
Structural damage to module	Single hull metal structure	Double hulled metal or composite with detectable fluid between	Composite hull with embedded fiber network to detect over-stress/leak conditions
Communications loss	Multi-satellite network operating at maximum data rate	On-board recorders to cover one satellite out	On-board recorders plus separate satellites configured to handle same type of traffic
Crew stranded	Original vehicle used for pick-up	Rescue-only vehicle attached	Launch and return vehicle attached *or* separate nearby facility accessible

Figure 9.18 Alternative approaches to handling space risks [6]. *Courtesy of International Technology Underwriters, Inc.*

using a repairable/serviceable satellite design for some systems. A few special cases indicated LCC savings of 30 to 50 percent. Other systems were found not to benefit at all from the serviceable concept. So on-orbit servicing is not for every program.

It might be argued that the potential cost savings achieved by employing maintenance and servicing is not the driving factor in deciding whether to adopt a service and repair philosophy for future satellite systems. Percentage wise, the cost savings may appear to be marginal compared with the cost of developing and operating specific satellite systems. However, when applying these cost savings in a mission model sense across a number of satellite systems, the absolute dollar savings can amount to billions. The exact amount of potential savings depends on a number of variables including the mission requirements imposed on future systems. For example, Reference 4 shows that servicing and repair options are more favorable to bigger and more expensive satellites. If the trend in satellite design breaks from historical precedent and tends to a "lightsat" design philosophy, little or no benefit could be gained from a maintainable design. In reality, some combination of both types of satellites will most likely be present in the inventory.

The cost benefit analyses in Reference 1 show that servicing provides substantial economic benefits using several scenarios. Sensitivities were performed to indicate the impact of changing key cost assumptions. Even with the changes in the cost assumptions, the benefits of servicing still indicate substantial savings. There is a significant trend toward larger and more complex satellites especially for the NASA Great Observatories science missions. This trend is expected to continue and the value of assets in space will increase accordingly. The prior results indicate, as they did in Reference 4, that the cost savings due to servicing are increasingly larger as the satellites become larger and more expensive.

Figure 9.19 presents a comparison of replacement costs and servicing costs as a function of satellite size. The difference between the two cost curves represents the economic benefits of servicing. The satellite costs are based upon a weight-based cost estimating relationship (CER) for satellites of nominal complexity (i.e. $100 million unit cost for a 907 kg (2,000 lb) satellite). Using a CER for higher complexity satellites such as surveillance satellites would further increase the economic benefits [1]. This figure illustrates increasing benefits realized as the satellites become larger and more expensive. The nominal case servicing cost curve represents the DoD only scenario and where ORU costs for the servicing represent 10 percent of the satellite costs. If the missions were reduced to 50 missions and the ORU costs increased to 20 percent of satellite costs, then the higher servicing cost curve results.

The impact of the high cost curve is to introduce a region where replacement is less costly than servicing. This occurs for satellites less than 907 kg (2,000 lb). Thus, the importance of this curve is to indicate that a break-even point can

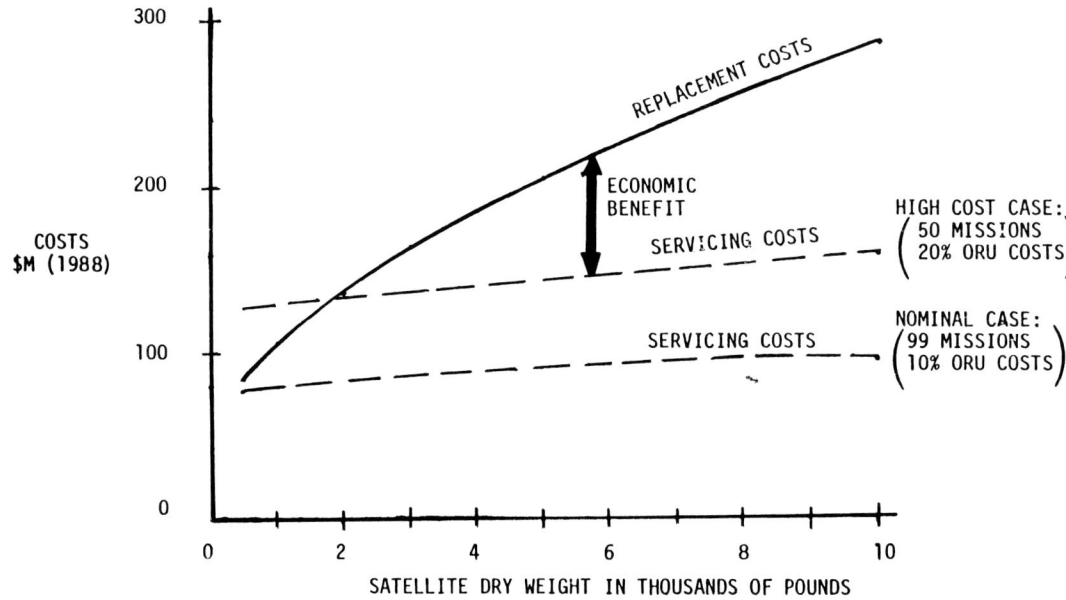

Figure 9.19 Comparison of replacement cost versus servicing costs as a function of satellite size. *Courtesy of TRW Space and Technology Group.*

occur at some point dependent upon the cost of satellites. The exact point at which the break-even point occurs depends upon the mission model and the servicing definition. Even though uncertainty will exist as to the exact point for break-even, the trend toward more complex, larger, and more expensive satellites leads to the conclusion that available servicing capabilities can substantially lower the total cost of the nation's assets in space.

Technology Changes

Future technological innovations may change some of the basic cost conclusions. Certain technological advances may reduce maintenance and servicing infrastructure requirements and simplify operational procedures causing a drastic reduction in infrastructure cost. Obviously there would be some doubts as to the validity and believability of results arrived at using this type of data. However, the potential impact of such developments in the future should not be ignored when evaluating maintainability issues even in today's environment.

One example of such technological innovations is the National Aero-Space Plane (NASP) which is postulated to be able to reach low Earth orbit at a payload delivery cost unprecedented in the history of spaceflight. Its increased operational flexibility should reduce logistics timelines measurable and allow easier access to space. Such a vehicle could itself service low altitude polar assets using manned EVA techniques already proven and alleviate the need for a robotic servicer completely (assuming minimal radiation hazards). Such a vehicle could also be used to access low inclination/low altitude orbits and ferry supplies to and from low Earth orbit or a space node.

Another example is the space node which may become more desirable once an advanced launch system (ALS) becomes operational. Not only could a node be placed in orbit with the large lift capability of an ALS, but the cost to orbit should be reduced appreciably. The ALS would also be useful to resupply missions to the space node. Such a space node is likely to be a useful commodity to both NASA and the military once the International Space Station Freedom is operational. Even though the two facilities would be distinct elements in space, there would be multiple opportunities for common use. As shown in References 1 and 4, a 28.5 degree orbit for such a space node would be useful for both NASA and DoD missions. Considering the potential for coattail missions, (such as fuel savings or hardware delivery,) an economy of scale can be exploited. In addition, the space basing of infrastructure elements at the node lends flexibility to mission planning.

Interorbit transportation system advances will also improve the viability of maintenance and servicing missions. Propulsion systems using advance technologies such as electric propulsion will greatly reduce fuel loads needed to be delivered to low Earth orbit as compared to storable or cryogenic systems. Such a reduction will reduce ground launch frequencies and upper stage design requirements. Such propulsion systems might also be attached to satellites themselves. In the event of a failure, the satellite could slowly spiral down to low earth orbit where it could be serviced at the space node or from a NASP type vehicle.

In addition, other hardware elements required to perform service and repair missions, such as a robotic servicer and refueler, are being studied seriously in the aerospace industry. NASA has taken the lead in several key technologies needed for the development of these systems. In some cases, such systems are mission-enabling for NASA. This will not be the case in most instances for military missions; however, the use of the technologies could obviously be beneficial if needed. The involvement of the military during the development of such systems could preclude costly retrofits and design modifications if such systems are adopted at later dates. The specific benefits of such systems are not quantifiable to the extent desirable at the present time.

Cost-Effectiveness

Like most analysis, the economic effectiveness of on-orbit servicing requires the balancing of several decision parameters. The major parameters are:

- Necessity—One fundamental question is whether it is absolutely essential to basic mission success to have a SAMS system. If a space system can be put in place and accomplish its mission without SAMS, then servicing is not absolutely essential. However, some systems, most notable SDI systems, will find that SAMS—type techniques are mission-enabling and that costs would otherwise become prohibitive.
- Satellite Mission Durability—This consideration is the "sine qua non" of servicing. If the basic mission of the satellite becomes obsolete before any benefit of extended service life is realized, then servicing to extend life has no merit.
- Required Near-Term Investment—The acquisition of the necessary infrastructure to implement servicing could require substantial early investment, without discernible payback for some time. It appears that, in the long term, SAMS is highly cost-effective, but there is a near-term question of affordability. There will not be offsetting cost savings during the buildup to an operational capability, and a national decision to invest in the future would have to be made, probably at the expense of continuing some near-term programs.
- Guaranteed SAMS Availability—One question near the front of the ultimate users' minds will concern the certainty of SAMS support when needed. The system must be sufficiently robust that there is no significant interruption in the availability of servicing or maintenance due to limited scope casualties. The SAMS system and logistic setup must tolerate equipment failure and keep functioning. Without guaranteed availability, a system operator would have to elect to "take care of himself" to ensure continuation of his basic mission. This approach will be particularly true of DoD/SDI users whose missions involve the national security.

Conclusions

Satellite servicing costing is a vast subject with many side issues, methodologies of approach, and expert economic viewpoints. This section has only attempted to define the important topics that enter into the total question of cost feasibility and affordability to implementing operational servicing.

The material of this section was taken, for the most part, from the USAF/NASA/SDIO sponsored SAMS study performed by TRW and Lockheed Missiles and Space Company [1–5]. In fact, there are many excellent treatments to the subject of evaluating costs associated with satellite servicing. References 8 through 20 are documents for suggested reading.

Two levels of decision making enter into the balancing strategy for implementing a national on-orbit space assembly, maintenance, and servicing (SAMS) capability. At the agency level (DoD, NASA, SDIO), these questions will be asked:

- Is a SAMS capability mission enabling? Must we have it?
- How much near-term investment is required?
- What are the runout costs for putting the SAMS infrastructure in place?
- Can we recover development costs?
- Will international and commercial space users avail themselves and therefore pay a user charge for SAMS missions?
- How will a robust SAMS capability impact spacecraft insurance premiums?
- What standards must be mandated?
- Which programs will be asked to consider incorporating servicing into their satellite design and operations?
- Which programs will be commanded to incorporate servicing into their satellite design and operations?

When the question of "To service or not to service?" is presented to a typical System Program Office (SPO) director, this second level decision maker will probably ask a series of questions that go something like this:

- Does it cost me more this year?
- Does it help me get to first/next launch?
- Is servicing availability guaranteed?
- Will my project be charged to develop servicing hardware?
- Will my project have to share the infrastructure with other programs and thereby incur scheduling problems?
- Can I do my mission without on-orbit servicing?

If the answers are not favorable, the SPO director will undoubtedly seek to avoid servicing. Near-term money is always a problem and the development of a new satellite inherently contains sufficient challenges that the SPO director is not moved to undertake anything which doesn't help get to the main goal.

In the final analysis, cost alone should not be the final or only discriminator used to evaluate the military, civilian, and commercial utility of maintainable satellites [4]. Chances are that if cost were the only parameter of importance, the United States would not have a civilian space program in the first place. Operating in space is an expensive undertaking. This is not to say that all cost issues should be completely ignored. Ideally a balance must be reached between mission requirements and cost. The time has come to expand the scope of on-orbit servicing analysis to investigate other issues of importance. For example, quick turnaround repair missions could increase constellation availability by using prepositioned space nodes containing appropriate ORUs. Satellite refueling systems could be used to replenish consumables on space test vehicles and/or operational satellites which may need periodic refueling (i.e., mission or maneuverability system tests). In addition, the overall survivability of space systems may be enhanced by ORU changeout and/or refueling, which provides another option to the ground-launched replacement and complete on-orbit sparing concepts. Once mission and cost data are evaluated together, sufficient information will be available to make hard decisions regarding the pros and cons of satellite maintainability. Such decisions probably should be reevaluated periodically as technologies evolve. It is expected that the concept of maintaining and servicing military satellites will be evolutionary rather than revolutionary.

Space maintenance and servicing strategies can save money over the life cycle of some satellite systems. However, an investment in the infrastructure is required prior to achieving this savings. If NASA and DoD together make the decision to commit to an infrastructure or to maintenance concepts which employ minimum infrastructure elements [5], many satellite systems would probably adopt the space repair and servicing concept. No one satellite user wants to have the first need (and thus cost) of the infrastructure development. However, the cost savings available through space repair and servicing, coupled with the mission enhancing possibilities mentioned above, could lead to a revolution in the way satellite systems are designed, produced, operated, and maintained in the future.

References

1. *Space Assembly, Maintenance and Servicing Study (SAMS) Final Report*, Volume I, Executive Summary, TRW No. SAMSS-196, Volume II, System Analysis, TRW No. SAMSS-195, Volume III, Design Concepts, TRW No. SAMSS-197, Volume IV, Concept Development Plan, TRW No. SAMSS-198, Volume V, Neutral Buoyancy Simulation, TRW No. SAMSS 199. TRW, Redondo Beach, CA, July 6, 1988.
   and
   Space Assembly, Maintenance, and Servicing Study (SAMS) Final Report, Volume I, Executive Summary, Volume II, System Analysis, Volume III, Design Concepts, Volume IV, Concept Development Plan, Volume V, Simulation Report. Lockheed Missiles and Space Company, Inc., Sunnyvale, CA, July 6, 1988.

2. Suttle, James H., Thomas E. Jee, and Kevin D. Wheaton. *Space Assembly, Maintenance and Servicing Analysis Model (SAMSA) Users Guide*. Tecolote Research, Inc., CR—0353, prepared for Department of the Air Force Headquarters Space Division, November 1, 1988.
3. Fong, F. K., G. N. Heydinger, M. E. Koscielski, J. E. Schmitz, J. R. Barnum, R. L. Nordli, R. L. Tomlinson, L. Wilcox. *Space Division Unmanned Spacecraft Cost Model*. Space Division/ACC, Los Angeles, CA, SD—TR-81-45, June 1981.
4. Suttle, James H., Thomas E. Jee, Stephen J. Stepanek, and Robert J. Curtis. *Space Assembly Maintenance and Servicing Study Independent Mission Cost Effectiveness Assessment*. Tecolote Research, Inc., and Science Applications International Corporation, CR—0307, prepared for Department of the Air Force Headquarters Space Division, November 1, 1988.
5. Jee, Thomas E., James H. Suttle, and Robert J. Curtis. *Space Assembly, Maintenance and Servicing Analysis Model, Minimum Investment Satellite Repair and Servicing Analysis*. Tecolote Research, Inc. and Science Applications International Corporation, CR—0371, Department of the Air Force Headquarters Space Division, December 7, 1988.
6. Higginbotham, John B., and Peter M. Stark. *Insuring Space Ventures*. Space Commerce Vol. 1, No. 1, p. 19, 1990.
7. Howe, John C., and Karen J. DeGraffenreid. *Program Risk Analysis*. TRW, April 1987.
8. Rysavy, Gordon. *A Guide for Evaluating Costs Associated with Satellite Servicing*. Satellite Services System Working Group, NASA JSC, June 1988.
9. *Satellite Servicing Cost Criteria*. Prepared by Science Applications International Corporation for Engineering Directorate, NASA JSC, Contract No. NAS9-17207, October 16, 1986.
10. Jordan, John W. *Servicing Transportation Scenarios*. Boeing Aerospace Company, Session II Satellite Servicing Studies and Planning, Satellite Servicing Workshop III, NASA Goddard Space Flight Center, June 9–11, 1987.
11. *On-Orbit Maintenance Study Phase II Space Logistics Support Concept Study Final Report*, Volume I, Executive Summary, Volume 2, Compendium Final Report, Volume 2, Annox: Monthly Status Reports and Executive Briefing, Volume 4, Classified Data Sets. ANSER, Arlington, VA, STDN 88-12, April 1988.
12. *Cost Effectiveness of On-Orbit Repair and Servicing Of Selected Satellite Programs*. Directorate of Aerospace Studies, AFCMD/SA, Kirtland AFB, NM, DAS TR 86-5, January 1987.
13. Smith, Scott. *On-Orbit Refueling—An Analysis of Potential Benefits*. SA-ALC/TIEO Space Assembly & Servicing Working Group Meeting #26, San Antonio Air Logistics Center, February 27, 1991.
14. *Preliminary Logistics Analysis Methodology Design Document*. Advanced Technology, Inc., El Segundo, CA, October 13, 1987.
15. Stepanek, Stephen S. *Integration and Assembly Savings to DoD and NASA Spacecraft Due to Modularity*. Tecolote Research, Inc., El Segundo, CA, CR-0283, March 1988.
16. *Pricing Options for the Space Shuttle*. Congressional Budget Office, Congress of the United States, March 1985.
17. *Spacecraft Partitioning and Interface Standardization*. FEDSIM, Washington, DC, Report No. 95603-02-USAF, February 1987.
18. *Servicer System Demonstration and Capability Development Final Technical Report*. Martin Marietta, Denver, CO, December 1987.
19. *Orbital Spacecraft Consumables Resupply System Preliminary Design Report*. Rockwell International, Downey, CA, Report No. STS 86-0268, July 1986.
20. Bloomquist, C., and W. Graham. *Analysis of Spacecraft On-Orbit Anomalies and Lifetimes*. Planning Research Corporation, Los Angeles, CA, PRC-R3579, February 1983.

Chapter 10
Road Map to On-Orbit Servicing

"The Administrator shall conduct a thorough and comprehensive study of satellite servicing with a view toward establishing national goals and objectives for utilizing such capabilities . . . The capital investment in space satellites and vehicles should be enhanced and protected by establishing a system of servicing, rehabilitation, and repair capabilities in orbit (hereinafter referred to as 'satellite servicing')"

National Aeronautics and Space Administration Authorization Act of 1988 (H.R. 2782), Title 1, Section 11

"Air Force policy is to (1) ensure that spacecraft maintenance options are considered in requirements definition, acquisition program management, and contractual documentation for those satellite programs wherein these options might be reasonably implemented, and (2) actively examine the utility of spacecraft maintenance options so as to avoid, wherever practicable, design actions which preclude on-orbit maintenance later in the spacecraft life cycle . . ."

Secretary of the Air Force Edward C. Aldridge, Jr., letter on spacecraft maintenance—September 1984

"Opportunities for deploying and maintaining systems in space should be vigorously pursued where they provide clear mission improvements and, or, cost reductions . . ."

Air Force Space Division Regulation 540-8, Space Servicing—October 1984

"DoD will vigorously pursue new support concepts . . . space support functions are those required to deploy and maintain military equipment and personnel in space . . . includes maintaining and sustaining space vehicles while on orbit . . ."

DoD Space Policy—March 1987

"SDI programs shall consider on-orbit servicing as a prime alternative in the life cycle support of space systems . . ."

SDIO Supportability Research Policy—October 1985

Steps to Operational Capability

On-orbit satellite servicing could be a routine operation in the 1995 to 2000 time period. The technology appears feasible, the economics look promising, but complex engineering and programmatic issues remain to be resolved.

Assuming the U.S. Space Station Freedom program proceeds as planned by NASA, we should witness, in the last decade of this century, the start of space-based spacecraft servicing that is motivated by economic and operational considerations.

Satellite servicing will probably progress, Figure 10.1, in three major phases of expansion and capability: concepts definition and validation, initial support capability (ISC), and full support capability (FSC) [1].

Phase I. Concepts Definition and Validation (1983–1995)

The learning process and early flight demonstration missions are in progress. Space Shuttle missions of 1983 through 1995 will be used to develop satellite servicing transportation, equipment, and techniques and to provide an educational basis for flight crewmembers and space technicians performing servicing tasks. On-orbit satellite servicing is, roughly, a 25 year program to improve space systems capability, flexibility, affordability, and responsiveness.

Phase I will feature: requirements analysis, advanced concept design studies, EVA servicing hardware/tools development, crew training for EVA servicing, ground-based simulations, cost-effectiveness assessments, development of serviceable spacecraft (HST, GRO, AXAF, OMV), shuttle-based early technology development missions for both EVA (manned) and autonomous (robotic) servicing technology development missions (TDMs), and planning for Phases II and III.

Specific capabilities in this phase will include:

- Space Assembly—Spacecraft checkout, deployment, calibration, and alignment and the mating of large, self-contained modules

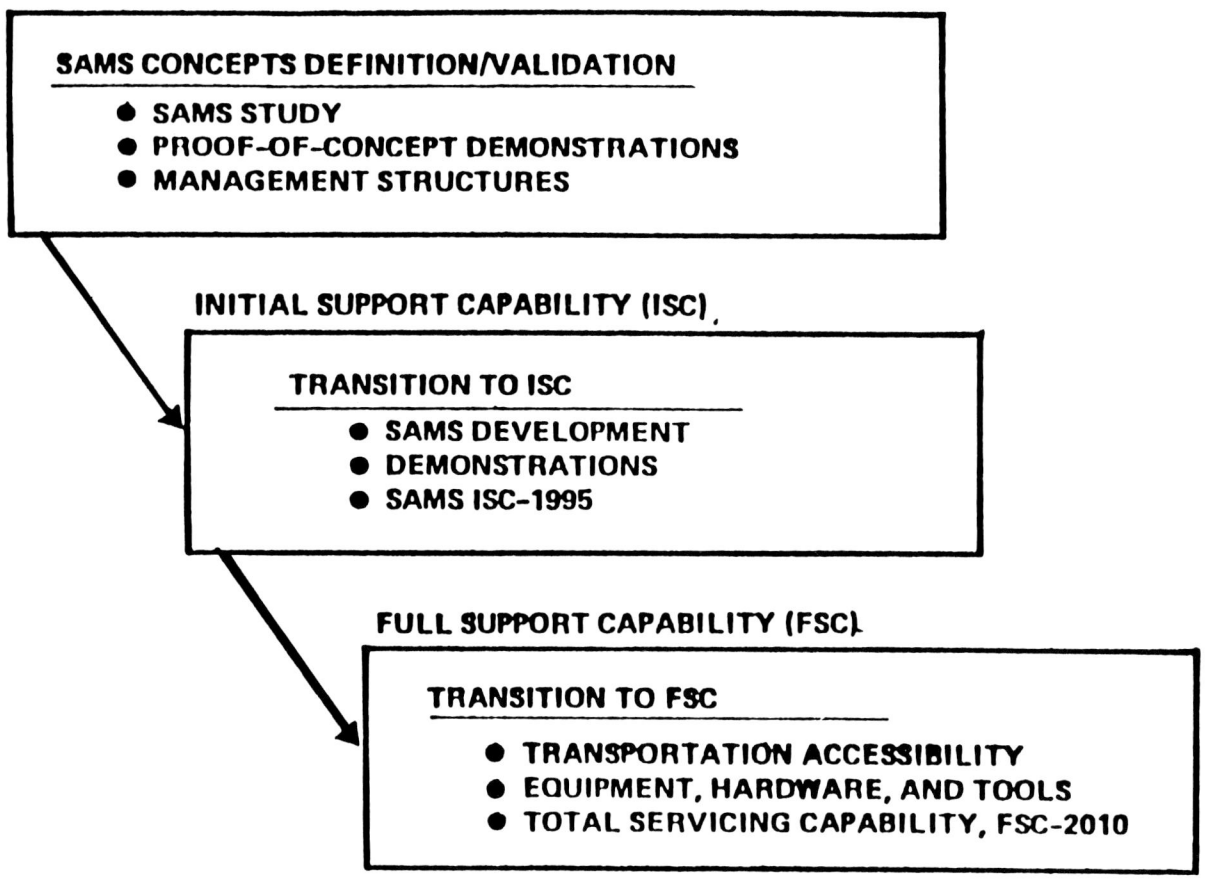

Figure 10.1 Space Assembly, maintenance, and servicing road-map. Courtesy of TRW.

- Space Maintenance—Observation, inspection, realignment, recalibration, simple ORU changeout, minor repairs, plus extensive satellite checkout, surface restoration, fault repair, complex module changeout, and early application of automation
- Space Servicing—Exchange of film packs, data types, batteries, propulsion tankage, monopropellant refueling, plus other fluid and gas container replacement

In Phase I proof-of-concept demonstrations will be conducted in ground laboratories and simulators and in space to verify equipment design ideas and build operational capability.

During this phase it is expected that government and industry management structures to guide and control the program will be put into place.

Phase II. Initial Support Capability (1996–2005)

The application of technology, Phase II, will take advantage of the presence of the early International Space Station Freedom in LEO (Shuttle supported) to conduct advanced tests and demonstrations and to focus the servicing technologies to specific program applications. This phase will feature the technology enhancements, hardware qualifications, and policy/legal actions to advance servicing from the concept stage to initial support capability (ISC)—a major goal in the national program.

During this phase, three major elements of the servicing infrastructure—the Space Station Freedom, an orbital maneuvering vehicle, and a telerobotic satellite servicer system—should be added to the national space capability, enabling servicing missions to meet a greatly expanded set of requirements, orbits, and time lines. Starting in the late 1990s, servicing tasks on or near the Space Station will be performed in a routine manner, repair tasks complexity will further increase, and in situ servicing will be initiated. Even geostationary orbital servicing missions will be feasible provided the orbital transfer vehicle is available with the requisite payload delivery and return capability.

Specific capabilities in this phase will include:

- Space Assembly—Assembly of complex space systems plus construction of large structures and joining them into space vehicle and orbit-to-orbit stages
- Space Maintenance—Damage repair, preplanned product improvement, depot type maintenance, large

alterations to spacecraft to allow mission objective changes, spacecraft self-diagnosis, and introduction of artificial intelligence technology
- Space Servicing—Bipropellant and cryogenics refueling, complete fluid management systems operations, and deployment of space-based tank farms and refueling depots for all types of fluids

About the year 2000, levels of automation will advance from the early 1990s flight demonstrations of supervised autonomy and teleoperation modes to more sophisticated robotic modes. The latter will incorporate machine intelligence support in diagnostics, troubleshooting, fault isolation and correction, and some levels of decision making.

Phase III Full Support Capability (2006–2015)

The third phase, starting about 2006, will begin an era of routine on-orbit servicing events where mature servicing equipment, developed crew skills, automated servicing with robotics and artificial intelligence software, space-based servicing infrastructure facilities (man and man-tended), advanced transportation systems, servicing of constellations of satellites on a single service call, and the existence of an advanced NASA Space Station will all be combined into a national capability that will greatly benefit NASA, military, and commercial and international space operations.

Specific capabilities in this phase will include:

- Space Assembly—Fabrication of components and unique materials and assembly into large systems
- Space Maintenance—Production line maintenance on constellations of satellites in pressurized facilities, complete spacecraft refurbishment, and utilization of space-based autonomy
- Space Servicing—Refueling of multiple spacecraft at the same time by robotics, and the emergence of a quick, "pit stop" capability

The above postulated time scale is meant as a scenario that is considered achievable. Yearly levels of government funding into the program, as well as technical events, will strongly modify the actual history and progress of the program. Another factor that could pace the program is the extent to which private industry perceives a potential for commercial ventures in servicing satellites on-orbit. Some companies—TRW, Lockheed, McDonnell Douglas, and Science Applications International Corporation—have quietly looked into this venture. Their results are company private and highly dependent on their assumptions of market value and their potential share of it.

National Program Plan

Based on specific design details, many spacecraft were identified in Chapter 9 as candidates for servicing operations in either the initial support capability time period (1996–2005) or the full support capability time period (2006–2015). The ISC time period spacecraft candidates must be examined for an evolutionary approach to servicing capability development. Design concepts must be developed to minimize operational cost and schedule impacts and implementation should be planned at spacecraft program block changes. The SAMS study [1] examined the parallel development of proposed spacecraft design concepts and SAMS hardware/tools and technology. Spacecraft changes should be planned for implementation only when the associated SAMS hardware/tools are available.

Certain SAMS hardware/tool approaches may adversely impact spacecraft design concepts and certain spacecraft design concepts may preclude SAMS hardware/tools. Existing and planned NASA SAMS hardware/tools and technologies must be examined for their impacts and the impacts on them from the spacecraft design concepts. The NASA SAMS hardware/tools and technologies require analysis so that DoD requirements/impacts can be developed for them. DoD spacecraft differ from NASA spacecraft in three areas: (1) Survivability—threat environment, shielding, hardening, demininhing, ASAT, (2) Security—classification, encryption, decryption, and (3) Mission Availability-operational effectiveness, mean time between critical failure, mean time to repair. The approach for the near-term spacecraft will establish the framework for the development of the SAMS requirements for the far-term spacecraft. The far-terrm spacecraft concepts and SAMS hardware must utilize the near-term concepts as a basis for their development.

The DoD has not, as yet, established firm servicing commitments to military satellite programs. Possible serviceable space systems in the near term are the: Global Positioning System, Defence Meteorological Space Program, Milstar, and Space-Based Radar. For the far term the advanced communication and surveillance programs and projects within the Strategic Defense Initiative are under study for the possible cost-effective applications of on-orbit servicing.

For the past 3 years NASA management and DoD leaders have been working on an affordable operational strategy that both organizations could factor into a national space systems servicing program. The major elements of this new servicing strategy are [2]:

1. Servicing of assets (where assets includes the Space Station as well as free flyers) in low Earth orbit as a starting point; then evolution to the servicing of satellites in polar orbits, later in high inclination orbits, and eventually in geostationary orbit.
2. The transition from EVA to remote servicing

The satellite servicing strategic plan must recognize [2]:

1. The reduced number of NSTS launches and the limited number of EVA hours

2. The value and accessibility of the space assets
3. The substantial investment in OMV and the FTS technology and early development
4. The need to demonstrate remote servicing at an early date in low Earth orbit
5. The degree to which DoD commits to servicing its assets
6. The increase in activity in LEO in the Space Station era
7. The need to support hazardous servicing requirements

The key considerations of a national space systems servicing program can be collected under ten principal elements: cost, servicing location, servicing functions, logistics, common hardware, transportation, standards, operations, technology, growth capability.

The key considerations will change with time and program maturity. Examples of key considerations under each element are:

1. *Cost*
   — Satellite servicing must be cost effective or program enabling.
   — Servicing pricing policy must be established for users in advance of spacecraft design and mission operations.
   — A national investment in the servicing infrastructure is required prior to achieving space program life cycle costs savings.

2. *Servicing Location*
   — Near-term strategy will address servicing of space assets in low Earth orbits.
   — The evolution of the satellite servicing strategy to polar, high inclination, or geostationary orbits will be a function of the cost benefits associated with the servicing of these assets in these orbits.
   — Users of serviceable space assets must locate these assets in orbits compatible with an operational servicer system.
   — The servicing of satellites at the Space Station Freedom will occur when a servicing capability is available at the Station and when such servicing is warranted as paced by user requirements and economics.

3. *Servicing Functions*
   — Satellite servicing is any activity performed on-orbit to assemble, maintain, repair, resupply, upgrade, deploy, retrieve, or return various space systems, satellites, or facilities.
   — How, when, and if each above function is performed on a space system depends on the technology status of the function, servicing hardware availability to accomplish the work, operational need for quick response, location of the serviceable spacecraft, and cost of the servicing mission versus spacecraft replacement cost.

4. *Logistics*
   — Replacement modules and fluids will be shipped to the Space Station through the Space Station logistics resupply system.
   — Supportability must be factored from the start of the space system design process.
   — Logistics and maintainability are integral in the total decision criteria of space systems life cycle cost analysis.

5. *Common Hardware*
   — Baselining and development of generic servicing equipment hardware and tools for classes of spacecraft will preclude each program office for having to procure its own set of hardware.
   — A designated government organization (NASA or Air Force) must own and issue (loan) the generic servicing hardware to user project offices.

6. *Transportation*
   — The operational satellite servicer system must be compatible with both the expendable launch vehicles (Delta, Titan) and the Space Shuttle.
   — An OMV and OTV are vital parts of the national servicing infrastructure.
   — The shortage of launch capacity created by the 1986 suspension of flights by the U.S. Shuttle and Titan, together with the stand-down of Europe's Ariane, has been replaced by a plethora of launchers being produced by many countries.
   — A goal should be to cut to one-tenth the cost per pound of placing payloads into low Earth orbit.

7. *Standards*
   — Servicing interfaces must meet previously recognized and agreed to interface standards.
   — Work should be accelerated to determine and define the requirements for robotic hardware/software and standard servicing interfaces for satellite servicing.

8. *Operations*
   — Safety of the crew and equipment is paramount in mission planning, time lining of events, and selection of infrastructure elements to accomplish on-orbit servicing objectives for all programs.
   — The low Earth orbit servicing capability should be space-based at the Space Station Freedom using an OMV and the operational version of a satellite servicer system (SSS).
   — Strong emphasis should be placed to identify compatible issues in the design of satellites and robotic servicing systems which promote a cooperative environment for blended EVA and automated servicing operations.
   — Development of orbital operational procedures,

through ground simulation of servicing operations, must be worked out for effective use of the Space Shuttle, OMV, OTV, satellite servicer, and manned operated mission control functions.

9. *Technology*
   — A satellite servicer system flight demonstration program will provide the confidence required to commit to a national operational remote servicer program. NASA had planned flight demonstrations with an OMV, FTS, satellite servicer system, and a target vehicle in the 1993–95 time period but these demonstrations are on hold pending NASA budgetary problems.
   — There is a need to quickly mature the critical technologies associated with satellite servicing. These technologies fall into categories of: man-in-space, spacecraft design, mission operations, and servicing equipment.
   — There is also a need to utilize ground testing and simulation facilities for developing a remote servicer capability.
   — Specific and directed studies must focus on: (1) Where are the high (early) technology payoffs? (2) Are they specific to users or generic? (3) How does sequence of technologies affect the solution via flight demonstrations? and (4) How much innovation is enough?

10. *Growth Capability*
    — Satellite servicing infrastructure, and the related hardware/tools inventory, can evolve with mission needs.
    — Almost all elements of the servicing infrastructure (Space Shuttle, Space Station, OMV, satellite servicer, etc.) have built into their respective programs a growth or extended capability. These separate growth plans should be integrated and factored into the national satellite servicing program.

With a desire to share its extensive expertise in technology, applications, and utilization of on-orbit servicing, NASA is supportive not only of the servicing needs of NASA users, but also those of the DoD, domestic commercial, and foreign users. The commercial Industrial Space Facility (ISF) and ESA's Eureca are examples of commercial and foreign spacecraft that are being designed to utilize NASA servicing capability. Although the initial phase of the joint study with the USAF (the Space Assembly, Maintenance and Servicing Study—SAMSS) did not identify near-term, firm, military satellite servicing requirements, NASA continues to work closely with the DoD in the area of satellite servicing. This joint effort takes place through a Memorandum of Agreement (MOA) designed to address on-orbit maintenance and repair [3]. This Memorandum of Agreement for joint DoD/NASA on-orbit maintenance and repair activities was signed in June 1986. Its purpose is to establish the collaborative relationship between NASA and the DoD to institutionalize on-orbit maintenance as a design option for current and future space systems. Under this MOA, NASA and the DoD participated in the SAMS study [1].

Servicing Technology Development

Chapter 2 contains a list of 30 technology items that will further enhance the development of satellite servicing. From the 30 technologies identified, 6 are considered [1] to be drivers from the standpoints of early payoff, relevance to a real or planned NASA/DoD/SDIO program, and belief that "good" proof-of-concept efforts can be developed to advance the technology. These 6 technologies are listed in Chapter 2.

Servicing Infrastructure Development System Requirements

Although the national servicing system concept will evolve as the mission requirements become more firm and better understood, several attributes of an infrastructure have been identified [1]. The system architecture must include:

1. Servicing facilities at Space Station Freedom
2. A reusable orbital transfer vehicle (ROTV), using cryogenic propellants
3. A facility for the on-orbit storage and handling of cryogenic propellants
4. A remotely piloted maneuvering vehicle (OMV) which can carry a servicing front end and appropriate spare modules for the serviced satellite
5. A propellant transfer system (OSCRS) which can service satellites with storable propellants (hydrazine and/or bipropellants)
6. A maintenance and servicing module with one or more servicing arms and stowage for replacement ORUs, adaptable to the OMV or ROTV
7. For later missions, a manned orbital transfer module (MOTM) which can be carried to a remote servicing location by the OMV

For deployment of SDI systems such as SSTS or midlife servicing of such a system, the infrastructure must include:

8. An assembly and overhaul station (AOS) in the same plane as the system to be deployed, with the capability to support a sizeable crew (up to 12) for a period of weeks
9. Co-orbital warehouses with an OMV/satellite servicer system combination to service high altitude platforms between overhauls
10. A propellant carrier capable of mating to the AOS and carrying sufficient propellants to service multiple SDI spacecraft
11. A reactor kick stage capable of placing a 2,722 kg (6,000 lb) reactor into a higher 18,530 km (10,000

nmi) storage orbit when the SDI spacecraft are overhauled

Depending on the nature of the mission model and the levels of activity, there may be further need for:

12. Man-tended platforms in polar or geosynchronous orbit to perform short duration servicing or assembly missions
13. A high energy upper stage (HEUS) for the heavy lift launch vehicle (HLLV)/advanced launch system (ALS) which will be required for direct emplacement of the SDI satellites and for manned or planetary/science missions

Regardless of the absolute numbers of satellites/systems to be serviced, or their locations, advanced transportation systems will also be needed which will reduce the cost of access to space by a significant factor. Reference 1 identified the need for a manned Shuttle capable of transporting 6 to 12 crew members and roughly 18,144 to 20,412 kg (40,000–45,000 lb) of equipment to a 463 km (250 nmi) orbit of 60 degrees inclination.

Tests and Demonstrations

The commitment to a flight program for demonstrating the on-orbit capability for servicing of satellites represents a significant milestone in the nation's endeavor to protect and efficiently use its orbital assets. The demonstration flights validate the servicing system approach particularly in autonomous rendezvous and docking, fluids resupply, and ORU changeout. From this beginning, the opportunities for reaping the benefits of serviceable space systems can be realized with the eventual development of a fully operational capability for on-orbit servicing.

NASA, with operational requirement inputs from the Air Force and the SDIO, is constructing a plan for a satellite servicing proof-of-concept program to establish a demonstrated servicing capability by the mid-1990s. The planning to date calls for ground-based tests and simulations, and flight tests and demonstrations. As elements of the plan and potential servicing concepts emerge they will be integrated into procurement actions for industry bid. Here is a preliminary set of 10 criteria for selecting and scheduling servicing concepts to be evaluated in a proof-of-concept set of flight demonstrations:

1. The concept, if validated, must be needed by NASA, Air Force, or SDIO for a specific servicing function on near-term missions.
2. The concept will support a number of missions; it must not be a single purpose action.
3. The concept must enable or enhance a servicing technology, as identified in Chapter 2.
4. The concept must have growth potential to far-term missions.
5. The cost of conducting the demonstration must be affordable. Affordability envelopes, therefore, must be defined.
6. Demonstration hardware must not be dead-ended but can be modified for use on operational servicing missions after the demonstration is completed.
7. Demonstration must be realistic allowing time for ground analysis, lab testing, simulator testing, crew training, and demonstration hardware fabrication and delivery.
8. Support equipment requirements must be realistic.
9. The concept demonstration must effectively use available transportation to orbit and return.
10. The concept, if validated, must meet all safety standards.

The above list is in no prioritized order; however, safety and cost would, of course, be at the top of any ranking. The demonstrations needing spaceflights for validation will be preceded by precursor activities in ground labs and test/simulator facilities.

A few examples of concept demonstrations are:

1. EVA orbital replacement unit changeout
2. EVA spacecraft refueling
3. OMV with servicer front-end kit conducting supervised automated (remote) ORU changeout and refueling
4. Telepresence demonstrations
5. Assembly, via EVA and robotics, of complicated space structures
6. On-orbit product improvement changes to a spacecraft
7. Automated inspection, test, and checkout
8. Productivity increase for man in space operations
9. Blended EVA/robotics demonstrations of servicing functions
10. Demonstration of response to an unplanned/unscheduled servicing demand

Ground-Based Tests and Simulations

NASA has a number of ground-based facilities to support on-orbit servicing operations and to demonstrate servicing technology developments. Ground test devices and simulators can reduce risk, provide development cost and schedule confidence, and demonstrate crew and hardware/software interfaces.

There are a number of facilities at the Goddard Space Flight Center (GSFC) and the Johnson Space Center (JFC) to support the on-orbit operations, maintenance, and growth of the flight telerobotic servicer (FTS). Prior to any of the FTS missions extensive crew training will be conducted by JSC at their facilities. A high-fidelity crew trainer will be installed at JSC for this activity. Also, a telerobot simulator will support crew training. This is a real-time, kinematic simulation of the telerobot. It will exercise all the work station interfaces, taking hand controller inputs and driving all the functions in the work station except force reflection. System kinematic information will be available for display, including graphic representations of all telerobot camera views.

Full-scale kinematic mockups of the telerobot will be available for testing in the JSC Weightless Environment Test Facility (WETF). These will be passive mockups that will be used to verify EVA interfaces and handling, such as manual release of the attachment mechanism, restowing, and transporting back through the Space Station hatch for repair. There will also be lightweight mockups for use on the JSC Manipulator Development Facility (MDF) to exercise the interfaces and operation with the Space Shuttle remote manipulator system.

The FTS Engineering Test System (ETS) will be housed at GSFC in the new Spacecraft Systems Development and Integration Facility (SSDIF). The ETS will serve as a high-fidelity test-bed for the development and testing of flight hardware and software. It will also be used to support on-orbit anomaly investigation.

The primary facility for monitoring the FTS operations and for the analysis, archiving, and trending of data will be the Engineering Support Center (ESC) at GSFC. Commands and software loads will be generated in the ESC and transmitted to the Space Station control center (SSCC) at JSC for relay to FTS. Similarly the downlink data comes to the ESC through the SSCC.

There will be a Software Development Facility (SDF) at GSFC for the maintenance of the flight software and for the generation of new code as new capabilities and new hardware come on line and as new task scenarios are developed.

The present GSFC robotics laboratory or Development, Integration, and Test Facility (DITFAC), as it is officially known, will support the FTS operations as an advanced development test-bed. Candidate hardware for future addition to the FTS will be tested and evaluated in the DITFAC before being selected for implementation. Such activities as tool development and evaluation and task element design and test will be conducted here. Currently the laboratory is involved in a number of activities in support of the FTS development, including the evaluation of hand controllers, orbital replaceable units (ORUs), task scenarios, camera locations, and work station designs.

It is necessary that all of these ground support facilities be interconnected with voice, video, and data links so that the operations support can be properly coordinated and managed. Work is currently underway to determine the requirements for these communications links.

Flight Tests and Demonstrations

To be cost-effective, the proof-of-concept flight tests and demonstrations should be coordinated with related NASA, Air Force, and SDIO operational servicing requirements. Present thinking is focused on the following near-term probable future on-orbit spacecraft operations where servicing functions will play a major role:

1. Performance extensions
2. Technology upgrading
3. Reconfiguration for product improvement
4. Spacecraft staging or buildup

Planned servicing intervals or contingency servicing operations are inherent in the above. Figure 10.2 shows some possible subsets of proof-of-concept (POC) candidates under the above four operations. A fifth operation which might emerge is debris control and space rescue. From a future operational context, the Figure 10.2 flight tests and demonstrations proof-of-concepts assume:

1. Integrated logistics, properly coordinated and programmed
2. Transportation capability to, from, and in space
3. Servicing and assembly support equipment availability
4. Ground base command, control, communications availability
5. Space qualified crews; maybe space-based
6. Space-based servicing facilities, stations, or platforms

The keys to servicing proof-of-concept selections include:

1. Cost-effective/high leverage/early benefits to users
2. High degree of relevance to technology plans and user programs
3. Experimental operational testing/ground simulations
4. Reduction of technical and programmatic user risks
5. Joint DoD/NASA/SDIO efforts where feasible
6. POC demonstration hardware which is affordable and applicable to operational missions

Proof-Of-Concept Candidate Ground Rules and Assumptions

As an operating ground rule in synthesis of servicing concepts, the POCs proposed have not used the juxtapositioning of man versus machine. Instead, the POC candidates should deal with the best mix of the two. Candidates proposed then may be of two major categories:

1. **PERFORMANCE EXTENSIONS**
   1. FLUIDS RESUPPLY
   2. MULTIPLE MODULE EXCHANGES
   3. SOLID CRYOGEN EXCHANGE
   4. SOLAR ARRAY REPLACEMENT
2. **TECHNOLOGY UPGRADING**
   1. SINGLE OR MULTIPLE MODULE EXCHANGE
   2. SENSOR/COMPONENT EXCHANGE
   3. SOLAR ARRAY (TECHNOLOGY) REPLACEMENT
3. **RECONFIGURATION FOR PRODUCT IMPROVEMENT**
   1. PAYLOAD - BOOM EXTENSION ON-ORBIT ADD-ONS
   2. SINGLE OR MULTIPLE MODULE EXCHANGE OR ADD-ONS
   3. ANTENNA, ARRAY, THERMAL, OR PROPULSION MODULE ADD-ONS
4. **SPACECRAFT STAGING/BUILDUP**
   1. STRUCTURAL ERECTION AND ASSEMBLY
   2. MODULE AND PAYLOAD ATTACHMENTS
   3. HARNESS ROUTING, ATTACHMENTS, AND CONNECTIONS
   4. FLUID, ELECTRICAL, POWER, THERMAL, AND SIGNAL CONSTRUCTION AND ASSEMBLY
   5. ANTENNA ERECTION
   6. ORBITAL CHECKOUT
   7. PROPULSION MODULE MATING

Figure 10.2 Proof of concept candidate demonstrations focused on probable future servicing operations.

1. *A blend of man and machine*—For example, large space structures assembly and erection, using a degree of automation in the erection process because of potential EVA hours limitations with final assembly and alignment being conducted by EVA astronauts.

2. *Dual/sequential mode of operation*—Automated or manned with either of the modes alternatively being the primary mode of operation with the other as the secondary or contingency mode of operation. This "flip-flop" capability can be very important. It permits flexibility in future operational choices. There are two major operational factors in which humans might not be considered:

— Natural or induced environments: natural radiation, nuclear space power radiation, high thermal rejection radiators, solar dynamic servicing, debris hazards

— Transportation/Logistics: manned transportation possibly not available, manifest contention of competing choices

An example of the first operational condition would be the conducting of a servicing task around a space nuclear power source where the operational solution would be a remotely manned (piloted) orbital maneuvering vehicle with either automated or man-in-the-loop servicer mission kits. An example of the second operational condition, assuming no manned geostationary-based or an intermittently manned capability, the manifest contention would be in payload versus "bus" mass allocations where the "bus" contains the space assembly maintenance and servicing SAMS capability.

Another major ground rule or assumption inherent in the proof-of-concept candidates is that standardization (at some degree or level) and modularity make sense at the systems level. In fact, standardization and modularity are facilitators to SAMS capability. It is assumed that there would be a family of modular sizes and a family of scalable connectors to meet system design and configuration flexibility concerns.

Other preliminary ground rules for POC candidates include:

1. That some of the new hardware (valves, connectors, etc.) which would be used in the POCs should be fully flight qualified (not dead-ended) and represent a validation of a standard or standards and that the design(s) would permit a multiple supplier base: for example, a zero leak refueling valve

2. That up to two buddy EVAs per NSTS flight can be committed to by the crew, e.g., nominal EVA 6 hours $\times$ 2 crew $\times$ 2 buddy EVAs = 24 EVA hours

3. That in a critical SDIO POC, a dual launch (ELV or CELV and STS) capability may be exercised

4. That an OMV is available to be utilized in appropriate POCs

Proof-of-Concept Functional Grouping and Logic

Figure 10.2 shows focused areas of POC candidates. Another categorization of POCs is to break them down into space assembly, maintenance, servicing, and support functions. This is done in Figure 10.3. POCs can now be separated into two parts—those related to SAMS systems which will *provide* the servicing functions and those related to the spacecraft systems which will *receive* the servicing functions. In many cases, a system level POC test/demonstration will contain subsets of tests that apply to both sides of the servicer/spacecraft interface. The two lists below suggest spacecraft design POCs and SAMS functional POCs.

1. ASSEMBLY
   - STRUCTURAL ERECTION
   - ATTACHMENTS
   - SUBSYSTEM ASSEMBLY
   - CHECKOUT
   - RECONFIGURATION

2. MAINTENANCE (ORBITAL REPLACEMENT UNITS (ORUs))
   - POWER
     - SOLAR ARRAY
     - SOLAR DYNAMIC
     - DISTRIBUTION/CONDITIONING
     - NUCLEAR
   - THERMAL
     - THERMAL PROTECTION SYSTEM (TPS, MLI, etc.)
     - RADIATORS
   - ATTITUDE CONTROL
     - IMUs
     - THRUSTERS
     - CONTROL OPTICS
   - PROPULSION
     - ENGINES
     - TANKS
     - LINES
   - COMMUNICATIONS
     - ANTENNAS
     - ELECTRONICS
   - PAYLOADS
     - INSTRUMENT PACKAGES
     - OPTICS CLEANING

3. SERVICING (CONSUMABLE RESUPPLY)
   - FUELS
     - HYDRAZINE
     - BI-PROPELLANT
     - PRESSURANTS (GAS)
     - CRYOGENS
   - BATTERIES
   - ORDNANCE

4. SUPPORT TECHNOLOGIES
   - EVA
     - ZERO LEAK EMU
     - ZERO PREBREATHE SUIT
     - ELECTRONIC DOCUMENTATION
     - VOICE ACTIVATED EQUIPMENT
     - GLOVES
   - ROBOTICS
     - SERVICERS
     - FREE FLYING TV INSPECTORS
     - CAPTURE/TOW DEVICES
     - TELEPRESENCE
     - TELEROBOTICS
     - RENDEZVOUS/DOCKING
     - PROXIMITY OPERATIONS
     - CONTAMINATION CONTROL
     - DEBRIS RECOVERY/DISPOSAL

Figure 10.3 Proof of concept demonstrations based on the three functions and the support technologies.

Example spacecraft design POCs are:

- Tests and demonstrations using a pathfinder test satellite vehicle. (Figure 10.4 is a possible vehicle that could be used for various POC tests/demonstrations. It is a nonoperational, lowcost pathfinder type spacecraft, employed to conduct concept validation activities close to the NSTS or the Space Station Freedom. It is only functional to the extent it has to be to conduct POC tests/demonstrations.)
- Proximity operations demonstrations
- Teleoperated function tests by direct link STS to satellite and indirect link via TDRS
- Laser docking tests
- Removable/replaceable thermal blanket demonstrations
- Secure real-time TV tests
- Demonstrations of membrane mirror deployment and phased array optics assembly
- Demonstration of built-in test equipment and diagnostics, fluid management and transfer devices/techniques, and checkout after servicing
- Tests of propulsion module mating, cryo refrigeration, EVA enhancement, and placement and routing of servicing platform utilities
- Demonstration of end-to-end checkout of an operational system in the STS cargo bay, with the crew ready to conduct EVA to restore the system to fully operational conditions before deployment
- Demonstration of the add-a-pod concept (See Appendix D.)

Example SAMS functional POCs are:

- Tests involving new automated work station systems, operator/hardware interfaces, and system architecture and integration for various control modes
- Tests to measure crew productivity, high dexterity gloves, and high pressure dual-gas spacesuits
- Evaluation of terrestrial robots in zero-g, telerobotic manipulators, handling and position aids, dexterous manipulators, end effectors, standard fittings, force feed back, and time delay
- Technology demonstrations of solar cell decontamination, use of space chemicals, cryogen transfer, and leak detection
- Demonstrations of optics inspection and diagnostics, tumbling satellite capture techniques, and surface repair kit vacuum deposition

Current Activities

Since on-orbit servicing is currently heavily dependent on human participation, only two national space programs, those of the United States and the C.I.S. can undertake such activities. Both the European and Japanese programs, however, are being directed toward the development of servicing capability as an integral activity within future manned space programs [3].

NASA satellite servicing activities are currently internally coordinated through the Office of Space Flight (OSF). This office is currently developing a Satellite Servicing Management Plan to establish agency policy and provide guidelines for implementation of servicing policy at

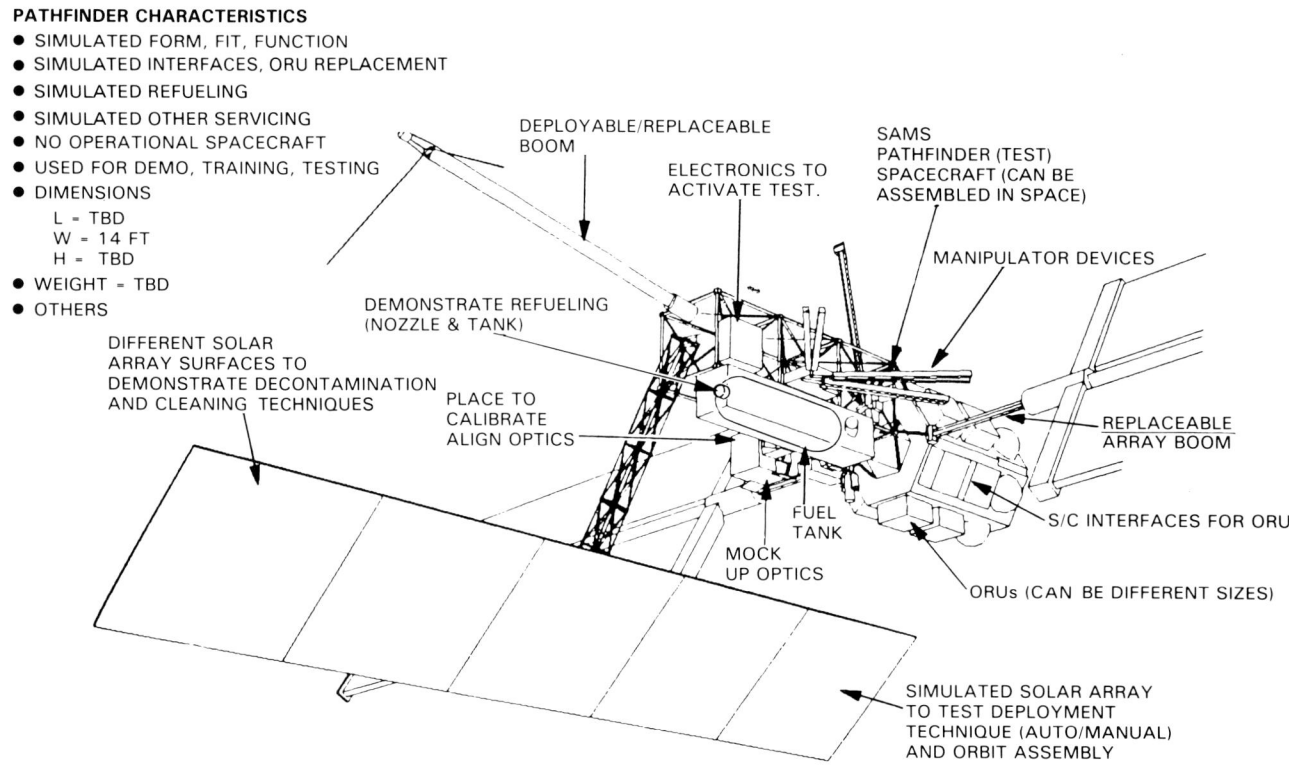

Figure 10.4 Pathfinder spacecraft for the conduct of on-orbit servicing tests and demonstrations of new servicing tools and techniques.

individual program levels within NASA and other U.S. agencies, in addition to providing for interaction with potential commercial users. In addition, the NASA Satellite Servicing Steering Committee provides coordination between the various NASA offices involved with servicing activities to ensure that the requirements of the various NASA user organizations are being met, and that servicing activities are not being duplicated. The Satellite Servicing Working Group, with representatives from all of the NASA field activities engaged in satellite servicing, periodically reviews the progress on servicing developments which are under way and provides input to NASA Headquarters.

For coordination with external parties involved in satellite servicing activities, NASA relies on two mechanisms. The NASA/DoD On-Orbit Maintenance Working Group established under the previously mentioned Memorandum of Agreement coordinates servicing activities between NASA and the DoD. External satellite servicing coordination is sponsored by the Office of Space Flight through periodic satellite servicing workshops. These workshops are the mechanism by which NASA communicates its servicing progress and future plans to industry and the greater user community [3].

NASA's Office of Space Science and Applications (OSSA) began emphasizing on-orbit servicing following the spectacular repair of the Solar Maximum Mission (SMM) satellite in the Space Shuttle bay in 1984. In addition to most payloads attached to Space Station Freedom, serviceable OSSA spacecraft include the four Great Observatories—the Hubble Space Telescope (HST), the Gamma Ray Observatory (GRO), the Advanced X-ray Astrophysics Facility (AXAF), and the Space InfraRed Telescope Facility (SIRTF)—plus the much smaller Extreme Ultra Violet Explorer(EUVE). Using the Goddard Space Flight Center's Multimission Modular Spacecraft (MMS), the EUVE platform will undergo the first-ever on-orbit payload changeout in 1994, to become the X-ray Timing Explorer. The TOPEX and the Upper Atmosphere Research Satellite (UARS), both—like Solar Max—also built around the MMS bus, as well as the Tropical Rain Measurement Mission and the Earth Observation System, are in principle expendable but designed to be serviceable [4].

On the other hand DoD and SDIO users of on-orbit servicing have not emerged to the point of incorporation of servicing into their spacecraft design decisions, although a number of spacecraft programs are considered strong candidates. Commercial users are thus far only watching servicing activities. Commercial communications satellites were temporarily serviceable in the mid-1980s, when the Space Shuttle was still allowed to handle such birds, but only when something went wrong at an early stage of the mission. Palapa B2, Westar VI, and Syncom IV/Leasat 3 were all rescued from useless LEO orbits where deployment faults had stranded them. ELV upper stage failures usually leave satellites in a geostationary transfer orbit, where their velocity near perigee is too high to permit external intervention. However, the salvage of TDRS-1, GStar-3, and Hipparcos through successful maneuvering with positioning thrusters has shown that even these are not necessarily hopeless cases. Though satellite recovery is not strictly servicing, the cliffhanger retrieval of the Long Duration Exposure Facility shows that NASA has lost none of its skills in this area [4].

Insurance

Insurance coverage is available for two separate phases of a satellite's life: launch and on-orbit operation. A leading underwriting institution has stated it would look favorably upon designs for improved repairability and retrievability, but cannot require satellite manufacturers to incorporate these features. Should these capabilities be included, they could produce reductions in insurance premiums for both phases. This could be achieved through the moderation of loss costs associated with possible failures, i.e., reducing what would otherwise be a total loss for an unrepairable and therefore unusable satellite to a partial loss. This would be accomplished by restoring all or a portion of the capability of such a satellite through on-orbit repair or retrieval [3].

The Palapa and Westar retrievals and the SYNCOM IV-3 on-orbit repair demonstrated to the insurance industry that the capability to service spacecraft in low Earth orbit accessible by Shuttle has been established. Consequently, one underwriter, as a result of this experience, has established in its insurance coverages the following requirement—that the insured exercise all available means within reason to salvage a satellite that has either failed to reach orbit or that has failed in orbit. To the extent that Shuttle-based servicing operations are available, the insured could have an increased ability to meet this requirement where satellites are stranded in low Earth orbit, or are capable of being de-orbited to low Earth orbit from a higher orbit. There is no current impact on insurance in the case of geosynchronous satellites or satellites unable to be placed in a Shuttle-accessible orbit since transportation for servicing is not yet available.

Proposed serviceable commercial space platforms could greatly impact future insurance coverages. This coverage is unique and subject, moreover, to negotiation between the parties. Certain basic coverages specifically tailored for these platforms are envisioned; for example, property coverage for asset values, liability coverage for damage to third parties, and liability coverage for product malfunctions. These insurance coverages will be related to the types of servicing to be conducted; therefore, their requirements will be specifically tailored to the ultimate use of the facility.

Insurance underwriters, when establishing rates, do analyze the partial loss versus total loss components. This information is calculated in terms of loss failure probabilities and in terms of monetary impact associated with projected failure scenarios. The addition of servicing enhancements can work toward reducing failure probabilities and the associated costs for failures. This would impact both launch insurance coverages to the extent those coverages included satellite initial operations and on-orbit coverages to the extent the satellites could be accessed.

The insurance issue, it should be noted, does not impact U.S. government spacecraft since all government payloads are self-insuring. Still, the potential for total loss to the government would be reduced if the spacecraft were serviceable in the event of failure [3].

Pricing

In August 1986, President Reagan directed that NASA shall no longer provide launch services for commercial and foreign payloads unless those payloads have unique, specific reasons to be launched aboard Shuttle. Accordingly, the National Space Transportation System will launch only payloads that are Shuttle unique or have national security or foreign policy implications [3].

NASA may still launch those payloads which conform to the above noted 1986 decision, including the provision of Shuttle-unique services for on-orbit spacecraft. Therefore, a pricing policy for both launching and servicing of commercial and foreign payloads is still a requirement.

NASA's experience to date in pricing satellite servicing for commercial customers consists of the Palapa B-2 and Westar VI retrievals on STS 51-A; the attempted repair of SYNCO IV-3 on STS 51-D and its subsequent successful repair on STS 51-I; and the retrieval and relaunch of Intelsat 6/F3 on the 47th flight of the Space Shuttle. These satellites had experienced upper stage failures immediately following their deployment from the Shuttle. The NASA charge for Palapa and Westar retrievals totaled $5.5M, the charge for SYNCOM repair totaled $8.5M, and Intelsat paid NASA $94M for servicing and relaunching the Intelsat 6/F3. At the time that services were priced for the Palapa, Westar, SYNCOM, and Intelsat satellites, NASA was strongly interested in demonstrating its Shuttle-based servicing capabilities. The servicing charges were determined on an additive cost basis. The costs which were included were those associated with mission planning, development of unique hardware, integration, training, and revisit/retrieval.

A proposed satellite servicing pricing policy presented in the NASA Report to Congress, December 1986, entitled "On-Orbit Service, Repair, and Recovery of Spacecraft Report," is based on pre-STS 51L data. This policy contains three costing elements: transportation, a tailored package of services, and additional optional services. This pricing policy is based on full cost recovery and included pro rate costs of transportation, the costs of using any servicing tools and capabilities (such as EVA, MMUs, OMVs, etc.), and the costs of any nonstandard optional services, including the full, rather than additive, costs of any mission-unique hardware. Satellite servicing missions priced on a full cost recovery basis will result in prices exceeding those charged for the prototype repair missions on STS 51-A, D, and I.

Previous pricing algorithms for Shuttle services, including satellite servicing, used factors which have changed significantly over the past several years. The two primary factors which will affect future Shuttle pricing are (1) a projected Shuttle flight rate which is reduced from earlier projections and (2) a recently announced and significant increase in downweight capability of the Shuttle Orbiter. Because Shuttle flight cost is a major element in pricing a retrieval mission, the current retrieval/revisit pricing policy is being reformulated to reflect these changes.

Policy has not yet been established for pricing of Space Station supplied servicing. It is NASA's intent, in accordance with the civil space policy, to seek full cost recovery for pricing of Shuttle servicing [3].

Outlook—A Summary of the Near Future

NASA has embraced satellite servicing. Various NASA spacecraft program offices have concluded that serviceable satellites would reduce the cost of in-space science. Serviceability means modularity and standardization. NASA has, therefore, with its contractors, dictated, for economic reasons, that their spacecraft and, indeed, also the Space Station Freedom, be designed with a certain amount of standard system level modules. NASA has also ordered that satellite servicing capability be incorporated into Phase II of the Space Station. Twelve NASA programs have used, use, or plan to use some modular system design. The Space Station Polar Orbiting Platform and the planned Multimission Modular Spacecraft based Explorer are totally modular, including modular removable experiment attachments. As a modular satellite becomes larger, the integration and test cost savings increase and the delta weight for modularity becomes less. At all satellite sizes, the potential increases for design cost savings by the use of standard system modules, while the inherent serviceability of modular design keeps open the use of contingency on-orbit servicing for premature failures.

Thoughts for the Immediate Future

On-orbit satellite servicing of NASA space systems will increase when science mission life requirements are extended beyond 8 to 10 years—the critical lifetime elements of the spacecraft will probably be the experiment/instrument modules. A wider range of programs, for instance communication and meteorology satellites, will consider servicing because remote servicing capabilities will be developed, demonstrated, and declared operational. Remote operations will extend the range of satellite servicing from low altitude and inclination Earth orbits with Space Shuttle manned EVA operations to orbits anywhere (polar, GEO, other) using expendable launch vehicles and OMVs with satellite servicer kits attached. The servicer kits will contain RMS and FTS derivatives for telerobotic operations.

In the far term, remote servicing technologies will play a big role on the lunar surface, in Mars exploration, and in planetary missions of the future.

The Air Force, SDIO, and commercial space operators are watching NASA developments in orbital servicing and conducting analyses of if, how, and when to implement servicing into their block changes or next generation space programs. Currently the Air Force is studying a low cost, mini-refueler vehicle while assessing how their extensive logistics capabilities might be used in servicing their spacecraft.

Goals and Objectives

Satellite servicing is a technologically evolving activity which has not yet attained final stages of development. Progressing from a contingency reaction to a baselined activity within many user programs, the intent of servicing is to extend operational life, enhance capabilities, and decrease system life-cycle costs. Servicing is currently constrained to Shuttle accessible orbits, but it will evolve to include Space Station based activities and remote operations with robots in support of permanent long-term operations in space. Developments are proceeding rapidly in the international arena; for example, (1) the Federal Republic of Germany will demonstrate robotic operations on an upcoming Spacelab flight; (2) Canada is developing the technology for a mobile servicing center on the Station; and (3) Japan is developing the technology for remote servicing from expendable launch vehicles.

In keeping with the information noted above, NASA proposes that the development of appropriate satellite servicing capabilities to enhance and protect national capital investments in space systems be considered and subsequently adopted as a national goal. With this in mind, NASA has presented the following agency objectives for utilizing satellite servicing [3]:

1. Develop the technology, hardware, tools, facilities, and infrastructure to meet projected NASA servicing requirements.

2. Continue to strive toward efficiency (cost-effectiveness) in on-orbit servicing support through the development of general purpose tools and common systems and subsystems.

3. Continue to evolve on-orbit servicing capabilities, including development of telerobotic and robotic servicing systems, to support servicing at remote sites.

4. Support servicing needs of DoD, domestic, commercial, and foreign space communities upon appropriate request.

5. Promote the spinoff of commercial services.

6. Develop and promote the use of servicing interface standards.
7. Stimulate the technology base in universities and industry.
8. Utilize commercially available on-orbit servicing to the fullest extent feasible, and avoid actions that may preclude or deter commercial space sector activities except as required by national security or public safety.

Satellite servicing should be a national level program. Acquisition costs will be high, which mandates the broadest possible user base and maximum possible standardization. These attributes are best achieved with national commitment and direction. The sponsoring organizations (USAF, SDIO, and NASA) should develop a unified policy and approach to servicing acquisition planning. There does not appear to be any fundamental technological roadblock to on-orbit assembly, maintenance, and servicing implementation. The driving technologies have been identified.

Satellite Servicing—What Is Needed

In order to attain a national operational capability to conduct servicing missions in the 1990s, the following actions are considered to be most pressing and absolutely necessary:

1. Development of baseline generic equipment and tools for EVA, robotic, and combination servicing functions.
2. Agreement by DoD, NASA, SDIO, and aerospace contractors to adhere to a set of standardized servicing interfaces.
3. Establishment of firm user service costing criteria, guidelines, and price lists for various servicing tasks.
4. Development of a servicing infrastructure that allows servicing functions to be performed at all orbital locations. The architecture of the infrastructure must include:
   — Low cost transportation to and from space, as well as in-space orbit-to-orbit transportation
   — Orbital replacement unit carriers
   — Robotic servicer kits attached to OMVs
   — Space-based support platforms that can be mantended
   — Refueling storage and tanker systems for satellite fluid resupply
   — Ground control centers dedicated to the command, control, and communication requirements of servicing operations in space
   — A servicing logistics system tuned to servicing mission operations
   — A trained cadre of space workers to carry out EVA and IVA servicing tasks

Conclusions and Summary

Space missions, whether manned or unmanned, are fundamentally difficult. Most demand large-scale undertakings that depend upon some of the world's most advanced technology. The Saturn V rocket required the integration of some 6 million components manufactured by thousands of separate contractors. Voyager 2 arrived at Neptune a mere 1 second behind its final updated schedule after a 12 year, (4.4 billion mile) flight, approaching within 5,559 km (3,000 nmi) of the planet's surface. The information to be gathered by the Earth Observing System could approach 10 trillion bits of information—about one Library of Congress—*per day*. The matter of human frailty is perhaps of even greater importance in the case of the Apollo program, some 400,000 people at some 20,000 locations were involved in its design, test, and operation [5].

Given the high cost of space operations, in both financial and human terms, and their profound impact on the prestige of the sponsoring country, no goal short of perfection is acceptable. Critical hardware on space systems can be very unforgiving on any form of simple or compound malfunction and certainly on outright failure. Satellites experiencing flaws after being placed in orbit are not readily "recalled" to the factory for repair or modification. Because of the intense interest in, and scrutiny of, the commendably open and visible civil space program of the United States, this type of problem evokes heavy public criticism directed at the government agency responsible for its development.

But now we have the option of on-orbit satellite servicing to save or extend the life of wornout or failed satellites, if the satellites' built-in systems for self-healing are not adequate to correct the flow. Satellite on-orbit servicing offers to the program office an alternative to satellite replacement. With servicing at planned intervals, a satellite program could also achieve potentially longer useful operational life, more mission flexibility, and lower life-cycle costs.

Three things are central to building a solid roadway for on-orbit servicing of space systems. First, the ability to access the space system with a servicing capability; second, the ability of space systems to be serviced; and third, the ability to create a National Servicing Management Policy and Plan which accommodates the interests and requirements of NASA, DoD, and commercial organizations. It is reasonable to assume we can get to any spacecraft which we originally put into orbit. Therefore, the real issue is being able to perform servicing functions on the orbiting spacecraft once rendezvous and docking have been accomplished. To assist program managers and planners in their cost/risk analysis as input to a formalized decision process to incorporate orbital servicing into their space system program, a comprehensive NASA/DoD national servicing capability should be acquired. Once a U.S. servicing capability is initially established, it should be incrementally expanded to include international partners as both operators of the servicing infrastructure and users of its functions.

This entire book, but specifically Chapter 10, has addressed the issues that form the bedrock of a roadway to satellite servicing. A checklist type of technical, programmatic, and economic milestones, related to successful orbital servicing, has been offered. To minimize the impact to the space system development program, decisions must be made at an early point in the conceptual design to incorporate, or not, serviceability design features into the system. A minimum of 24 months prior to system preliminary design review (PDR) is required to effectively assess the required design considerations for servicing infrastructure and transportation compatibility. Additionally, a national baseline servicing infrastructure program must be implemented as a separate entity to ensure its availability to support the space programs when they become space operational.

Finally this thought—to achieve an international orbital servicing and support capability, many key events are needed. As indicated in this book, on-orbit servicing supportability design considerations and guidelines must be created, documented, and updated such that the designs of both the "servicer" and "servicee" systems can evolve in concert with each other. However, prior to these efforts, a joint NASA/DoD decision on whether or not to fully support their space assets on-orbit is required.

With that decision, other key events will take place.

References

1. *Space Assembly, Maintenance and Servicing Study (SAMS)* Final Report, Volume I, Executive Summary, TRW No. SAMSS-196, Volume II, System Analysis, TRW No. SAMSS-195, Volume III, Design Concepts, TRW No. SAMSS-197, Volume IV, Concept Development Plan, TRW No. SAMSS-198, Volume V, Neutral Buoyancy Simulation, TRW No. SAMSS 199. TRW, Redondo Beach, CA, July 6, 1988.
   and
   Space Assembly, Maintenance, And Servicing Study (SAMS) Final Report, Volume I, Executive Summary, Volume II, System Analysis, Volume III, Design Concepts, Volume IV, Concept Development Plan, Volume V, Simulation Report. Lockheed Missiles and Space Company, Inc., Sunnyvale, CA, July 6, 1988.
2. Levin, George M. *A Draft Strategic Plan for NASA Satellite Servicing.* Satellite Servicing Workshop IV, Session IV: Future Opportunities, NASA/JSC, June 21–23, 1989.
3. *Satellite Servicing—a NASA Report to Congress.* NASA Office of Space Flight, Washington, DC, March 1, 1988.
4. Chenard, Stephane. *Can Satellite Servicing Pay?* Space Markets Publication, 1/1990.
5. *Summary and Principal Recommendations of the Advisory Committee on the Future of U.S. Space Program.* (Norman Augustine Report) December 10, 1990.

Appendix A

Acronyms and Glossary

Acronyms

Every aerospace engineering discipline has its list of acronyms and abbreviations. Satellite servicing is no exception. Where practical, I have attempted to define the acronyms in the text. They are collected in this appendix as an aid to quick reference.

ACCESS	Assembly Concept for Construction of Erectable Space Structures	CRT	Cathod Ray Tube
		CRU	Contingency Replacement Unit
ACS	Attitude Control System	CSC	University of Colorado, Center for Space Construction
AI	Artificial Intelligence		
AIAA	American Institute of Aeronautics and Astronautics	CTV	Cargo Transfer Vehicle
		DARPA	Defense Advanced Research Projects Agency
ALS	Advanced Launch System		
Ao	Operational availability	DFI	Development Flight Instrumentation
AOS	Assembly and Overhaul Station	DITFAC	Development, Integration and Test Facility at NASA GSFC
ARC	Ames Research Center		
ASAT	Anti-Satellite	DMSP	Defense Meteorological Satellite Program
ASE	Aerospace Support Equipment		
ASME	American Society of Mechanical Engineers	DoD	Department of Defense
		DOF	Degrees of Freedom
ATE	Automatic Test Equipment	DRM	Design Reference Mission
AWACS	Advance Warning Aircraft Communication System	DSCS	Defense Space Communication System
		DSP	Defense Support Program
AXAF	Advance X-Ray Astrophysics Facility	EASE	Experimental Assembly of Structures in Extravehicular Activity
BITE	Built-In Test Equipment		
BSTS	Boost Surveillance and Tracking System	ECS	Environmental Control System
C^3	Communications, Command, Control	EDO	Extended Duration Orbiter
CA	Carrier Assembly	ELV	Expendable Launch Vehicle
CAD	Computer Aided Design	EMST	External Maintenance Solutions Team
CCD	Contract Control Documentation	EMTT	External Maintenance Task Team
C&DH	Communications and Data Handling	EMU	Extravehicular Maneuvering Unit
CDM	Command and Data Management	EOS	Earth Observing System
CELV	Commercial Expendable Launch Vehicles	EP-PED	Explorer Platform
		EPOP	European Polar Orbiting Platforms
CER	Cost Estimating Ratio	ESA	European Space Agency
CETA	Crew Equipment Translation Aid	ESC	Engineering Support Center at NASA GSFC
CI	Controlled Interface		
CITE	Cargo Integrated Test Equipment	ESGP	Earth Survey Science Platform
CMG	Control Moment Gyro	ETS	Engineering Test System at NASA GSFC
C/O	Checkout/changeout		
COSEMS	Comprehensive Operational Support Evaluation Model for Space	EURECA	European Recoverable Carrier
		EUVE	Extreme Ultraviolet Explorer

263

EVA	Extravehicular Activity	LDR	Large Deployable Reflector
FEL	First Element Launch	LDS	Lasar Docking Sensor
FLTSATCOM	Fleet Satellite Communications System	LEO	Low Earth Orbit
FOC	Final Operational Capability	LHTP	Liquid Helium Transfer Project
FSC	Full Support Capability	LMSC	Lockheed Missiles & Space Company
FSC	Fairchild Space Company	LOS	Line of Sight
FSED	Flight System Engineering Development	LTV	LTV Aerospace and Defense Company
FSS	Flight Support System	M&S	Maintenance and Servicing
FTS	Flight Telerobotic Servicer	MDF	Manipulator Development Facility
GBL	Ground Based Laser	MDSSC	McDonnell Douglas Space Systems Company
GCC	Ground Control Console	MEB	Main Electronics Box
GEO	Geosynchronous Earth Orbit	MEO	Medium Earth Orbit
GNC	Guidance Navigation Control	MILSTAR	Military, Strategic, Tactical, and Relay
GPS	NAVSTAR Global Positioning System	MLI	Multi-Layer Insulation
GRO	Gamma Ray Observatory	MMC	Martin Marietta Corporation
GSE	Ground Support Equipment	MMD	Mean Mission Duration
GSFC	Goddard Space Flight Center	MMS	Multimission Modular Spacecraft
GSTDN	Ground Space Tracking and Data Network	MMU	Manned Maneuvering Unit
GTO	Geosynchronous Transfer Orbit	MOA	Memorandum of Agreement
H/T	Hardware/tools	MOTV	Manned Orbital Transfer Vehicle
HEO	High Earth Orbit	MPESS	Mission Peculiar Equipment Support Structure
HEUS	High Energy Upper Stage	MPS	Materials Processing in Space
HGA	High Gain Antenna	MRS	Module Retention System
HGAS	High Gain Antenna System	MSC	Mobile Servicing Center
HLLV	Heavy Lift Launch Vehicle	MSFC	Marshall Space Flight Center
HOSC	Huntsville Operations Support Center	MSS	Mobile Servicing System
HR	House Rule	MST	Module Support Tool
HRM	High Resolution Mirror	MTC	Man-Tended Capability
HRMA	High Resolution Mirror Assembly	MTFF	Man-Tended Free Flyer
HST	Hubble Space Telescope	NASA	National Aeronautics and Space Administration
ICD	Interface Control Document	NASDA	Japanese Civilian Space Agency
IEEE	Institute of Electrical and Electronic Engineers	NASP	National Aero-Space Plane
IOC	Initial Operational Capability	NAV	Navigation
IOSS	Integrated Orbital Servicing System	NLS/SSTO	National Launch System/Single Stage to Orbit
IR&D	Independent Research and Development	NOAA	National Oceanic and Atmospheric Administration—Dept. of Commerce
IRAS	Infrared Astronomical Satellite	NPB	Neutral Particle Beam
ISAS	In Space Assembly and Servicing Facility	NPOP	NASA Polar Orbiting Platforms
ISC	Initial Support Capability	NROSS	Navy Remote Ocean Surveillance System
ISF	Industrial Space Facility	NSTS	National Space Transportation System
ISO	International Standards Organization	O&S	Operations & Support
IUS	Inertial Upper Stage	OAO	Orbital Astronomy Observatory
IVA	Intravehicular Activity	OBC	On-board Computer
JEM	Japanese Experimental Module	OCC	Operations Control Center
JPL	Jet Propulsion Laboratory of California Institute of Technology	OMV	Orbital Maneuvering Vehicle
JPOP	Japanese Polar Orbiting Platform	OPC	Open Cherry Picker
JSC	Johnson Space Center	OPF	Orbiter Processing Facility
KEW	Kinetic Energy Weapon Spacecraft	ORS	Orbital Refueling System
KSC	Kennedy Space Center	ORU	Orbit Replacement Unit
LaRS	Langley Research Center	OSCRS	Orbital Spacecraft Consumables Resupply System
LASS	Large Amplitude Space Simulator		
LCC	Life Cycle Cost		
LDEF	Long Duration Exposure Facility		

Appendix A

OSF	Office of Space Flight, NASA Headquarters
OSL	Orbital Solar Laboratory
OTS	Orbital Telescope System
OTV	Orbital Transfer Vehicle
PBS	Payload Berthing System
PCAD	Pointing Control and Aspect Determination
PDR	Preliminary Design Review
PEEK	Alumina/Poly-ether-ether-ketone
PLSS	Primary Life Support System
PM	Propulsion Module
PMC	Permanently Manned Capability
POC	Proof of Concept
POCC	Payload Operational Control Center
POP	Polar Orbiting Platforms
PPD	Propulsion and Power Division at NASA JSC
PRC	Planning Research Corporation
PROD.DEV.	Product Development
R&D	Research and Development
RCS	Reaction Control System
RCTU	Remote Command and Telemetry Units
RDV	Rendezvous/Docking Vehicle
RF	Radio Frequency
RGDM	RMS Grapple Docking Mechanism
RMS	Remote Manipulator System
RMU	Remote Maneuvering Unit
ROTV	Reuseable Orbital Transfer Vehicle
RSSS	Reconfigurable Satellite Servicing System
S/C	Spacecraft
SA	Space Assembly
SAMS	Space Assembly, Maintenance, and Servicing
SAMSS	Space Assembly, Maintenance & Servicing Study
SAR	Space Antenna Radar
SAS	Solar Array System
SASS	Space Assist Support System
SASWG	Space Assembly and Servicing Working Group
SBI	Space Based Interceptor
SBL	Space Based Laser
SBR	Space Based Radar
SBSP	Space Based Support Platform
SDI	Strategic Defense Initiative
SDIO	Strategic Defense Initiative Organization
SDS	Strategic Defense System
SEA	Servicer Equipment Assembly
SHOOT	Superfluid Helium On-Orbit Transfer
SIC	Standard Interface Connector
SIRTF	Space Infrared Telescope Facility
SLC	Submarine Laser Communications Satellite
SM	Support Module
SMM	Solar Maximum Mission
SMRM	Solar Maximum Repair Mission
SPA	Servicer Position Arm
SPO	System Program Office
SRMS	Shuttle Remote Manipulator System
SRV	Short Range Vehicle
SS	Space Station
SSC	Science Support Center
SSCC	Space Station Control Center at NASA JSC
SSDIF	Spacecraft Systems Development and Integration Facility at NASA GSFC
SSE	Servicing Support Equipment
SSF	Space Station Freedom
SSS	Satellite Servicer System
SSS/FD	Satellite Servicer System Flight Demonstration
SSTS	Space Surveillance and Tracking System
STD	Standard
STAS	Space Transportation Architecture Study
STS	Space Transportation System (Space Shuttle)
TAGS	Test and Graphics System
TDM	Technology Development Missions
TDRSS	Tracking and Data Relay Satellite System
THURIS	The Human Role In Space
TMG	Thermal/Micrometeroid Garment
TOPEX	Ocean Topography Experiment
TPC	Total Program Cost
TPDM	Three Point Docking Mechanism
TRMM	Tropical Rainfall Measurement Mission
TT&C	Telemetry Tracking and Control
TV	Television
UARS	Upper Atmospheric Research Satellite
UHF	Ultrahigh Frequency
USAF	United States Air Force
USCM	Unmanned Spacecraft Cost Model
UST	Universal Servicer Tool
VAFD	Vandenberg Air Force Base
VTE	Variable Thrust Engines
WBS	Work Breakdown Structure
WETF	Weightless Environment Test Facility
WF/PC	Wide Field/Planetary Camera
WTR	Western Test Range
XPOP	Extra Polar Orbiting Platforms
XTE	X-Ray Timing Explorer
ZPB	Zero Pre-Breathe
ZS	Zenith Star

Glossary

Certain words or phrases in the satellite servicing discipline require definition or explanation for better understanding of on-orbit servicing operations.

ARTIFICIAL INTELLIGENCE: That branch of computer science concerned with the design and implementation of programs which make complicated decisions, learn or become more adept at making decisions, interact with humans in a way natural to humans, and in general, behave in a manner typically considered the mark of intelligence.

AUGMENTED TELEOPERATOR: A teleoperator with sensing and computation capability that can carry out portions of a desired operation without requiring detailed operator control.

AUTOMATION: Automation is the use of machines to effect initiation, control, modification, or termination of system/subsystem processes in a predefined or modeled set of circumstances. The implication is that little or no further human intervention is needed in performing the operation. The terms *hard automation* and *flexible automation* define subsets of automation.

AUTONOMY: The ability to function as an independent unit or element, over an extended period of time, performing a variety of actions necessary to achieve predesignated objectives, while responding to stimuli produced by integrally contained sensors.

BERTHING: The linkup of one orbiting object with another, wherein the closing energy is provided in a closely controlled fashion by an intermediate mechanism (e.g., RMS/MRMS) attached between the two.

CO-ORBITING VEHICLES: In regard to the Space Station, any vehicle with the same average orbital period, inclination, and node as the Station and which maintains its orbit track along the Station's orbit track.

CREW ACTIVITY PLAN (CAP): The schedule of activities to be performed by the various Station crewmembers. The CAP can be in either summary or detailed format, depending on operational requirements.

CUPOLA: An appendage located on a Space Station interconnect node, affording its occupants hemispherical viewing capability from a single vantage point.

DEPLOYABLE STRUCTURE: Space system structure that is delivered to orbit in collapsed or telescoped form. Generally assumed to contain stored energy devices to allow extension or self-assembly with a minimum of crew intervention.

DOCKING: The linkup of one orbiting object to another, wherein their kinetic energy brings the objects into contact.

END-EFFECTOR (end of arm tooling): Tools such as drills, cutting edges, or screw drivers attached to the end of a reboot arm to do specific jobs.

ERECTABLE STRUCTURE: Space system structure that is delivered to orbit in kit form which must be manually assembled by the crew via RMS, MSC/MRMS, or EVA.

EVA: Extravehicular activity: Operations performed by crewmembers wearing space suits outside a habitable environment.

EXPERT SYSTEM: An expert or knowledge-based system is one that stores, processes, and utilizes a significant amount of information about a particular domain of knowledge to solve problems or answer questions pertaining to that domain. The system is able to perform at the level of an experienced human practitioner working in that domain of knowledge.

EXTRAVEHICULAR MOBILITY UNIT (EMU): The protective garment worn for EVA. Includes life-support and communications equipment but not the maneuvering unit.

FREE-FLYER: A spacecraft that may require servicing by the Station or an OMV but is not associated with one of the platforms. Free-flyers may have their own translation capability or require an OMV for orbit maneuvers.

FUNCTION: A separate and distinct action required to achieve a given objective to be accomplished by the use of hardware, computer programs, personnel, facilities, procedural data, or a combination thereof.

HOOKS AND SCARS: Design features that permit the future addition of software (hooks) and hardware (scars).

IN SITU: Literally means in place. Refers to performing maintenance and servicing of platforms or free-flyers on location in orbit, not at the Station. In situ operations require the delivery of maintenance or servicing elements via OMV or shuttle.

INTEGRATED LOGISTICS SYSTEM (ILS): A disciplined approach to the activities necessary to cause support consider-

Appendix A

ations to be integrated into system and equipment design; developed support requirements that are consistently related to design and to each other, acquire the required support, and provide the required support during the operational phase at minimum cost.

INTERFACE: The point or area where a relationship exists between two or more parts, systems, programs, persons, or procedures where physical and functional compatibility is required.

INTRAVEHICULAR ACTIVITY (IVA): Operations performed by crewmembers within a habitable environment.

MANIPULATOR: A mechanism, usually consisting of a series of segments, jointed or sliding relative to one another, for the purpose of grasping and moving objects usually in several degrees of freedom. It may be remotely controlled by a computer or by a human.

MANIPULATOR FOOT RESTRAINT (MFR): An end-effector adapter that allows an EVA crewmember to be attached to the end of the Station manipulator and to have both hands free for EVA tasks.

MANNED MANEUVERING UNIT (MMU): A multipurpose astronaut mobility unit allowing short-range free flights.

MOBILE SERVICE CENTER: The Station's mobile RMS facility, consisting of one or more Canadian-provided manipulator systems and a U.S.-provided transport mechanism.

OPEN LOOK CONTROL METHOD: One joint (internal state information is used). Control achieved by driving control actuators with a sequence of preprogrammed signals without measuring actual system response and closing the feedback loop.

ORBITAL MANEUVERING VEHICLE (OMV): A potential propulsion stage (reusable) capable of transporting payloads between low Earth orbits and performing a number of different missions.

ORBITAL REPLACEMENT UNIT (ORU): Any item the replacement of which constitutes organizational maintenance.

ORBITAL TRANSFER VEHICLE (OTV): A potential propulsion stage (reusable) that will be capable of transporting payloads from low Earth orbits to higher energy orbits.

PROXIMITY OPERATIONS: The operation of one spacecraft in the vicinity of another with the relative positions stabilized and the rate small enough to preclude the requirements for re-rendezvous.

REMOTE MANIPULATOR SYSTEM (RMS): A 16.8 meter (55 foot) teleoperated arm on the STS for spacecraft deployment and proximity manipulation.

ROBOT: A device which is capable of autonomous, unsupervised operation; usually applied to systems capable of responding constructively to outside stimulus while carrying out rather complex operations within fairly broad, predefined limits. May, but will not necessarily be anthropomorphic. (Please note that devices such as the remote manipulator system of the Space Shuttle Orbiter are not robots!)

SAFE HAVEN: A volume or number of dedicated volumes or functional capabilities within the envelope of the Space Station that can provide a safe, protected environment for members of the crew in the possible event of anomaly situations.

SPACE ASSEMBLY: The on-orbit joining, construction, or fabrication of spacecraft, space systems, or structures to include the deployment of solar arrays, antennas, and other appendages into their operational configurations. On-orbit assembly occurs before a space system becomes operational.

SPACECRAFT/SATELLITE: The combination of mission specific equipment and space vehicle bus subsystems and support hardware which collectively are capable of autonomous operation in space.

SPACE MAINTENANCE: Any on-orbit activity performed for the purpose of extending the operation use or life of a space system, but not including consumables replenishment. On-orbit maintenance is performed after a space system is operational. Preventive space maintenance can include observation, inspection, surface restoration, realignment, recalibration, test, and checkout. Corrective space maintenance includes all actions performed as a result of a system failure to restore that system to a specified condition such as repair of faults or components and replacement of failed modules or subassemblies.

SPACE SERVICING: The on-orbit resupply and replenishment of consumables and expendables to keep a space system in operating condition, but not including preventative or corrective maintenance tasks. A servicing function is not considered a failure correction even when equipment shutdown occurs because a supply of consumables is exhausted. On-orbit servicing is performed after a space system is operational, although hardware developed for servicing may be used to load initial supplies of consumables and expendables during the process of space assembly. Space servicing may involve direct transfer of fluids including both liquids and pressurized gases, exchange of fluid tankage, and exchange of packaged consumables such as batteries or film.

SUPERVISORY CONTROL: The control of the robot system including both internal sections and synchronization with the external environment with an operator or possibly a higher level computer. This enables a higher level of robot decision making.

SUPPORT EQUIPMENT (SE): Items required to maintain equipment and systems in effective operating condition under various environments (ground and orbit). Support equipment includes general and special purpose vehicles, power units, stands, test equipment, tools, and test benches needed to facilitate or sustain maintenance actions, detect or diagnose malfunctions, or monitor the operational status of equipment and systems.

TELEOPERATION ("REMOTE OPERATION"): Use of remotely controlled sensors and actuators allowing a human to operate equipment even though the human presence is removed from the work site. Refers to controlling the motion of a complex piece of equipment such as a mechanical arm, rather than simply turning a device on or off from a distance. The human is provided with some information feedback (visual display or voice) that enables him to safely and effectively operate the equipment by remote control.

TELEOPERATOR: A device which performs functions such as mechanical manipulations or signal detection (e.g., removing a bolt or performing an IR scan) under the direct, manual control of a human operator while that human operator is physically displaced from the location of the teleoperator. That displacement may be only a matter of a few meters or many millions of kilometers. The essential criterion is that the operator has no direct mechanical connection to the teleoperator so that the responses of the end effector on the teleoperator are induced by signals evoked by the operator. (Please note that the remote manipulator system is a teleoperator!)

TELEPRESENCE ("REMOTE PRESENCE"): The ability to transfer a human's sensory perceptions, e.g., visual, tactile, to a remote site for the purpose of improved teleoperation performance. At the work site, the manipulators have the dexterity to allow the operator to perform normal human functions. At the control station, the operator receives sufficient quantity and quality of sensory feedback to provide a feeling of actual presence at the worksite.

Appendix B

SAMS Study Summary

As stated in Chapter 1, the Space Assembly, Maintenance, and Servicing (SAMS) study was performed in 1986–87 by two contractor teams headed by the TRW Space and Technology Group and the Lockheed Missiles and Space Company [1, 2]. The government customer agency for the two contractors was the Department of the Air Force, Headquarters Space Division, Air Force Systems Command, Los Angeles, California.

Sponsored jointly by the Air Force Space Division, the Strategic Defense Initiative Organization, and the National Aeronautics and Space Administration, the SAMS study addressed the requirements, concepts, and planning for full implementation of a U.S. space, maintenance, and servicing capability by the early 2000s. The study was conducted over the 16 month timeframe from March 1, 1986, through June 12, 1987. TRW was supported by the team member subcontractor work performed by McDonnell Douglas Astronautics Company, Grumman Space Systems, Booz Allen and Hamilton, and Advanced Technology, Inc. Lockheed also had a strong subcontractor team consisting of Boeing Aerospace, Honeywell Corporation, the Illinois Institute of Technology, Carnegie-Mellon University, and Life Support Systems, Inc.

Captain Joseph T. Wong, Air Force Space Division, was the SAMS study contract project officer. He was effectively assisted in study management by Mr. Rod Lochmann of The Aerospace Corporation and Mr. Robert Curtis of the Science Applications International Corporation. Mr. James H. Suttle of Tecolote Research, Inc. provided cost analysis information to the contractor teams and monitored their progress in generating life cycle cost benefits for selected servicing options and infrastructures.

The purpose of the SAMS study, which was a Phase I Definition effort, was to provide an alternative to satellite on-orbit replacement by using space assembly, maintenance, and servicing functions to achieve the mission lifetime, capability, and flexibility of space systems. Where SAMS was determined to be cost-effective, the study then defined and established servicing infrastructure capabilities to improve and enhance space systems affordability and operational responsiveness. Military, civilian, and commercial space programs were included in study tasks. SAMS study definitions are shown on Figure B.1.

Study Background, Tasks, and Flow

The SAMS study pursued the DoD and NASA objective of making efficient use of technologies to increase the mission effectiveness and to reduce the cost of space operations. To maximize this objective, SAMS capability should logically evolve from past and current baseline knowledge of spacecraft, transportation systems, servicing hardware and tools, mission performance, and the optimum mix of crew skills with telerobotic devices to a new, higher level, baseline in the mid 1990s. Figure B.2 depicts the background for a SAMS program.

Both TRW and the Lockheed Missiles and Space Company performed the study according to the same statement of work described by the task flow of Figures B.3, B.4, and B.5.

Design Reference Missions

The study sponsors, USAF/SDIO/NASA, provided the contractors with five design reference missions (DRMs) as a means of exercising the SAMS study process for realistic conditions. From these DRMs, integrated SAMS program requirements were generated and scenarios written for the spacecraft to be serviced, the hardware/tools to do the servicing tasks, and the space/ground infrastructure to support a SAMS program. The detailed scenarios enabled analysis of SAMS on a system basis by defining common equipment items and basic principals of on-orbit servicing operations.

SAMS Architecture

The study laid out a SAMS systems architecture as an integral part of an international space operations infrastructure. This architecture included:

1. Servicing facilities at Space Station Freedom
2. A reusable orbital transfer vehicle (ROTV), using cryogenic propellants

ASSEMBLY (OCCURS BEFORE A SPACE SYSTEM BECOME OPERATIONAL)
- CONSTRUCTION
- JOINING
- DEPLOYMENT
- MATING OF SPACECRAFT TO STAGES

MAINTENANCE (PERFORMED AFTER A SPACE SYSTEM IS OPERATIONAL)
- OBSERVATION
- INSPECTION AND ASSESSMENT
- SURFACE RESTORATION
- REALIGNMENT/RECALIBRATION
- REPAIR
- REPLACEMENT OF MODULES, PAYLOADS, SUB-ASSEMBLIES
- CONTAMINATION REMOVAL
- CHECK-OUT AND TEST
- PRE-PLANNED PRODUCT IMPROVEMENT (P^3I)

SERVICING (PERFORMED AFTER A SPACE SYSTEM IS OPERATIONAL)
- REPLENISHMENT OF FLUIDS AND PRESSURIZED GASES
- RESUPPLY OF EXPENDABLES SUCH AS FILM, TAPE, BATTERIES

OTHER
- DEBRIS CAPTURE/CONTAINMENT
- EMERGENCY OPERATIONS AND RESCUE
- RETRIEVAL/RETURN

Figure B.1 SAMS study definitions.

3. A facility for the on-orbit storage and handling of cryogenic propellants
4. A remotely piloted maneuvering vehicle (OMV), which can carry a servicing front end and appropriate spare modules for the serviced satellite
5. A propellant transfer system (OSCRS), which can service satellites with storable propellants (hydrazine and/or bipropellants)
6. A teleoperated satellite servicer system (SSS), with dual servicing arms and stowage for fuel for ORUs replacement, adaptable to the OMV or ROTV
7. For later missions, a manned orbital transfer module (MOTM), which can be carried to a remote servicing location by the OMV

For deployment of SDI systems such as SSTS, or midlife servicing of such a system, the architecture must include:

1. An assembly and overhaul station (AOS) in the same plane as the system to be deployed, with the capability to support a sizeable crew (up to 12) for a period of weeks
2. Co-orbital warehouses with OMV/SSS combination to service high altitude platforms between overhauls
3. A propellant carrier capable of mating to the AOS and carrying sufficient propellants to service multiple SDI spacecraft
4. A reactor kick stage capable of placing a 3,262 kg (6,000 lb) reactor into a higher 18,530 km (10,000 nmi) storage orbit when the SDI spacecraft are overhauled

Depending on the nature of the mission model and the levels of activity, there may be further need for:

1. Man-tended platforms in polar or geosynchronous orbit to perform short duration servicing or assembly missions
2. A high energy upper stage (HEUS) for the heavy lift launch vehicle (HLLV)/advanced launch system (ALS) (Required for direct emplacement of the SDI satellites and for manned or planetary/science missions.)

Study Merits

Here, in summary, are the standout features of the SAMS study. They represent an integrated consolidation of the efforts of both contractor teams.

SAMS Requirements—The study examined mission, system, and spacecraft requirements across a broad spectrum of

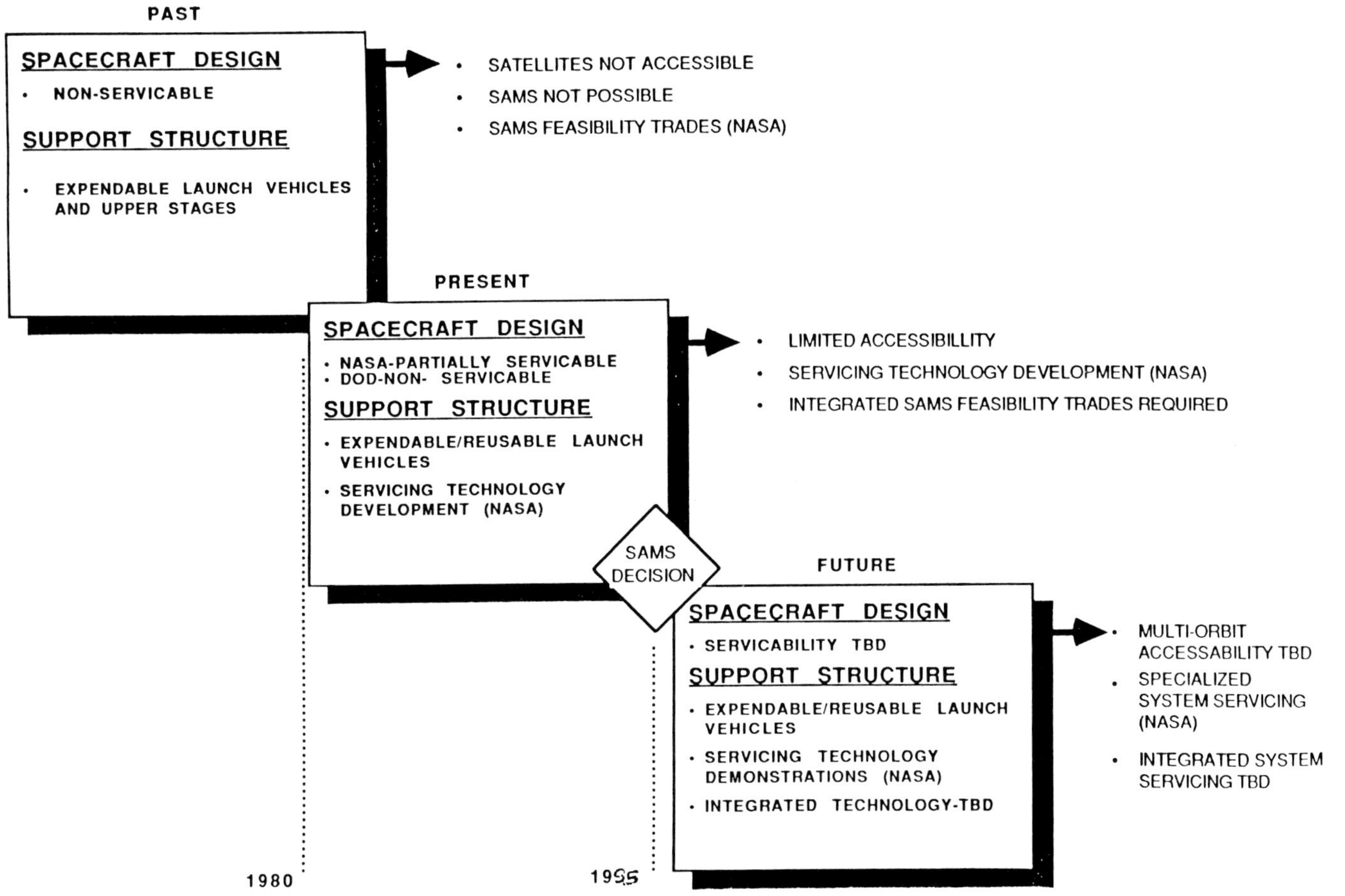

Appendix B

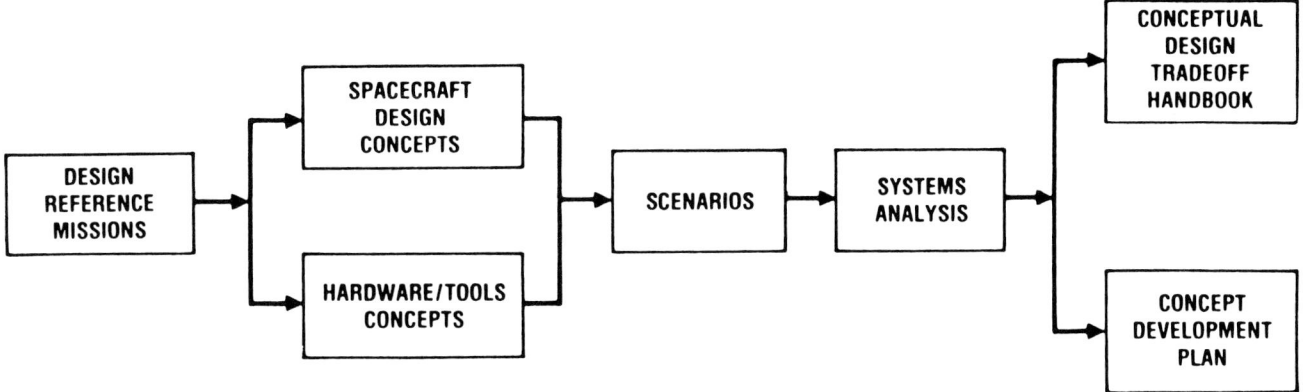

Figure B.2 Background for on-orbit satellite assembly, maintenance, and servicing operations.

Figure B.3 SAMS study task flow.

DoD, SDI, NASA, and commercial on-orbit servicing missions. The DRM study structure and methodology allowed the identification of satellite system design and cost drivers.

Spacecraft Design—Spacecraft design concepts were built on current and past NASA efforts such as the HST, OMV, and AXAF. Emphasis was placed on spacecraft modularity, standardization, and design commonality as these items would be impacted by satellite servicing. An attempt was made to adopt as many industry standards as possible. Reference 3, which is in the form of a designer's handbook, contains a good summary of SAMS spacecraft design concepts.

Hardware/Tools—Reference 4 was used to provide list of

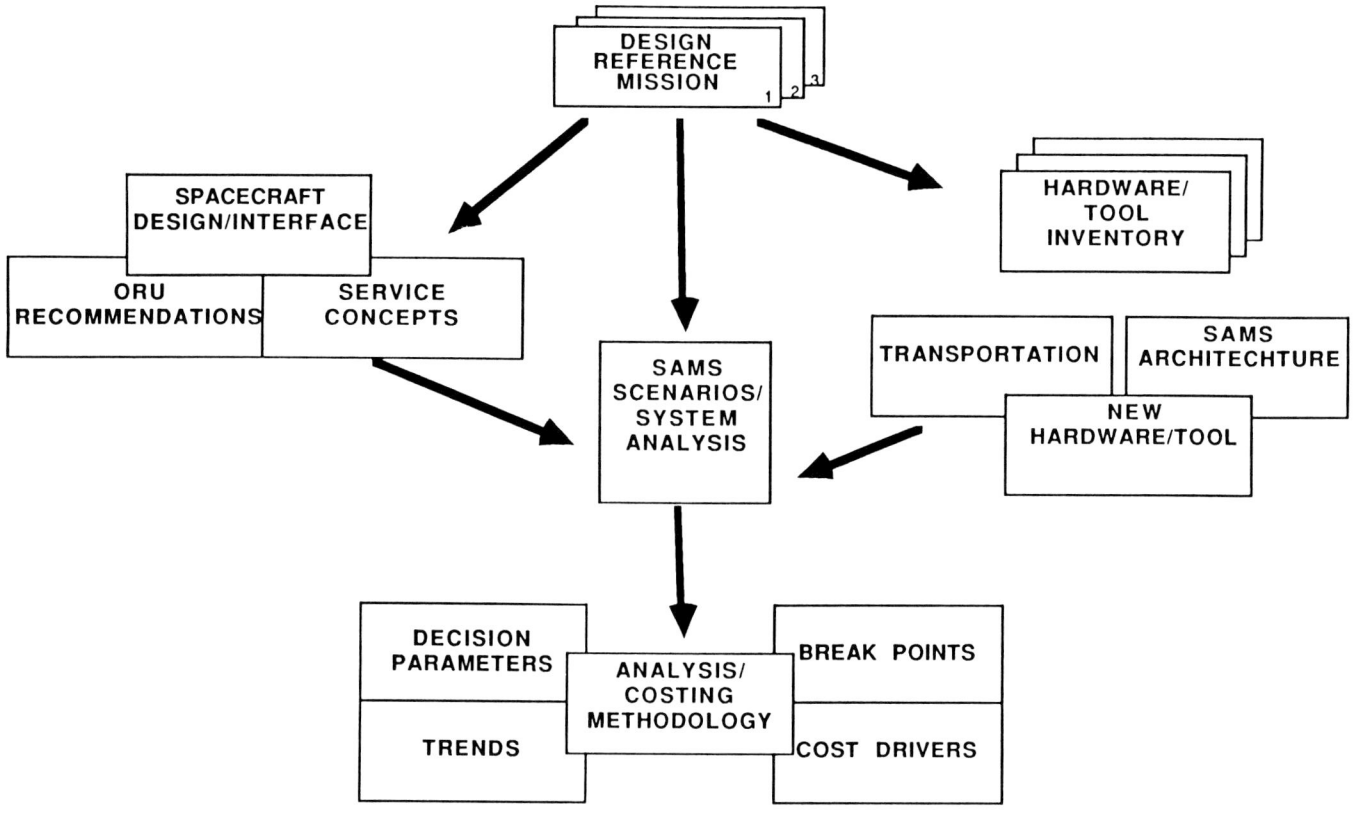

Figure B.4 SAMS study interrelationships.

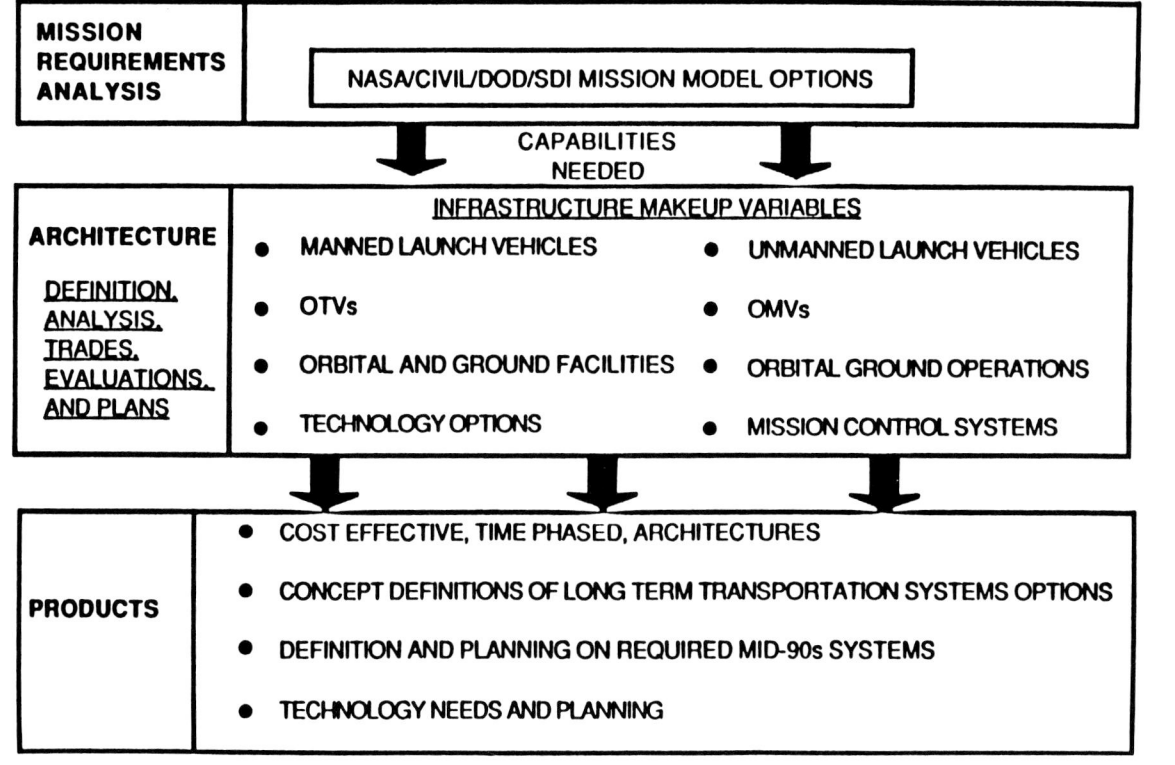

Figure B.5 SAMS study requirements, arcitecture, and products.

Appendix B

AREA	FOCAL POINT
REQUIREMENTS	ROD LOCHMANN, AEROSPACE CORP.
S/C DESIGN	LT COL CHARLES BROWN, AFSC/SD (CWX)
HARDWARE/TOOLS	GORDON RYSAVY, NASA/JSC (EX2)
GOV'T ASTRONAUT REVIEW BOARD	LT COL JERRY ROSS, NASA/JSC (CB)
SAMS SCENARIOS/ SYSTEM ANALYSIS	MAJ LOUISE JACKSON, AFSC/SD (XR) / DR. N. MARZWELL / NASA/JPL
LOGISTICS SUPPORTABILITY	COL JAMES GRAHAM, SDIO/SY / MAJ NEAL ELY, AFSC/SD/ (ALI)
TECHNOLOGY	GEORGE LEVIN, HQ NASA/MT
STANDARDIZATION	LT COL GEORGE SAWAYA AFSC/SD (ALI)

Figure B.6 SAMS study advisory panel.

current hardware and tools planned for HST and AXAF and manifested on current Shuttle flights. Unique servicing tools were designed for a broad spectrum of missions. Satellite servicer system concepts were suggested using the maximum amount of current NASA technology.

Mission Scenarios—SAMS study activities identified sequences and established timelines for both manned and robotic servicing missions. Top level mission equipment was specified for specific scenarios. From the scenarios, servicing architecture was turned into operational infrastructures and commonality of elements identified.

Systems Analysis—The study established a methodology to assess SAMS cost-effectiveness for satellite systems. Attempts were made to show where satellite systems could benefit by servicing as an alternative to replacement. SAMS cost trends, breakpoints and cost drivers were established and potential benefits, both quantitative and qualitative, identified for specific programs within the design reference missions.

Concept Development Plan—A roadmap and development plan was generated for selected SAMS concepts. A prioritized development schedule was generated for both NASA and military programs, and required technologies were synthesized from requirements and scenarios. The technologies were divided into four categories:

1. Man-in-Space—The study identified the rapidly evolving technologies critical to the use of the crew. Examples are the higher pressure space suit; helmet-mounted, voice-controlled alphanumeric and graphic display systems; improved, higher pressure gloves; and a non-outgassing, regenerable, portable life support system for the new suit. A new conformal airlock which will almost eliminate loss of pressurizing gas and allow very rapid transits to and from vacuum was also recommended for development. These advances will support broader options for use of EVA in conjunction with or as a backup or substitute for robotics and teleoperation until that technology catches up.

2. Spacecraft Design—Examples of how SAMS has identified currently available as well as needed spacecraft design improvements were documented in the scenarios for the DRMs. The key point is that these changes (such as standardization, modularization) are needed and will pay for themselves even if there is no SAMS implementation. That may be the single most important new information to come out of the study, because it can lead to creation of a SAMS-friendly

environment that need not be charged to SAMS. That environment will include the SPOs, the design offices of their contractors, and ultimately, the space architecture that they create.

3. SAMS Equipment—The study showed what equipment is needed in order to implement SAMS. It also showed that, perhaps surprisingly, much of that equipment has already been developed and used successfully in actual SAMS-type missions in space. Perhaps the most important item of equipment identified in the study was the satellite servicer system.

4. Mission Equipment—The DRM scenarios, which include pre- and post-flight processing, identified the associated equipment. Very little new ground equipment chargeable to SAMS is needed, other than what would be required for any new space system.

SAMS Advisory Panel

In order to provide the government sponsoring agency managers and the contractor teams with an overview of study progress and products, an advisory panel was formed, Figure B.6. This panel met from time to time with the contractors and study managers to review the work and plan future SAMS activities.

Related Studies

In addition to the hardware development spacecraft programs related to satellite servicing that were ongoing such as the HST, OMV, GRO, AXAF, and SSF, a number of definition studies were used for input information to the SAMS study. These are:

1. Space Transportation Architecture Study, sponsored by AFSC/SD (CFP)
2. Standardization Study, sponsored by AFSC/SD (YO)
3. On-Orbit Maintenance Repair Study (Policy), sponsored by HQ USAF/LEY
4. Orbital Spacecraft Consumables Resupply System Study, sponsored by NASA/JSC
5. Logistics Integration Study, sponsored by SDIO/SY
6. SDI Architecture Studies, sponsored by SDIO
7. KEW/DEW/Sensors Studies, sponsored by SDIO/AF/Army

References

1. Waltz, D. M. *Space Assembly Maintenance and Servicing Study (SAMS) Final Report*, Volume I, Executive Summary, TRW No. SAMSS-196, Volume II, System Analysis, TRW No. SA SS-195, Volume II, Design Concepts, TRW No. SAMSS-197, Volume IV, Concept Development Plan, TRW No. SAMSS-198, Volume V, Neutral Buoyancy Simulation, TRW No. SAMSS-199. TRW, Redondo Beach, CA, July 6, 1988.
2. Patterson, C. D., and T. E. Styczynski. *Space Assembly, Maintenance and Servicing (SAMS) Study Final Report*, Volume I, Executive Summary, Volume II, System Analysis, Volume III, Design Concepts, Volume IV, Concept Development Plan, Volume V, Simulation Report. Lockheed Missiles and Space Company, Inc., Sunnyvale, CA.
3. Chappelear, N. N., and T. R. Danielson. *Space Assembly, Maintenance and Servicing (SAMS) Design Concepts Handbook*. Advanced Systems, Astronautics Division, Lockheed Missiles and Space Company, Inc., LMSC-F223506, July 1, 1988.
4. *EVA* Catalog Tools and Equipment. NASA JSC-20466 Rev. A, April 1989.

Appendix C
On-Orbit Servicing Checklist

The on-orbit satellite servicing technical and programmatic information in this book has been placed in the form of a recommended checklist for:

- Spacecraft designers
- Mission operations planners
- Logistics and support organizers
- Servicer hardware/tools/equipment designers

Spacecraft Designers

1. Treat the design of a serviceable spacecraft, the servicer to be used, and servicing mission hardware as a systems element of the total program.
2. Group, if possible, spacecraft components with similar on-orbit lifetimes.
3. Design ORUs for cost-effective production and test. Trades indicate that if the ORUs cost 50 percent or more of the satellite replacement costs, servicing is nonviable.
4. Choose ORUs from a standard "stable" or catalog for the generic spacecraft functions such as attitude control, electric power, command, and telemetry. This offers advantages of shared spares inventory, with consequent reduced cost, lowered concern for "shelf life," and greater spares turnover.
5. Don't customize an ORU design unless it is absolutely unavoidable.
6. Provide for easy EVA and robotic access to spacecraft docking ports, ORUs for changeout, refueling intakes, film replacement, appendage deployment mechanisms, and inspection of critical components.
7. Make maximum use of built-in test equipment and automated test equipment. This lowers preparation and test time, which is a significant cost driver.
8. Use a standardized spacecraft structure instead of ORU subsystems, if available.
9. Design the spacecraft and ORUs for standardized mechanical, electrical, fluid, optical, thermal, and data/communications interfaces. The NASA/JSC Space Assembly and Servicing and Working Group (SASWG) has an Interface Standards Committee organized according to the six functional areas named above. Further, nine Interface Standards Subcommittees have been formed to work on standards for: hex head bolts and sockets, flight releasable grapple fixtures, grasping/berthing/docking devices, electrical connectors, fluid connectors, fiber optic connectors, utility connectors, and replaceable thermal insulation.
10. Don't overspecify components or mission life requirements.
11. Design the spacecraft and its components and servicing hardware to tolerate the space environment, including particulate contamination, natural radiation, thermal effects, and electromagnetic interfaces and spacecraft charging.
12. Provide a complete and total spacecraft configuration control system.

Mission Operations Planners

1. Plan multiple spacecraft servicings per mission. This reduces the cost per unit serviced.
2. Allow sufficient time for in-space maneuvering so that the most efficient maneuvers can be used, thereby saving spacecraft fuel.
3. Make maximum use of robotic and automated systems for remote servicing. Use manned EVA activities only where the unique skills of the crew are really required.
4. Keep to the minimum the amount of real time commanding and activity monitoring required for the mission.
5. Avoid overly complex servicing scenarios and too crowded timelines. Allow sufficient time to repeat procedures or to recover from contingencies.
6. Never bring back anything into the gravity well that can be stored on orbit.

7. Plan missions to minimize trash and debris. Plan to dispose of unavoidable debris.
8. Make maximum use of existing tools, facilities, and servicing elements.
9. Plan missions to minimize causing the servicing activity to produce induced contamination, effluents, thermal impacts, and electromagnetic interference.
10. Develop various mission segment sequences of events prior to their need so that for quick reaction, contingency servicing flights, these "canned" segments can be used to shorten the mission planning task.
11. As an economic measure, share ground payload processing, crew training, mission control, and ORU refurbishment facilities between NASA, DoD, and commercial users as schedule and security considerations permit.

Logistics and Support Organizers

1. Develop a standard catalog of servicing logistics elements.
2. Set up a "lending bank" of standard ORUs from which users can draw as needed for servicing.
3. Develop a highly automated inventory control system which tracks each item of servicing equipment and spares in the entire system. Be able to identify by type, serial number, location, and condition from any node in the system. The inventory control system should be organized on a user priority basis, such that missions of national importance could obtain limited or singular assets in preference to less critical users.
4. Develop highly automated spares activation and checkout procedures. ORUs/spares should be designed to require the minimum of attention and test prior to use. Maximum use of built-in test equipment and automated test equipment is essential.
5. "Pipeline" spares for critical servicing equipment. Any breakdown in servicing availability will ripple through the system with potentially disastrous results.
6. Optimize spares distribution, based on number of users in each servicing orbital regime and predicted use rates. Update the usage modeling based on actual servicing data.

7. To the maximum extent possible, design to use existing and/or planned facilities and tools.
8. Normalize logistic support to current and future systems where possible.
9. Reduce sole source contracting for current space support activities and create a competitive environment for logistic support of space systems.
10. Set up and maintain a configuration and management control system specifically for operations and logistics support.
11. Insist that logistics inputs be made early in the requirements analysis of new programs.
12. Identify logistics infrastructure requirements and potential shortfalls.
13. Continually upgrade the normalization process for logistic support based on mission effectiveness, potential cost savings, and resource availability.

Servicer Hardware/Tools/Equipment Designers

1. Integrate the servicing functional capabilities into the overall SAMS system.
2. Include sufficient contingency capability into all servicing hardware.
3. Define the spacecraft-servicer interface in detail—a drive toward standardization and commonality.
4. Develop servicing hardware to the same reliability values as the spacecraft.
5. Make sure the servicer operations are relatively time efficient.
6. Ensure that the servicing system is compatible with evolving and alternative transportation systems, including orbit-to-orbit transportation vehicles.
7. Design the servicing system to have minimum impact on the spacecraft configuration and design. Where there is a choice, build complexity into the servicing hardware.
8. Develop servicing hardware so that it is reusable for cost-effectiveness.
9. Where possible, design the servicing hardware to be reconfigurable to meet specific mission needs.

Appendix D

Add-a-Pod Concept

A number of on-orbit satellite maintenance, repair, and servicing design and operational concepts have been suggested in the ten chapters of this book. Most focus on the idea of replacing failed or obsolete components designated as orbital replacement units (ORUs). Various infrastructure elements, that is, OMVs, satellite servicers, tankers, and space support platforms, have been identified to support these concepts.

Another possible concept for maintenance and servicing, but given a smaller amount of attention by government and industry than the ORU concept, is an autonomous on-orbit satellite maintenance and servicing system. Such a system is called the add-a-pod technique [1]. The basic feature of this concept is to design a spacecraft in such a way that a module could be flown to and docked with a failed spacecraft and then reconfigured internally to use the attached replacement parts in lieu of failed components. The module would then become a permanent fixture on the spacecraft, with both elements operating as one unit. Failed components in the spacecraft would be disconnected via telemetry and software and the new components in the added-on module switched on. This concept precludes the need for replacement of ORUs as in conventional servicing scenarios, thus simplifying the repair mission. Rendezvous and docking maneuvers would be performed by a small, simple, low-cost, short life "cargo bus" attached to the add-a-pod module. Rendezvous and docking would be performed with active participation by both the spacecraft and the cargo bus or by the cargo bus alone, depending on the design configuration of the system [1].

Add-a-pod concepts have been studied by TRW, LMSC, RCA, and Fairchild Space Company.

Concept Discussion

In a top level review of the add-a-pod option by Air Force Systems Command (Space Division), Tecolote Research, Inc. and Science Applications International Corporation [1], these general observations were made and are discussed in the following paragraphs. It appeared that no potential showstoppers were inherent in the concept or the various proposed designs. An autonomous rendezvous and docking capability required for flight operations has been demonstrated in laboratory simulations in the United States and were performed regularly by the Soviet Union. The OMV program also developed man-in-the-loop techniques for rendezvous and docking but is not totally autonomous. An advanced data management system, required to switch components in and out of operation, is a viable technology. Basic spacecraft design philosophies, however, will require a different "mindset" compared to current design practices. Such a change is not as severe as required for ORU type designs. The actual interface designs which have been proposed appear to be functionally straightforward and technically valid. But as with any design, including ORU types, the technique will require test and validation.

The add-a-pod concept allows the cargo bus to be launched directly into the orbit of interest at the appropriate time in a launch window. The cargo bus which has internal guidance, navigation, and control capabilities then performs the required phasing maneuvers to rendezvous and dock with the degraded satellite. The concept appears valid for access to all LEO satellites, including potential SDI weapon platforms. The exact applicability and feasibility of repair missions using this concept depends on the altitude/payload tradeoffs for various boosters. The concept can be expanded to allow access to satellites in high energy orbits (GEO and MEO). This could be done by adding a propulsion system to the cargo bus that would perform a circularization and plane change at the desired altitude. This implies that a heavier launch vehicle will be required to lift the increased payloads, but does not make the concept infeasible.

The spacecraft receiving the add-a-pod does not need to be structurally modularized, only electrically modularized with the software in the central computer units capable of turning off any of the modules and transferring those functions to any of several spare connections on the data bus. The scar to the spacecraft is the docking provisions for the pod, and the electrical bus accommodations. Figure D.1 is from the LMSC study. Two possible configurations are

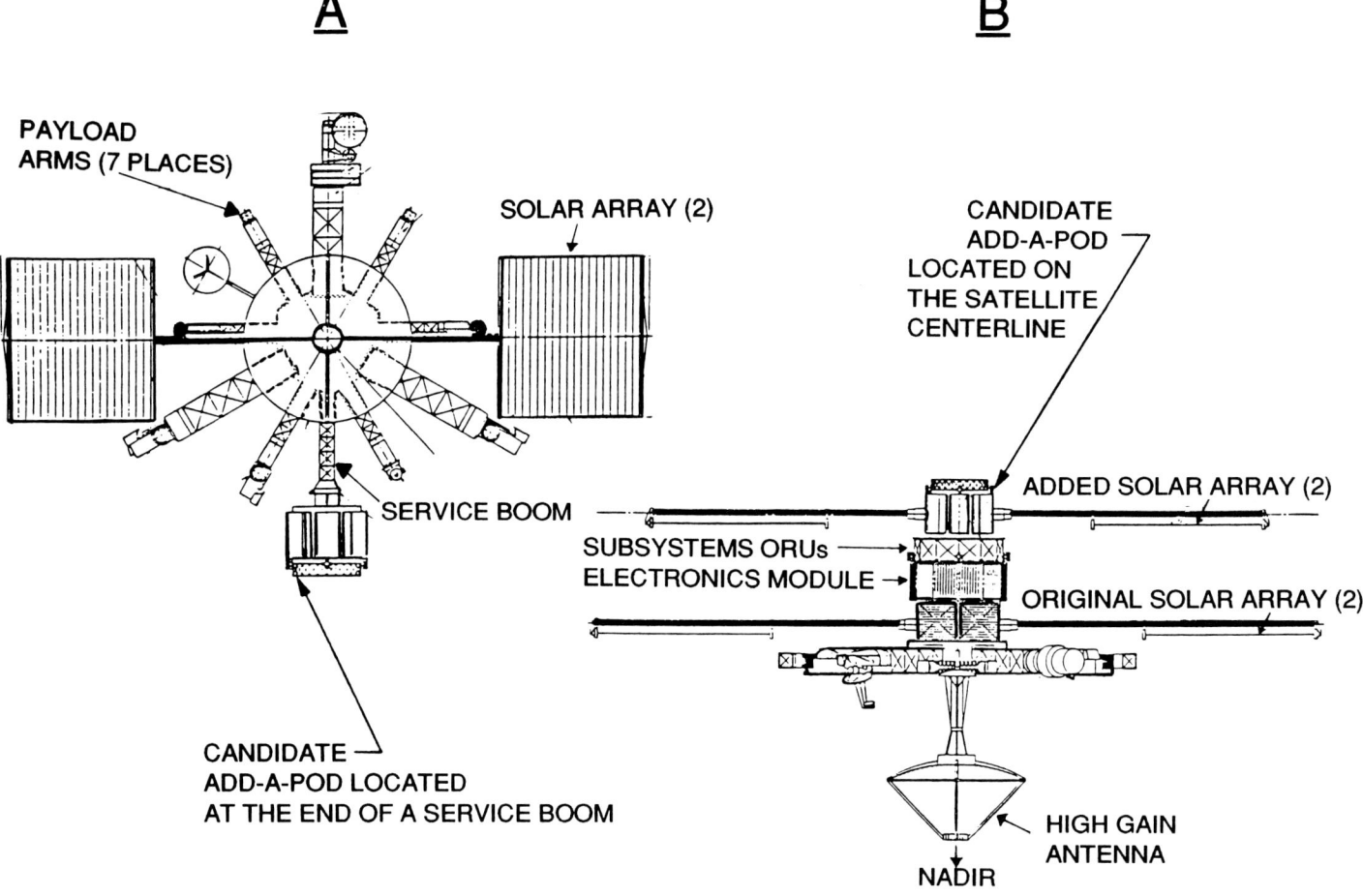

Figure D.1 Two possible add-a-pod configurations on a typical satellite. *Courtesy of Lockheed Missiles and Space Company.*

shown. The add-a-pod location at the end of a boom, Figure D.1a, could incorporate new or replacement sensors, as well as other ORUs. The on center line location, Figure D.1b, incorporates replacement or added solar arrays.

The transportation problems are addressed by making only a one way trip with a smaller package. The pod would have to include the systems necessary to control itself from launch through rendezvous and docking, as well as communications, electrical power, and thermal control for the same period. Those systems will not be separated from the platform, but the cost of the nonrecovered equipment should be small compared to the transportation costs saved. This entire add-a-pod transportation may be accomplished by an expendable launch vehicle.

Advantages and Disadvantages

There are a number of advantages and disadvantages to the add-a-pod system [1].

Advantages:

1. It minimizes physical design impacts on original spacecraft design.
2. A one way repair mission simplifies operations.
3. It precludes the need for an reusable orbital transfer vehicle (ROTV)/OMV and a servicer.
4. It allows assembly of large structures on orbit.
5. It requires minimum infrastructure for maintenance and servicing missions and can be adopted by individual program offices as needed (i.e., it does not require infrastructure amortization).
6. The potential for mission success may be increased (assuming adequate fault isolation) due to less complex maintenance procedures.
7. System program offices may find this concept attractive since such a system does not impact baseline designs as much as ORU type servicing systems.

Disadvantages:

1. If the original spacecraft has a critical failure in systems required for guidance and navigation systems (which are required to be active), a repair mission is not possible.
2. The spacecraft attitude control system must be capa-

ble of handling varying spacecraft mass and moments of inertia.

3. Multiple satellite visits per mission are not possible using common equipment.
4. An upper stage propulsion system must be used for high energy missions and potential debris problems with spent stages must be addressed.
5. Satellite retrieval is not possible.
6. Replacement module spacecraft may have a different physical design layout than original spacecraft.
7. The system may be less flexible in handling on-orbit contingencies than ORU maintenance and servicing systems.
8. It may limit the number of repair missions possible to a specific satellite.
9. Multiple pod missions may increase radar or optical signatures which could adversely affect survivability.

Conclusions

Given the above qualifiers, the autonomous on-orbit spacecraft assembly, maintenance, and servicing concept does seem to have merit as a near-term option for achieving an on-orbit repair capability. Such a concept, however, may not offer the flexibility and growth potential of an ORU-type system but can be performed with less supporting infrastructure and potentially less cost. Baseline satellite designs, as with ORU type designs, must be changed to allow effective utilization of the concept but not to the extent required for ORU type maintenance. The infrastructure required to support such a repair system would be minimal. This is because only a launch vehicle (and upper stage for high energy missions) would be required to deliver the replacement modules to the degraded satellite.

The add-a-pod concept can be applied to both hardware repair and refueling scenarios. Since the payload portion of the cargo bus is a "dead weight" to the cargo bus system itself, either fuel or hardware (or a combination of both) can be delivered to orbit. The standard interface proposed for the system is designed to allow for both electrical and fluid connections between the original spacecraft and the cargo bus. All interconnects are not necessarily used for each mission. It is felt that a cargo bus configuration containing the hardware required for refueling may be slightly heavier and more complex than pure "electronic" pods. This is because of the complex plumbing and fluid transfer systems required for refueling missions.

There are a number of questions that must be addressed before a decision can be made as to whether this type of maintenance and servicing system is beneficial. Since the replacement module is physically attached to the original spacecraft, it is possible that only one repair mission can be performed per satellite over its lifetime. As mentioned earlier, this may be due to ACS limitations or an eventual field-of-view blockage of sensors, radiators, solar cells, etc., as the spacecraft grows. However, some early studies have indicated that this problem can be adequately addressed by using auspicious designs. Also, an add-a-pod system may not be able to evolve to a more flexible (i.e., real time contingency operations) ORU type system due to different design requirements. This point also is subject to debate. The overall question that remains is whether an add-a-pod repair system is cheaper or more expensive in the long run or life cycle than other postulated concepts.

Reference

1. Jee, Thomas E., James H. Suttle, Tecolote Research, Inc., and Robert J. Curtis. Science Applications International Corporation, *Space Assembly, Maintenance and Servicing Analysis Model Minimum Investment Satellite Repair and Servicing Analysis.* Prepared for Department of the Air Force, Headquarters Space Division, Report No. CR-0371, December 7, 1988.

Index

ACCESS, 2, 15, 29–30, 75–79
ADA programming language, 58
Add-a-Pod, 277–279
Advanced Land Observing System, 19
Advanced Launch System, 3–4, 40, 52, 62–63, 246, 254, 270
Advanced Technology, Inc., 5, 58, 75, 133, 269
Advanced Warning Aircraft Communications System, 19
Aerospace Corporation, The, 9, 66, 269, 273
AIAA, 114, 133
Aircraft Industries Association, 133
Air Force Plans and Advanced Programs Organization, 51
Air Force Systems Command, Space Division, 234–235, 237–238, 242, 249, 269, 273–274, 277
Akers, Thomas D., 175–176
Aldridge, Edward C., 249
Ames Research Center, 160–163
Apollo, 3, 13, 14, 37, 106, 261
Apt, Jerome, 157
Ariane, 51–52, 62–63
ARINC Research Corporation, 133
Artificial intelligence, 29, 33–34
ASAT, 178, 251
ASME, 114
Assembly and Overhaul Station, 5, 253, 270
ASTROMAG, 207
Atlantis, 29, 155, 157
Atlas Agena, 139
Atlas Centaur, 52, 62–63, 132
Attitude control system, 11–13, 56, 126–128
Automated Structural Assembly Laboratory, 207
Automatic Test Equipment (ATE), 54
AXAF, 3, 13, 15, 19, 32, 35, 37, 42, 46, 64, 66, 99, 123, 127, 136, 139, 142, 146–155, 213, 215, 230, 249, 258, 273–274

Ball, David, 133
Ballistic coefficient, 100
Bean, Alan, 10
Bell Aerospace Systems Division, 146
Berthing cradle, 12
Boeing Aerospace, 5, 146, 269
Booz Allen & Hamilton, 5, 269

Brandenstein, Daniel C., 175
Breakpoints for servicing, 41
Brown, Charles, 273
BSTS, 32, 237–238
Built-In Test Equipment, 54

Cameron, Kenneth D., 157
Canada, 51, 75, 83, 260
Cargo transfer vehicle, 168
Carnegie-Mellon University, 5, 269
Carr, Gerald, 11
Cepollina, Frank, 139
Cernan, Eugene A., 174
Challenger, 11–13, 174
Chilton, Kevin P., 175
Cislunar space, 3
Civil Needs Data Base, 22
Cleve, Mary, 75
Columbus, 51, 215
Commonwealth of Independent States, 9, 16, 257
Compressed Gas Association, 134
Compton, Arthur, 155
Computer-aided design, 13
COMSAT, 32
Configuration Control Board, 135
Conrad, Charles "Pete", 10
Coolant system, 9, 14
Co-Orbital Warehouses and Tankers, 53
Coronograph/polarimeter instrument, 11–12
COSEMS, 58–61
COSTAR, 143
Cost drivers, 41, 47
Covert, Eugene, 175
Crippen, Robert, 12
Critical design review, 64
Cryogen resupply, 18, 41, 45–47, 53, 56
Curtis, Robert, 269

Data Collection and Location System, 19
Davis, Bob, 133
Deep Space Network, 154
Delta launch vehicle, 11, 52, 62–63, 92–93, 96–97, 132, 139, 141
Depots, 4, 45–46, 53
Development flight instrumentation, 13
Development Integration and Test Facility, 207, 255
Discovery, 157

DMSP, 32, 66, 215, 237–238, 251
Dolinsky, Shlomo, 133
DSCS, 32, 66, 237
DSP, 66, 237

Earth Observing System, 19, 32, 35, 37, 66, 87, 91–92, 123–124, 127, 132, 139, 164–167, 213, 215, 258
Earth viewing satellites, 19
EASE, 2, 15, 29–30, 75–79
Eastman Kodak Corporation, 146
Edwards Air Force Base, 133, 174
Electromagnetic contamination, 7
Electromagnetic interference, 7
Electronic Industries Association, 134
Ely, Neal, 273
Endeavour, 2, 174–176
Energia, 62–63
Equatorial Science Platform, 19
ESGP, 123, 215
Eureca, 66, 215, 253
European Space Agency, 51, 64, 75, 79, 82, 84, 143, 164
EVA Catalog of Tools and Equipment, 185–190
Expendable launch vehicles, 3–4, 33, 38, 59–60, 63–65, 72–73, 79, 83, 92–97, 165, 197, 242
Explorer, 136
Extended duration orbiter, 71
Extravehicular maneuvering unit, 134, 185–186
Extreme UV Explorer, 19, 258

Fairchild Corporation, 199–200, 218–220, 222–224, 277
Far UV Spectroscopy Explorer, 19
Fisher, William J., 188–190
FLEETSATCOM, 32, 66, 237–238
Flight Support System, 11, 38, 141, 148, 152, 160–161
Flight Telerobotics System, 42, 46–47, 179, 202–207, 209, 213, 252, 255
Free Flying Imaging Radar Experiment, 19
Fuel cell, 13
Fuel transfer, 15–16, 32, 38, 41–45, 56, 223–225, 248

Garriott, Owen, 9–10
GBL, 32

Index

Gemini, 14, 37, 75, 106, 185
General Dynamics Corporation, 242
General Electric, 58, 93, 132, 141
General Research Corporation, 58, 66
GEO Platform, 32, 215
Geopotential Research Mission, 19
Geosynchronous Earth orbit, 21–22, 32–33, 38, 40, 42, 45, 87
Germany, 260
Gibson, Edward, 11
Goddard Space Flight Center, 11, 13, 16, 113, 121, 132, 133, 136, 139–143, 148, 155, 157, 160–164, 202–203, 207, 229, 254–255
Goldin, Daniel, 155
GPS, 32, 53, 66, 106, 109, 237–238, 251
Graham, James, 273
Gravitational force, 87
GRO, 2–3, 16, 19, 32, 35, 37, 42, 63–64, 66, 99, 123, 134, 136, 139, 142, 146, 155–159, 215, 218, 230, 242, 249, 258, 274
Ground control console, 106–107
Ground Station Tracking and Data Network, 140, 147
Grumman Space Systems, 5, 82, 141, 147, 269
G Star-3, 259

Haddad, Al, 133
Hart, Terry, 12
Heavy lift launch vehicle, 40, 52, 254
Hermes, 51, 62–63
Herschel, Sir William, 158
High Energy Upper Stage, 52–53, 254, 270
HIMSS, 123
HIRIS, 127
Honeywell Corporation, 5, 269
Hotol, 62–63
HRMA, 146, 149, 151
Hubble, Edwin P., 142
Hubble Space Telescope, 3, 13, 16, 18, 32, 35, 42, 46, 63–64, 66, 99, 136, 139–145, 167, 169, 187, 215, 230, 249, 258, 273–274
Hughes Danbury Optical Systems, Inc., 13, 146
Human Role in Space Study, 70–71
Huntsville Operations Support Center, 147

IEEE, 114, 133–134
ILC Dover, Inc., 183–188, 190, 213
Illinois Institute of Technology, 5, 269

Independent Research & Development, 18, 48
Industrial Space Facility, 60, 253
Inertial Upper Stage, 21, 28, 125–126
Infrared Astronomical Satellite, 160
In-Space Assembly/Servicing (ISAS) Facility, 33, 58–59
Integration, 1, 18, 23, 32, 54
Intelligent Systems Research Laboratory, 207
Intelsat 6/F3, 2, 174–176, 259
Inter-module connector, 165
International Standards Organization, 133
International Technology Underwriters, Inc., 242, 245
IOSS, 198–199, 201–202, 220–221, 226
IRAS, 163

Jackson, Louise, 273
Japan, 51, 75, 79–81, 83, 257
Japanese Space Development Agency, 164
Jet Propulsion Laboratory, 107, 133, 143, 164, 166, 203, 207–208, 273
Johnson Space Center, 12, 107, 187–188, 203, 209, 213, 217–218, 229, 254–255, 273
Johnson Space Center Space Assembly and Servicing Working Group (SASWG), 49, 133–134, 144, 175

Kennedy Space Center, 9–11, 71, 144, 172, 174, 208, 229
Keplerian motion, 87
Kerwin, Joseph, 10
KEW, 32, 66, 274
Kohrs, Richard, 83

Landsat, 11, 66, 141, 160, 183, 215, 229
Langley Research Center, 58, 207
Large amplitude space simulator, 171
Large Deployable Reflector, 19, 35, 37, 66
Laser docking sensor, 46–47, 107–109
LDEF, 2, 258
LDR, 123, 215
Leasat, 66
Leasecraft, 66, 215
Ledford, Otto, 133
Lerner, Eric J., 155
Levin, George, 273
Lewis Research Center, 82
Life cycle cost, 4, 39, 47, 48, 231–238, 240–241, 245, 261, 269
Life Support Systems, Inc., 5, 269
Liquid Helium Transfer Project, 162

Lochmann, Rod, 269, 273
Lockheed Missiles and Space Company, 5, 13, 33, 52, 65–66, 75, 122, 129, 133, 146, 167–168, 190, 197, 213, 131, 238, 247, 251, 269, 277–278
Logistic system, 23, 45–46, 135
Long March, 62–63
Lousma, Jack, 10
Lower Atmosphere Research Satellite, 19
LTV Corporation, 168
Lunar rover, 14

Magellan, 208
Magnetic fields, 11
Magnetic Monitoring Mission, 19
Main electronic box, 11–12
Manned maneuvering unit, 11, 30, 32, 38, 42, 185
Man-tended capability, 80
Man-tended free flyer, 84
Mars, 29, 37, 74, 84
Marshall Space Flight Center, 13, 64, 70, 75, 93, 120, 146–147, 154, 161, 168, 169–173, 203
Martin Marietta Corporation, 58, 129, 133, 167–168, 199–201, 203–204, 206, 209, 212, 218, 221, 225–227, 242
Marzwell, Dr. Neville, 273
McDonnell Douglas Space Systems Company, 5, 58, 68–71, 81–82, 107, 109, 176, 190, 195–196, 242, 251, 269
Melnick, Bruce E., 175
Mercury Program, 185
Meteoroid shield, 9–10
Microgravity, 27, 76, 84
Microwave antenna, 9
Military space missions, 17
Milstar, 32, 66, 237–238, 251
Mir (C.I.S. Space Station), 85
Mission Control Center, 11
Mission peculiar equipment support structure, 162
MIT, 107, 175
Mobile servicing center, 28
Mobile servicing system, 5, 84
Modularized spacecraft, 4, 6, 9, 119–120, 125, 129
Module service tool, 46–47, 154
Momentum wheel, 11
Moon, 29, 37, 74
Multimission modular spacecraft, 124, 132–133, 139, 142, 152, 155, 157, 160, 167, 209, 215, 229, 258, 260

NASA Space Systems Technology Report, 38
National Aero-Space Plane, 52, 62–63, 246
National Fluid Power Association, 134
National Launch System, 62, 72
National Oceanic and Atmospheric Administration, 22, 34, 164
National pricing policy, 4
National Space Transportation System: advanced servicing support, 53, 257; EVA activities, 117, 157; servicing support, 4, 54, 148, 162, 185, 201, 209; Space Shuttle vehicle, 2, 62, 143, 229, 242; transportation operations, 71, 125, 132, 139, 161, 169, 251, 259
Navy Remote Ocean Satellite System, 19, 66
Nelson, George, 11–12
Neptune, 261
Neutral-buoyancy water tank, 29, 135, 172, 209
NIMBUS, 140
NPB, 32, 66
NRSS, 32

Ocean Microwave Package, 19
O'Connor, Bryan, 75
OMNIPLAN Corporation, 213
OMNISTAR, 66
OMV short range vehicle, 170–171
On-Orbit servicing platforms and work stations (manned, man-tended, or automated), 4–5, 7, 26, 52, 214–215, 236, 240, 254; *Also see* Space Assist Support System, Tank farms, Space-based support platform, Assembly and Overhaul Station, In-Space Assembly/Servicing Facility
Optical Society of America, 134
Orbiter processing facility, 172
Orbiting Astronomy Observatory, 140
Orbiting Solar Laboratory, 19, 35
OSCRS, 197, 199, 213, 217–221, 224, 253, 270, 274

Palapa B2, 2, 15, 37, 42, 175, 183, 258–259
Payload Operational Control Center, 11, 147, 162
P^{3}I, 4, 45
Pioneer, 208
Planning Research Corporation, 57, 60–61
Pogue, William, 11

Polar meteorological satellite, 19
Polar orbiting platform, 19, 37, 42, 87, 91–97, 103–104, 132, 164
Polar orbiting satellite, 5
Pressurized module, 5, 7, 81
Price, Charles R., 188–190
Primary life support system, 185
Progress, 16, 106
Proteus, 136
Proton, 62–63

Radarsat, 66, 215
Radioactive thermoelectric generator (RTG), 56
Radtke, Robert, 133
Rate gyro, 9, 14
RCA, 277
Reaction control system, 10, 81
Rendezvous and docking, 14–17, 71, 109
RMS grapple docking mechanism, 169, 171, 201
Robotics Industries Association, 134
Robotics Systems Development Branch/JSC, 188
Robotics Working Group, 190
Rockwell International Corporation, 174, 199–200, 216–218, 229
Ross, Jerry, 29, 75–76, 157, 273
Rotating service structure, 172
RSSS, 215
Rysavy, Gordon, 273

SAIC, 238, 251, 269, 277
Salyut, 16–17
Satellite Control Facility, 19
Satellite servicer system, 13, 30, 32, 34, 52, 57, 197, 208–213, 252, 270
Satellite servicing issues, 43, 48
Satellite Servicing Steering Committee (NASA), 258
Satellite Servicing Working Group, 258
Saturn launch vehicle, 9–10
Savitskaya, Svetlana, 175
Sawaya, George, 273
SBI, 35, 123, 237–238
SBL, 32, 66
SBR, 237–238, 251
Schmitt, Harrison H., 174
Science Support Center, 147
Servicing functions, 21–22, 25
Servicing models, 17
Shaw, Brewster, 75
Sieck, Robert, 155

Simon, William E., 189–190
SIRTF, 19, 32, 46, 64, 66, 99, 123, 136, 142, 157–163, 215, 258
Skylab, 3, 9–11, 13–14, 37, 75, 135, 142, 168, 190
SLC, 32, 66
Slow spin mode, 11
Society of Automotive Engineers, 133–134
Solar array system, 10, 14–15, 124
Solar Corona Diagnostic Mission, 19
Solar flare, 11
Solar Maximum Mission, 2–3, 11–13, 15, 17, 37, 42, 113, 132, 134–135, 140, 142, 152, 161, 169, 175, 229, 258
Solar Terrestrial Observatory, 19
Soviet Union, 6
Soyuz, 16, 106
Space Assist Support System (SASS), 52, 58–60, 139
Space-based support platform, 52, 58–60
Space Exploration Initiative, 111
Spacelab, 15, 37, 42
Space Station Assembly Planning Group, 82
Space Station Control Center and Payload Operations Integrations Center, 80
Space suits, 16, 27, 32, 34, 82, 183, 185
Space support equipment, 148
Space Transportation Architecture Study, 22
Spartan, 215
SPAS-01, 15
Spring, Sherwood, 29, 75–76
SSTS, 32, 35, 66, 123, 127, 215, 237–238, 253, 270
Stafford, Tom, 175
STP, 66
Strategic Defense System: architecture, 60, 269–270; engineering/system integration, 58, 260; operations, 57, 238, 247, 249, 255, 277; program, 56, 208–209, 229, 253–254, 261, 273–274; servicing, 59, 139, 183, 230, 251, 271
Stytle, Leo, 173
Sullivan, Kathryn, 175
Sun, 11
Sunshield, 14
Sunspot, 11
Superfluid Helium On-Orbit Transfer Project, 46–47, 217–218
Surveyor, 14, 208
Suttle, James H., 269

Index

SYNCOM 4-F5, 2, 37, 175
SYNCOM IV-3, 2, 15, 42, 258–259

Tank farms, 4, 81
Target acquisition sequence, 108
TDRSS, 66, 106–107, 109, 111, 140, 143, 147, 149, 167, 221, 257, 259
Teal, Ruby, 66
Tecolote Research Corporation, 61, 238, 269, 277
Teleoperation, 3, 5, 8, 13, 15
Text and graphics system, 13
Thompson, Al, 133
Thornton, Kathryn C., 175–176
Threepoint docking mechanism, 169
Thuot, Pierre J., 175
Titan launch vehicle, 52, 62–63, 92, 95–97, 132, 139, 163, 169, 175
TOPEX, 66, 123, 215, 258
Topography experiment, 19
Tracor Applied Sciences, 133
TRMM, 123, 215, 258
TRW Inc.: modularity, 120–121, 128, 130, 173; planning, 53, 150–251; program risks, 243–244; requirements, 34, 36, 65, 124–127, 216; serviceable spacecraft, 35, 123, 146–159, 167–172; servicing costs, 48–49, 131, 231–233, 238, 246; servicing spectrum, 31, 37, 67, 114, 194, 239; studies, 5, 40, 52, 58, 64, 66, 132, 141, 208–209, 247, 269, 277; technology, 41–45, 88–105, 108–110, 210–213

United Technologies Chemical Systems Division, 175
University of Colorado, Center for Space Construction, 74
Unmanned spacecraft cost model, 232
Unpressurized servicing facility, 5, 7
Upper atmosphere research satellite, 19, 35, 66, 123, 141, 160
Upper stages, 3–4
USAF B-1, 19
U.S.S. *New Orleans*, 10
UV scanning polychromator spectroheliometer, 11

Van Allen radiation belt, 163
Vandenberg AFB, 33, 63, 229
van Hoften, James, 12
Vela, Rodolfo Neri, 75
Viking, 168, 208
VLBI, 66
Voyager, 208, 261

Walker, Charles, 75
WASS, 6
Weitz, Paul, 10
Westar VI, 1, 15, 37, 42, 175, 183, 258–259
Western Test Range, 87, 165, 172, 229
White light coronagraph, 11
Windsat, 19
Wong, Joseph T., 269

X-ray polychromator instrument, 13
X-Ray Spectrographic Telescope, 11
X-Ray Timing Explorer, 19, 258

Zenith Star, 35, 123, 139, 167, 209, 213, 215